E S S E N T

MATHEMATICS

AN INTERACTIVE APPROACH

Linda Pulsinelli   Patricia Hooper

Department of Mathematics, Western Kentucky University

MACMILLAN PUBLISHING COMPANY   NEW YORK

COLLIER MACMILLAN, Inc.   TORONTO
MAXWELL MACMILLAN INTERNATIONAL PUBLISHING GROUP

NEW YORK   OXFORD   SINGAPORE   SYDNEY

Macmillan Publishing Company
866 Third Avenue, New York, New York 10022

Collier Macmillan Canada, Inc.

Library of Congress Cataloging-in-Publication Data

Pulsinelli, Linda Ritter.
    Essential mathematics: An interactive approach / Linda
Pulsinelli, Patricia Hooper.
        p.   cm.
    ISBN 0-02-357170-5
    1. Mathematics.   I. Hooper, Patricia (Patricia I.)   II. Title.
QA39.2.P83   1991
510—dc20   89-27693
                                                            CIP

Preface

As its title indicates, *Essential Mathematics: An Interactive Approach* is intended to teach students fundamental concepts and skills by involving them continually in the learning process, allowing them to see their own progress and therefore motivating them to continue.

The vocabulary level of a book of this type is crucial to its success and we have attempted to word our explanations in simple, straightforward language. The writing style is informal without being condescending, and important ideas are summarized in table form rather than paragraph form whenever appropriate.

Our experience with students at this level leads us to believe that they must practice each new skill as soon as it has been presented. For this reason, the text includes several unique features that are designed to provide maximum reinforcement.

Each chapter in the book follows the same basic structure.

Motivational Applied Program

At the beginning of each chapter we have presented an applied problem that can be solved after the student has mastered the skills in that chapter. Its solution appears within the chapter.

Explanations

We have tried to avoid the "cookbook" approach by including a straightforward and readable explanation of each new concept. Realizing that students at the introductory level become easily bogged down in reading lengthy explanations, we have attempted to make our explanations as brief as possible without sacrificing rigor.

Highlighting

Definitions, properties, and formulas are highlighted in boxes throughout the book for easy student reference. In most cases a rephrasing of a generalization in words accompanies the symbolic statement, and it is also highlighted in a box.

Examples

Immediately following the presentation of a new idea, several completely worked-out examples appear together with several partially worked-out examples to be finished by the student. These examples are completed correctly at the end of each section and the student is advised to check his or her work immediately.

Trial Runs

Sprinkled throughout each section are several short Trial Runs, a list of six or eight problems to check on the student's grasp of a new skill. The answers appear at the end of the section.

Exercise Sets

Each section concludes with an extensive Exercise Set in which each odd-numbered problem corresponds closely to the following even-numbered problem.

Stretching the Topics
At the end of each Exercise Set there are several problems designed to challenge the students by extending to the next level of difficulty the skills learned in the section.

Checkups
Following each Exercise Set, a list of about 10 problems checks on the student's mastery of the most important concepts in the section. Each Checkup problem is keyed to comparable examples in the section for restudy if necessary.

Problem Solving
One section of almost every chapter involves switching from words to mathematics. By including such a section in each chapter we are attempting to treat problem solving as a natural outgrowth of acquiring arithmetic skills.

Chapter Summaries
Each chapter concludes with a summary in which the important ideas are again highlighted, in tables when possible. New concepts are presented in symbolic form and verbal form, accompanied by a typical example.

Speaking the Language of Mathematics; Writing About Mathematics
Following the summary, we have included a group of sentences to be completed *with words* and questions to be answered *in sentences* by the student. Students (especially those in self-paced programs) often lack the opportunity to ''speak and write about mathematics.'' We hope that these short sections will help them develop a better mathematics vocabulary and become more comfortable expressing themselves in mathematical terms.

Review Exercises
A list of exercises reviewing all the chapter's important concepts serves to give the student an overview of the content. Each problem is keyed to the appropriate section and examples.

Practice Test
A Practice Test is included to help the student prepare for a test over the material in the chapter. Once again, each problem is keyed to the appropriate chapter section and examples.

Sharpening Your Skills
Finally, we have included a short list of exercises that will provide a cumulative review of concepts and skills from earlier chapters. Retention seems to be a very real problem with students at this level and we hope that these exercises will serve to minimize that problem. Each cumulative review exercise is keyed to the appropriate chapter and section.

Throughout the book we have adhered to a rather standard order of topics, making an attempt to connect new concepts to old ones whenever appropriate. This modified spiraling technique is designed to help students maintain an overview of the content. Success in future courses seems to us to hinge on students' seeing mathematics as a logical progression of ideas rather than a set of unrelated skills to be memorized and forgotten.

This book follows the four basic operations of arithmetic through the whole numbers, fractional numbers, and decimal numbers in Chapters 1–7. Then the important topic of ratio and proportion is introduced (Chapter 8) and is carried through the discussion of percents and their applications (Chapters 9 and 10). A brief look at statistics (including percentiles) follows in Chapter 11. Chapters 12 and 13 contain the topics of measurement and geometry, using American and metric units of measure. Chapter 13 also includes a short discussion of the Pythagorean Theorem. Chapters 14 and 15 provide an introduction to algebra. Although the chapter on calculators (Chapter 16) has been placed at the end of the book, its sections are constructed to allow for their use along with earlier chapters. At whatever point an instructor wishes to include calculator options, he or she may introduce the first section of Chapter 16 and refer to the remaining sections at the appropriate times during the course.

The answers to the odd-numbered exercises in the Exercise Sets appear in the back of the book together with answers for *all* items in Stretching the Topics, Checkups, Speaking the Language of Mathematics, Review Exercises, Practice Tests, and Sharpening Your Skills.

More assistance for students and instructors can be found among the supplementary materials that accompany this book.

Instructor's Manual with Test Bank

The Instructor's Manual contains the answers for all exercises in the Exercise Sets and Stretching the Topics. In addition there are eight Chapter Tests (four open-ended and four multiple choice) for each chapter and four Final Examinations (two open-ended and two multiple choice). Answers to these tests and examinations also appear in the Instructor's Manual. The chapter tests are also available on computer disks.

Student's Solutions Manual

The Student's Solutions Manual, written by Rebecca Stamper, contains step-by-step solutions for the even-numbered exercises in the Exercise Sets and for *all* items in the Review Exercises, Practice Tests, Sharpening Your Skills, and exercises involving word problems. Using the same style as appears in the text, these solutions emphasize the procedure as well as the answer.

Video Tapes

A series of video tapes (each 20 to 30 minutes in length) provides explanations for many topics in the course.

Computerized Test Generator

A set of computer-generated tests is available for producing tests of any length for each chapter. Cumulative tests and final exams may also be constructed.

Interactive Software Program

This computer-assisted program is available for use with Apple and IBM systems and is a series of lessons including problems at differing levels of difficulty.

Acknowledgments

The writing of this book would not have been possible without the assistance of many people. We express our appreciation to our typist Maxine Worthington, to Rebecca Stamper of Western Kentucky University and Kathy Rodgers of the University of Southern Indiana for carefully working all our problems, to our families for tolerating our obsessive work schedules, to our editors Gary Ostedt and Bob Pirtle for their enthusiastic support, and to Ed Burke and the production staff at Hudson River Studio for their expert guidance and efficient supervision.

We also thank our reviewers for their careful scrutiny and helpful comments.

David A. Petrie, Cypress College
Robert Levine, Community College of Allegheny Co., Boyce Campus
Cynthia Miller, Georgia State University
Deann Christianson, College of the Pacific

Robert A. Nowlan, Southern Connecticut State University
Brenda B. Allen, Georgia College
Robert Kaiden, Lorain County Community College

Deborah Hale, Thomas Nelson Community
   College
James Magliano, Union County College
Shirley Markus, University of Louisville
Mrs. Ara B. Sullenberger, Tarrant County
   Junior College, South Campus

Lucy C. Thrower, Francis Marion College
Carole Welch, Eastern Kentucky
   University
John Snyder, Sinclair Community College
John T. Gordon, Georgia State University

The cover painting, Weavers of Time, was executed in acrylic, gauze, and dyes by Frances
Wells, a Kentucky artist and teacher. Her prints, oils, and watercolors appear in numerous
galleries and in corporate and private collections throughout the United States.

Contents

E S S E N T I A L
MATHEMATICS

AN INTERACTIVE APPROACH

Adding and Subtracting Whole Numbers

Leroy has been on a diet for 6 months. He has kept the following record of his monthly weight loss or gain.

October—lost 5 pounds   January—lost 3 pounds

November—lost 7 pounds   February—lost 8 pounds

December—gained 4 pounds   March—gained 2 pounds

If Leroy weighed 192 pounds before dieting, what was his weight at the end of March?

In this chapter we will learn to use arithmetic to solve this problem. Arithmetic is the most fundamental branch of mathematics. Whether a student wishes to continue the study of mathematics through algebra or merely wants to be able to perform calculations in daily life, an understanding of the arithmetic of numbers is essential for success.

In this chapter we learn how to

1. Recognize whole numbers.
2. Use the number line.
3. Understand place value.
4. Add whole numbers.
5. Subtract whole numbers.
6. Switch from words to whole numbers.

1.1 Working with Whole Numbers

Recognizing Whole Numbers

The numbers that we use every day of our lives to answer the question "How many?" are called the **whole numbers.** See if you can fill each blank in these statements with a whole number.

1. There are _____ feet in a yard.
2. There are _____ eggs in a dozen.
3. There are _____ vowels in the word pizza.
4. There are _____ states in the United States.
5. There are _____ cents in a dollar.
6. There have been _____ female U.S. Presidents.

Check your work on page 8. ▶

Now that you have seen examples of whole numbers, perhaps you will agree that the set of whole numbers can be written as

> Whoie numbers: {0, 1, 2, 3, 4, 5, . . .}

The dots at the end of the list show that the set continues indefinitely in the same pattern. We know, therefore, that numbers such as 12, 100, and 365 belong to the set of whole numbers.

Many numbers that you encounter in everyday life do *not* belong to the set of whole numbers. For instance, fractions such as $\frac{1}{2}$, $\frac{2}{3}$, and $\frac{8}{5}$ do *not* represent whole numbers. Likewise, decimal numbers such as 1.5, 0.25, and 5.67 do *not* represent whole numbers. Fractions and decimal numbers are certainly important types of numbers, but we shall consider them later in our study of arithmetic.

▐▐▐▶ Trial Run

Fill in each blank with the correct whole number.

_____ 1. There are _____ days in a week.

_____ 2. There are 5280 feet in _____ mile(s).

_____ 3. Water freezes at _____ °F.

_____ 4. There are _____ inches in a yard.

_____ 5. There are _____ girls in the Vienna Boys' Choir.

Answers are on page 9.

Using the Number Line

To picture the whole numbers, we can use a **number line.** We draw a line and choose a zero point and a length to represent 1 unit. Then all points spaced 1 unit apart to the right of zero are labeled with the whole numbers in order.

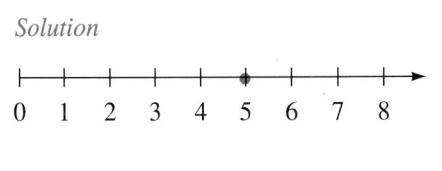

The arrow shows that this line goes on indefinitely, so that numbers such as 36 and 5280 also correspond to points on this line.

To illustrate the position of a whole number on the number line, we put a solid dot at the point corresponding to that number. This is called **plotting a point** on the number line.

Example 1. Plot 5 on the number line.

Solution

```
|---+---+---+---+---+---•---+---+---+--->
0   1   2   3   4   5   6   7   8
```

Example 2. Name the whole number corresponding to each point on the number line.

Solution

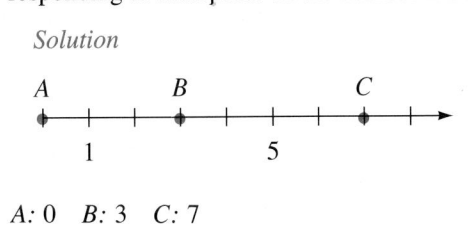

A: 0   B: 3   C: 7

The number line also gives us a handy way of **comparing** numbers. If we look at a number, say 3, we notice that all the numbers to the right of 3 are larger than 3. All numbers to the left of 3 are smaller than 3. For instance, 5 is greater than 3 because 5 lies to the right of 3 and 0 is less than 3 because 0 lies to the left of 3.

We may also write these statements using symbols.

> < means "is less than."
> > means "is greater than."

Such statements are called **inequalities.**

Statement	Number Line	Symbols
5 is greater than 3	0 1 2 3 4 5 6 7 8 9 10	5 > 3
9 is greater than 3	0 1 2 3 4 5 6 7 8 9 10	9 > 3
2 is less than 3	0 1 2 3 4 5 6 7 8 9 10	2 < 3
0 is less than 3	0 1 2 3 4 5 6 7 8 9 10	0 < 3

Note that if 5 > 3, we may also say that 3 < 5. Likewise, if 0 < 3 we may also say that 3 > 0.

Example 3. Place a < or > symbol between the numbers.

$$10 \underline{\quad} 4 \qquad 0 \underline{\quad} 1$$
$$5 \underline{\quad} 6 \qquad 101 \underline{\quad} 99$$

Solution

10 > 4 because 10 lies to the right of 4.
5 < 6 because 5 lies to the left of 6.
0 < 1 because 0 lies to the left of 1.
101 > 99 because 101 lies to the right of 99.

We shall see later that the number line is a useful tool in learning to add and subtract whole numbers.

⫸ Trial Run

Plot the following whole numbers on the number line.

1. 6   **2.** 2   **3.** 0   **4.** 8

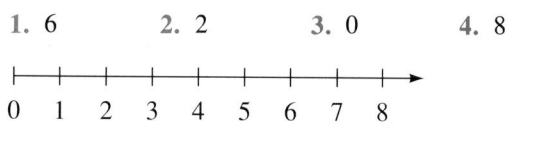

Name the whole number corresponding to each point.

5. *A*   **6.** *B*   **7.** *C*

Place a < or > symbol between the numbers.

8. 31 _____ 17   **9.** 1000 _____ 1001   **10.** 2997 _____ 3000

Answers are on page 9.

Understanding Place Value

Our number system is based upon the number 10. Why? Probably because man has 10 fingers. Every whole number that we write is made up of one or more **digits,** and every digit must be a whole number from the following set.

Digits: {0, 1, 2, 3, 4, 5, 6, 7, 8, 9}

When we write a two-digit number such as 26, we say "twenty-six," but what do we mean? The number 26 actually stands for 2 tens and 6 ones. Similarly, if we write the two-digit number 62, we say "sixty-two," but we mean 6 tens and 2 ones. To read the number 153, we say "one hundred fifty-three" which we can translate into 1 hundred and 5 tens and 3 ones. Likewise, we can give meaning to numbers having 4 digits, 5 digits, and so on. Let's see if a table can be used to switch from number form to word form.

Number Form	Word Form	Meaning
47	forty-seven	4 tens and 7 ones
74	seventy-four	7 tens and 4 ones
80	eighty	8 tens and 0 ones
400	four hundred	4 hundreds and 0 tens and 0 ones
915	nine hundred fifteen	9 hundreds and 1 ten and 5 ones
1853	one thousand, eight hundred fifty-three	1 thousand and 8 hundreds and 5 tens and 3 ones

Notice how commas are used after 3 digits to make the word form of the number easier to read. In naming whole numbers of 4 digits or more, it is customary to read the digits in the groups separated by commas. For instance, we read 624,305 as "six hundred twenty-four thousand, three hundred five" rather than as "six hundred thousands, two ten thousands, four thousand, three hundred five." The second form for reading the number is not incorrect, but it sounds terribly awkward.

The rightmost digit of a whole number tells us how many ones to use for the number. The next digit to the left tells us how many tens to use. The next digit tells how many hundreds to use, and so on. We therefore call the rightmost digit the **ones digit** and the position it occupies is called the **ones place.** We call the next digit to the left the **tens digit** and the position it occupies is called the **tens place.** The next digit to the left is called the **hundreds digit.** It occupies the **hundreds place.**

Notice that the value of each place is just 10 times as large as the value of the place immediately to the right. Consider the whole number 2375.

$$\underset{\substack{\uparrow \\ \text{thousands} \\ \text{place}}}{2} \quad \underset{\substack{\uparrow \\ \text{hundreds} \\ \text{place}}}{3} \quad \underset{\substack{\uparrow \\ \text{tens} \\ \text{place}}}{7} \quad \underset{\substack{\uparrow \\ \text{ones} \\ \text{place}}}{5}$$

In this number

2 occupies the thousands place, so 2 is the thousands digit.
3 occupies the hundreds place, so 3 is the hundreds digit.
7 occupies the tens place, so 7 is the tens digit.
5 occupies the ones place, so 5 is the ones digit.

We can summarize the meaning of each place in a whole number with the following diagram.

_____ _____ _____ , _____ _____ _____ , _____ _____ _____   Places

hundred millions | ten millions | millions | hundred thousands | ten thousands | thousands | hundreds | tens | ones

Example 4. Write the word form and the meaning of 12,836.

Solution. 12,836 is twelve thousand, eight hundred thirty-six.

12,836 means 1 ten thousand and 2 thousands and 8 hundreds and 3 tens and 6 ones.

You try Example 5.

Example 5. In the whole number 6735

(a) The thousands digit is _____ .

(b) The ones digit is _____ .

(c) The tens digit is _____ .

(d) The hundreds digit is _____ .

Check your work on page 8. ▶

Example 6. The population of the United States in 1970 was 203,235,298. Write the word form for this whole number.

Solution. 203,235,298 is two hundred three million, two hundred thirty-five thousand, two hundred ninety-eight.

⑉➡ Trial Run

_____ **1.** Write the word form for 21,342.

_____ **2.** Write the number form for thirty thousand twelve.

_____ **3.** Write the meaning of 203,506.

_____ **4.** In the whole number 17,296 name the thousands digit.

_____ **5.** In the whole number 736,052 name the tens digit.

_____ **6.** Gene purchased a used car that had been driven 73,296 miles. Write the word form for this whole number.

Answers are on page 9.

Rounding Whole Numbers

In everyday conversation, we often use approximate values of whole numbers. For instance, if the attendance at a ball game were 11,231, the announcer might say that the attendance was "about 11,000." Or, if you paid $14,595 for your new car, you might say that you had paid "about $14,600" for the car. In each of these cases, the actual number was **rounded** to a particular place.

To round a whole number such as 27 to the *tens* place, we must decide whether that number is closer to 20 (2 tens) or 30 (3 tens). Perhaps the number line will help us decide.

```
  ←—+——+——+——+——+——+——+——+——+——+——+——+——+——→
    19  20  21  22  23  24  25  26  27  28  29  30  31
                                    ↑
```

It is clear that 27 is closer to 30 than to 20, so rounding to the tens place

$$27 \doteq 30$$

The symbol $\doteq$ means "is approximately equal to".

You try Example 7.

Example 7. Use the number line to round 52 to the tens place.

Solution

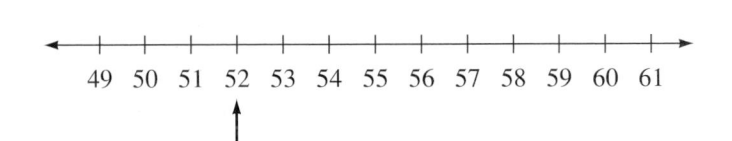

We see that 52 is closer to _____ than to _____ . Therefore, rounding to the tens place, 52 $\doteq$ _____ .

Check your work on page 8. ▶

Whole numbers can be rounded to any place we choose, but the number-line method can become awkward when rounding to higher places. Instead, we may use the following method.

Rounding Whole Numbers

1. Locate the rounding place in the number.
2. Look at the digit in the next place to the right of the rounding place.
3. If the digit in the next place is less than 5, keep the same digit in the rounding place and replace all digits to the right of the rounding place with zeros.
4. If the digit in the next place is 5 or greater, increase the digit in the rounding place by 1 and replace all digits to the right of the rounding place with zeros.
5. State that the original number is approximately equal to ($\doteq$) the rounded number.

Example 8. Round 3832 to the thousands place.

Solution

3832   Locate the rounding place.
↑

3832   Look at the next place to the right.
↑

4000   Since 8 is greater than 5, increase 3 to 4 in the thousands place and replace digits to the right with zeros.

Rounded to the thousands place, 3832 $\doteq$ 4000.

Example 9. Round 213,978 to the ten thousands place.

Solution

213,978   Locate the rounding place.
↑

213,978   Look at the next place to the right.
↑

210,000   Since 3 is less than 5, keep 1 in the ten thousands place and replace digits to the right with zeros.

Rounded to the ten thousands place, 213,978 $\doteq$ 210,000.

You complete Example 10.

Example 10. Round 7,522,123 to the millions place.

Solution

7,522,123   Locate the rounding place.
↑

7,522,123   Look at the next place to the right.
　↑

_____   Since the digit in the next place is ____ , the digit in the rounding place will be ____ .

Rounded to the millions place, 7,522,123 ≐ ____ .

Check your work on page 9. ▶

⫸ Trial Run

_____ **1.** Use a number line to round 76 to the tens place.

Use the rules to round each whole number to the indicated place.

_____ **2.** 21,502 to the hundreds place.

_____ **3.** 21,502 to the thousands place.

_____ **4.** 736,052 to the hundred thousands place.

_____ **5.** Gina paid $83,750 for her new home. Round the price to the nearest thousand dollars.

Answers are on page 9.

▶ Examples You Completed

1. There are 3 feet in a yard.
2. There are 12 eggs in a dozen.
3. There are 2 vowels in the word pizza.
4. There are 50 states in the United States.
5. There are 100 cents in a dollar.
6. There have been 0 female U.S. Presidents.

Example 5. In the whole number 6735
　(a) The thousands digit is 6.
　(b) The ones digit is 5.
　(c) The tens digit is 3.
　(d) The hundreds digit is 7.

Example 7. Use the number line to round 52 to the tens place.

Solution

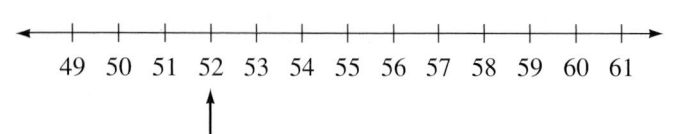

We see that 52 is closer to 50 than to 60. Therefore, rounding to the tens place, 52 ≐ 50.

Example 10. Round 7,522,123 to the millions place.

Solution

7,522,123   Locate the rounding place.
↑

7,522,123   Look at the next place to the right.
 ↑

8,000,000   Since the digit in the next place is 5, the digit in the rounding place will be 8.

Rounded to the millions place, 7,522,123 ≐ 8,000,000.

Answers to Trial Runs

page 2   1. 7   2. 1   3. 32   4. 36   5. 0

page 4   1. – 4.

5. A-1   6. B-5   7. C-9   8. >   9. <   10. <

page 6   **1.** 21,342 is twenty-one thousand, three hundred forty-two.   **2.** 30,012
3. 203,506 means 2 hundred thousands and 3 thousands and 5 hundreds and 6 ones.
4. The thousands digit is 7.   **5.** The tens digit is 5.
6. Seventy-three thousand, two hundred ninety-six.

page 8   1. 80   2. 21,500   3. 22,000   4. 700,000   5. $84,000

EXERCISE SET 1.1

Fill in the blanks with the correct whole number.

_____ **1.** There are _____ inches in a foot.

_____ **2.** During leap year, February has _____ days.

_____ **3.** A baseball team has _____ players.

_____ **4.** On Memorial Day, all auto racing fans would like to be at the Indianapolis _____ .

Plot each whole number on the number line.

5. 1   **6.** 12   **7.** 0   **8.** 5

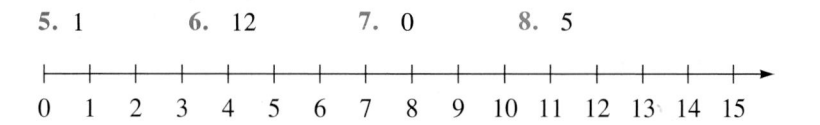

Name the whole number corresponding to each point.

9. _A_   **10.** _B_   **11.** _C_   **12.** _D_

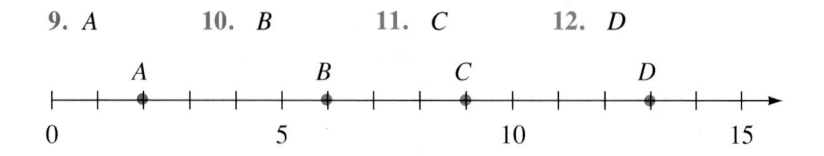

Place a < or > symbol between the numbers.

13. 99 _____ 100   **14.** 2 _____ 0

15. Write the word form for 47,592.

_____

16. Write the word form for 300,920.

_____

17. Write the number form for fifty-six thousand, seven hundred four. _____

18. Write the number form for four hundred sixty thousand, eight hundred. _____

19. Write the meaning of 93,005.

_____

20. Write the meaning of 310,723.

_____

21. Write the meaning of 1,003,576.

_____

22. Write the meaning of 4007.

_____

_____ 23. In the whole number 157,365 name the hundreds digit.

_____ 24. In the whole number 4,072,650 name the hundred thousands digit.

_____ 25. In the whole number 725,000 name the ten thousands digit.

_____ 26. In the whole number 31,416 name the tens digit.

27. Charley farms 5236 acres of land. Write the word form for this whole number.

_____

28. Canada has a total area of 3,694,863 square miles. Write the word form for this whole number.

_____

29. The number of injuries to schoolchildren in one city during the school year on school grounds was 7175. Write the word form for this whole number.

_____

30. In Coppertown, the number of traffic accidents caused by drunken driving was 461. Write the word form for this whole number.

_____

Round each whole number to the indicated place.

_____ 31. 38; tens place   _____ 32. 61; tens place

_____ 33. 573; tens place   _____ 34. 708; tens place

_____ 35. 1549; hundreds place   _____ 36. 871; hundreds place

_____ 37. 2699; thousands place   _____ 38. 26,499; thousands place

_____ 39. 32,480; ten thousands place   _____ 40. 50,500; ten thousands place

_____ 41. $570,698; hundred thousands place   _____ 42. $1,287,300; hundred thousands place

☆ Stretching the Topics _____

_____ 1. Write the word form for 11,101,001.

_____ 2. Write the meaning of 205,360,007.

_____ 3. In a certain whole number the thousands digit is one half the hundreds digit and the tens digit is twice the hundreds digit. If the sum of the hundreds digit and the units digit is 8 and the units digit is 6, write the whole number.

Check your answers in the back of your book.

If you can do **Checkup 1.1,** you are ready to go on to Section 1.2.

✓ **CHECKUP 1.1**

Fill in the blanks with the correct whole number.

_____ 1. There are _____ months in a year.

_____ 2. The national speed limit for rural interstate highways is _____ miles per hour.

3. Write the word form for 30,170.

   _____

_____ 4. Write the number form for twenty-six million two.

5. Write the meaning of 73,452.

   _____

6. Write the meaning of 150,000.

   _____

_____ 7. In the whole number 73,857 name the ones digit.

_____ 8. In the whole number 350,000 name the thousands digit.

9. The number of potential female voters in a country is 42,407,943. Write the word form for this whole number.

   _____

_____ 10. The enrollment in technical programs at a certain community college has increased this past year by 5360. Round this whole number to the hundreds place.

Check your answers in the back of your book.

If You Missed Problems:	You Should Review Examples:
1, 2	—
3–6	4
7, 8	5
9	6
10	8–10

1.2 Adding Whole Numbers

If you have a collection of 6 cassettes and you purchase 5 more cassettes, how do you figure out how many cassettes you have altogether? Probably you would *not* line them up and start counting: 1, 2, 3, 4, and so on. Certainly this method would give you the correct answer, but you learned long ago to use **addition** to find out the answer to such a problem.

We turn our attention now to sharpening our addition skills. Remember that when we add numbers, the result is called the **sum.**

Adding One-Digit Numbers

Earlier we used the number line to picture the positions of the whole numbers. The number line is sometimes useful in performing addition. For instance, to find the sum of 5 and 2, we can first locate the point corresponding to 5 and then count 2 more units to the *right,* ending up at 7.

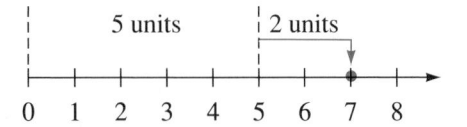

You are probably right if you think it is easier to learn the addition fact $5 + 2 = 7$ than to use a number line to find this sum. However, the number line does provide us with a picture of an addition problem.

To find the total number of cassettes in the previous problem, we must *add* 6 cassettes and 5 cassettes. We may write this problem in either of two ways: horizontally or vertically. Students often prefer the vertical method, but the horizontal method saves space. We must learn to find sums in either way. Of course the sum of 6 and 5 is 11, no matter which method we use.

$$\begin{array}{cc}
\textit{horizontally} & \textit{vertically} \\
6 + 5 = 11 & \begin{array}{r} 6 \\ +5 \\ \hline 11 \end{array}
\end{array}$$

If you have never mastered the basic addition facts for all the one-digit numbers, now is the time to learn them. You cannot hope to succeed with the rest of your work in arithmetic unless you know these facts.

Perhaps the following addition table will help. To find the sum of two numbers, locate the first number in the far left column and locate the second number in the top row. Follow the first number across its row and the second number down its column. The sum of 6 and 5 appears where the 6-row meets the 5-column.

+	0	1	2	3	4	5	6	7	8	9
0	0	1	2	3	4	5	6	7	8	9
1	1	2	3	4	5	6	7	8	9	10
2	2	3	4	5	6	7	8	9	10	11
3	3	4	5	6	7	8	9	10	11	12
4	4	5	6	7	8	9	10	11	12	13
5	5	6	7	8	9	10	11	12	13	14
6	6	7	8	9	10	11	12	13	14	15
7	7	8	9	10	11	12	13	14	15	16
8	8	9	10	11	12	13	14	15	16	17
9	9	10	11	12	13	14	15	16	17	18

You should not memorize the construction of this table, but you must learn the addition facts contained in it. Practice those sums that do not come to you automatically by reciting them aloud and writing them down. Try Example 1.

Example 1. Fill in the table.

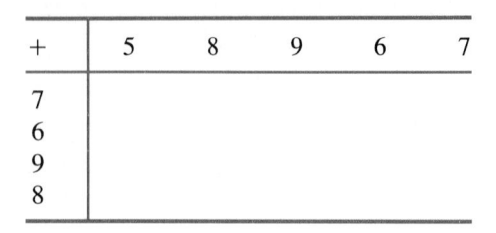

+	5	8	9	6	7
7					
6					
9					
8					

Check your table on page 19. ▶

To add more than two numbers in a sum, we may do the addition in any order. Some students like to proceed down the column of numbers, while others like to proceed up the column. The result will be the same. For instance, either

$$
\begin{array}{c}
6 \\
2 \\
+3
\end{array}
\left.\begin{array}{c} \\ \end{array}\right\} 8 \text{or}
\begin{array}{c}
6 \\
2 \\
+3
\end{array}
\left.\begin{array}{c} \\ \end{array}\right\} 6 5
$$

$$11 11$$

will yield the same answer.

Example 2. Find the sum 6 + 8 + 2.

Solution

$$
\begin{array}{ll}
\underline{6 + 8} + 2 & \text{or} \quad 6 + \underline{8 + 2} \\
= 14 + 2 & = 6 + 10 \\
= 16 & = 16
\end{array}
$$

IIII➡ Trial Run

Find the sums.

_____ 1. 5 + 9   _____ 2. 6 + 7   _____ 3. 9 + 8

_____ 4. 4 + 3 + 6   _____ 5. 7 + 2 + 6   _____ 6.   7
          +9

_____ 7.   5   _____ 8.   6   _____ 9.   5
         1   3   3
    +8   +4   1
                                            +6

Answers are on page 19.

Adding Larger Whole Numbers

So far we have discussed finding sums of one-digit numbers. Let's recall what one-digit numbers mean as we find the sum of 3 and 4.

$$\begin{array}{r} 3 \text{ ones} \\ +4 \text{ ones} \\ \hline 7 \text{ ones} \end{array}$$

To find a sum such as 12 + 35, we are really adding

$$\begin{array}{rr} 1 \text{ ten and 2 ones} & 12 \\ +3 \text{ tens and 5 ones} & +35 \\ \hline 4 \text{ tens and 7 ones} & 47 \end{array}$$

On the right we found this sum without the word meanings by first adding the digits in the ones place and then adding the digits in the tens place.

This method of adding digits in the same place will continue to work when we try to add three-digit numbers, four-digit numbers, and so on. Just remember to line up like digits beneath each other in writing down the problem.

Example 3. Find the sum of 966 and 30.

Solution

$$\begin{array}{r} 966 \\ +\ 30 \\ \hline 996 \end{array}$$

You try Example 4.

Example 4. Find the sum 32 + 821 + 14.

Solution

$$\begin{array}{r} 32 \\ 821 \\ +\ 14 \\ \hline \end{array}$$

Check your work on page 19. ▶

Notice how we carefully line up the ones digits, tens digits, and hundreds digits beneath each other. Then we begin our addition with the ones digits and move to the left.

⭢ Trial Run

Find the sums.

_____ 1. 13 + 5

_____ 2. 26 + 42

_____ 3. 53 + 245

_____ 4. 165 + 821

_____ 5. $\begin{array}{r} 832 \\ +107 \\ \hline \end{array}$

_____ 6. $\begin{array}{r} 231 \\ 42 \\ +515 \\ \hline \end{array}$

_____ 7. $\begin{array}{r} 18 \\ 20 \\ +51 \\ \hline \end{array}$

_____ 8. $\begin{array}{r} 6342 \\ +1057 \\ \hline \end{array}$

Answers are on page 19.

Suppose we wish to find the sum 47 + 6. Beginning with the ones digits, we find that 7 + 6 = 13. But 13 is 1 ten and 3 ones. Only the 3 belongs in the ones place of our answer, and the 1 belongs in the tens place. Our thinking goes like this.

		1 ten	1 ten	1
47	4 tens and 7 ones	4 tens and 7 ones	4 tens and 7 ones	47
+ 6	+ 6 ones	+ 6 ones	+ 6 ones	+ 6
	13 ones	3 ones	5 tens and 3 ones	53

Of course you recall that we do *not* write down these five steps in finding such a sum. Instead we simply remember to regroup or **carry** the 1 to the tens place when the sum of the ones digits is 13. The carrying process was shown in the last step above. Whenever the sum of a column of digits is more than 9, you must carry to the next place to the left. Try Example 5.

Example 5. Find the sum 38 + 49.

Solution

$$\begin{array}{r} 38 \\ +49 \\ \hline \end{array}$$

Check your work on page 19. ▶

Example 6. Add 189 and 270.

Solution

$$\begin{array}{r} 1 \\ 189 \\ +270 \\ \hline 459 \end{array}$$

Now can you see why it is important that you know your basic addition facts? Let's try two more examples with three numbers so you can see that the method does not change.

Example 7. Find

22,834 + 9,210 + 16,975.

Solution

$$\begin{array}{r} 12\ 1 \\ 22{,}834 \\ 9{,}210 \\ 16{,}975 \\ \hline 49{,}019 \end{array}$$

Now complete Example 8.

Example 8. Find 67 + 235 + 574.

Solution

$$\begin{array}{r} 1 \\ 67 \\ 235 \\ 574 \\ \hline 6 \end{array}$$

Check your work on page 19. ▶

⫸ Trial Run

Find the sums.

_____ 1. 9 + 18

_____ 2. 52 + 38

_____ 3. 26 + 35 + 79

_____ 4. 89 + 236 + 429

_____ 5.   2378
              +4729

_____ 6.   5086
                239
              +7985

_____ 7.   13,745
              +95,376

_____ 8.   36,286
                45,795
              +23,021

Answers are on page 19.

▶ **Examples You Completed** _____

Example 1

+	5	8	9	6	7
7	12	15	16	13	14
6	11	14	15	12	13
9	14	17	18	15	16
8	13	16	17	14	15

Example 4. Find the sum 32 + 821 + 14.

Solution

$$
\begin{array}{r}
32 \\
821 \\
+\ 14 \\
\hline
867
\end{array}
$$

Example 5. Find the sum 38 + 49.

Solution

$$
\begin{array}{r}
1 \\
38 \\
+49 \\
\hline
87
\end{array}
$$

Example 8. Find 67 + 235 + 574.

Solution

$$
\begin{array}{r}
11 \\
67 \\
235 \\
574 \\
\hline
876
\end{array}
$$

Answers to Trial Runs _____

page 16   **1.** 14   **2.** 13   **3.** 17   **4.** 13   **5.** 15   **6.** 16   **7.** 14   **8.** 13   **9.** 15

page 17   **1.** 18   **2.** 68   **3.** 298   **4.** 986   **5.** 939   **6.** 788   **7.** 89   **8.** 7399

page 18   **1.** 27   **2.** 90   **3.** 140   **4.** 754   **5.** 7107   **6.** 13,310   **7.** 109,121   **8.** 105,102

EXERCISE SET 1.2

Find the sums.

_____ 1. 6 + 2   _____ 2. 5 + 4

_____ 3. 8 + 7   _____ 4. 9 + 6

_____ 5. 4 + 7   _____ 6. 8 + 9

_____ 7. 3 + 2 + 8   _____ 8. 6 + 3 + 4

_____ 9. 9 + 7 + 3   _____ 10. 8 + 9 + 2

_____ 11.　　6   _____ 12.　　3
　　　　　　　　+8   　　　　　　+9

_____ 13.　　7   _____ 14.　　8
　　　　　　　　3   　　　　　　1
　　　　　　　　+8   　　　　　　+7

_____ 15.　　3   _____ 16.　　4
　　　　　　　　6   　　　　　　9
　　　　　　　　+8   　　　　　　+6

_____ 17. 14 + 5   _____ 18. 16 + 2

_____ 19. 23 + 15   _____ 20. 42 + 57

_____ 21. 38 + 81   _____ 22. 36 + 92

_____ 23. 271 + 827   _____ 24. 356 + 642

_____ 25. 171 + 928   _____ 26. 123 + 574

_____ 27. 95 + 203   _____ 28. 78 + 520

_____ 29.　　273   _____ 30.　　392
　　　　　　　　+514   　　　　　　+706

_____ 31.　　423   _____ 32.　　502
　　　　　　　　204   　　　　　　315
　　　　　　　　+ 52   　　　　　　+ 81

_____ 33.　　33   _____ 34.　　13
　　　　　　　　41   　　　　　　71
　　　　　　　　+52   　　　　　　+24

_____ 35.　　65   _____ 36.　　51
　　　　　　　　203   　　　　　　107
　　　　　　　　+ 21   　　　　　　+ 20

_____ 37.　　3247   _____ 38.　　6542
　　　　　　　　+4521   　　　　　　+1327

—————— 39. 2030
              +7829

—————— 40. 5642
              +4040

—————— 41. 15 + 8

—————— 42. 6 + 19

—————— 43. 29 + 32

—————— 44. 58 + 63

—————— 45. 57 + 39 + 70

—————— 46. 80 + 79 + 68

—————— 47. 39 + 276 + 583

—————— 48. 56 + 479 + 231

—————— 49. 32 + 45 + 64 + 97

—————— 50. 86 + 23 + 40 + 58

—————— 51. 4839
              +7148

—————— 52. 9762
              +3495

—————— 53. 3654
              +7346

—————— 54. 4893
              +6107

—————— 55. 7392
               364
              +1269

—————— 56. 3689
               240
              +7149

—————— 57. 15,369
              +28,236

—————— 58. 23,976
              +94,678

—————— 59. 52,769
              36,703
              +40,205

—————— 60. 78,320
              22,721
              +63,582

—————— 61. 765
              803
               94
              279
             +650

—————— 62. 223
              452
               87
              365
             +248

—————— 63. 5460
              7531
              8205
             + 463

—————— 64. 2860
              1945
               720
             +9735

☆ Stretching the Topics

Find the sums.

—————— 1. 387,273 + 284,007 + 89,710 + 379 + 3542

—————— 2. 48 + 9 + 18 + 29 + 58 + 37 + 69 + 89 + 8 + 79 + 98 + 89

—————— 3. 1001 + 10,001 + 101 + 110,001 + 10,000,000 + 101,110,001

Check your answers in the back of your book.

If you can find the sums in **Checkup 1.2,** you are ready to go on to Section 1.3.

✔ CHECKUP 1.2

Find the sums.

_____ 1. 9 + 8

_____ 2. 6 + 9 + 3

_____ 3. 42 + 27

_____ 4. 279 + 38 + 973

_____ 5. 808
    + 757

_____ 6. 4895
    + 9605

_____ 7. 7196
    307
    + 5587

_____ 8. 52,379
    28,786
    + 10,203

_____ 9. 103
    5769
    + 89

_____ 10. 463 + 1245 + 53 + 279

Check your answers in the back of your book.

If You Missed Problems:	You Should Review Examples:
1, 2	2
3	3, 4
4–10	5–8

1.3 Subtracting Whole Numbers

Subtracting Using Basic Addition Facts

If you go to the store with $9 and spend $4, you might like to figure out how much money you have left without counting. Such a problem involves the operation of **subtraction** and we recall that the answer in a subtraction problem is called the **difference.**

We can use a number line to work a subtraction problem. In our example we wish to find the difference when $4 is subtracted from $9 and we write the difference as $9 - 4$.

To do this subtraction using the number line, we start at the point corresponding to 9 and then count 4 units to the *left,* ending up at 5.

Therefore, we see that

$$\text{horizontally} \qquad \text{vertically}$$

$$9 - 4 = 5 \quad \text{or} \quad \begin{array}{r} 9 \\ -4 \\ \hline 5 \end{array}$$

and we conclude that you are left with $5.

Notice that this subtraction statement corresponds to an addition statement. We know that

$$9 - 4 = 5 \quad \text{because} \quad 9 = 5 + 4$$

In fact, *every* subtraction statement corresponds to an addition statement. For example,

$$10 - 3 = 7 \quad \text{because} \quad 10 = 7 + 3$$
$$5 - 1 = 4 \quad \text{because} \quad 5 = 4 + 1$$
$$12 - 9 = 3 \quad \text{because} \quad 12 = 3 + 9$$

We can always "check" a subtraction problem by writing the corresponding addition statement.

Example 1. Find each difference and check.

$$9 - 2 = 7 \quad \text{because} \quad 9 = 7 + 2$$
$$16 - 7 = 9 \quad \text{because} \quad 16 = 9 + 7$$
$$11 - 3 = 8 \quad \text{because} \quad 11 = 8 + 3$$

You complete Example 2.

Example 2. Fill in the missing numbers.

$$15 - 10 = \underline{\quad} \quad \text{because} \quad 15 = \underline{\quad} + 10$$
$$6 - 4 = \underline{\quad} \quad \text{because} \quad 6 = \underline{\quad} + 4$$
$$8 - \underline{\quad} = 1 \quad \text{because} \quad 8 = 1 + \underline{\quad}$$
$$10 - \underline{\quad} = 4 \quad \text{because} \quad 10 = 4 + \underline{\quad}$$
$$\underline{\quad} - 3 = 11 \quad \text{because} \quad \underline{\quad} = 11 + 3$$

Check your work on page 27. ▶

As you can see, it's easier to learn the basic subtraction facts if you already have learned the basic addition facts. Again, the key is lots of practice.

If you wonder about checking a subtraction problem written vertically, look at these examples.

$$\left.\begin{array}{r} 9 \\ -4 \\ \hline 5 \end{array}\right\} \; 5 + 4 = 9 \; \checkmark \qquad \left.\begin{array}{r} 12 \\ -8 \\ \hline 4 \end{array}\right\} \; 4 + 8 = 12 \; \checkmark$$

Once again we check a subtraction problem by writing the corresponding addition statement. You complete Example 3.

Example 3. Subtract and check.

$$\left.\begin{array}{r} 18 \\ -9 \\ \hline \end{array}\right\} \; \underline{} + 9 = 18 \; \checkmark \qquad \left.\begin{array}{r} 16 \\ -9 \\ \hline \end{array}\right\} \; \underline{} + 9 = 16 \; \checkmark$$

$$\left.\begin{array}{r} 11 \\ -7 \\ \hline \end{array}\right\} \; \underline{} + 7 = 11 \; \checkmark \qquad \left.\begin{array}{r} 13 \\ -8 \\ \hline \end{array}\right\} \; \underline{} + 8 = 13 \; \checkmark$$

Check your work on page 27. ▶

⫸ Trial Run

Fill in the missing numbers.

_____ 1. $8 - 3 = $ _____   because $8 = $ _____ $ + 3$

_____ 2. $12 - 5 = $ _____   because $12 = $ _____ $ + 5$

_____ 3. $13 - $ _____ $ = 6$ because $13 = 6 + $ _____

_____ 4. $17 - $ _____ $ = 9$ because $17 = 9 + $ _____

Subtract.

_____ 5. $\begin{array}{r} 11 \\ -5 \\ \hline \end{array}$ _____ 6. $\begin{array}{r} 16 \\ -7 \\ \hline \end{array}$ _____ 7. $\begin{array}{r} 13 \\ -8 \\ \hline \end{array}$ _____ 8. $\begin{array}{r} 12 \\ -9 \\ \hline \end{array}$

Answers are on page 28.

Subtracting Numbers with More Digits

Remember that in adding numbers with more than one digit we carefully lined up beneath each other the ones digits, the tens digits, and so on. Then we *added* the ones digits, the tens digits, and so on.

In subtraction we proceed in the same way, lining up the digits beneath each other and then subtracting in each place from *right to left*.

$$\begin{array}{r} 684 \\ -372 \\ \hline 312 \end{array} \qquad \text{CHECK:} \qquad \begin{array}{r} 372 \\ 312 \\ \hline 684 \; \checkmark \end{array}$$

Example 4. Subtract and check.

$$
\begin{array}{r} 938 \\ -221 \\ \hline 717 \end{array} \Big\}
\quad
\begin{array}{r} 221 \\ +717 \\ \hline 938 \ \checkmark \end{array}
\qquad
\begin{array}{r} 815 \\ -\ 11 \\ \hline 804 \end{array} \Big\}
\quad
\begin{array}{r} 11 \\ +804 \\ \hline 815 \ \checkmark \end{array}
\qquad
\begin{array}{r} 579 \\ -201 \\ \hline 378 \end{array} \Big\}
\quad
\begin{array}{r} 201 \\ +378 \\ \hline 579 \ \checkmark \end{array}
$$

How do we proceed in a subtraction problem such as

$$
\begin{array}{r} 35 \\ -18 \end{array}
$$

where we must subtract 8 from 5 in the ones place? We must regroup or **borrow** a 1 from the tens digit. In other words, we borrow 1 ten (which we can now think of as 10 ones) and we now have 15 ones instead of 5 ones. Our tens digit changes from 3 to 2. Recall the meaning of place value as we illustrate the borrowing process.

$$
\begin{array}{l} 3 \text{ tens and } 5 \text{ ones} \\ -1 \text{ ten and } 8 \text{ ones} \\ \hline \end{array}
\qquad
\begin{array}{l} 2 \text{ tens and } 1 \text{ ten and } 5 \text{ ones} \\ -1 \text{ ten} \qquad\qquad\quad \text{and } 8 \text{ ones} \\ \hline \end{array}
\qquad
\begin{array}{l} 2 \text{ tens and } 15 \text{ ones} \\ -1 \text{ ten and } \ \ 8 \text{ ones} \\ \hline 1 \text{ ten and } \ \ 7 \text{ ones} \end{array}
$$

We show our borrowing in the following way.

$$
\begin{array}{r} 35 \\ -18 \\ \hline \end{array}
\qquad
\begin{array}{r} {\scriptstyle 2\ 15} \\ \cancel{35} \\ -18 \\ \hline \end{array}
\qquad
\begin{array}{r} {\scriptstyle 2\ 15} \\ \cancel{35} \\ -18 \\ \hline 17 \end{array}
$$

We first subtracted 8 from 15, then 1 from 2. We can check our answer by the usual method (18 + 17 = 35).

Let's try a few more examples.

Example 5. Find 523 − 219 and check.

Solution. We cannot subtract 9 from 3, so we must borrow 1 ten from the tens place.

$$
\begin{array}{r} {\scriptstyle 1\ 13} \\ 5\cancel{2}\cancel{3} \\ -219 \\ \hline 304 \end{array} \Big\}
\quad
\begin{array}{r} 219 \\ +304 \\ \hline 523 \ \checkmark \end{array}
$$

Example 6. Find 729 − 354 and check.

Solution. There is no trouble here in the ones place, but in the tens place we cannot subtract 5 from 2. From the hundreds place we borrow 1 hundred (or 10 tens). The tens digit becomes 12 and the hundreds digit becomes 6.

$$
\begin{array}{r} {\scriptstyle 6\ 12} \\ \cancel{7}\cancel{2}9 \\ -354 \\ \hline 375 \end{array} \Big\}
\quad
\begin{array}{r} 354 \\ +375 \\ \hline 729 \ \checkmark \end{array}
$$

In some cases it is necessary to borrow more than once in a subtraction problem. The bookkeeping in such a problem can be confusing at first, so let's use several steps to find $852 - 479$.

Borrow 1 ten. Subtract ones.	Borrow 1 hundred. Subtract tens.	Subtract hundreds.	CHECK:
4 12	7 14 12	7 14 12	
8 5 2	8 5 2	8 5 2	479
− 4 7 9	− 4 7 9	− 4 7 9	+ 3 7 3
3	7 3	3 7 3	852 √

You carefully complete Example 7.

Example 7. Find $2625 - 1158$ and check.

Solution

		1 15	5 11 15	
	2625	2 6 2 5	2 6 2 5	
	− 1158	− 1 1 5 8	− 1 1 5 8 ⎱	1158
		7	7 ⎰	+
				2625 √

Check your work on page 28. ▶

Example 8. Find $16,543 - 9876$ and check.

Solution

		3 13	4 13 13	5 14 13 13	
16,543	16,5 4 3	16,5 4 3	1 6,5 4 3	9876
− 9,876	− 9,876	− 9,876	− 9,876 ⎱	+ 6667
	7	6 7	6,6 6 7 ⎰	16,543 √

What happens if we wish to borrow 1 from the next place but there is a 0 in that place? In such a case we must move *one more* place to the left and borrow 1. Look at this example.

	Borrow 1 hundred.	Borrow 1 ten. Subtract ones.	Subtract tens. Subtract hundreds.	CHECK:
		9	9	
	5 10	5 10 14	5 10 14	
604	6 0 4	6 0 4	6 0 4	38
− 38	− 3 8	− 3 8	− 3 8	+ 566
		6	5 6 6	604 √

Now carefully complete Example 9.

Example 9. Find $4023 - 785$ and check.

Solution. Let's show our steps.

		1 13	3 10 1 13	9 3 10 1 13	
	4023	40 2 3	4 0 2 3	4 0 2 3	785
	− 785	− 785	− 7 8 5	− 7 8 5 ⎱	+
		8	8	8 ⎰	4023 √

Check your work on page 28. ▶

Example 10. Find $1001 - 859$ and check.

Solution

$$
\begin{array}{c}
\overset{10}{\cancel{1}}001 \\
-859
\end{array}
\qquad
\begin{array}{c}
\overset{9}{\cancel{10}}10 \\
\cancel{1}\cancel{0}\cancel{0}1 \\
-859
\end{array}
\qquad
\left.
\begin{array}{c}
\overset{99}{\cancel{10}\cancel{10}}11 \\
\cancel{1}\cancel{0}\cancel{0}\cancel{1} \\
-859 \\
\hline
142
\end{array}
\right\}
\begin{array}{r}
859 \\
+142 \\
\hline
1001 \checkmark
\end{array}
$$

Example 11. Find $3000 - 1983$ and check.

Solution

$$
\begin{array}{c}
\overset{2\ 10}{\cancel{3}\cancel{0}}00 \\
-1983
\end{array}
\qquad
\begin{array}{c}
\overset{9}{2\cancel{10}}10 \\
\cancel{3}\cancel{0}\cancel{0}0 \\
-1983
\end{array}
\qquad
\left.
\begin{array}{c}
\overset{9\ 9}{2\cancel{10}\cancel{10}}10 \\
\cancel{3}\cancel{0}\cancel{0}\cancel{0} \\
-1983 \\
\hline
1017
\end{array}
\right\}
\begin{array}{r}
1983 \\
+1017 \\
\hline
3000 \checkmark
\end{array}
$$

▥➡ Trial Run

Subtract and check.

_____ 1. $\begin{array}{r} 837 \\ -213 \\ \hline \end{array}$
_____ 2. $\begin{array}{r} 65 \\ -29 \\ \hline \end{array}$
_____ 3. $\begin{array}{r} 526 \\ -398 \\ \hline \end{array}$

_____ 4. $\begin{array}{r} 5726 \\ -2948 \\ \hline \end{array}$
_____ 5. $\begin{array}{r} 403 \\ -56 \\ \hline \end{array}$
_____ 6. $\begin{array}{r} 3000 \\ -294 \\ \hline \end{array}$

Answers are on page 28. ▶

▶ Examples You Completed

Example 2

$$
\begin{array}{llll}
15 - 10 = 5 & \text{because} & 15 = 5 + 10 \\
6 - 4 = 2 & \text{because} & 6 = 2 + 4 \\
8 - 7 = 1 & \text{because} & 8 = 1 + 7 \\
10 - 6 = 4 & \text{because} & 10 = 4 + 6 \\
14 - 3 = 11 & \text{because} & 14 = 11 + 3
\end{array}
$$

Example 3

$$
\left.\begin{array}{r} 18 \\ -9 \\ \hline 9 \end{array}\right\} 9 + 9 = 18 \checkmark
\qquad
\left.\begin{array}{r} 16 \\ -9 \\ \hline 7 \end{array}\right\} 7 + 9 = 16 \checkmark
$$

$$
\left.\begin{array}{r} 11 \\ -7 \\ \hline 4 \end{array}\right\} 4 + 7 = 11 \checkmark
\qquad
\left.\begin{array}{r} 13 \\ -8 \\ \hline 5 \end{array}\right\} 5 + 8 = 13 \checkmark
$$

Example 7

$$
\begin{array}{r}
2625 \\
-1158 \\
\hline
\end{array}
\qquad
\begin{array}{r}
{}^{1\,15} \\
26\cancel{2}\cancel{5} \\
-1158 \\
\hline
7
\end{array}
\qquad
\begin{array}{r}
{}^{5\,11\,15} \\
2\cancel{6}\cancel{2}\cancel{5} \\
-1158 \\
\hline
1467
\end{array}
\left.\right\}
\begin{array}{r}
1158 \\
+1467 \\
\hline
2625\ \checkmark
\end{array}
$$

Example 9. Find $4023 - 785$ and check.

Solution. Let's show our steps.

$$
\begin{array}{r}
4023 \\
-\ 785 \\
\hline
\end{array}
\qquad
\begin{array}{r}
{}^{1\,13} \\
402\cancel{3} \\
-\ 785 \\
\hline
8
\end{array}
\qquad
\begin{array}{r}
{}^{3\,10\,1\,13} \\
4\cancel{0}\cancel{2}\cancel{3} \\
-\ 785 \\
\hline
8
\end{array}
\qquad
\begin{array}{r}
{}^{9} \\
{}^{3\,10\,11\,13} \\
4\cancel{0}\cancel{2}\cancel{3} \\
-\ 785 \\
\hline
3238
\end{array}
\left.\right\}
\begin{array}{r}
785 \\
+3238 \\
\hline
4023\ \checkmark
\end{array}
$$

Answers to Trial Runs

page 24   1. 5   2. 7   3. 7   4. 8   5. 6   6. 9   7. 5   8. 3

page 27   1. 624   2. 36   3. 128   4. 2778   5. 347   6. 2706

EXERCISE SET 1.3

Fill in the missing numbers.

1. $13 - 5 =$ _____ because $13 =$ _____ $+ 5$ 2. $17 - 9 =$ _____ because $17 =$ _____ $+ 9$

3. $13 - 7 =$ _____ because $13 =$ _____ $+ 7$ 4. $7 - 3 =$ _____ because $7 =$ _____ $+ 3$

5. $12 -$ _____ $= 7$ because $12 = 7 +$ _____   6. $15 -$ _____ $= 8$ because $15 = 8 +$ _____

7. $9 -$ _____ $= 4$ because $9 = 4 +$ _____   8. $11 -$ _____ $= 5$ because $11 = 5 +$ _____

Subtract and check.

————— 9. $\begin{array}{r} 17 \\ -\ 8 \\ \hline \end{array}$ ————— 10. $\begin{array}{r} 13 \\ -\ 6 \\ \hline \end{array}$ ————— 11. $\begin{array}{r} 10 \\ -\ 6 \\ \hline \end{array}$

————— 12. $\begin{array}{r} 12 \\ -\ 7 \\ \hline \end{array}$ ————— 13. $\begin{array}{r} 19 \\ -11 \\ \hline \end{array}$ ————— 14. $\begin{array}{r} 23 \\ -12 \\ \hline \end{array}$

————— 15. $\begin{array}{r} 95 \\ -76 \\ \hline \end{array}$ ————— 16. $\begin{array}{r} 71 \\ -48 \\ \hline \end{array}$ ————— 17. $\begin{array}{r} 70 \\ -29 \\ \hline \end{array}$

————— 18. $\begin{array}{r} 60 \\ -42 \\ \hline \end{array}$ ————— 19. $\begin{array}{r} 31 \\ -25 \\ \hline \end{array}$ ————— 20. $\begin{array}{r} 93 \\ -85 \\ \hline \end{array}$

————— 21. $\begin{array}{r} 629 \\ -432 \\ \hline \end{array}$ ————— 22. $\begin{array}{r} 837 \\ -654 \\ \hline \end{array}$ ————— 23. $\begin{array}{r} 825 \\ -707 \\ \hline \end{array}$

————— 24. $\begin{array}{r} 685 \\ -509 \\ \hline \end{array}$ ————— 25. $\begin{array}{r} 790 \\ -248 \\ \hline \end{array}$ ————— 26. $\begin{array}{r} 620 \\ -253 \\ \hline \end{array}$

————— 27. $\begin{array}{r} 800 \\ -471 \\ \hline \end{array}$ ————— 28. $\begin{array}{r} 500 \\ -332 \\ \hline \end{array}$ ————— 29. $\begin{array}{r} 22{,}365 \\ -\ 6{,}587 \\ \hline \end{array}$

————— 30. $\begin{array}{r} 34{,}472 \\ -\ 9{,}548 \\ \hline \end{array}$ ————— 31. $\begin{array}{r} 47{,}157 \\ -46{,}679 \\ \hline \end{array}$ ————— 32. $\begin{array}{r} 36{,}345 \\ -35{,}649 \\ \hline \end{array}$

————— 33. $\begin{array}{r} 55{,}207 \\ -31{,}808 \\ \hline \end{array}$ ————— 34. $\begin{array}{r} 47{,}306 \\ -24{,}909 \\ \hline \end{array}$ ————— 35. $\begin{array}{r} 264{,}686 \\ -121{,}687 \\ \hline \end{array}$

	36.	457,524 −243,529		37.	50,000 −45,305		38.	90,000 −63,705

	39.	389,426 −128,717		40.	225,365 −174,458

☆ Stretching the Topics _____

Subtract and check.

_____ 1. 32,272,101 − 18,346,897

_____ 2. 1,000,000 − 879,376

_____ 3. 101,110,001 − 95,678,327

Check your answers in the back of your book.

If you can find the differences in **Checkup 1.3,** you are ready to go on to Section 1.4.

✓ **CHECKUP 1.3**

Fill in the missing numbers.

1. $17 - 9 = $ _____ because $17 = $ _____ $+ 9$

2. $12 - $ _____ $= 7$ because $12 = 7 + $ _____

Subtract and check.

_____ 3. $\begin{array}{r} 89 \\ -34 \\ \hline \end{array}$ _____ 4. $\begin{array}{r} 71 \\ -55 \\ \hline \end{array}$

_____ 5. $\begin{array}{r} 234 \\ -196 \\ \hline \end{array}$ _____ 6. $\begin{array}{r} 5000 \\ -3876 \\ \hline \end{array}$

_____ 7. $\begin{array}{r} 36{,}246 \\ -18{,}398 \\ \hline \end{array}$ _____ 8. $\begin{array}{r} 871{,}503 \\ -306{,}725 \\ \hline \end{array}$

Check your answers in the back of your book.

If You Missed Problems:	You Should Review Examples:
1, 2	2
3	4
4, 5	5, 6
6–8	7–11

1.4 Switching from Words to Numbers

One of the main reasons for gaining skills in arithmetic is to be able to use those skills in performing calculations that are important in our daily lives. Even if we plan to use a handheld calculator to find a sum or difference, we must first decide whether it is appropriate to add or subtract! Most everyday mathematical problems appear in word form.

1. If James deposits $37 in his checking account and his current balance is $168, what will be his new balance?
2. If Audra has already paid $1283 in federal income taxes throughout the year and her 1040 form shows that she should have paid $1311, has she overpaid or underpaid her taxes? By how much?
3. If the odometer in Carlo's car reads 32186 at the start of a trip and 32723 at the end of the trip, how far has Carlo traveled?

Before beginning a mathematics problem stated in words, we must first stop and ask ourselves whether it is necessary to add or subtract. There are some key phrases that can help us decide.

Word Phrase	Number Phrase
3 plus 2	$3 + 2$
4 and 6	$4 + 6$
sum of 17 and 13	$17 + 13$
7 more than 11	$11 + 7$
8 increased by 9	$8 + 9$
15 added to 12	$12 + 15$
15 minus 10	$15 - 10$
9 take away 3	$9 - 3$
difference between 82 and 17	$82 - 17$
9 less 5	$9 - 5$
6 less than 14	$14 - 6$
12 decreased by 3	$12 - 3$
30 subtracted from 53	$53 - 30$

Example 1. Marcia scored 81 points on her first history test. She improved that score by 8 points on her second test. What was Marcia's score on her second test?

Solution. Since Marcia *improved* her score, she must have *increased* it. By how much? By 8 points. We must find 81 points increased by 8 points so we *add*.

$$81 + 8 = 89$$

On her second test, Marcia scored 89 points.

Example 2. If Audra has paid $1283 in federal taxes but her tax bill is $1311, by how much has she underpaid her taxes?

Solution. Audra must find the *difference* between what she has paid and what she owes. We must *subtract*.

$$\begin{array}{r} \$1311 \\ -\ 1283 \\ \hline \$\ \ 28 \end{array}$$

Audra owes the IRS $28.

You complete Example 3.

Example 3. If James deposits $37 in his checking account and his current balance is $168, what will be his new balance?

Solution. Since James is *depositing* money in his checking account, we must _____ to find his new balance.

$168
___
$

James now has a balance of $_____ .

Check your work on page 37. ▶

Example 4. Now suppose that James writes one check for $17 and another check for $49. What is the total amount of these checks? What is James's new balance?

Solution. To find the total amount of these checks, we must *add*.

$$\begin{array}{r} \$17 \\ +\ 49 \\ \hline \$66 \end{array}$$

James has written checks totaling $66. From his balance of $205 he must *subtract* the amount of the checks he has written.

$$\begin{array}{r} \$205 \\ -\ 66 \\ \hline \$139 \end{array}$$

James now has a balance of $139.

The distance around the outside of a figure is called its **perimeter.** If we wish to find the perimeter of a figure, we must find the sum of the lengths of its sides. Consider these figures and their perimeters (*p*).

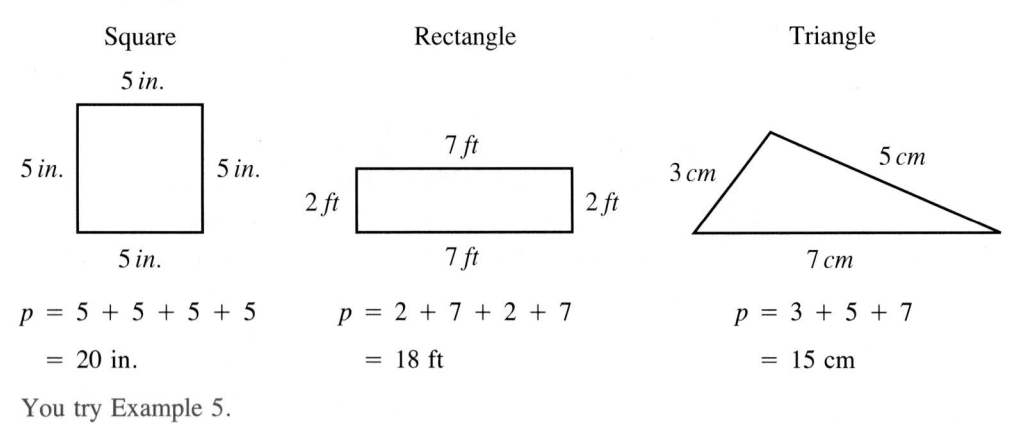

$$p = 5 + 5 + 5 + 5$$
$$= 20 \text{ in.}$$

$$p = 2 + 7 + 2 + 7$$
$$= 18 \text{ ft}$$

$$p = 3 + 5 + 7$$
$$= 15 \text{ cm}$$

You try Example 5.

Example 5. Find the number of feet of fencing needed to go around the garden illustrated.

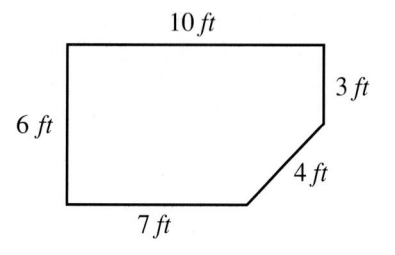

Solution. We must find the _____ of the garden.

$$p = \underline{\quad} + \underline{\quad} + \underline{\quad} + \underline{\quad} + \underline{\quad}$$
$$= \underline{\quad} \text{ ft}$$

We need _____ feet of fencing.

Check your work on page 37. ▶

⇒ Trial Run

_____ 1. If the distance from Louisville to Bowling Green on Route 65 is 123 miles and the distance from Bowling Green to Nashville on Route 65 is 68 miles, find the distance from Louisville to Nashville on Route 65.

_____ 2. A stereo can be purchased at Record Rack for $326. The same stereo costs $295 at Tape Town. Which store offers the better deal? By how much?

_____ 3. The winner in the student body presidential election had 1579 votes and the loser had 738 votes. Find the number of persons who voted in the election.

_____ 4. The first summer Steve worked for Computer Comp, Inc., he earned $1275. The next summer his salary was $1500. What was Steve's raise for the second summer of work?

_____ 5. A room is 15 feet long and 14 feet wide. How many feet of wallpaper border will be required for the room?

Answers are on page 37.

Sometimes it is necessary to add *and* subtract within the same problem. Let's consider such a situation as we solve the problem stated at the beginning of this chapter.

Example 6. Leroy has been on a diet for 6 months. He has kept the following record of his monthly weight loss or gain.

October—lost 5 pounds	January—lost 3 pounds
November—lost 7 pounds	February—lost 8 pounds
December—gained 4 pounds	March—gained 2 pounds

If Leroy weighed 192 pounds before dieting, what was his weight at the end of March?

Solution. We must *subtract* Leroy's losses and *add* his gains. Let's first compute his *total* loss and his *total* gain.

$$\text{Loss: } 5 + 7 + 3 + 8 = 23 \text{ pounds}$$

$$\text{Gain: } \qquad 4 + 2 = 6 \text{ pounds}$$

If he weighed 192 pounds when he began his diet, we must *subtract* his total loss: $192 - 23 = 169$. Then we *add* his total gain: $169 + 6 = 175$. Leroy weighed 175 pounds at the end of March.

You complete Example 7.

Example 7. During the past five years, the following changes occurred in enrollment at Hometown College.

Year #1	increased by 230
Year #2	decreased by 73
Year #3	decreased by 12
Year #4	increased by 115
Year #5	increased by 36

If enrollment was 7672 at the beginning of the five-year period, what was enrollment at the end of the five years?

Solution. First we find the *total* increase and the *total* decrease.

increase	decrease
230	73
115	+12
+ 36	

Starting with the original enrollment, we must _____ the total increase and _____ the total decrease:

At the end of the five years, enrollment was _____ .

Check your work on page 37. ▶

⟫ Trial Run

_____ **1.** In December Sandra charged purchases of $25, $38, and $19. On the first of the next month she paid $50 on her account. How much did she still owe for December purchases?

_____ **2.** In order to reach their favorite vacation spot, the Johnsons went 1029 miles by plane. Then they went 135 miles by jeep and hiked the rest of the way. If the total trip was 1180 miles, how far did they hike?

_____ **3.** During the past month, Clifton made deposits of $50, $110, and $85 to his account, which had a balance at the beginning of the month of $250. If he also wrote checks for $123, $96, $15, and $78, what was his balance at the end of the month?

_____ **4.** A weekend's sales of burgers at Snack Shack were as follows.

	Friday	Saturday
Regular hamburgers	536	509
Deluxe hamburgers	483	395
Super Snack Burger	246	302

On which day were the most burgers sold? How many more?

_____ **5.** At the beginning of the semester, Potter Hall had 678 residents. At the end of the first quarter, 38 residents moved out and 47 new residents moved in. At the end of the second quarter, 53 moved out and 28 moved in. By the end of the third quarter, 19 more residents had moved out. How many residents were living in Potter Hall at the end of the third quarter?

Answers are below.

▶ Examples You Completed

Example 3. If James deposits $37 in his checking account and his current balance is $168, what will be his new balance?

Solution. Since James is *depositing* money in his checking account, we must *add* to find his new balance.

$$
\begin{array}{r}
\$168 \\
+\ \ \ 37 \\
\hline
\$205
\end{array}
$$

James now has a balance of $205.

Example 5

Solution. We must find the *perimeter* of the garden.

$$p = 6 + 7 + 4 + 3 + 10$$
$$= 30 \text{ ft}$$

We need 30 feet of fencing.

Example 7

Solution. First we find the *total* increase and the *total* decrease.

increase	decrease
230	73
115	$+12$
$+\ 36$	85
381	

Starting with the original enrollment, we must *add* the total increase and *subtract* the total decrease.

$$
\begin{array}{r}
7672 \\
+\ 381 \\
\hline
8053
\end{array}
\qquad
\begin{array}{r}
8053 \\
-\ \ \ 85 \\
\hline
7968
\end{array}
$$

At the end of the five years, enrollment was 7968.

Answers to Trial Runs

page 35   **1.** 191 miles   **2.** Tape Town; $31   **3.** 2317   **4.** $225   **5.** 58

page 36   **1.** $32   **2.** 16 miles   **3.** $183   **4.** Friday; 59   **5.** 643

EXERCISE SET 1.4

_____ 1. Last month the Danhauers used 1218 kilowatt-hours of electricity at their home and 5387 kilowatt-hours at their business. Find the total number of kilowatt hours of electricity used by the Danhauers last month.

_____ 2. A piece of chocolate cake has 428 calories and a glass of milk has 166 calories. If Jane has cake and milk for an afternoon snack, how many calories will she consume?

_____ 3. The U.S. Postal Service employee figures for a recent year show 212,561 city carriers and 50,309 rural carriers. Find the total number of persons carrying the mail.

_____ 4. A baseball diamond is a square 90 feet on each side. Find the total distance a batter will run if she hits a home run.

_____ 5. In 1970 the average yearly earnings of a salesperson were $6970. By 1980, the average earnings had increased by $2315. What were the average yearly earnings of a salesperson in 1980?

_____ 6. If a full-size tennis court is 78 feet long and 36 feet wide, how many feet of fence will be needed to enclose the tennis court?

_____ 7. The elevation above sea level of Chicago is 583 feet. If Butte is 5538 feet above sea level, find the difference between the elevations of Chicago and Butte.

_____ 8. In one city during a year, 95,740 persons used the public library's Bookmobile. The next year 120,000 used the Bookmobile. How many more persons used the Bookmobile during the second year?

_____ 9. Last month Bonnie spent $187 for groceries, $242 for rent, and $85 for utilities. What was the total amount of her living expenses?

_____ 10. On a business trip, Gay's car mileage gauge changed from 21,623 to 22,007. How many miles can Gay claim for travel on her company's expense account?

_____ 11. The attendance at Plaza I Theater for three nights was 796, 929, and 986.
_____ The attendance at Plaza II was 856, 897, and 964 for the same three nights. Which theater had the larger attendance? By how much?

The Recreation and Parks Department for a certain city gave the following report for the past year. Use these figures to answer questions 12 through 16.

Activity	Participants	Spectators
Athletics	15,525	234,799
Arts and crafts	23,789	50,256
Dramatics	7,836	115,769

_____ **12.** Find the total number of persons involved in athletic events.

_____ **13.** How many more spectators than participants were there at the dramatic events?

_____ **14.** Find the total number of spectators during the year.

_____ **15.** Find the total number of participants during the year.

_____ **16.** How many more spectators than participants were there during the year?

_____ **17.** The Stallions football team gained 6178 yards last year. If 3926 yards were gained in passing, how many yards were gained in rushing?

_____ **18.** Coletta's bank balance is $783. She writes checks for $84, $75, and $13. What is her new balance?

_____ **19.** Find the unknown length of the line segment in the figure at the right.

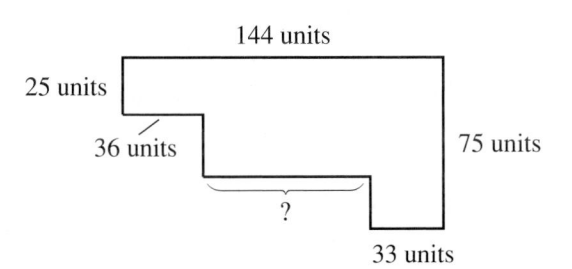

_____ **20.** During the past week, the Pizza Shack took in $1873. Supplies cost $920, salaries for the workers totaled $768, and advertising cost $53. How much profit or loss did the Pizza Shack show for the week?

☆ Stretching the Topics

_____ **1.** To install a strip of wood trim above the paneling in the recreation room with the illustrated floor plan, how many feet of trim are needed?

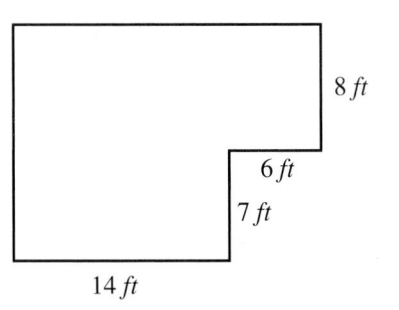

_____ **2.** At a sale after Christmas, Fran bought a coat regularly priced at $169 for $54, _____ a pair of boots regularly priced at $87 for $35, a sweater regularly priced at _____ $48 for $17, and a pair of jeans regularly priced at $46 for $29. What was the total of the regular prices for the items Fran purchased? What was the total of the sale prices? How much did Fran save by purchasing the items on sale?

Check your answers in the back of your book.

If you can solve the problems in **Checkup 1.4,** you are ready to do the **Review Exercises** for Chapter 1.

✓ CHECKUP 1.4

_____ 1. Wesley and Wanda share a paper route to help pay their tuition. One day Wesley delivered 87 papers and Wanda delivered 113. How many papers are delivered on their route?

_____ 2. When it was announced that a new automobile assembly plant was opening, there were 12,568 applicants for jobs. If there were 8579 women applicants, how many men applied for jobs?

_____ 3. On the Blazers' offensive football team, the right end weighs 245 pounds, the right tackle 239 pounds, the center 206 pounds, the left tackle 252 pounds, and the left end 245 pounds. Find the total weight of the Blazers' offensive line.

_____ 4. The following is an inventory list of the stock on one wall of Day's Bookstore.

	On Hand January 1	Bought in January	Sold in January
Fiction	526	148	278
Biography	198	52	113

Find the number of books on hand on February 1.

_____ 5. Roger plans to enclose his triangular-shaped garden with a fence. If his garden has the given dimensions, how many feet of fencing should he buy?

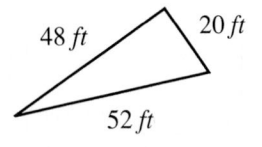

48 ft   20 ft   52 ft

Check your answers in the back of your book.

If You Missed Problems:	You Should Review Examples:
1	1
2	2
3	3, 5
4	6, 7
5	5

Summary

In this chapter we discussed the set of **whole numbers,** {0, 1, 2, 3, . . .} and the operations of addition and subtraction. We learned to use a **number line** to plot whole numbers and compare them.

Symbols	Statement	Meaning	Number Line
$4 > 2$	4 is greater than 2	4 lies to the right of 2	0 1 2 3 4 5
$0 < 3$	0 is less than 3	0 lies to the left of 3	0 1 2 3 4 5

We also discussed the **place value** of the digits in a whole number and practiced writing whole numbers in several forms.

Number Form	Word Form	Meaning
83	eighty-three	8 tens and 3 ones
5900	five thousand, nine hundred	5 thousands and 9 hundreds and 0 tens and 0 ones

We learned to **round** whole numbers to a particular place by looking at the digit in the next place to the right of the rounding place. If the digit in the next place is *less* than 5, we keep the same digit in the rounding place. If the digit in the next place is *5 or greater,* we add 1 to the digit in the rounding place. In either case, we replace all digits to the right of the rounding place with zeros.

$$\text{Rounded to the tens place,} \qquad 7352 \doteq 7350$$

$$\text{Rounded to the hundreds place,} \qquad 7352 \doteq 7400$$

$$\text{Rounded to the thousands place,} \qquad 7352 \doteq 7000$$

We discovered that the basic addition facts for one-digit numbers were all we needed to memorize in order to be able to perform addition with any whole numbers. The method of **carrying** was introduced to help us add larger numbers.

To help us understand subtraction, we related each subtraction statement to its corresponding addition statement. This idea also provided us with a method for checking subtraction problems.

Subtraction Statement		Addition Statement
$11 - 4 = 7$	because	$11 = 7 + 4$
$6 - 0 = 6$	because	$6 = 6 + 0$
$5 - 5 = 0$	because	$5 = 0 + 5$

To help us subtract a digit from a smaller digit, we learned to use the process of **borrowing** from the next digit to the left.

Finally, we used our addition and subtraction skills to switch from word statements to whole-number statements, paying attention to key phrases that would help us decide whether to add or subtract.

❏ Speaking the Language of Mathematics ────────

Complete each sentence with an appropriate word or phrase.

1. The set {0, 1, 2, 3, . . .} is the set of ──────── ──────── .

2. We know that 7 > 3 because 7 lies to the ──────── of 3 on the number line.

3. The set {0, 1, 2, 3, 4, 5, 6, 7, 8, 9} is the set of ──────── .

4. In the whole number 329, the 3 occupies the ──────── place, the 2 occupies the ──────── place, and the 9 occupies the ──────── place.

5. In an addition problem, the answer is called the ──────── . In a subtraction problem, the answer is called the ──────── .

6. For the subtraction statement 17 − 9 = 8, the corresponding addition statement is ──────── .

7. The word phrase "10 increased by 11" corresponds to the number phrase ──────── .

8. The distance around the outside of a geometric figure is called its ──────── .

❏ Writing About Mathematics ────────

Write your response to each question in complete sentences.

1. Using the ideas of place value, tell why the number 7359 does not have the same value as the number 9537.

2. Write four word expressions that could be translated into the number expression 7 + 5.

3. Explain in your own words the process used to find the difference shown here.

$$\begin{array}{r} 1002 \\ -\ 345 \\ \hline 657 \end{array}$$

REVIEW EXERCISES for Chapter 1

Place a < or > symbol between the numbers.

1. 27 _____ 17   **2.** 2009 _____ 2010

    3. Write the meaning of 3784.

    _____

_____ **4.** Write the number form for ten thousand sixty-one.

_____ **5.** In the whole number 750,245, name the hundreds digit. Then round the number to the hundreds place.

_____ **6.** In the whole number 6,587,286, name the ten thousands digit. Then round the number to the ten thousands place.

    7. At Southern College, the total attendance for basketball games during the season was 208,407. Write the word form for this whole number.

    _____

Find the sums.

_____ **8.** 7 + 9 + 6   _____ **9.** 35 + 42

_____ **10.**   376   _____ **11.**   48
             +423   279
                                       +164

_____ **12.** 7839   _____ **13.** 16,728 + 9722
            +4097

_____ **14.** 38,729   _____ **15.**   846
              57,421   29
          +30,704   1342
                                 756
                           + 219

Fill in the missing numbers.

_____ **16.** 35 − 17 = _____ because 35 = _____ + 17.

_____ **17.** 21 − _____ = 9 because 21 = 9 + _____ .

Subtract and check.

_____ **18.** 15   _____ **19.** 35
             − 7   −23

_____ 20.   57
                − 19

_____ 21.   80
                − 57

_____ 22.   734
               − 208

_____ 23.   900
               − 236

_____ 24.   8726
               − 2324

_____ 25.   10,000
               − 7053

_____ 26.   305,206
               − 277,098

_____ 27.   150,200
               − 54,876

_____ 28. The drive-in theater sold 810,315 tickets one season. Total ticket sales for the next year were 615,708. Find the decrease in ticket sales.

_____ 29. Administrative salaries for a county school system are as follows: Superintendent $52,780; Director of Instruction $42,680; Finance Director $40,680; Pupil Personnel Director $45,785; and Curriculum Specialist $50,728. Find the total amount budgeted for salaries of the administrators.

_____ 30. Lucian's bank balance is $1782. He writes checks for $178, $305, and $98. He also makes deposits of $325 and $200. What is his new balance?

Check your answers in the back of your book.

If You Missed Exercises:	You Should Review Examples:	
1, 2	Section 1.1	3
3, 4		4
5, 6		5, 8–10
7		6
8	Section 1.2	2
9, 10		3, 4
11–15		5–8
16, 17	Section 1.3	2
18		3
19		4
20–22		5–8
23–27		9–11
28	Section 1.4	2
29		3–5
30		6, 7

If you have completed the **Review Exercises** and corrected your errors, you are ready to take the **Practice Test** for Chapter 1.

PRACTICE TEST for Chapter 1

		SECTION	EXAMPLE
_____	1. Use a < or > symbol to compare the numbers 1020 and 1002.	1.1	3
	2. Write the word form for 306,719.	1.1	4
_____	3. In the whole number 78,350 name the hundreds digit. Then round the number to the hundreds place.	1.1	5, 8–10
_____	4. Find the sum of 553 and 42.	1.2	3

Find the sums.

_____	5. 903 + 678 + 529	1.2	8
_____	6. 5984 + 5069	1.2	6
_____	7. 8342 + 209 + 13,206	1.2	7
_____	8. 97,325 68,782 + 30,201	1.2	7
_____	9. Fill in the missing number. _____ − 7 = 8 because _____ = 7 + 8	1.3	2

Subtract and check.

_____	10. 839 − 427	1.3	4
_____	11. 726 − 459	1.3	5, 6
_____	12. 4668 − 3279	1.3	7
_____	13. 25,721 − 18,736	1.3	8
_____	14. 5079 − 3687	1.3	9
_____	15. 3000 − 1843	1.3	10

	SECTION	EXAMPLE

_____ 16. In the Ladies Golf Tourney, Patti won the first round with a score of 73 for 18 holes. In the next round her score increased by 13 strokes. What was Patti's score for the second round?

SECTION 1.4   EXAMPLE 1

_____ 17. A total of 683 gobblers were taken statewide during the spring turkey season. The top counties in the state were Christian with 65, Harlan with 517 and Letcher with 41. How many turkeys were taken in all the other counties combined?

SECTION 1.4   EXAMPLE 4

_____ 18. In one series of plays, the Chicago Bears completed a 12-yard pass, a 3-yard pass, and an 8-yard run. Then the quarterback was sacked for a 6-yard loss and the team received a 15-yard penalty. What was the team's total gain or loss?

SECTION 1.4   EXAMPLE 6, 7

Check your answers in the back of your book.

2

Multiplying and Dividing Whole Numbers

Compute the number of square yards of carpeting needed to cover the floor of the illustrated rooms.

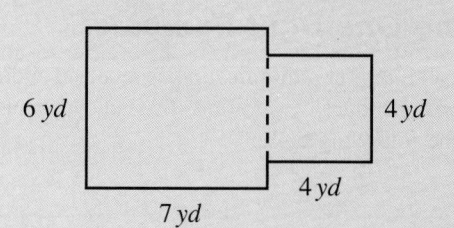

In order to solve this problem, we need more than the operations of addition and subtraction. We must turn our attention to the further topics of multiplication and division. In this chapter we learn how to

1. Multiply whole numbers.
2. Divide whole numbers.
3. Write whole numbers as products.
4. Switch from words to numbers.

2.1 Multiplying Whole Numbers

If Sam wishes to buy 6 movie tickets costing $3 each, how could you figure out how much money he'll need to make his purchase? You could certainly add up the prices of the tickets, one by one, but it seems that there must be a better way to perform this repeated addition. Of course the new operation we need is **multiplication.**

In this example we agreed that we could find Sam's total by repeated addition of the ticket prices: $3 + 3 + 3 + 3 + 3 + 3 = 18$. But we can find the total cost of 6 tickets costing $3 each more quickly by multiplying 6 times $3. Since $6 \times 3 = 18$, 6 tickets will cost $18.

We see that multiplication is another way to perform repeated addition. For instance

$$3 \times 5 = 5 + 5 + 5 = 15$$

$$4 \times 1 = 1 + 1 + 1 + 1 = 4$$

$$3 \times 0 = 0 + 0 + 0 = 0$$

In performing multiplication, the numbers being multiplied together are called **factors** and the answer is called the **product.**

$$3 \quad \times \quad 6 = \quad 18$$

factors   product

Multiplying One-Digit Numbers

Although we can interpret multiplication as repeated addition, it would be very time consuming to compute every product by addition. Instead, you must recall the basic multiplication facts appearing in the following table:

×	0	1	2	3	4	5	6	7	8	9
0	0	0	0	0	0	0	0	0	0	0
1	0	1	2	3	4	5	6	7	8	9
2	0	2	4	6	8	10	12	14	16	18
3	0	3	6	9	12	15	18	21	24	27
4	0	4	8	12	16	20	24	28	32	36
5	0	5	10	15	20	25	30	35	40	45
6	0	6	12	18	24	30	36	42	48	54
7	0	7	14	21	28	35	42	49	56	63
8	0	8	16	24	32	40	48	56	64	72
9	0	9	18	27	36	45	54	63	72	81

To find a product such as 8×7, we must locate the 8-row and the 7-column and where they meet. Here we see that $8 \times 7 = 56$.

From this table, it is possible to make some interesting observations. Notice that in the 0-row and in the 0-column all the entries are 0. In other words,

> If 0 is multiplied times any number, then the product is 0.

We also note from the 1-row and the 1-column that

> If 1 is multiplied times any number, the product is that number.

Thus we agree that

$$0 \times 6 = 0 \qquad\qquad 1 \times 6 = 6$$
$$17 \times 0 = 0 \qquad\qquad 17 \times 1 = 17$$
$$0 \times 2396 = 0 \qquad 1 \times 2396 = 2396$$

You complete Example 1.

Example 1. Fill in the table.

×	5	7	9	6	8
6					
8					
9					
7					

Check your work on page 54. ▶

As with addition, we may write multiplication problems either horizontally or vertically.

$$3 \times 5 = 15 \quad \text{or} \quad \begin{array}{r} 5 \\ \times 3 \\ \hline 15 \end{array}$$

Your success in multiplication depends upon your learning the multiplication facts for all the one-digit numbers. Whether we write the problems horizontally or vertically, the answer is based upon the facts in the table. Those facts *must* be memorized before you continue.

Ⅲ➡ Trial Run

Find the products.

_____ 1. $\begin{array}{r} 5 \\ \times 4 \end{array}$ _____ 2. $\begin{array}{r} 7 \\ \times 9 \end{array}$

_____ 3. $\begin{array}{r} 6 \\ \times 5 \end{array}$ _____ 4. $\begin{array}{r} 7 \\ \times 1 \end{array}$

_____ 5. 4×7 _____ 6. 8×8

_____ 7. 9×10 _____ 8. 6×9

Answers are on page 54.

Multiplying Larger Whole Numbers

To multiply a one-digit number times a number with more than one digit, we must recall the meaning of the digits in a whole number. For instance,

$$2 \times 431 \quad \text{means} \quad 2 \times (4 \text{ hundreds and } 3 \text{ tens and } 1 \text{ one})$$

$$\text{means} \quad 2 \times 4 \text{ hundreds and } 2 \times 3 \text{ tens and } 2 \times 1 \text{ one}$$

$$\text{means} \quad 8 \text{ hundreds and } 6 \text{ tens and } 2 \text{ ones}$$

$$\text{means} \quad 862$$

Therefore $2 \times 431 = 862$. Of course we can find this product more easily just by multiplying 2 times each of the digits in 431.

Example 2. Multiply 4×102.

Solution

$$
\begin{array}{r}
102 \\
\times \quad 4 \\
\hline
408
\end{array}
$$

Now you try Example 3.

Example 3. Multiply 3×2103.

Solution

$$
\begin{array}{r}
2103 \\
\times \quad 3
\end{array}
$$

Check your work on page 54. ▶

What happens when we multiply whole numbers and obtain a place holder larger than 9? Consider the product 2×36.

$$2 \times 36 \quad \text{means} \quad 2 \times (3 \text{ tens and } 6 \text{ ones})$$

$$\text{means} \quad 2 \times 3 \text{ tens and } 2 \times 6 \text{ ones}$$

$$\text{means} \quad 6 \text{ tens and } 12 \text{ ones}$$

$$\text{means} \quad 6 \text{ tens and } 1 \text{ ten and } 2 \text{ ones}$$

$$\text{means} \quad 7 \text{ tens and } 2 \text{ ones}$$

So $2 \times 36 = 72$

We can find this product in another way. First we multiply 2×6 and then we multiply 2×30. Then we add those products.

$$
\begin{array}{r}
3\,6 \\
\times \quad 2 \\
\hline
1\,2
\end{array}
\qquad
\begin{array}{r}
3\,6 \\
\times \quad 2 \\
\hline
1\,2 \\
6\,0 \\
\hline
7\,2
\end{array}
$$

An even shorter method of multiplication allows us to **carry** digits from one place to the next place, adding them to the product obtained in that place. For instance,

$$
\begin{array}{r}
36 \\
\times\ 2
\end{array}
\quad \text{becomes} \quad
\begin{array}{r}
\overset{1}{3}6 \\
\times\ 2 \\
\hline
2
\end{array}
\quad \text{becomes} \quad
\begin{array}{r}
\overset{1}{3}6 \\
\times\ 2 \\
\hline
72
\end{array}
$$

Notice that we multiplied 2×3 (in the tens place) to obtain 6 tens. Then we added the carried 1 ten to the 6 tens to get 7 tens.

Let's try another example in which we must carry more than once. The procedure is still the same.

Example 4. Multiply 6×534.

Solution. Let's show our steps.

$$
\begin{array}{ccc}
2 & 22 & 22 \\
534 & 534 & 534 \\
\times6 & \times6 & \times6 \\
\hline
4 & 04 & 3204
\end{array}
$$

⫸ **Trial Run**

Find the products.

_____ 1. $\begin{array}{r} 57 \\ \times\ 6 \\ \hline \end{array}$ _____ 2. $\begin{array}{r} 30 \\ \times\ 7 \\ \hline \end{array}$

_____ 3. $\begin{array}{r} 87 \\ \times\ 9 \\ \hline \end{array}$ _____ 4. $\begin{array}{r} 307 \\ \times\ 6 \\ \hline \end{array}$

_____ 5. 37×6 _____ 6. 72×8

_____ 7. 812×8 _____ 8. 3600×5

Answers are on page 54.

When we wish to multiply two numbers, both larger than one digit, we must be very careful to line them up with the digits in the same places beneath each other. For instance, to multiply 21×434, we proceed in the following way.

	Multiply 1 times each digit in 434.	Multiply 2 times each digit in 434.	Add the two products.
$\begin{array}{r} 434 \\ \times\ 21 \\ \hline \end{array}$	$\begin{array}{r} 43\ 4 \\ \times\ 2\ 1 \\ \hline 43\ 4 \end{array}$	$\begin{array}{r} 43\ 4 \\ \times\ 2\ 1 \\ \hline 43\ 4 \\ 868 \end{array}$	$\begin{array}{r} 434 \\ \times\ 21 \\ \hline 434 \\ 868 \\ \hline 9114 \end{array}$

Therefore $21 \times 434 = 9114$.

Notice that in each multiplication the rightmost digit in the product is lined up directly beneath the digit we are multiplying by. If any carrying is necessary in our multiplication, we proceed as before.

Example 5. Multiply 26×307.

Solution. Let's show our steps one more time.

$$
\begin{array}{ccc}
\begin{array}{r} 3\ 0\ 7 \\ \times\ 2\ 6 \\ \hline 18\ 4\ 2 \end{array} &
\begin{array}{r} 3\ 0\ 7 \\ \times\ 2\ 6 \\ \hline 18\ 4\ 2 \\ 61\ 4 \end{array} &
\begin{array}{r} 307 \\ \times\ 26 \\ \hline 1842 \\ 614 \\ \hline 7982 \end{array}
\end{array}
$$

The product is 7982.

Example 6. Multiply 563×896.

Solution

$$
\begin{array}{r}
896 \\
\times 563 \\
\hline
2688 \\
5376 \\
4480 \\
\hline
504448
\end{array}
$$

The product is 504,448.

In Chapter 1 we learned that to add more than two numbers we could perform the addition in any order. To multiply more than two numbers, we may also do our multiplication in any order. For instance,

$$2 \times 3 \times 5 = 2 \times 15 \qquad \text{or} \qquad 2 \times 3 \times 5 = 6 \times 5$$
$$= 30 \qquad\qquad\qquad\qquad = 30$$

The order in which you multiply does not change your answer.

You complete Example 7.

Example 7. Multiply $2 \times 9 \times 3 \times 4$.

Solution

$$2 \times 9 \times 3 \times 4 = 2 \times 9 \times 12$$
$$= \underline{} \times \underline{}$$
$$= \underline{}$$

Check your work on page 54. ▶

▰▶ Trial Run

Find the products.

_____ 1. 34×56

_____ 2. 625×57

_____ 3. 8007×63

_____ 4. 358×526

_____ 5. 3127×465

_____ 6. 589×306

_____ 7. $8 \times 9 \times 13$

_____ 8. $210 \times 8 \times 15$

Answers are on page 54.

Using Shortcuts in Multiplication

In finding products of whole numbers ending in one or more zeros, notice the wasted time and effort spent in multiplying by those zeros.

$$
\begin{array}{ll}
(1) \quad
\begin{array}{r}
723 \\
\times 200 \\
\hline
000 \\
000 \\
1446 \\
\hline
144600
\end{array}
&
(2) \quad
\begin{array}{r}
86 \\
\times 4000 \\
\hline
00 \\
00 \\
00 \\
344 \\
\hline
344000
\end{array}
\end{array}
$$

Product: 144,600   Product: 344,000

In problem (1) the 2 zeros at the end of the factor 200 contributed 2 zeros at the end of the product. In problem (2) the 3 zeros at the end of the factor 4000 contributed 3 zeros at the end of the product. In finding such products, perhaps we should consider a shortcut that allows us to cut down on the amount of writing. Let's rework the same problems in a slightly different way.

$$
\begin{array}{ccc}
(1) & 723 \\
 & \underline{\times\quad 200} \\
 & 144600
\end{array}
\qquad
\begin{array}{cc}
(2) & 86 \\
 & \underline{\times\ 4000} \\
 & 344000
\end{array}
$$

Notice that we never actually multiplied by zero at all. We simply allowed the ending zeros from the factor to appear in the product. Of course our answers are the same as before.

Example 8. Use a shortcut to multiply 15000 × 477.

Solution

$$
\begin{array}{r}
477 \\
\underline{\times\ 15000} \\
2385000 \\
\underline{477\qquad\ } \\
7155000
\end{array}
$$
    Product: 7,155,000

Notice from this example that when we multiply by a nonzero digit, the rightmost digit of the product is still lined up beneath the digit we are multiplying by. If you keep that in mind, you will not become confused about the position of each line in the sum.

Now that we have discussed a shortcut to use when multiplying by a number ending in one or more zeros, perhaps you are wondering if there's a way to cut down on writing in products such as

$$
\begin{array}{cr}
(1) & 134 \\
 & \underline{\times 207} \\
 & 938 \\
 & 000 \\
 & \underline{268\quad} \\
 & 27738
\end{array}
\qquad
\begin{array}{cr}
(2) & 2631 \\
 & \underline{\times 4003} \\
 & 7893 \\
 & 0000 \\
 & 0000 \\
 & \underline{10524\quad} \\
 & 10531893
\end{array}
$$

Product: 27,738   Product: 10,531,893

Once again, the lines of zeros seem to be unnecessary. Indeed they can be omitted *if* you remember to line up the rightmost digit of each of the other products beneath the digit that you are multiplying by.

$$
\begin{array}{cr}
(1) & 134 \\
 & \underline{\times 207} \\
 & 938 \\
 & \underline{268\quad} \\
 & 27738
\end{array}
\qquad
\begin{array}{cr}
(2) & 2631 \\
 & \underline{\times 4003} \\
 & 7893 \\
 & \underline{10524\quad} \\
 & 10531893
\end{array}
$$

You try Example 9.

Example 9. Use shortcuts to multiply 20400 × 12372.

Solution

$$
\begin{array}{r}
12372 \\
\underline{\times\quad 20400}
\end{array}
$$

Product: _____

Check your work on page 54. ▶

⇒ Trial Run

Find the products.

_____ 1. 318×10

_____ 2. 949×1000

_____ 3. 805×12

_____ 4. 543×100

_____ 5. 306×900

_____ 6. 2007×58

_____ 7. 683×8010

_____ 8. 5000×39

Answers are below.

▶ Examples You Completed

Example 1. Fill in the table.

Solution

$\times$	5	7	9	6	8
6	30	42	54	36	48
8	40	56	72	48	64
9	45	63	81	54	72
7	35	49	63	42	56

Example 3. Multiply 3×2103.

Solution

$$\begin{array}{r} 2103 \\ \times \quad 3 \\ \hline 6309 \end{array}$$

Example 7. Multiply $2 \times 9 \times 3 \times 4$.

Solution

$$2 \times 9 \times 3 \times 4 = 2 \times 9 \times 12$$
$$= 2 \times 108$$
$$= 216$$

Example 9. Use shortcuts to multiply 20400×12372.

Solution

$$\begin{array}{r} 12372 \\ \times \quad 20400 \\ \hline 4948800 \\ 24744 \quad\quad \\ \hline 252388800 \end{array}$$

Product: 252,388,800

Answers to Trial Runs

page 49 1. 20   2. 63   3. 30   4. 7   5. 28   6. 64   7. 90   8. 54

page 51 1. 342   2. 210   3. 783   4. 1842   5. 222   6. 576   7. 6496   8. 18,000

page 52 1. 1904   2. 35,625   3. 504,441   4. 188,308   5. 1,454,055   6. 180,234   7. 936
8. 25,200

page 54 1. 3180   2. 949,000   3. 9660   4. 54,300   5. 275,400   6. 116,406   7. 5,470,830
8. 195,000

EXERCISE SET 2.1

Find the products.

—————— **1.**   8
              ×7

—————— **2.**   6
              ×9

—————— **3.**   7
              ×6

—————— **4.**   8
              ×6

—————— **5.** 8 × 1

—————— **6.** 6 × 1

—————— **7.** 7 × 7

—————— **8.** 6 × 6

—————— **9.** 3 × 0

—————— **10.** 9 × 0

—————— **11.** 34
              × 2

—————— **12.** 21
              × 5

—————— **13.** 50
              × 3

—————— **14.** 80
              × 4

—————— **15.** 38
              × 7

—————— **16.** 59
              × 8

—————— **17.** 276
              × 7

—————— **18.** 329
              × 6

—————— **19.** 4500
              ×   9

—————— **20.** 3700
              ×   4

—————— **21.** 79 × 43

—————— **22.** 86 × 39

—————— **23.** 456
              × 53

—————— **24.** 527
              × 24

—————— **25.** 397
              ×245

—————— **26.** 783
              ×326

—————— **27.** 3695
              × 372

—————— **28.** 4157
              × 643

—————— **29.** 5 × 9 × 12

—————— **30.** 6 × 8 × 13

—————— **31.** 16 × 5 × 7

—————— **32.** 21 × 6 × 5

—————— **33.** 112 × 3 × 15

—————— **34.** 212 × 4 × 13

—————— **35.** 428 × 10

—————— **36.** 523 × 100

—————— **37.** 302 × 14

—————— **38.** 506 × 19

—————— **39.** 407 × 800

—————— **40.** 608 × 700

—————— **41.** 3008 × 93

—————— **42.** 7006 × 83

—————— **43.** 721 × 3020

—————— **44.** 625 × 1050

—————— **45.** 48 × 7000

—————— **46.** 72 × 6000

—————— **47.** 30,000 × 259

—————— **48.** 50,000 × 638

☆ Stretching the Topics

Find the products.

_____ 1. $38 \times 29 \times 54 \times 500$

_____ 2. $29,426 \times 5024$

_____ 3. $429 \times 512 \times 2000$

Check your answers in the back of your book.

If you can find the products in **Checkup 2.1,** you are ready to go on to Section 2.2.

✓ **CHECKUP** 2.1

Find the products.

—————— 1. 9
          ×7

—————— 2. 51
          × 8

—————— 3. 700
          × 6

—————— 4. 64
          × 5

—————— 5. 594
          × 4

—————— 6. 9 × 7 × 15

—————— 7. 89 × 26

—————— 8. 429 × 318

—————— 9. 709 × 58

—————— 10. 256 × 800

Check your answers in the back of your book.

If You Missed Problems:	You Should Review Examples:
1–3	1–3
4, 5	4
6	7
7–9	5, 6
10	8, 9

2.2 Dividing Whole Numbers

Suppose that Ann has a total of $15 to last until payday, 5 days from now. She would like to know how she can spend the same amount each day and not run out of money until payday. Solving Ann's problem requires another new operation: the operation of **division.** Unless you wished to guess the amount in each part, you would probably try to solve this problem by dividing 5 into 15.

We write that division as

$$5\overline{)15}^{\,3} \quad \text{or} \quad 15 \div 5 = 3$$

and we conclude that Ann can spend $3 each day until payday.

To solve this division problem we had to remember a multiplication fact. We know that

$$15 \div 5 = 3 \quad \text{because} \quad 15 = 5 \times 3$$

In fact, *every division statement corresponds to a multiplication statement.* For example,

$$12 \div 3 = 4 \qquad \text{because} \qquad 12 = 3 \times 4$$
$$18 \div 2 = 9 \qquad \text{because} \qquad 18 = 2 \times 9$$
$$35 \div 7 = 5 \qquad \text{because} \qquad 35 = 7 \times 5$$

Indeed all the basic division facts follow from the basic multiplication facts that we have just mastered.

In a division statement the number being divided is called the **dividend;** the number doing the dividing is called the **divisor;** and the answer is called the **quotient.** Let's label the parts in our division statement.

$$
\begin{array}{c}
\text{quotient} \longrightarrow 3 \\
15 \div 5 = \overset{\downarrow}{3} \quad \text{or} \quad 5\overline{)15} \\
\uparrow \quad \uparrow\!\text{---divisor---}\uparrow \;\; \uparrow \\
\text{---------dividend---------}
\end{array}
$$

We may now summarize the connection between a division statement and its corresponding multiplication statement.

> If
> dividend ÷ divisor = quotient
> then
> dividend = divisor × quotient

Example 1. Find 42 ÷ 6 and check using multiplication.

Solution

$$42 \div 6 = 7 \qquad \text{CHECK:} \qquad 42 = 6 \times 7$$

Notice in each of these examples that the divisor could be divided *exactly* into the dividend with a whole number quotient. When this occurs, we say that the dividend is **divisible** by the divisor. For instance,

$$42 \text{ is divisible by } 6 \qquad \text{because} \qquad 42 \div 6 = 7$$

We close this section with a comment about **zero** (0) as a dividend or a divisor. You should agree that

$$0 \div 1 = 0 \quad \text{because} \quad 0 = 1 \times 0$$
$$0 \div 13 = 0 \quad \text{because} \quad 0 = 13 \times 0$$

In fact we may say:

> If 0 is divided by any nonzero divisor, the quotient is zero.

Now consider a quotient such as $13 \div 0$. Can we say that

$$13 \div 0 = 0? \quad \text{No, because} \quad 13 \neq 0 \times 0$$
$$13 \div 0 = 13? \quad \text{No, because} \quad 13 \neq 0 \times 13$$

In fact there is no correct quotient, and we conclude that

> It is not possible to divide *by* 0.

⫸ Trial Run

Divide and check.

_____ 1. $54 \div 9$ _____ 2. $36 \div 4$

_____ 3. $30 \div 5$ _____ 4. $72 \div 8$

_____ 5. $24 \div 0$ _____ 6. $2 \div 1$

_____ 7. $49 \div 7$ _____ 8. $0 \div 3$

Answers are on page 66.

Working with Remainders

Suppose one number does *not* divide *exactly* into another number. How do we perform divisions such as $13 \div 5$ or $29 \div 4$ or $71 \div 9$? In such problems, we choose a quotient by guessing the largest number of times the divisor will "go" into the dividend. Then we multiply the chosen quotient times the divisor and subtract the product from the dividend. The number left over is called the **remainder.**

For instance to divide 13 by 5, we think $5 \times 2 = 10$ and $5 \times 3 = 15$. Since 5×3 exceeds our dividend of 13, we must choose 2 as our quotient.

$$
\begin{array}{r}
2 \\
5\overline{)13} \\
10 \\
\hline
3
\end{array}
$$

       Multiply 2×5 and subtract.

       Remainder

So we say that "13 divided by 5 equals 2, with a remainder of 3." We can write our answer as 2 R3.

To use multiplication to check a division problem with a remainder, we must multiply the divisor times the quotient *and then add* the remainder; the result should be the dividend. We check our problem 13 ÷ 5 = 2 R3 as follows:

$$(5 \times 2) + 3 \overset{?}{=} 13$$

$$10 + 3 \overset{?}{=} 13$$

$$13 = 13 \;\checkmark$$

Now we must rewrite our statement concerning the connection between division and multiplication to include a possible remainder.

> If
> dividend ÷ divisor = quotient with remainder
> then
> dividend = (divisor × quotient) + remainder

Example 2. Find 29 ÷ 4 and check.

Solution

$$\begin{array}{r} 7 \\ 4\overline{)29} \\ \underline{28} \\ 1 \end{array}$$

So 29 ÷ 4 = 7 R1.

CHECK:

$$29 \overset{?}{=} (4 \times 7) + 1$$

$$\overset{?}{=} 28 + 1$$

$$29 = 29 \;\checkmark$$

You try Example 3.

Example 3. Find 71 ÷ 9 and check.

Solution

$$9\overline{)71}$$

So 71 ÷ 9 = _____ R _____.

CHECK:

$$71 \overset{?}{=} (9 \times \underline{\quad}) + \underline{\quad}$$

$$\overset{?}{=} \underline{\quad} + \underline{\quad}$$

$$71 = \underline{\quad} \;\checkmark$$

Check your work on page 65. ▶

How can you tell when you have *not* chosen the correct quotient? To choose incorrectly means that your quotient is too large or too small, but how will that show up in your work?

Suppose we wish to find 40 ÷ 6. Let's pretend that we choose some incorrect quotients.

Pretend we think the quotient is 7.   Pretend we think the quotient is 5.

$$\begin{array}{r} 7 \\ 6\overline{)40} \\ \underline{42} \end{array}$$
$$\begin{array}{r} 5 \\ 6\overline{)40} \\ \underline{30} \\ 10 \end{array}$$

Clearly, we have *exceeded our dividend* of 40, so we know we have made an error.

In this case, we have *not* exceeded our dividend, but the *amount left after subtracting is larger than the divisor.*

Now you work the problem correctly.

Example 4. Find 40 ÷ 6.

Solution

$$6\overline{)40}$$

$$40 \div 6 = \underline{\hspace{1cm}} R\underline{\hspace{1cm}}$$

Check your work on page 65. ▶

From our work here, you should be able to recognize the signals for an incorrect quotient choice.

> **1.** If the product of the divisor times the quotient is larger than the dividend, then the chosen quotient is too *large.*
> **2.** If the amount left after subtracting the product of the divisor times the quotient from the dividend is equal to or larger than the divisor, then the chosen quotient is too *small.*

With practice in division, you will learn to spot these signals automatically and to change your quotient accordingly.

⫸ **Trial Run**

Find the quotients and remainders.

    _____ **1.** 31 ÷ 7   _____ **2.** 15 ÷ 2

    _____ **3.** 50 ÷ 8   _____ **4.** 73 ÷ 9

    _____ **5.** 45 ÷ 6   _____ **6.** 19 ÷ 5

    _____ **7.** 29 ÷ 3   _____ **8.** 38 ÷ 4

Answers are on page 66.

Dividing with Larger Whole Numbers

To divide using larger whole numbers in the dividend merely requires that we repeat the same process more than once. We do not try to compute immediately the final quotient in a problem such as $1368 \div 3$. Instead, we consider the first part of the dividend into which the divisor can be divided. Here we must consider the digits 13. We try 4 as the first digit of our quotient, being certain to write the 4 above the last digit of the part of the dividend being used. We multiply 4×3 and subtract as before. The process continues in the following way.

	Bring down 6.	Bring down 8.
Divide 3 into 13.	Divide 3 into 16.	Divide 3 into 18.
Multiply and subtract.	Multiply and subtract.	Multiply and subtract.

$$
\begin{array}{r} 4 \\ 3\overline{)1368} \\ \underline{12} \\ 1 \end{array}
\qquad
\begin{array}{r} 45 \\ 3\overline{)1368} \\ \underline{12} \\ 16 \\ \underline{15} \\ 1 \end{array}
\qquad
\begin{array}{r} 456 \\ 3\overline{)1368} \\ \underline{12} \\ 16 \\ \underline{15} \\ 18 \\ \underline{18} \\ 0 \end{array}
$$

Since there are no more digits in the dividend to bring down and there is nothing left after our last subtraction, we conclude that 3 divides into 1368 exactly 456 times, with no remainder: $1368 \div 3 = 456$. Of course we can check our answer using multiplication.

$$
\begin{array}{r} 456 \\ \times\quad 3 \\ \hline 1368 \ \surd \end{array}
$$

Let's try another division problem, including the steps along the way. Let's find $2598 \div 7$.

$$
\begin{array}{r} 3 \\ 7\overline{)2598} \\ \underline{21} \\ 4 \end{array}
\qquad
\begin{array}{r} 37 \\ 7\overline{)2598} \\ \underline{21} \\ 49 \\ \underline{49} \\ 0 \end{array}
\qquad
\begin{array}{r} 371 \\ 7\overline{)2598} \\ \underline{21} \\ 49 \\ \underline{49} \\ 08 \\ \underline{7} \\ 1 \end{array}
$$

Our answer is 371 R1. Notice that we did not become disturbed when 0 appeared in the subtraction. We merely brought down the next digit and divided again.

Example 5. Find $6327 \div 9$.

Solution

$$
\begin{array}{r} 703 \\ 9\overline{)6327} \\ \underline{63} \\ 02 \\ \underline{0} \\ 27 \\ \underline{27} \\ 0 \end{array}
$$

Therefore $6327 \div 9 = 703$.

You try Example 6.

Example 6. Find $8431 \div 3$.

Solution

$$
\begin{array}{r} 3\overline{)8431} \\ \underline{} \\ \underline{} \\ \underline{} \\ \underline{} \end{array}
$$

Therefore $8431 \div 3 = $ _____ R _____ .

Check your work on page 66. ▶

⫸ Trial Run

Divide and check.

_____ 1. 87 ÷ 4   _____ 2. 963 ÷ 3

_____ 3. 359 ÷ 5   _____ 4. 4291 ÷ 7

_____ 5. 3625 ÷ 4   _____ 6. 1924 ÷ 6

_____ 7. 3920 ÷ 8   _____ 8. 3002 ÷ 9

Answers are on page 66.

The process we have just discussed is called **long division.** Now we must consider some long division problems in which both the divisor and the dividend consist of more than one digit. You will be glad to know that the process is not very different.

To begin to find a quotient such as 609 ÷ 29 we must move over in the dividend until we arrive at a number larger than the divisor of 29. Since 6 is not large enough, we consider 60 and ask how many times 29 will divide into 60. We choose 2, and write the 2 above the last digit of the part of the dividend being used. From then on, the steps for completing the division are the same steps used earlier.

```
        21          CHECK:
    29)609             29
       58            ×21
       29             29
       29             58
        0            609 √
```

Once again we checked our answer by multiplying the divisor times the quotient.

Let's try another problem and find 37,218 ÷ 63. Notice that we must move all the way over to 372 in the dividend because neither 3 nor 37 is larger than the divisor of 63.

```
         590              CHECK:
    63)37218                 590
       315                 ×  63
       571                  1770
       567                  3540
        48                 37170
         0            +       48    Add remainder
        48  Remainder     37218 √
```

Therefore, 37,218 ÷ 63 = 590 R48.

Notice in this problem that we did not forget the last 0 in the quotient. Once we place our first digit in the quotient, we must be sure that there is exactly one digit above each of the remaining digits in the dividend.

Example 7. Find $700,305 \div 41$.

Solution

$$
\begin{array}{r}
17080 \\
41\overline{)700305} \\
\underline{41} \\
290 \\
\underline{287} \\
33 \\
\underline{0} \\
330 \\
\underline{328} \\
25 \\
\underline{0} \\
25
\end{array}
$$

Therefore $700,305 \div 41 = 17,080$ R25.

You try Example 8.

Example 8. Find $830,624 \div 516$.

Solution

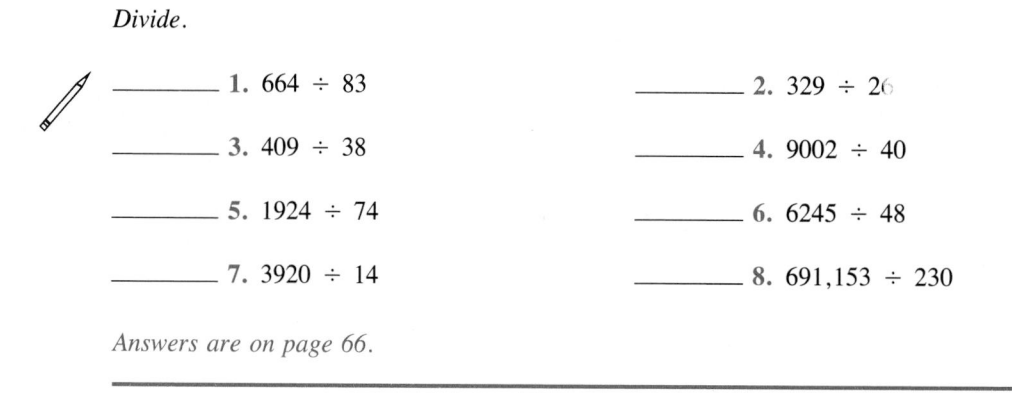

$$516\overline{)830624}$$

Therefore $830,624 \div 516 =$ _____ R_____ .

Check your work on page 66. ▶

From these examples you should see that you must be careful when zeros occur in the dividend, divisor, or quotient. When your divisor will not divide into a quantity, be sure to enter a 0 in the proper place in your quotient. In such situations taking shortcuts is not a good idea.

||||➡ **Trial Run**

Divide.

_____ **1.** $664 \div 83$

_____ **2.** $329 \div 26$

_____ **3.** $409 \div 38$

_____ **4.** $9002 \div 40$

_____ **5.** $1924 \div 74$

_____ **6.** $6245 \div 48$

_____ **7.** $3920 \div 14$

_____ **8.** $691,153 \div 230$

Answers are on page 66.

▶ **Examples You Completed**

Example 3. Find $71 \div 9$ and check.

Solution

$$
\begin{array}{r}
7 \\
9\overline{)71} \\
\underline{63} \\
8
\end{array}
$$

So $71 \div 9 = 7$ R8.

CHECK: $71 \overset{?}{=} (9 \times 7) + 8$

$\overset{?}{=} 63 \quad + 8$

$71 = 71 \;\checkmark$

Example 4. Find $40 \div 6$.

Solution

$$
\begin{array}{r}
6 \\
6\overline{)40} \\
\underline{36} \\
4
\end{array}
$$

$40 \div 6 = 6$ R4

Example 6. Find $8431 \div 3$.

Solution

$$
\begin{array}{r}
2810 \\
3\overline{)8431} \\
6 \\
\hline
24 \\
24 \\
\hline
03 \\
3 \\
\hline
01 \\
0 \\
\hline
1
\end{array}
$$

Therefore $8431 \div 3 = 2810$ R1.

Example 8. Find $830{,}624 \div 516$.

Solution

$$
\begin{array}{r}
1609 \\
516\overline{)830624} \\
516 \\
\hline
3146 \\
3096 \\
\hline
502 \\
0 \\
\hline
5024 \\
4644 \\
\hline
380
\end{array}
$$

Therefore $830{,}624 \div 516 = 1609$ R380.

Answers to Trial Runs

page 60 1. 6   2. 9   3. 6   4. 9   5. It is not possible to divide by 0.   6. 2   7. 7   8. 0

page 62 1. 4 R3   2. 7 R1   3. 6 R2   4. 8 R1   5. 7 R3   6. 3 R4   7. 9 R2   8. 9 R2

page 64 1. 21 R3   2. 321   3. 71 R4   4. 613   5. 906 R1   6. 320 R4   7. 490
8. 333 R5

page 65 1. 8   2. 12 R17   3. 10 R29   4. 225 R2   5. 26   6. 130 R5   7. 280
8. 3005 R3

EXERCISE SET 2.2

Find each quotient and check.

_____ 1. 63 ÷ 7

_____ 4. 72 ÷ 9

_____ 7. 64 ÷ 8

_____ 10. 5 ÷ 1

_____ 13. 60 ÷ 9

_____ 16. 52 ÷ 6

_____ 19. 84 ÷ 6

_____ 22. 67 ÷ 5

_____ 25. 715 ÷ 9

_____ 28. 2112 ÷ 8

_____ 31. 245 ÷ 35

_____ 34. 576 ÷ 0

_____ 37. 9315 ÷ 23

_____ 40. 3912 ÷ 65

_____ 43. 9000 ÷ 26

_____ 46. 4104 ÷ 513

_____ 49. 16,519 ÷ 718

_____ 52. 37,800 ÷ 700

_____ 2. 42 ÷ 6

_____ 5. 0 ÷ 3

_____ 8. 81 ÷ 9

_____ 11. 54 ÷ 7

_____ 14. 25 ÷ 3

_____ 17. 19 ÷ 3

_____ 20. 98 ÷ 7

_____ 23. 324 ÷ 6

_____ 26. 649 ÷ 7

_____ 29. 3427 ÷ 4

_____ 32. 416 ÷ 52

_____ 35. 3219 ÷ 87

_____ 38. 9682 ÷ 47

_____ 41. 3065 ÷ 15

_____ 44. 8000 ÷ 52

_____ 47. 5400 ÷ 450

_____ 50. 28,738 ÷ 821

_____ 3. 40 ÷ 8

_____ 6. 0 ÷ 4

_____ 9. 8 ÷ 1

_____ 12. 65 ÷ 8

_____ 15. 36 ÷ 5

_____ 18. 15 ÷ 2

_____ 21. 89 ÷ 7

_____ 24. 280 ÷ 8

_____ 27. 2748 ÷ 6

_____ 30. 6735 ÷ 9

_____ 33. 739 ÷ 0

_____ 36. 3726 ÷ 54

_____ 39. 7652 ÷ 85

_____ 42. 7074 ÷ 35

_____ 45. 1901 ÷ 312

_____ 48. 9490 ÷ 730

_____ 51. 11,600 ÷ 400

☆ Stretching the Topics _____

Find the quotients.

_____ 1. 38,151,750 ÷ 1365

_____ 2. 2,633,400 ÷ 872

_____ 3. 160,820 ÷ 205

Check your answers in the back of your book.

If you can find the quotients in **Checkup 2.2,** you are ready to go on to Section 2.3.

✔ **CHECKUP 2.2**

Find each quotient and check.

_____ 1. 63 ÷ 7

_____ 2. 82 ÷ 5

_____ 3. 435 ÷ 6

_____ 4. 3862 ÷ 9

_____ 5. 525 ÷ 21

_____ 6. 0 ÷ 42

_____ 7. 5239 ÷ 13

_____ 8. 4215 ÷ 70

_____ 9. 9840 ÷ 15

_____ 10. 23,560 ÷ 380

Check your answers in the back of your book.

If You Missed Problems:	You Should Review Examples:
1	1
2	2, 3
3, 4	5, 6
5–10	7, 8

2.3 Writing Whole Numbers as Products

Most of us would agree that we can write the whole number 12 as a product of whole numbers in several ways. For instance,

$$12 = 1 \times 12$$
$$12 = 2 \times 6$$
$$12 = 4 \times 3$$
$$12 = 2 \times 2 \times 3$$

We might also agree that we can write the whole number 13 only as a product of the whole numbers 1 and 13.

$$13 = 1 \times 13$$

A whole number greater than 1 (such as 13) that can *only* be written as a product of whole numbers using the factors 1 and itself is called a **prime number.** A whole number (such as 12) that is *not* a prime number is called a **composite number.** Zero and 1 are the only whole numbers that are *not* considered to be prime or composite. Every whole number greater than 1 must be either prime or composite.

Let's take a look at the first 20 nonzero whole numbers and decide whether or not each is a prime number. Remember, if we are able to write a whole number as a product of any other whole numbers, then it is *not* a prime number.

1 is neither prime nor composite.		$11 = 1 \times 11$	11 is prime
$2 = 1 \times 2$	2 is prime	$12 = 3 \times 4$	12 is not prime
$3 = 1 \times 3$	3 is prime	$13 = 1 \times 13$	13 is prime
$4 = 2 \times 2$	4 is not prime	$14 = 2 \times 7$	14 is not prime
$5 = 1 \times 5$	5 is prime	$15 = 3 \times 5$	15 is not prime
$6 = 2 \times 3$	6 is not prime	$16 = 4 \times 4$	16 is not prime
$7 = 1 \times 7$	7 is prime	$17 = 1 \times 17$	17 is prime
$8 = 2 \times 4$	8 is not prime	$18 = 3 \times 6$	18 is not prime
$9 = 3 \times 3$	9 is not prime	$19 = 1 \times 19$	19 is prime
$10 = 2 \times 5$	10 is not prime	$20 = 4 \times 5$	20 is not prime

From this list we notice that 2 is the only *even* prime number. Every other even number is *not* prime because every even number can always be written as 2 times some other whole number.

Now that we have seen a few prime numbers, we can practice writing some composite numbers as products of whole numbers that are prime numbers. For instance, we agreed earlier that

$$12 = 1 \times 12 \quad \text{or} \quad 12 = 2 \times 6 \quad \text{or} \quad 12 = 4 \times 3 \quad \text{or} \quad 12 = 2 \times 2 \times 3$$

In which of these versions have we written 12 as a product of *prime* numbers? Only in the last version: $12 = 2 \times 2 \times 3$. Notice that each of the other products contains a whole number factor that can be rewritten as a product of primes. For instance,

$$12 = 2 \times 6 \quad \text{becomes} \quad 12 = 2 \times 2 \times 3 \quad \text{because} \quad 6 = 2 \times 3$$

and similarly

$$12 = 4 \times 3 \quad \text{becomes} \quad 12 = 2 \times 2 \times 3 \quad \text{because} \quad 4 = 2 \times 2$$

After writing every non-prime factor as a product of prime numbers, we see that there is only one way to write 12 as a product of prime number factors.

$$12 = 2 \times 2 \times 3$$

Note that the order in which the factors are written is not important. In fact this observation is true for every whole number that is not prime.

> Every whole number that is not prime can be written in just one way as a product of prime number factors.

For work in Chapter 3, it will be important for us to learn to write whole numbers as products of prime number factors. This process is called **factoring a whole number into a product of primes** or finding the prime factorization. Suppose we practice this skill and write 75 as a product of prime factors.

$$75 = 3 \times 25$$
$$= 3 \times 5 \times 5$$

When a factor appears more than once in a product, there is a shorthand way of writing that product.

Product Form	New Form	Word Form
3×3	3^2	3 squared
5×5	5^2	5 squared
$2 \times 2 \times 2$	2^3	2 cubed
$7 \times 7 \times 7 \times 7$	7^4	7 to the fourth power
$3 \times 3 \times 3 \times 3 \times 3$	3^5	3 to the fifth power

In each of these expressions, the small raised number is called an **exponent** and the other number is called the **base.**

> An exponent tells the number of times that the base is to be used as a factor in a product.

We can now rewrite 12 and 75 as products of prime factors using exponents.

$$12 = 2 \times 2 \times 3 \qquad 75 = 3 \times 5 \times 5$$
$$= 2^2 \times 3 \qquad\qquad = 3 \times 5^2$$

Often you will not see immediately the prime factors of a larger composite number. To begin, think of any whole number that divides exactly into the original number and use that number as one factor. Continue this process until you have found all the prime factors.

For example, it will take several steps to factor the whole number 120.

$$120 = \underbrace{10}\quad \times \quad \underbrace{12}$$
$$= 2 \times 5 \times \quad 4 \times 3$$
$$= 2 \times 5 \times \underbrace{2 \times 2} \times 3$$
$$120 = 2 \times 2 \times 2 \times 3 \times 5$$
$$= 2^3 \times 3 \times 5$$

Notice that in the last step we wrote the factors in order, from smallest to largest. This is not absolutely necessary, but it is customary.

Perhaps you are thinking that you would not have begun factoring 120 as 10 × 12, and you are wondering whether you would have ended up with the same prime factors. Let's try factoring 120 again.

$$120 = 2 \quad \times \quad 60$$
$$= 2 \quad \times \quad 6 \quad \times \quad 10$$
$$= 2 \times 2 \times 3 \times 2 \times 5$$
$$120 = 2 \times 2 \times 2 \times 3 \times 5$$
$$= 2^3 \times 3 \times 5$$

Indeed, our prime factors are the same even though we "followed a different path."

Here are a few hints that may help you get started when factoring less familiar whole numbers.

Divisibility Tests

If a whole number ends in 0, 2, 4, 6, or 8, then 2 must be a factor of that whole number. Such numbers are called **even** numbers.

If a whole number ends in 0 or 5, then 5 must be a factor of that whole number.

If a whole number ends in 0, then 10 must be a factor of that whole number.

If the sum of a whole number's *digits* is divisible by 3, then 3 must be a factor of that whole number.

Example 1. Factor 184 into a product of prime factors.

Solution. Since 184 ends in 4, we know that 2 is a factor of 184. To find the factor to go with 2, we can divide 184 by 2.

$184 = 2 \times 92$	Note that 92 is even.
$= 2 \times 2 \times 46$	Rewrite 92 as 2 × 46; note that 46 is even.
$= 2 \times 2 \times 2 \times 23$	Rewrite 46 as 2 × 23; note that 23 is prime.

We have found the prime factors. We write $184 = 2^3 \times 23$.

You try Example 2.

Example 2. Write 105 as a product of prime numbers.

Solution. Since 105 ends in 5, we know one of its factors is 5.

$$105 = 5 \times \underline{} \qquad\qquad 5\overline{)105}$$
$$= 5 \times \underline{} \times \underline{}$$

Check your work on page 72. ▶

Example 3. Write 234 as a product of prime factors.

Solution

Observation	Conclusion	Factorization
The number 234 ends in 4.	A factor of 234 is 2.	$234 = 2 \times 117$
The sum of the digits of 117 (1 + 1 + 7 = 9) is divisible by 3.	A factor of 117 is 3.	$234 = 2 \times 3 \times 39$
The sum of the digits of 39 (3 + 9 = 12) is divisible by 3.	A factor of 39 is 3.	$234 = 2 \times 3 \times 3 \times 13$
All factors are prime numbers.	The factorization is complete.	$234 = 2 \times 3^2 \times 13$

➡ Trial Run

Write each of the following numbers as a product of prime factors.

_____ **1.** 42   _____ **2.** 60   _____ **3.** 288

_____ **4.** 315   _____ **5.** 198   _____ **6.** 1575

Answers are below.

▶ Example You Completed

Example 2. Write 105 as a product of prime numbers.

Solution. Since 105 ends in 5, we know one of its factors is 5.

$$105 = 5 \times 21$$
$$= 5 \times 3 \times 7$$

$$\begin{array}{r} 21 \\ 5\overline{)105} \\ \underline{10} \\ 05 \\ \underline{5} \\ 0 \end{array}$$

Answers to Trial Run

page 72   **1.** $2 \times 3 \times 7$ **2.** $2^2 \times 3 \times 5$ **3.** $2^5 \times 3^2$ **4.** $3^2 \times 5 \times 7$ **5.** $2 \times 3^2 \times 11$
6. $3^2 \times 5^2 \times 7$

EXERCISE SET 2.3

Classify each number as prime or composite. If the number is composite, write it as a product of prime numbers.

_____ 1. 15

_____ 2. 21

_____ 3. 27

_____ 4. 8

_____ 5. 19

_____ 6. 23

_____ 7. 12

_____ 8. 50

_____ 9. 1000

_____ 10. 1125

_____ 11. 183

_____ 12. 177

_____ 13. 91

_____ 14. 51

_____ 15. 173

_____ 16. 139

_____ 17. 105

_____ 18. 385

_____ 19. 1512

_____ 20. 882

☆ Stretching the Topics _____

_____ 1. List the even prime numbers.

_____ 2. Write 6048 as a product of prime factors.

_____ 3. When 520 and 7371 are written as products of prime factors, what is the only prime factor that they have in common?

Check your answers in the back of your book.

If you can complete **Checkup 2.3,** then you are ready to go on to Section 2.4.

✓ CHECKUP 2.3

Classify each number as prime or composite. If the number is composite, write it as a product of prime numbers.

_____ 1. 50

_____ 2. 45

_____ 3. 68

_____ 4. 171

_____ 5. 37

_____ 6. 104

_____ 7. 216

_____ 8. 133

_____ 9. 708

_____ 10. 840

Check your answers in the back of your book.

If You Missed Problems:	You Should Review Examples:
1–10	1–3

2.4 Switching from Words to Numbers

After learning how to add and subtract whole numbers in Chapter 1, we discovered how to change problems stated in words into number problems. Now that we have gained skill with multiplication and division, we shall again turn our attention to using those skills to solve problems stated in words.

Solving Multiplication Problems

As with addition and subtraction statements, there are certain key word phrases that tell us that we should multiply.

Word Phrase	Number Phrase
twice as much as 7	2×7
3 times as large as 10	3×10
19 doubled	2×19
5 times as great as 6	5×6
42 tripled	3×42

Example 1. During July, the Backers' electric bill was twice what it was during April. If their electric bill was $47 in April, what was their bill in July?

Solution. To find twice a number, we must multiply by 2.

$$\begin{array}{r} 47 \\ \times\ 2 \\ \hline 94 \end{array}$$

The Backers' electric bill in July was $94.

You complete Example 2.

Example 2. Five years ago there were 76 entrants in the Hometown Marathon. This year there were 4 times as many entrants. How many people entered the marathon this year?

Solution. To find 4 times as many as some number, we must multiply by _____ .

$$\begin{array}{r} 76 \\ \times\ \underline{} \end{array}$$

This year there were _____ entrants.

Check your work on page 83. ▶

In working with geometric figures we must use multiplication to compute **area.** The area of any geometric figure (square, rectangle, triangle, and so on) measures the amount of surface enclosed by the figure. The area is shaded for each of the following figures.

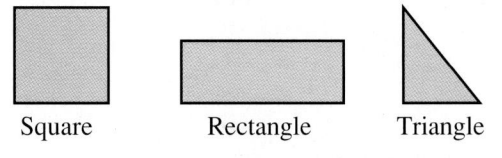

Square   Rectangle   Triangle

Consider first a **square** that measures 1 inch on each side. To find the area of this square, we want to measure the shaded region.

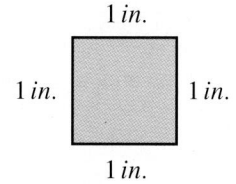

1 *in.*

1 *in.*   1 *in.*

1 *in.*

We say that the area of this region is 1 square inch (or 1 sq in.). In fact this is how we define 1 square inch.

```
1 sq in. = area of a square with 1-in. sides
```

Similarly,

```
1 sq ft = 1 square foot
        = area of a square with 1-ft sides
1 sq yd = 1 square yard
        = area of a square with 1-yd sides
1 sq m = 1 square meter
        = area of a square with 1-m sides
1 sq cm = 1 square centimeter
        = area of a square with 1-cm sides
```

To measure the area of a square with 3-inch sides, let's look at these figures.

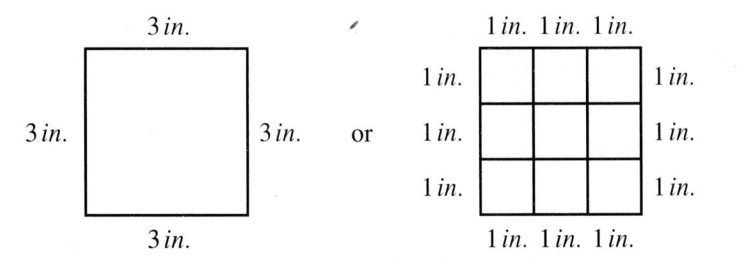

As we can see, the area we seek is actually made up of 9 smaller squares, each having an area of 1 square inch. Therefore the area of our square is 9 square inches.

A shorter way of computing the area of this square is to *multiply* the length of one side (3 in.) times the length of another side (3 in.)

$$\text{Area} = \text{side} \times \text{side} \qquad \text{(Area of square)}$$

$$\text{Area} = 3 \times 3 \text{ sq in.}$$

$$= 9 \text{ sq in.}$$

Note that area is always measured in *square units* (sq in., sq ft, sq cm, and so on).

You complete Example 3.

Example 3. Find the area of a square room that measures 11 feet on each side.

Solution

$$\text{Area} = \text{side} \times \text{side}$$

$$\text{Area} = \underline{\hspace{1cm}} \times \underline{\hspace{1cm}} \text{ sq ft}$$

$$= \underline{\hspace{1cm}} \text{ sq ft}$$

Check your work on page 83. ▶

What about the area of a *rectangle*? Let's look at a rectangle with length 4 inches and width 2 inches.

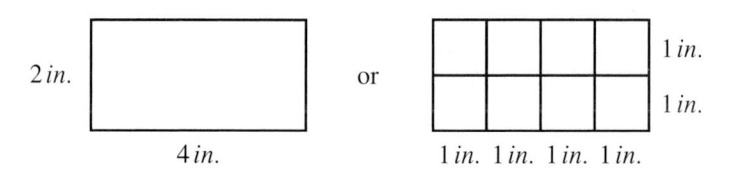

We can see that the area contains 8 little squares, each with area 1 square inch, so the area of our rectangle must be 8 square inches.

To compute this area, you should agree that we could *multiply* the length (4 in.) times the width (2 in.).

$$\text{Area} = \text{length} \times \text{width} \qquad \text{(Area of rectangle)}$$

$$\text{Area} = 4 \times 2 \text{ sq in.}$$

$$= 8 \text{ sq in.}$$

You complete Example 4.

Example 4. Find the area of a piece of drawing paper with length 12 inches and width 9 inches.

Solution

$$\text{Area} = \text{length} \times \text{width}$$

$$= \underline{\hspace{1cm}} \times \underline{\hspace{1cm}} \text{ sq in.}$$

$$= \underline{\hspace{1cm}} \text{ sq in.}$$

Check your work on page 84. ▶

Now let's solve the problem stated at the beginning of the chapter.

Example 5. Compute the number of square yards of carpeting needed to cover the floor of the illustrated rooms.

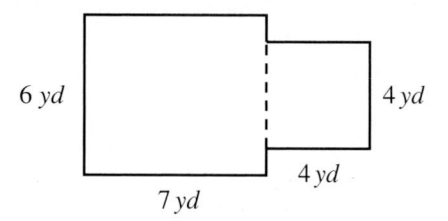

Solution. We must compute the area of the rectangle and the area of the square, and then find the sum of those areas.

Rectangle: Area = length × width   Square: Area = side × side

= 7 × 6 sq yd   = 4 × 4 sq yd

= 42 sq yd   = 16 sq yd

$$\text{Total: } A = \text{Area of rectangle} + \text{Area of square}$$
$$= 42 + 16 \text{ sq yd}$$
$$= 58 \text{ sq yd}$$

We need 58 square yards of carpeting.

Trial Run

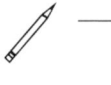

_______ 1. The round-trip distance from Paul's home to his job is 17 miles. If Perry travels twice as far to and from his job as Paul does, how far does Perry travel each day?

_______ 2. It is estimated that the cost of attending college will triple in the next 20 years. If the tuition at State University is presently $6238 per semester, what will the estimated tuition be in 20 years?

_______ 3. A square plot of land is 32 feet on each side. What is the area of the plot?

_______ 4. How many square feet are in a piece of plywood that is 4 feet by 6 feet?

_______ 5. Find the area of the patio at the right.

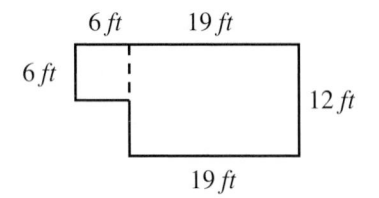

Answers are on page 84.

Solving Problems Involving "per" or "each"

One kind of problem that often occurs involves some sort of **rate,** such as miles per hour, dollars per hour, price per item, revolutions per minute, and so on. Each of these rates is described as *something per something else* and tells us how many of the first thing we can expect for every *one* of the second thing. For instance, an mpg rating of "18 miles per gallon" means we can expect to travel 18 miles on *one* gallon of gas.

Example 6. Clyde's car gets 18 miles per gallon of gasoline. How far can he travel if he has 11 gallons of gas in the tank?

Solution. Clyde can travel *11 times as far* with 11 gallons of gas as he can with 1 gallon of gas. We find how far he can travel by *multiplying*.

```
  18
 ×11
  18
 18
 198
```

Clyde can travel 198 miles on 11 gallons of gas.

Example 7. If John has worked 17 hours this week at a rate of $4 per hour, what amount has he earned?

Solution. Again we expect his pay for 17 hours to be *17 times* as much as his pay for *one* hour. We find his pay by *multiplying*.

$$\begin{array}{r} 17 \\ \times\ 4 \\ \hline 68 \end{array}$$

John has earned $68 for 17 hours of work.

Now you try Example 8.

Example 8. The label on a bag of marshmallows reads ''23 calories per marshmallow.'' If Gertie eats 13 marshmallows, how many calories has she consumed?

Solution. Gertie has consumed _____ times as many calories as she would have if she had eaten just *one* marshmallow. We find her total by _____ .

Gertie consumed _____ calories by eating 13 marshmallows.

Check your work on page 84. ▶

⫸ Trial Run

_____ 1. If Tom averages 55 miles per hour driving home from college at the end of the semester, how far will he have traveled after driving for 6 hours?

_____ 2. A room at the Happy Travelers Motel near the beach costs $52 per day. If Holly spends 5 days of her vacation there, find her total motel bill.

_____ 3. The workers at the carpet factory lose an average of 52 hours per week because of illness. How many hours would be lost in 4 weeks?

_____ 4. Sam pays rent of $325 per month for his apartment. How much will he pay in rent for 12 months?

_____ 5. A taxpayer is allowed a $1000 deduction per dependent. If a taxpayer claims 5 dependents, what would be the total deduction?

Answers are on page 84.

We learned to begin with a rate "something *per* one of something else" and to compute a total by *multiplying*. Now we would like to learn to reverse that process. For instance, if Marie spends a total of $36 for 9 basketball tickets, we might like to figure out the cost *per* ticket. It seems logical that we should *divide* the total cost by the number of tickets.

$$36 \div 9 = 4$$

So the cost is $4 per ticket. We can check our answer by thinking, if one ticket costs $4, how much will 9 tickets cost? Of course the answer is $9 \times \$4 = \$36$.

Example 9. If Hazel drove her car at the same speed for 6 hours and traveled a total of 312 miles, how fast was she driving (in miles per hour)?

Solution. Since we know the total miles driven in 6 hours, we divide that total by 6 to find the miles driven in 1 hour.

$$\begin{array}{r} 52 \\ 6\overline{)312} \\ \underline{30} \\ 12 \\ \underline{12} \\ 0 \end{array}$$

Hazel was driving at the rate of 52 miles per hour (mph).

Other word problems that must be solved using division involve a total that is to be separated into a certain number of equal parts. (Ann's payday problem mentioned earlier in this chapter is an example of such a problem.) To find the size of each equal part, we divide the total by the number of parts.

You complete Example 10.

Example 10. Mr. Honeywell left a 1570-acre farm to be divided equally among his 5 sons. What was each son's share of the farm?

Solution. We must divide the total (_____) by the number of equal parts (_____).

$$5\overline{)1570}$$

Each son received _____ acres.

Check your work on page 84. ▶

⫸ Trial Run

1. A beef roast weighs 40 ounces. If the roast is to serve 8 persons, how many ounces will there be in each serving?

2. A bowler has a total score of 1395 for 9 games. Find the average score per game.

3. A canning factory has 1248 cans of green beans ready for shipping. If 24 cans are packed in each carton, how many cartons will be needed?

4. If a total of 5022 persons attended the production of "My Fair Lady" during its 6-night run at Harrison Theater, find the average nightly attendance.

_____ **5.** The disc jockey at WCKT has 192 minutes of playing time for records. If playing time for a record averages 3 minutes, how many records can he play during the time he is on the air?

Answers are on page 84.

Solving Problems Involving Several Operations

In discussing areas earlier we mentioned triangles but did not compute their area. Here we shall briefly discuss finding the area of one particular kind of triangle, a **right triangle.** A right triangle is a 3-sided figure with one square corner (also called a **right angle,** denoted by ⌐).

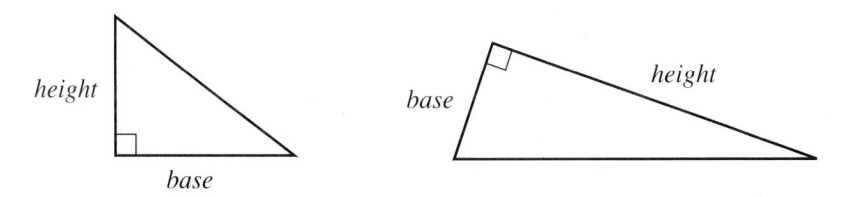

We label the sides that meet in the right angle the **base** and the **height.** Since a right triangle is just half a rectangle, it seems sensible to find the area by *dividing* the area of the rectangle by 2.

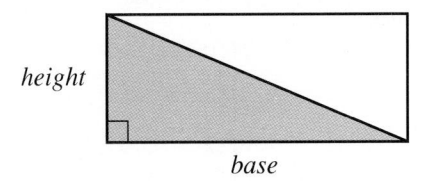

Area = (base × height) ÷ 2

To find the area of a triangle, we first multiply the base times the height. Then we divide that product by 2. In fact, this formula will give the area of *any* triangle for which you know the height and base. For instance, consider this triangle.

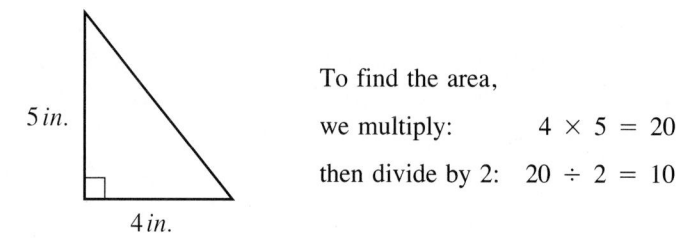

To find the area,

we multiply: 4 × 5 = 20

then divide by 2:   20 ÷ 2 = 10

The area is 10 square inches.

You try Example 11.

Example 11. Find the area of a sail in the shape of a triangle with base 11 feet and height 14 feet.

Solution. First we multiply the base times the height.   Then we divide by 2.

$$\begin{array}{r} 14 \\ \times 11 \\ \hline \end{array}$$

$$2\overline{)}$$

The area of the sail is _____ square feet.

Check your work on page 84. ▶

Another kind of problem that requires more than one operation involves finding **averages.**

To find the **average** for a set of values, we add them together and then divide the sum by the number of values in the set.

Example 12. If Marie's latest 5 bowling scores were 110, 121, 130, 112, and 122 points, what is Marie's bowling average?

Solution

First we add the scores.   Then we divide the sum by 5.

$$\begin{array}{r} 110 \\ 121 \\ 130 \\ 112 \\ +122 \\ \hline 595 \end{array}$$

$$\begin{array}{r} 119 \\ 5\overline{)595} \\ \underline{5} \\ 09 \\ \underline{5} \\ 45 \\ \underline{45} \\ 0 \end{array}$$

Marie's average score is 119 points.

You complete Example 13.

Example 13. Rudy checked the price of Levi's jeans at 8 stores in his town and found that the prices were $26, $15, $19, $21, $23, $29, $18, and $17. Find the average price for Levi's jeans in Rudy's town.

Solution. To find the average for this group of prices, we must add all the prices together and then divide the sum by the number of prices that Rudy checked.

$$
\begin{array}{r}
26 \\
15 \\
19 \\
21 \\
23 \\
29 \\
18 \\
+17 \\
\end{array}
\qquad 8\overline{)}
$$

The average price for Levi's is $____ .

Check your work on page 84. ▶

⟹ Trial Run

_____ **1.** Find the area of a triangle with base 36 feet and height 15 feet.

_____ **2.** In the first 5 basketball games of the season, Clarence scored 24, 32, 19, 26, and 24 points. What was his scoring average going into the sixth game?

_____ **3.** On the first 6 days of racing at Churchill Downs, a lucky bettor won $25, $274, $6, $19, $52, and $32. Find his average winnings per day.

_____ **4.** Wilma has a right triangle flower bed whose base is 18 feet and whose height is 12 feet. Find its area.

Answers are on page 84.

▶ Examples You Completed

Example 2. Five years ago there were 76 entrants in the Hometown Marathon. This year there were 4 times as many entrants. How many people entered the marathon this year?

Solution. To find 4 times as many as some number, we must multiply by 4.

$$
\begin{array}{r}
76 \\
\times\ 4 \\
\hline
304 \\
\end{array}
$$

This year there were 304 entrants.

Example 3. Find the area of a square room that measures 11 feet on each side.

Solution

$$\text{Area} = \text{side} \times \text{side}$$
$$\text{Area} = 11 \times 11 \text{ sq ft}$$
$$= 121 \text{ sq ft}$$

Example 4. Find the area of a piece of drawing paper with length 12 inches and width 9 inches.

Solution

$$\text{Area} = \text{length} \times \text{width}$$
$$= 12 \times 9 \text{ sq in.}$$
$$= 108 \text{ sq in.}$$

Example 8 (*Solution*). Gertie has consumed *13 times* as many calories as she would have if she had eaten just *one* marshmallow. We find her total by *multiplying*.

$$
\begin{array}{r}
23 \\
\times 13 \\
\hline
69 \\
23 \\
\hline
299
\end{array}
$$

Gertie consumed 299 calories by eating 13 marshmallows.

Example 10 (*Solution*). We must divide the total (1570) by the number of equal parts (5).

$$
\begin{array}{r}
314 \\
5)\overline{1570} \\
\underline{15} \\
07 \\
\underline{5} \\
20 \\
\underline{20} \\
0
\end{array}
$$

Each son received 314 acres.

Example 11. Find the area of a sail in the shape of a triangle with base 11 feet and height 14 feet.

Solution. First we multiply the base times the height.

$$
\begin{array}{r}
14 \\
\times 11 \\
\hline
14 \\
14 \\
\hline
154
\end{array}
$$

Then we divide by 2.

$$
\begin{array}{r}
77 \\
2)\overline{154} \\
\underline{14} \\
14 \\
\underline{14} \\
0
\end{array}
$$

The area of the sail is 77 square feet.

Example 13. Rudy checked the price of Levi's jeans at 8 stores in his town and found that the prices were $26, $15, $19, $21, $23, $29, $18, and $17. Find the average price for Levi's jeans in Rudy's town.

Solution. To find the average for this group of prices, we must add all the prices together and then divide the sum by the number of prices that Rudy checked.

$$
\begin{array}{rr}
26 & \\
15 & 21 \\
19 & 8)\overline{168} \\
21 & \underline{16} \\
23 & 08 \\
29 & \underline{8} \\
18 & 0 \\
+17 & \\
\hline
168 &
\end{array}
$$

The average price for Levi's is $21.

Answers to Trial Runs

page 78 1. 34 miles   2. $18,714   3. 1024 square feet   4. 24 square feet   5. 264 square feet

page 79 1. 330 miles   2. $260   3. 208 hours   4. $3900   5. $5000

page 80 1. 5 ounces   2. 155   3. 52 cartons   4. 837 persons   5. 64

page 83 1. 270 square feet   2. 25 points   3. $68   4. 108 square feet

EXERCISE SET 2.4

_____ 1. Last year the county fair beauty pageant had only 14 entrants. If the number of entrants tripled this year, how many contestants were there in this year's pageant?

_____ 2. Linda wishes to make 5 times as many meatballs for her party as her recipe makes. If her recipe calls for 3 pounds of ground beef, how many pounds of ground beef will she need?

_____ 3. A recipe for pancakes calls for 6 tablespoons of butter. If the recipe is tripled, how many tablespoons of butter will be needed?

_____ 4. When grilling chicken wings, one should allow 2 pounds of wings per person. How many pounds of chicken wings should Eddie buy if he plans to feed 17 persons at his cookout?

_____ 5. On a map, 1 inch represents a distance of 36 miles. Find the actual distance from Evansville to Pensacola if the distance on the map is 17 inches.

_____ 6. On a scale drawing of the floor plan of a house, 1 inch represents 6 feet. Find how many inches on the scale drawing would be used to represent the width of the front of the house if the actual width is 72 feet.

_____ 7. If Earl's new car gets approximately 23 miles per gallon of gasoline, how far can he expect to travel on a 16-gallon tank of gasoline?

_____ 8. The maximum number of calories per pound needed each day by an eighteen-year-old female is 31. If Mary is eighteen and weighs 110 pounds, what would be the maximum number of calories she should consume each day?

_____ 9. On a trip of 854 miles, an automobile averages 14 miles per gallon of gasoline. How many gallons of gasoline will be necessary for the trip?

_____ 10. Mr. Lukar's weekly salary is $444. If he works 37 hours per week, how much does he earn per hour?

_____ 11. At one point the Grand Canyon is about 24,200 feet wide. If sound waves travel about 1100 feet per second, how long does it take a shout to travel from one side of the canyon to the other?

_____ 12. If a piece of wood 72 inches long is divided into 9 equal pieces, how long is each piece?

_____ 13. Jo Ann bought a box of sugar that weighed 24 pounds. If the sugar was in 3-pound bags, how many bags of sugar were in the box?

_____ 14. If one gallon of paint covers about 300 square feet, how many gallons will be needed to paint 25,500 square feet?

_____ 15. Ms. Tapp has a square garden that is 52 feet on each side. Find the area of the garden.

_____ 16. A rectangular plot of land 52 feet by 36 feet is to be shared equally by 4 families for vegetable gardens. How many square feet of garden space will each family have?

_____ 17. The roof of a garage consists of 4 triangular-shaped sections. The base of each triangle is 24 feet and the height is 14 feet. What is the area of the roof?

_____ 18. On the first four days of their vacation, the McNarys drove 323 miles, 374 miles, 256 miles, and 427 miles. What was their average mileage per day?

_____ 19. Monty lives 23 miles from his place of work. How many miles does he drive to and from work in 5 days?

_____ 20. Each window in Ella's house is 64 inches long. If she allows 8 inches on each drapery for hems, how many yards of material will she need for 4 pairs of draperies? (1 yard = 36 inches.)

☆ Stretching the Topics _____

_____ 1. The estate of Mr. Rich is equally divided among his five nieces. If the estate includes a house worth $150,000 and 3 automobiles worth $11,000 each and debts of $9000 and $29,000, how much will each niece inherit?

_____ 2. How much fabric will be needed to construct the kite illustrated?

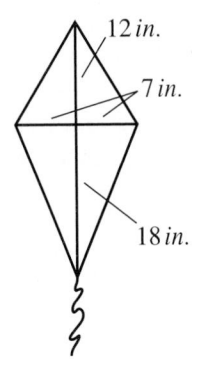

12 in.

7 in.

18 in.

Check your answers in the back of your book.

If you can do the problems in **Checkup 2.4,** you are ready to do the **Review Exercises** for Chapter 2.

✓ CHECKUP 2.4

_____ 1. If $138 is deducted from Lou's salary each month for taxes, what will be the total of his tax deduction for 12 months?

_____ 2. Ms. Churchill buys coffee by the case for her restaurant. She orders 5 cases at one time. If each case contains 8 cans of coffee and each can contains 3 pounds of coffee, how many pounds of coffee did she order?

_____ 3. The city maintains a triangular lot as a small park. Find the area of the park if the base of the triangle is 220 feet and the height is 150 feet.

_____ 4. If the attendance figures for the first 3 basketball games this season were 8369, 7574, and 9506, what was the average attendance for the first 3 games?

_____ 5. The mileage reading on Scott's car when he left for vacation was 23,528. When he returned, the reading was 25,400. If he used 104 gallons of gasoline, find the average number of miles per gallon for Scott's car on the trip.

Check your work in the back of your book.

If You Missed Problems:	You Should Review Examples:
1, 2	1, 2
3	11
4	12, 13
5	6, 12

Summary

In this chapter we learned to use multiplication to represent repeated addition. The numbers being multiplied are called **factors** and the result of multiplication is called the **product.** As with addition, we discovered that the products of one-digit numbers are all that need to be memorized for performing any multiplication.

We learned that we could use shortcuts to multiply factors containing zeros as digits *if* we were careful to always line up the rightmost digit of each product beneath the digit we were multiplying by.

$$
\begin{array}{r} 823 \\ \times\ \ 4000 \\ \hline 3292000 \end{array}
\qquad
\begin{array}{r} 1461 \\ \times 2005 \\ \hline 7305 \\ 292200 \\ \hline 2929305 \end{array}
$$

In performing division, we learned to name the parts of every division statement.

$$
\underset{\underset{\text{dividend}}{\uparrow}}{14} \ \div\ \underset{\underset{\text{divisor}}{\uparrow}}{2} \ = 7 \leftarrow \text{quotient}
$$

Then we discovered that we could relate every division statement to its corresponding multiplication statement.

Division Statement		*Multiplication Statement*
$14 \div 2 = 7$	because	$14 = 2 \times 7$
$0 \div 5 = 0$	because	$0 = 5 \times 0$

We noted that 0 divided by any nonzero number is equal to 0, but division *by* 0 is impossible.

When a number cannot be divided *exactly* into another number, we learned to use **remainders.**

Division Statement		*Multiplication Statement*
$15 \div 7 = 2\ \text{R}1$	because	$15 = (7 \times 2) + 1$
$9 \div 5 = 1\ \text{R}4$	because	$9 = (5 \times 1) + 4$

Then we learned to use **long division** to perform more complicated divisions.

To help us in future chapters, we discussed the process of rewriting a whole number as a product of prime number factors. We noted that a **prime number** is a whole number greater than 1 whose only whole number factors are itself and 1. If a whole number greater than 1 is *not* prime, then we call it a **composite number.** Every composite number can be rewritten in one way as a product of prime number factors.

$$
\begin{array}{l}
\underset{\text{number}}{\text{composite}} \rightarrow 84 = 2 \times 42 \\
\qquad\qquad\quad = 2 \times 2 \times 21 \\
\qquad\qquad\quad = 2 \times 2 \times 3 \times 7 \leftarrow \text{prime number factors} \\
\qquad\qquad\quad = 2^2 \times 3 \times 7
\end{array}
$$

Finally we practiced switching from word statements to whole number statements requiring the operations of multiplication and/or division. Again we learned to look for key phrases to help us decide what operation to perform. In particular, we looked at the **areas** of several familiar geometric figures.

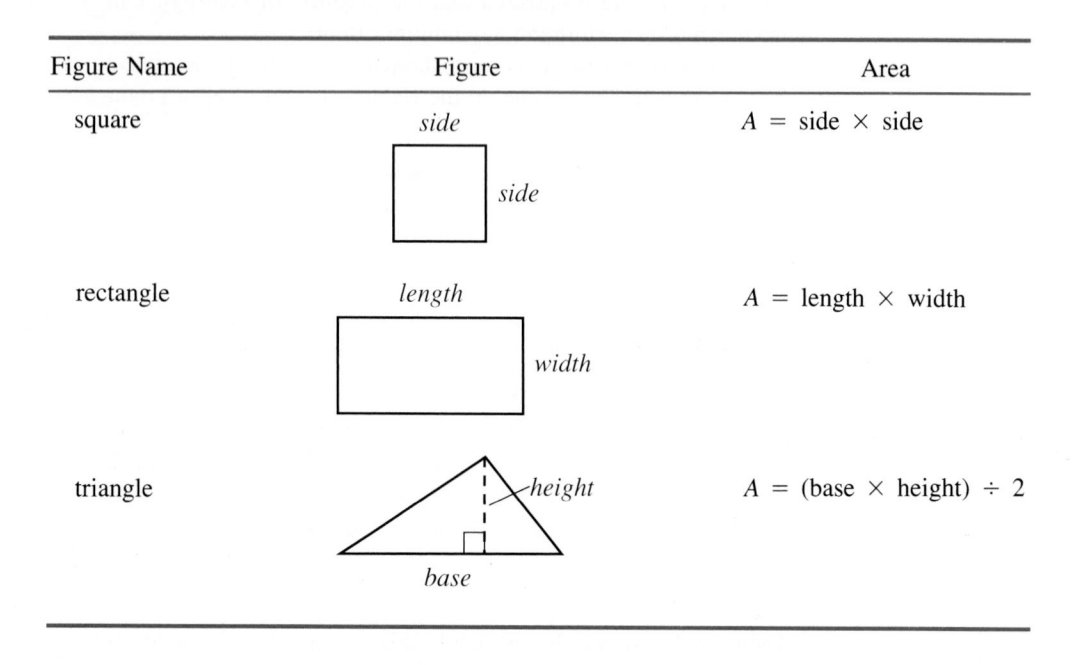

Figure Name	Figure	Area
square		$A = \text{side} \times \text{side}$
rectangle		$A = \text{length} \times \text{width}$
triangle		$A = (\text{base} \times \text{height}) \div 2$

We also solved several problems involving **rates** and **averages**.

❏ Speaking the Language of Mathematics

Complete each sentence with the appropriate word or phrase.

1. In a multiplication problem, the numbers being multiplied are called _____ and the answer is called the _____ .

2. If 0 is multiplied times any number, the product is _____ .

3. In a division problem, the number being divided is called the _____ , the number you are dividing by is called the _____ , and the answer is called the _____ .

4. For the division statement $24 \div 6 = 4$, the corresponding multiplication statement is _____ .

5. If a number does not divide *exactly* into another number, then the leftover number is called the _____ .

6. For the division statement $20 \div 3 = 6$ R2, the corresponding multiplication statement is _____ .

7. A whole number greater than 1 whose only whole number factors are itself and 1 is called a _____ number. Every whole number greater than 1 is either _____ or _____ .

8. When we measure the amount of surface enclosed by a geometric figure, we are finding its _____ .

9. To find the average of a set of 5 numbers, we first _____ those numbers and then _____ by 5.

△ Writing About Mathematics

Write your response to each question in complete sentences.

1. State, in words, how you would apply the divisibility tests to find the prime factors of 420.

2. Explain how a division problem with a remainder can be checked.

3. Why do you think it is important to memorize the multiplication facts for all one-digit numbers?

REVIEW EXERCISES for Chapter 2

Multiply.

_____ 1. 127 × 5

_____ 2. 4 × 1203

_____ 3. 894 × 7

_____ 4. 36 × 308

_____ 5. 729 × 326

_____ 6. 8 × 9 × 3 × 7

_____ 7. 596 × 700

_____ 8. 47 × 1200

_____ 9. 431 × 402

_____ 10. 30,500 × 3678

Divide and check.

_____ 11. 72 ÷ 9

_____ 12. 28 ÷ 0

_____ 13. 107 ÷ 8

_____ 14. 5762 ÷ 7

_____ 15. 15,252 ÷ 41

_____ 16. 11,040 ÷ 23

_____ 17. 709,652 ÷ 236

_____ 18. 32,500 ÷ 25

_____ 19. 225,287 ÷ 167

_____ 20. 178,300 ÷ 500

Classify each number as prime or composite. If the number is composite, write it as a product of prime numbers.

_____ 21. 182

_____ 22. 271

_____ 23. 72

_____ 24. 600

_____ 25. During July the Greens' water bill tripled what it was in March. If their bill in March was $28, what was it during July?

_____ 26. Find the amount of waxed paper needed to line the bottom of a rectangular cake pan that is 9 inches wide and 13 inches long.

_____ 27. If Ricky works 2 hours in the morning at $3 per hour and 4 hours in the afternoon at $5 per hour, how much does he earn in 1 day?

_____ 28. Lanita drove at a rate of 55 miles per hour for 3 hours and at a rate of 50 miles per hour for 2 hours. How far did Lanita drive?

_____ 29. Mr. Robinson sold 1542 bushels of soybeans for $6168. How much per bushel was he paid for the soybeans?

_____ 30. If Baron's monthly bills were $325 for rent, $75 for utilities, and $200 for groceries, find his average weekly living expenses. (Assume a 4-week month.)

Check your answers in the back of your book.

If You Missed Exercises:	You Should Review Examples:	
1–3	Section 2.1	2–4
4, 5		5, 6
6		7
7–10		8, 9
11	Section 2.2	1
12–14		2–6
15–20		7, 8
21–24	Section 2.3	1–3
25	Section 2.4	1, 2
26		4
27, 28		7, 8
29		9, 10
30		12, 13

If you have completed the **Review Exercises** and corrected your errors, you are ready to take the **Practice Test** for Chapter 2.

PRACTICE TEST for Chapter 2

Multiply.

		SECTION	EXAMPLE
_____ 1.	8×3807	2.1	3
_____ 2.	509×38	2.1	5
_____ 3.	986×321	2.1	6
_____ 4.	$9 \times 2 \times 3 \times 5$	2.1	7
_____ 5.	756×1300	2.1	8
_____ 6.	$67,005 \times 36,724$	2.1	8, 9

Divide and check.

_____ 7.	$37 \div 8$	2.2	2, 3
_____ 8.	$5621 \div 7$	2.2	5
_____ 9.	$42,483 \div 9$	2.2	6
_____ 10.	$605,207 \div 26$	2.2	7
_____ 11.	$758,910 \div 615$	2.2	8
_____ 12.	$205,036 \div 410$	2.2	8

Classify each number as prime or composite. If the number is composite, write it as a product of prime numbers.

		SECTION	EXAMPLE
_____ 13.	84	2.3	1–3
_____ 14.	73	2.3	1–3
_____ 15.	441	2.3	1–3
_____ 16.	For his garden, Irvin bought 8 boxes of tomato plants at $6 per box. How much did Irvin spend for the plants?	2.4	1, 2
_____ 17.	Find the area of a rectangular room that measures 15 feet by 18 feet.	2.4	3, 4
_____ 18.	If Randy's car gets 21 miles per gallon, how far can he travel on 16 gallons of gas?	2.4	6
_____ 19.	Find the area of a yacht's triangular mainsail if the base is 80 feet long and the height is 164 feet.	2.4	11
_____ 20.	During 5 days, Malcolm jogged 2, 5, 4, 6, and 3 miles. Find the average number of miles Malcolm jogged each day.	2.4	12, 13

Check your answers in the back of your book.

Working with Fractions

In making the mixture for hamburgers at the Student Center Grill, the cooks may use 2 ounces of filler in every 16 ounces of mixture. The rest of the 16 ounces must be beef. What fractional part of the mixture must be beef?

Now that we have become familiar with whole numbers and have learned how to perform the four basic operations with whole numbers, we turn our attention to another set of numbers that are useful in many everyday situations. These new numbers are called **fractions.**

In this chapter we learn how to

1. Work with proper fractions.
2. Work with improper fractions.
3. Change mixed numbers and fractions.
4. Reduce fractions.
5. Switch from words to fractions.

The methods for using a calculator to solve problems of the types encountered in Chapters 3–7 are discussed in Sections 16.1 and 16.2 of Chapter 16.

3.1 Understanding Fractions

Suppose we cut a cake into 8 equal pieces and Martha eats one of the pieces. How can we represent that part of the cake that has been eaten? We can say that Martha has eaten 1 piece out of a total of 8 pieces or we can use mathematical shorthand and write that Martha has eaten $\frac{1}{8}$ of the cake.

Similarly, if 5 pieces of the cake have been eaten, we may write that as $\frac{5}{8}$ of the cake. The remaining 3 pieces can be described as $\frac{3}{8}$ of the cake.

These new numbers that we are using are called **fractions.** In our example, $\frac{1}{8}$, $\frac{5}{8}$, and $\frac{3}{8}$ were fractions that we used to compare some part of the cake to the whole cake. We must learn to work with several kinds of fractions and we shall consider each type in turn.

Understanding Proper Fractions

We could have illustrated our cake example with the following drawings in which a cake has been cut into 8 equal pieces. In each case, the shaded portion represents the part described by the fraction below it.

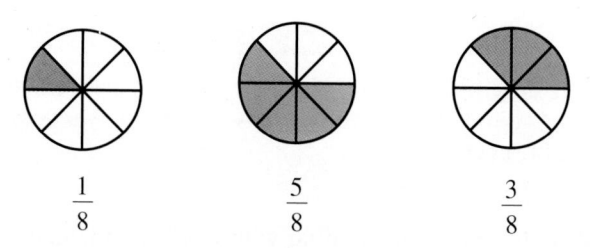

$$\frac{1}{8} \qquad\qquad \frac{5}{8} \qquad\qquad \frac{3}{8}$$

Every fraction is made up of two numbers and a fraction bar. The number above the fraction bar is called the **numerator** and the number below the fraction bar is called the **denominator.**

$$\text{fraction bar} \rightarrow \frac{5}{8} \begin{array}{l} \leftarrow \text{numerator} \\ \leftarrow \text{denominator} \end{array}$$

Here, the denominator (8) tells us that the whole cake has been divided into 8 equal portions, and the numerator (5) tells us that we are looking at 5 of those 8 portions. Once again, the fraction $\frac{5}{8}$ describes the relationship between the part and the whole.

$$\frac{\text{numerator}}{\text{denominator}} = \frac{\text{part}}{\text{whole}}$$

Example 1. Use a fraction to describe the shaded part of the figure.

Solution. The figure is divided into 4 equal portions, and 1 portion is shaded.
The fraction we want is

$$\frac{\text{part}}{\text{whole}} = \frac{1}{4}$$

Example 2. If 10 girls are enrolled in a class of 29 students, what fractional part of the class is girls?

Solution. We must consider the relationship

$$\frac{\text{part}}{\text{whole}} = \frac{10}{29}$$

Therefore, we say that $\frac{10}{29}$ of the class is girls.

You try Example 3.

Example 3. Shade the part of the ruler described by the fraction $\frac{11}{12}$.

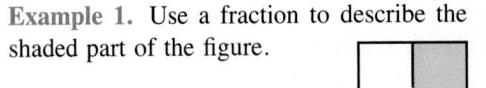

| 1 | 2 | 3 | 4 | 5 | 6 | 7 | 8 | 9 | 10 | 11 | 12 |

Check your work on page 107. ▶

Let's pause for a moment and be sure we know how to read fractions.

Fraction	Words
$\frac{1}{2}$	one half
$\frac{2}{3}$	two thirds
$\frac{1}{4}$	one fourth
$\frac{3}{5}$	three fifths
$\frac{4}{11}$	four elevenths
$\frac{10}{29}$	ten twenty-ninths
$\frac{7}{100}$	seven hundredths

Notice the *-ths* attached to the number appearing in the denominator (except when the denominator is 2 or 3).

Example 4. Read the following fractions.

$$\frac{7}{8}, \frac{113}{1000}, \frac{5}{9}$$

Solution

$\frac{7}{8}$ is read "seven eighths."

$\frac{113}{1000}$ is read "one hundred thirteen thousandths."

$\frac{5}{9}$ is read "five ninths."

So far we have discussed fractions in which the numerator was smaller than the denominator. Such fractions are called **proper fractions.**

$\frac{1}{4}$ is a proper fraction.

$\frac{10}{29}$ is a proper fraction.

$\frac{11}{12}$ is a proper fraction.

> **Proper Fraction.** If the numerator of a fraction is smaller than its denominator, the fraction is called a proper fraction.

⇒ Trial Run

_____ 1. Use a fraction to describe the shaded part of the figure.

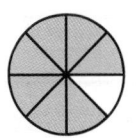

_____ 2. In a psychology class of 120 students there are 79 freshmen. What fractional part of the class is freshmen?

_____ 3. Billy Joe budgets $13 of his weekly salary for gasoline. If his salary is $235 per week, what fractional part of his salary does he budget for gasoline?

_____ 4. Raetta's diet allows her to have 1200 calories per day. If she uses 325 calories for breakfast, what fractional part of her day's allowance has she used?

_____ 5. Mr. Greenwell planted 36 fruit trees in his orchard. If he planted 11 peach trees, what fractional part of the orchard is peach trees?

_____ 6. Write the fraction $\frac{5}{14}$ in words.

Answers are on page 108.

Understanding Fractions That Are Whole Numbers

Let's return a minute to the cake that was divided into 8 equal pieces and think of a fraction we could use to describe the shaded part.

All 8 pieces have been shaded, so the fraction we need is

$$\frac{\text{part}}{\text{whole}} = \frac{8}{8}$$

But we could also describe this shaded part as 1 whole cake. So we must agree that $\frac{8}{8} = 1$.

Similarly, if we consider the following figure,

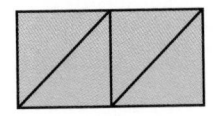

we should agree that all 4 of the 4 equal portions have been shaded. The shaded part represents $\frac{4}{4}$ of the figure. But we can also say that we have shaded 1 whole figure. Therefore, $\frac{4}{4} = 1$.

In general, we observe that

> If the numerator is equal to the denominator in a fraction (and they are not 0), then the fraction equals 1.

Example 5. Find the missing part in each statement.

$$\frac{157}{157} = \underline{\quad ? \quad} \qquad \frac{76}{?} = 1 \qquad \frac{?}{2} = 1$$

Solution. A fraction is equal to 1 when the numerator equals the denominator. Therefore

$$\frac{157}{157} = 1 \qquad \frac{76}{76} = 1 \qquad \frac{2}{2} = 1$$

Now consider the cakes in this drawing, each of which is cut into 8 pieces.

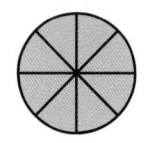 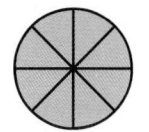

If we are asked to describe how many *eighths* have been shaded, we must respond $\frac{16}{8}$. But we also see that 2 whole cakes have been shaded, so we agree that $\frac{16}{8} = 2$.

Look at the following group of squares, where each square has been divided into 2 equal portions.

 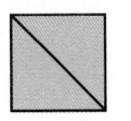 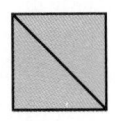 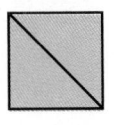 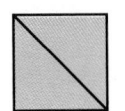

How many *halves* have been shaded? We count 10 halves or $\frac{10}{2}$, but we also see that 5 whole squares are shaded. We conclude that $\frac{10}{2} = 5$.

These examples illustrate the fact that some fractions represent whole numbers. They are *not* proper fractions. In fact, they look more like *division* problems that come out exactly, with no remainder.

$$\tfrac{16}{8} = 2 \quad \text{and} \quad 16 \div 8 = 2$$

$$\tfrac{10}{2} = 5 \quad \text{and} \quad 10 \div 2 = 5$$

Indeed if the numerator of a fraction is *not* smaller than its denominator, it may help you to think of that fraction as a division problem. If the denominator divides into the numerator exactly (with no remainder) then the original fraction can be written as a *whole number*.

Example 6. Find the missing part in each statement

$$\frac{39}{13} = \underline{\quad ? \quad} \qquad \frac{?}{5} = 8$$

$$\frac{63}{9} = \underline{\quad ? \quad} \qquad \frac{48}{?} = 16$$

Solution

$$\frac{39}{13} = 3 \qquad \frac{40}{5} = 8$$

$$\frac{63}{9} = 7 \qquad \frac{48}{3} = 16$$

You try Example 7.

Example 7. Write a fraction and a whole number to describe the shaded region.

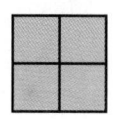 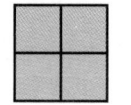 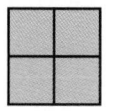 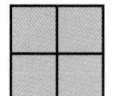

Solution. Each square has been separated into fourths. The shaded region represents _____ or _____ .

Check your work on page 107. ▶

We note here that every *whole number* can be rewritten as a fraction with that whole number as its numerator and a denominator of 1. For instance,

$$3 = \tfrac{3}{1} \qquad 173 = \tfrac{173}{1}$$

We mention also the value of a fraction with 0 in the numerator or denominator.

Zero in a Fraction. If the numerator of a fraction is 0 (and the denominator is not 0), the value of the fraction is 0.

Example: $\tfrac{0}{5} = 0$

If the denominator of a fraction is 0, the fraction has no meaning.

Example: $\tfrac{7}{0}$ has no meaning

These ideas are consistent with the ideas of division with 0 discussed in Section 2.2.

⇒ Trial Run

Find the missing part in each statement.

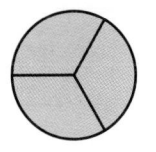

_____ 1. $\dfrac{7}{7} = \dfrac{?}{}$ _____ 2. $\dfrac{?}{15} = 1$ _____ 3. $\dfrac{21}{7} = \dfrac{?}{}$

_____ 4. $\dfrac{54}{?} = 6$ _____ 5. $\dfrac{72}{8} = ?$ _____ 6. $\dfrac{8}{?} = 8$

_____ 7. $\dfrac{?}{1} = 5$ _____ 8. $\dfrac{0}{7} = ?$ _____ 9. $\dfrac{?}{10} = 0$

_____ 10. Write a fraction and a whole number to represent the shaded region.

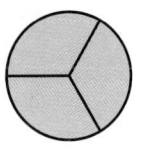 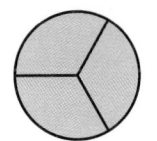 

Answers are on page 108.

Understanding Improper Fractions and Mixed Numbers

Recall that a proper fraction has a numerator that is smaller than its denominator. When this is *not* true, a fraction is called an **improper fraction.**

> **Improper Fraction.** If the numerator of a fraction is *not* smaller than its denominator, the fraction is called an improper fraction.

If the denominator of an improper fraction can be divided into its numerator exactly, we have seen that the improper fraction can be written as a whole number. Now we must consider improper fractions in which the denominator does *not* divide exactly into the numerator. For instance,

$\frac{5}{2}$ is an improper fraction.

$\frac{11}{3}$ is an improper fraction.

$\frac{39}{17}$ is an improper fraction.

We can illustrate improper fractions of this type by using the cakes that have been cut into 8 pieces.

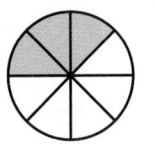

One way that the shaded portion can be described is by using the improper fraction $\frac{11}{8}$. But the drawing also helps us see that the shaded region is

1 whole cake and $\frac{3}{8}$ of a cake

or

1 and $\frac{3}{8}$ cake

or

$1\frac{3}{8}$ cake

A number such as $1\frac{3}{8}$ that is made up of a whole number part (1) and a fractional part ($\frac{3}{8}$) is called a **mixed number.** Every improper fraction can be written as a whole number or as a mixed number. Every whole number or mixed number can be written as an improper fraction. For instance, in the following drawing we can represent the shaded portion as $\frac{9}{4}$ or $2\frac{1}{4}$.

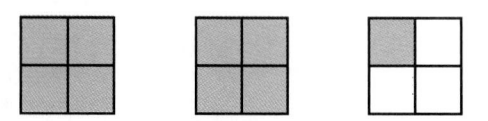

Example 8. Every window in Victoria's house contains 12 panes. If Victoria has washed 41 panes, represent the number of windows she has washed using an improper fraction and then using a mixed number.

Solution. If Victoria has washed 41 panes, then she has washed

$\frac{41}{12}$ windows

or $3\frac{5}{12}$ windows

Perhaps an illustration of Victoria's windows will help you see why these numbers are correct.

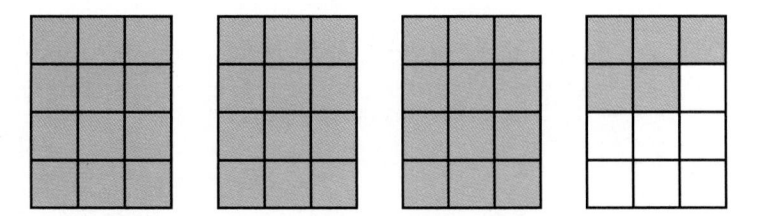

⫸ Trial Run

_____ 1. Every carton contains 6 compartments for bottles of catsup. If 23 compartments are filled with bottles of catsup, represent the number of cartons filled using an improper fraction and then a mixed number.

_____ 2. Each pen in Carl's barn will hold 5 pigs. If he puts 18 pigs in pens, represent the number of pens used with an improper fraction and then a mixed number.

_____ 3. Each box will hold 10 pieces of homemade candy. If Lynn puts 49 pieces of candy in the boxes, represent the number of boxes of candy with an improper fraction and then a mixed number.

Answers are on page 108.

Changing Between Improper Fractions and Mixed Numbers

It would certainly be simpler if we could find a way to change back and forth between improper fractions and mixed numbers without using a shaded drawing. We have already seen how to make such a change when an improper fraction represents a whole number. In such cases we used division. Recall that

$$\frac{20}{4} = 5 \quad \text{because} \quad 20 \div 4 = 5$$

Let's see if we can use division to change other improper fractions to mixed numbers. Consider the cake example in which we looked for a mixed number to represent the improper fraction $\frac{11}{8}$. Suppose we divide the denominator (8) into the numerator (11).

$$\begin{array}{r} 1 \\ 8\overline{)11} \\ \underline{8} \\ 3 \end{array}$$

The answer here is 1 R3. In other words, $\frac{11}{8}$ is 1 whole cake with 3 parts left over. But since each of these leftover parts is $\frac{1}{8}$ of another whole cake, we have $\frac{3}{8}$ of a cake left over. Therefore

$$\frac{11}{8} = 1\frac{3}{8}$$

Let's use the cake example to create a general method to change an improper fraction to a mixed number.

To change an improper fraction to a mixed number

1. Divide the denominator into the numerator.
2. The whole number part of the mixed number is the quotient.
3. The fraction part of the mixed number is the remainder written over the divisor.

Example 9. Change $\frac{41}{12}$ to a mixed number.

Solution. We must divide 12 into 41.

$$\begin{array}{r} 3 \text{ R5} \\ 12\overline{)41} \\ \underline{36} \\ 5 \end{array}$$

Since $41 \div 12 = 3$ R5, we know $\frac{41}{12} = 3\frac{5}{12}$.

You try Example 10.

Example 10. Change $\frac{83}{7}$ to a mixed number.

Solution. We must divide 7 into 83.

$$7\overline{)83}$$

Since $83 \div 7 = \underline{\qquad}$ R $\underline{\qquad}$, we know $\frac{83}{7} = \underline{\qquad}$.

Check your work on page 108. ▶

Now that we have learned to change improper fractions to mixed numbers, perhaps we can tackle the method for changing mixed numbers to improper fractions. Let's consider the mixed number $2\frac{1}{4}$ and remember that this mixed number came from some improper fraction by dividing the denominator into the numerator. The whole number part (2) was the quotient in the division, and the fraction part ($\frac{1}{4}$) was the remainder (1) over the divisor (4). Our division problem must have looked like this:

$$\begin{array}{r} 2 \text{ R1} \\ 4\overline{)?} \end{array}$$

The missing part is the dividend. But we learned in Chapter 2 that in every division problem it must be true that

$$\text{dividend} = (\text{divisor} \times \text{quotient}) + \text{remainder}$$

So the dividend here must be

$$? = (4 \times 2) + 1$$
$$? = 8 + 1$$
$$? = 9$$

Our division problem becomes

$$\begin{array}{r} 2 \text{ R1} \\ 4\overline{)9} \\ \underline{8} \\ 1 \end{array}$$

In other words $9 \div 4 = 2\frac{1}{4}$ or $\frac{9}{4} = 2\frac{1}{4}$ and we know that we can say that the mixed number $2\frac{1}{4}$ equals the improper fraction $\frac{9}{4}$.

Let's review what we have just done and see if there is a quicker way to change from mixed numbers to improper fractions.

$$\text{quotient} \rightarrow 2\frac{1}{4} \quad = \quad \frac{9}{4}$$

remainder ↓ (over numerator), dividend ↓ (over numerator); divisor ↑ (under denominator), divisor ↑ (under denominator)

To change a mixed number to an improper fraction

1. Multiply the denominator of the fraction part of the mixed number times the whole number part of the mixed number. Add the numerator of the fraction part of the mixed number. This will be the numerator of the improper fraction.
2. The denominator of the fraction part of the mixed number will be the denominator of the improper fraction.

Let's try this method and change $8\frac{2}{7}$ to an improper fraction.

$$\text{Numerator:} \quad (7 \times 8) + 2 \qquad \text{Denominator: } 7$$
$$= \quad 56 \quad + 2$$
$$= \quad 58$$

$$\text{So} \quad 8\frac{2}{7} = \frac{58}{7}$$

You complete Example 11.

Example 11. Change $10\frac{6}{13}$ to an improper fraction.

Solution

Numerator: $(13 \times 10) + 6$ Denominator: _____

$= \underline{\quad\quad} + 6$

$= \underline{\quad\quad}$

So $10\frac{6}{13} = \underline{\quad\quad}$

Check your work on page 109. ▶

⫸ **Trial Run**

_____ **1.** Change $\frac{35}{6}$ to a mixed number.

_____ **2.** Change $5\frac{3}{4}$ to an improper fraction.

_____ **3.** Change $\frac{139}{15}$ to a mixed number.

_____ **4.** Change $21\frac{3}{5}$ to an improper fraction.

_____ **5.** Change $35\frac{1}{2}$ to an improper fraction.

_____ **6.** Change $\frac{273}{10}$ to a mixed number.

Answers are on page 109.

▶ **Examples You Completed**

Example 3. Shade the part of the ruler described by the fraction $\frac{11}{12}$.

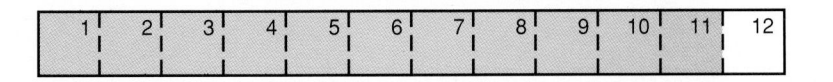

Example 7 (*Solution*). Each square has been separated into fourths. The shaded region represents $\frac{16}{4}$ or 4.

Example 10. Change $\frac{83}{7}$ to a mixed number.

Solution. We must divide 7 into 83.

$$\begin{array}{r} 11 \text{ R6} \\ 7\overline{)83} \\ \underline{7} \\ 13 \\ \underline{7} \\ 6 \end{array}$$

Since $83 \div 7 = 11$ R6,
we know $\frac{83}{7} = 11\frac{6}{7}$.

Example 11. Change $10\frac{6}{13}$ to an improper fraction.

Solution

$$\text{Numerator:} \quad (13 \times 10) + 6 \qquad \text{Denominator: 13}$$

$$= \quad 130 \quad + 6$$

$$= \quad 136$$

$$\text{So} \quad 10\frac{6}{13} = \frac{136}{13}$$

Answers to Trial Runs

page 100   1. $\frac{7}{8}$ 2. $\frac{79}{120}$ 3. $\frac{13}{235}$ 4. $\frac{325}{1200}$ 5. $\frac{11}{36}$ 6. five fourteenths

page 103   1. 1`   2. 15   3. 3   4. 9   5. 9   6. 1   7. 5   8. 0   9. 0   10. $\frac{9}{3}$ or 3

page 104   1. $\frac{23}{6}$ or $3\frac{5}{6}$ cartons   2. $\frac{18}{5}$ or $3\frac{3}{5}$ pens   3. $\frac{49}{10}$ or $4\frac{9}{10}$ boxes

page 107   1. $5\frac{5}{6}$ 2. $\frac{23}{4}$ 3. $9\frac{4}{15}$ 4. $\frac{108}{5}$ 5. $\frac{71}{2}$ 6. $27\frac{3}{10}$

EXERCISE SET 3.1

Use a fraction to describe the shaded part of each figure.

_____ **1.**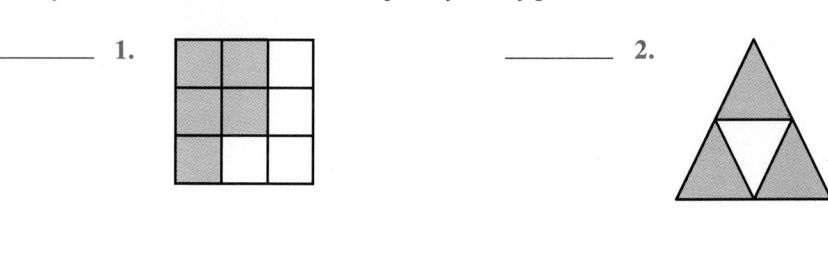

_____ **2.**

_____ **3.**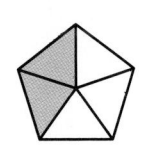

_____ **4.**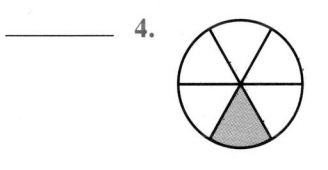

_____ **5.** On Mr. Waller's farm of 625 acres, he planted 486 acres of corn. What fractional part of the farm is planted in corn?

_____ **6.** At the camp for Special Olympians there were 384 campers during the second week of June. If 151 of the campers were girls, what fractional part of the campers were girls?

_____ **7.** Ms. Lovern earns $1283 per month and $385 is withheld. What fractional part of her salary is withheld?

_____ **8.** A family of four with a monthly income of $1577 spends $450 a month on rent. What fractional part of their income is spent on rent?

_____ **9.** There were 28 basketball teams in the city league this year and 15 of them were men's teams. What fractional part of the league were *women's* teams?

_____ **10.** One year St. Louis won 67 of the 154 baseball games they played. What fractional part of their games did they *lose*?

_____ **11.** Of 250 households surveyed, 123 had microwave ovens. What fractional part of those households surveyed owned microwave ovens?

_____ **12.** When George looked over his record collection, he noticed that out of 115 records he had 58 with a country and western label. What fractional part of his collection is country and western?

Find the missing part in each statement.

_____ **13.** $\dfrac{35}{5} = ?$

_____ **14.** $\dfrac{63}{9} = ?$

_____ **15.** $\dfrac{?}{8} = 7$

_____ **16.** $\dfrac{?}{6} = 9$

_____ **17.** $\dfrac{72}{?} = 8$

_____ **18.** $\dfrac{81}{?} = 9$

_____ **19.** $\dfrac{42}{7} = ?$

_____ **20.** $\dfrac{21}{3} = ?$

_____ **21.** $\dfrac{?}{1} = 12$

_____ **22.** $\dfrac{?}{8} = 1$

Represent each of the following using an improper fraction and then a mixed number.

_____ **23.** If each carton holds a dozen (12) eggs, how many cartons will be needed for 95 eggs?

_____ **24.** If each row in the theater has 26 seats, how many rows will be needed to seat 73 persons?

_____ **25.** If each bookshelf will hold approximately 25 books, how many shelves will be used for 138 books?

_____ **26.** If each box will hold 36 doughnuts, how many boxes will be needed for 475 doughnuts?

_____ **27.** If 12 calves can be fed at each feeder, how many feeders will be needed for 101 calves?

_____ **28.** If each case contains 24 soft drinks, how many cases should be purchased so that there will be 493 soft drinks at the picnic?

Change each improper fraction to a mixed number.

_____ 29. $\dfrac{87}{10}$ _____ 30. $\dfrac{123}{5}$

_____ 31. $\dfrac{79}{4}$ _____ 32. $\dfrac{110}{3}$

_____ 33. $\dfrac{14}{5}$ _____ 34. $\dfrac{17}{2}$

_____ 35. $\dfrac{95}{3}$ _____ 36. $\dfrac{77}{10}$

Change each mixed number to an improper fraction.

_____ 37. $8\dfrac{3}{5}$ _____ 38. $19\dfrac{1}{2}$

_____ 39. $1\dfrac{2}{3}$ _____ 40. $2\dfrac{3}{4}$

_____ 41. $19\dfrac{7}{10}$ _____ 42. $17\dfrac{4}{5}$

_____ 43. $45\dfrac{3}{4}$ _____ 44. $23\dfrac{3}{10}$

☆ Stretching the Topics _____

_____ 1. Over a period of 2 months Eudena's weight changed from 150 pounds to 137 pounds. What fractional part of her original weight did she lose?

_____ 2. (a) Change $273\frac{49}{100}$ to an improper fraction.

   (b) Change $\frac{8246}{25}$ to a mixed number.

_____ 3. If each bag contains 3 pounds of sugar and each carton contains 10 bags, how many cartons will be needed to package 161 pounds of sugar? Write your answer as an improper fraction and then as a mixed number.

Check your answers in the back of your book.

If you can complete **Checkup 3.1,** then you are ready go on to Section 3.2.

✓ CHECKUP 3.1

_____ **1.** Use a fraction to describe the shaded part of the figure.

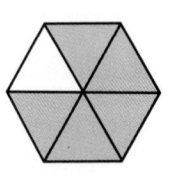

_____ **2.** If 15 males are enrolled in a class of 38 students, what fractional part of the class is females?

_____ **3.** If 3 pounds of ground beef are used to make a 5-pound meat loaf, what fractional part of the meat loaf is ground beef?

Find the missing part in each statement.

_____ **4.** $\dfrac{45}{9} = ?$

_____ **5.** $\dfrac{77}{?} = 11$

_____ **6.** If each box of saltwater taffy holds 36 pieces, how many boxes will be needed to box 257 pieces? Write your answer as an improper fraction and then as a mixed number.

_____ **7.** Change $\frac{50}{7}$ to a mixed number.

_____ **8.** Change $\frac{15}{4}$ to a mixed number.

_____ **9.** Change $7\frac{3}{5}$ to an improper fraction.

_____ **10.** Change $8\frac{4}{9}$ to an improper fraction.

Check your answers in the back of your book.

If You Missed Problems:	You Should Review Examples:
1	1, 3
2, 3	2
4, 5	5, 6
6	8
7, 8	9, 10
9, 10	11

3.2 Reducing Fractions

Let's return to Victoria's windows and look at one of them more closely.

Recall that each window contains 12 panes, so each pane represented $\frac{1}{12}$ of the window.

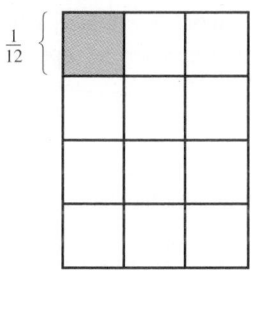

We might also think of the window as being divided into 6 equal portions. In this case, every 2 panes would represent $\frac{1}{6}$ of the window.

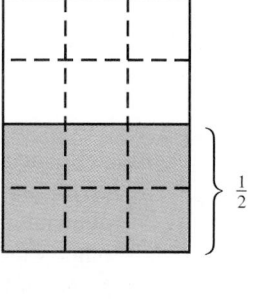

Or we could divide the window into 4 equal parts. Then every 3 panes would represent $\frac{1}{4}$ of the window.

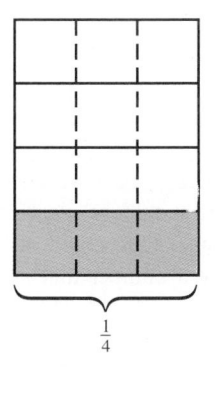

Or we could divide the window into 2 equal parts and every 6 panes would represent $\frac{1}{2}$ of the window.

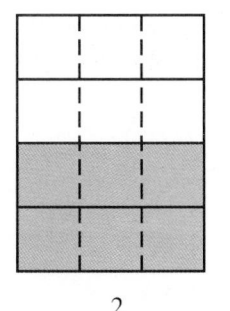

Now consider the shaded parts in each of the following windows.

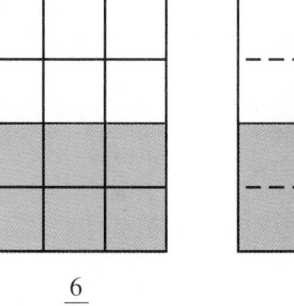 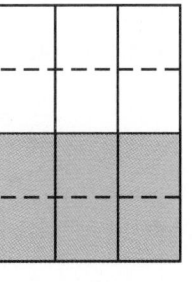 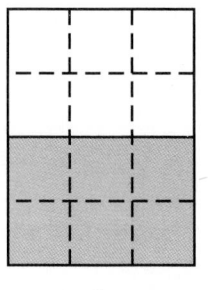

$$\frac{6}{12} \qquad \frac{3}{6} \qquad \frac{2}{4} \qquad \frac{1}{2}$$

In each case, *half* the window is shaded. In other words, all the fractions illustrated are equal to the fraction $\frac{1}{2}$.

$$\frac{6}{12} = \frac{1}{2} \qquad \frac{3}{6} = \frac{1}{2} \qquad \frac{2}{4} = \frac{1}{2}$$

We say that $\frac{1}{2}$ is the **simplest form** of all these fractions. A fraction is in simplest form when its numerator and denominator are not divisible by any whole number except one. But how do we change a fraction to its simplest form (also called **lowest terms**)? The process used to change a fraction to its lowest terms is called **reducing a fraction.**

Reducing Fractions to Simplest Form. To reduce a fraction, we may *divide* the numerator and denominator by the same number (except 0).

Suppose we use this method to reduce $\frac{6}{12}$. We see that both 6 and 12 are divisible by 6, so we divide the numerator and denominator by 6.

$$\frac{6}{12} = \frac{6 \div 6}{12 \div 6} = \frac{1}{2}$$

Notice that $\frac{6}{12}$ does indeed reduce to $\frac{1}{2}$.

Let's try the same procedure with $\frac{3}{6}$. Noticing that 3 and 6 are both divisible by 3, we divide the numerator and denominator by 3.

$$\frac{3}{6} = \frac{3 \div 3}{6 \div 3} = \frac{1}{2}$$

Once again we see that $\frac{3}{6}$ reduces to $\frac{1}{2}$.

We can reduce $\frac{2}{4}$ to $\frac{1}{2}$ by dividing the numerator and denominator by 2.

$$\frac{2}{4} = \frac{2 \div 2}{4 \div 2} = \frac{1}{2}$$

Reducing fractions is not a difficult process, once you decide by what number you should divide the numerator and denominator. It should be clear that you would like to divide the numerator and denominator by some number that would divide *exactly* into both of them. In other words, you would like to divide by a number that is a **factor** of the numerator *and* a *factor* of the denominator.

For instance, in reducing $\frac{6}{12}$ we knew that

$$\frac{6}{12} = \frac{1 \times 6}{2 \times 6}$$

We chose to divide the numerator and denominator by 6 because 6 was a factor of the numerator and denominator. We can indicate that we are dividing the numerator and denominator by 6 with a slash through each 6. The 1 that appears shows that $6 \div 6 = 1$.

$$\frac{6}{12} = \frac{1 \times 6}{2 \times 6}$$

$$= \frac{1 \times \overset{1}{\cancel{6}}}{2 \times \underset{1}{\cancel{6}}}$$

$$= \frac{1 \times 1}{2 \times 1}$$

$$\frac{6}{12} = \frac{1}{2}$$

Let's illustrate this procedure as we try to reduce $\frac{10}{35}$. From Chapter 2 we recall that numbers ending in 0 or 5 contain the factor 5. We write the numerator and denominator as products of factors and then divide by any factor that the numerator and denominator have in common.

$$\frac{10}{35} = \frac{2 \times 5}{7 \times 5}$$

$$= \frac{2 \times \overset{1}{\cancel{5}}}{7 \times \underset{1}{\cancel{5}}}$$

$$= \frac{2 \times 1}{7 \times 1}$$

$$\frac{10}{35} = \frac{2}{7}$$

Let's try another problem and reduce the improper fraction $\frac{150}{66}$. Since both numbers are even, we know they contain the factor 2.

$$\frac{150}{66} = \frac{75 \times 2}{33 \times 2}$$

$$= \frac{75 \times \overset{1}{\cancel{2}}}{33 \times \underset{1}{\cancel{2}}}$$

$$\frac{150}{66} = \frac{75}{33}$$

Are we finished reducing? Do the numerator (75) and the denominator (33) contain any common factor? Yes, both are divisible by 3.

$$\frac{150}{66} = \frac{75}{33}$$

$$= \frac{25 \times \overset{1}{\cancel{3}}}{11 \times \underset{1}{\cancel{3}}}$$

$$\frac{150}{66} = \frac{25}{11}$$

Since 11 is a prime number but 11 is not a factor of 25, we know we have reduced $\frac{150}{66}$ to lowest terms.

Example 1. Reduce $\frac{30}{105}$ to lowest terms.

Solution

$$\frac{30}{105} = \frac{6 \times 5}{21 \times 5}$$

$$= \frac{6 \times \cancel{5}^{1}}{21 \times \cancel{5}_{1}}$$

$$= \frac{6}{21}$$

$$= \frac{2 \times \cancel{3}^{1}}{7 \times \cancel{3}_{1}}$$

$$\frac{30}{105} = \frac{2}{7}$$

Some people prefer to decide what factors the numerator and denominator have in common after looking at *all* the factors of the numerator and denominator. To use this method, we first write the numerator and denominator as products of prime factors (as we learned in Section 2.3 of Chapter 2). Then we look for common factors.

For instance, to work Example 1 by this method, we write

$$\frac{30}{105} = \frac{2 \times 3 \times 5}{3 \times 5 \times 7}$$

Then we divide numerator and denominator by common factors.

$$\frac{30}{105} = \frac{2 \times \cancel{3}^{1} \times \cancel{5}^{1}}{\cancel{3}_{1} \times \cancel{5}_{1} \times 7}$$

$$\frac{30}{105} = \frac{2}{7}$$

This method may seem shorter and indeed it is, *if* you have mastered writing whole numbers as products of prime number factors. Perhaps you now understand why we spent time on that topic in Chapter 2.

Example 2. Reduce $\dfrac{270}{198}$ to lowest terms.

Solution

$$\frac{270}{198} = \frac{2 \times 3 \times 3 \times 3 \times 5}{2 \times 3 \times 3 \times 11}$$ Factor numerator and denominator.

$$= \frac{\cancel{2}^{1} \times \cancel{3}^{1} \times \cancel{3}^{1} \times 3 \times 5}{\cancel{2}_{1} \times \cancel{3}_{1} \times \cancel{3}_{1} \times 11}$$ Divide numerator and denominator by common factors.

$$\frac{270}{198} = \frac{15}{11}$$ Multiply remaining factors in the numerator; multiply remaining factors in the denominator.

Remember, the process used to reduce fractions is to divide the numerator and denominator by all common factors. Perhaps the best rules to remember are these.

Reducing a Fraction

1. If you spot a common factor, then divide the numerator and denominator by that factor.
2. If you do not spot a common factor, write the numerator and denominator as products of prime number factors. Then divide the numerator and denominator by all common factors. A fraction has been reduced to lowest terms when the numerator and denominator contain no more common factors (except 1).
3. Multiply the remaining factors in the numerator. Multiply the remaining factors in the denominator.

You try Example 3.

Example 3. Reduce $\dfrac{400}{5700}$ to lowest terms.

Solution. $\dfrac{400}{5700} =$

Check your work on page 118. ▶

Example 4. Reduce $\dfrac{30}{77}$ to lowest terms.

Solution. $\dfrac{30}{77} = \dfrac{2 \times 3 \times 5}{7 \times 11}$

Since the numerator and denominator contain no common factor, the fraction cannot be reduced. In lowest terms

$$\frac{30}{77} = \frac{30}{77}$$

⫸ **Trial Run**

Reduce each fraction to lowest terms.

_____ 1. $\dfrac{18}{24}$

_____ 2. $\dfrac{35}{77}$

_____ 3. $\dfrac{360}{1260}$

_____ 4. $\dfrac{49}{81}$

_____ 5. $\dfrac{165}{396}$

_____ 6. $\dfrac{200}{750}$

Answers are on page 118.

▶ Example You Completed

Example 3. Reduce $\dfrac{400}{5700}$ to lowest terms.

Solution

$$\frac{400}{5700} = \frac{4 \times 100}{57 \times 100}$$

$$= \frac{4 \times \overset{1}{\cancel{100}}}{57 \times \underset{1}{\cancel{100}}}$$

$$= \frac{4}{57}$$

Answers to Trial Run

page 117   1. $\dfrac{3}{4}$ 2. $\dfrac{5}{11}$ 3. $\dfrac{2}{7}$ 4. $\dfrac{49}{81}$ 5. $\dfrac{5}{12}$ 6. $\dfrac{4}{15}$

EXERCISE SET 3.2

Reduce each fraction to lowest terms.

_____ 1. $\dfrac{6}{9}$ _____ 2. $\dfrac{10}{25}$ _____ 3. $\dfrac{15}{96}$

_____ 4. $\dfrac{4}{18}$ _____ 5. $\dfrac{5}{15}$ _____ 6. $\dfrac{6}{18}$

_____ 7. $\dfrac{3}{24}$ _____ 8. $\dfrac{3}{21}$ _____ 9. $\dfrac{22}{33}$

_____ 10. $\dfrac{14}{35}$ _____ 11. $\dfrac{9}{12}$ _____ 12. $\dfrac{24}{36}$

_____ 13. $\dfrac{25}{100}$ _____ 14. $\dfrac{50}{100}$ _____ 15. $\dfrac{12}{36}$

_____ 16. $\dfrac{17}{34}$ _____ 17. $\dfrac{20}{36}$ _____ 18. $\dfrac{63}{81}$

_____ 19. $\dfrac{56}{90}$ _____ 20. $\dfrac{135}{210}$ _____ 21. $\dfrac{30}{126}$

_____ 22. $\dfrac{42}{60}$ _____ 23. $\dfrac{52}{143}$ _____ 24. $\dfrac{77}{121}$

_____ 25. $\dfrac{49}{63}$ _____ 26. $\dfrac{9}{33}$ _____ 27. $\dfrac{49}{100}$

_____ 28. $\dfrac{36}{49}$ _____ 29. $\dfrac{210}{770}$ _____ 30. $\dfrac{150}{210}$

_____ 31. $\dfrac{108}{162}$ _____ 32. $\dfrac{72}{168}$ _____ 33. $\dfrac{28}{63}$

_____ 34. $\dfrac{98}{162}$ _____ 35. $\dfrac{165}{364}$ _____ 36. $\dfrac{182}{255}$

_____ 37. $\dfrac{315}{297}$ _____ 38. $\dfrac{84}{104}$ _____ 39. $\dfrac{270}{330}$

_____ 40. $\dfrac{80}{260}$

☆ Stretching the Topics _____

Reduce each fraction to lowest terms.

_____ 1. $\dfrac{2520}{8316}$

_____ 2. $\dfrac{1095}{1752}$

_____ 3. $\dfrac{224}{336}$

Check your answers in the back of your book.

If you can reduce the fractions in **Checkup 3.2,** you are ready to go on to Section 3.3.

✓ CHECKUP 3.2

Reduce each fraction to lowest terms.

_____ 1. $\dfrac{9}{54}$ _____ 2. $\dfrac{16}{28}$

_____ 3. $\dfrac{25}{125}$ _____ 4. $\dfrac{33}{121}$

_____ 5. $\dfrac{144}{216}$ _____ 6. $\dfrac{150}{315}$

_____ 7. $\dfrac{84}{189}$ _____ 8. $\dfrac{408}{468}$

_____ 9. $\dfrac{750}{980}$ _____ 10. $\dfrac{576}{864}$

Check your answers in the back of your book.

If You Missed Problems:	You Should Review Examples:
1, 2, 3	1, 2
4	4
5, 7, 8, 10	1, 2
6, 9	3

3.3 Switching from Words to Fractions

In our earlier work with fractions, we have already seen several instances in which fractions were used to change word phrases into number phrases. One of the key phrases that we should learn to spot is "out of."

For instance, if 3 out of every 10 graduates of Hometown High School attend college, we may use the fraction $\frac{3}{10}$ to express the fractional part of the graduating class attending college. Here we see that the whole class has been divided into 10 equal portions and 3 of those portions represent the part attending college.

$$\frac{\text{part}}{\text{whole}} = \frac{3}{10}$$

Example 1. If \$30 of Mark's \$480 monthly check is withheld for Social Security, what fractional part is withheld?

Solution

$$\frac{\text{part}}{\text{whole}} = \frac{30}{480}$$

We must remember to reduce our fractions whenever possible.

$$\frac{30}{480} = \frac{3 \times \overset{1}{\cancel{10}}}{48 \times \underset{1}{\cancel{10}}}$$

$$= \frac{3}{48}$$

$$= \frac{1 \times \overset{1}{\cancel{3}}}{16 \times \underset{1}{\cancel{3}}}$$

$$\frac{30}{480} = \frac{1}{16}$$

So $\frac{1}{16}$ of Mark's check is withheld for Social Security.

Example 2. At the church spaghetti supper, 75 children's tickets were sold. If a total of 200 tickets were sold, what fractional part were children's tickets?

Solution

$$\frac{\text{part}}{\text{whole}} = \frac{75}{200}$$

$$\frac{75}{200} = \frac{3 \times \overset{1}{\cancel{25}}}{8 \times \underset{1}{\cancel{25}}}$$

$$= \frac{3}{8}$$

Therefore, $\frac{3}{8}$ of the tickets were children's tickets.

Suppose we wanted to find out what fractional part of the spaghetti supper tickets in Example 2 were *adults'* tickets. Before we can decide, we must find out how many adults' tickets were sold by subtracting the number of *children's* tickets from the *total* number of tickets.

$$\text{number of adults' tickets} = 200 - 75$$
$$= 125$$

Now we can use a fraction to compare the number of adults' tickets (125) to the total number of tickets (200).

$$\frac{\text{part}}{\text{whole}} = \frac{125}{200}$$

$$= \frac{5 \times \overset{1}{\cancel{25}}}{8 \times \underset{1}{\cancel{25}}}$$

$$= \frac{5}{8}$$

So $\frac{5}{8}$ of the tickets were adults' tickets.

Example 3. Tom is painting a fence 80 feet long for his aunt. If he has painted 15 feet so far, what fractional part of the fence has *not* been painted?

Solution. If he has painted 15 of the 80 feet, then he has *not* painted $80 - 15 = 65$ feet. Now we can use a fraction to compare the number of feet of unpainted fence (65) to the total number of feet of fence (80).

$$\frac{\text{part}}{\text{whole}} = \frac{65}{80}$$

$$= \frac{13 \times \overset{1}{\cancel{5}}}{16 \times \underset{1}{\cancel{5}}}$$

$$= \frac{13}{16}$$

So $\frac{13}{16}$ of the fence is unpainted.

Now try to solve the problem stated at the beginning of the chapter.

Example 4. In making the mixture for hamburgers at the Student Center Grill, the cooks may use 2 ounces of filler in every 16 ounces of mixture. The rest of the 16 ounces must be beef. What fractional part of the mixture must be beef?

Solution. First we compute the number of ounces of beef.

$$16 - 2 = \underline{\quad\quad}$$

Now we compare the amount of beef to the total amount of mixture.

$$\frac{\text{part}}{\text{whole}} = \frac{\quad}{16}$$

$$= \underline{\quad\quad}$$

So _____ of the mixture must be beef.

Check your work below. ▶

Example 5. Ricardo usually works 40 hours per week. Last week he worked 46 hours. What fractional part of a usual workweek did Ricardo work last week?

Solution. We must use a fraction to compare the hours Ricardo worked last week to the hours he usually works.

$$\frac{46}{40} = \frac{23 \times \cancel{2}^{1}}{20 \times \cancel{2}_{1}}$$

$$= \frac{23}{20}$$

Ricardo worked $\frac{23}{20}$ or $1\frac{3}{20}$ of a usual workweek.

▶ Example You Completed ────────────────

Example 4 (*Solution*). First we compute the number of ounces of beef.

$$16 - 2 = 14$$

Now we compare the amount of beef to the total amount of mixture.

$$\frac{\text{part}}{\text{whole}} = \frac{14}{16}$$

$$= \frac{7}{8}$$

So $\frac{7}{8}$ of the mixture must be beef.

EXERCISE SET 3.3

——————— 1. If there are 90 peach trees in Drury's orchard of 270 fruit trees, what fractional part of the orchard is peach trees?

——————— 2. There are 24 hours in a day. If Mr. Garcia works 8 hours a day, what fractional part of a day does he work?

——————— 3. On a baseball team, 5 of the starting 9 players have batting averages over .300. What fractional part of the team is batting over .300?

——————— 4. Barney lives 15 blocks from where he works. If he has to walk 3 blocks to catch a bus, what fractional part of the total distance does he walk?

——————— 5. Of the 6 people in Lana's family, only she and her sister like to play monopoly. What fractional part of the family likes to play monopoly?

——————— 6. For a concert, 1725 tickets were sold in advance and the rest were sold at the door. If 5000 tickets were sold for the concert, what fractional part of the tickets were sold in advance?

——————— 7. Ingrid is making party hats for the 30 guests invited to her brother's birthday party. If she has finished 18 hats, what fractional part of the hats does she still have to make?

——————— 8. The distance from Gary's hometown to college is 126 miles. If only 60 miles can be driven on interstate highways, what fractional part of the distance is driven on other roads?

——————— 9. Allied Construction Company has been hired to pave 150 miles of highway. If they have completed 85 miles, what fractional part of the distance remains to be paved?

——————— 10. On a football team of 11 players, 8 of the players weigh over 200 pounds. What fractional part of the team weighs 200 pounds or less?

——————— 11. In making punch for a New Year's party, Buffy uses 3 quarts of club soda for every 9 quarts of punch. The rest must be fruit juice. What fractional part of the punch must be fruit juice?

——————— 12. Wesley has space for 48 rows in his garden. If he plants 6 rows of green beans, what fractional part of his garden can be planted in other vegetables?

——————— 13. Beth agreed to baby-sit for 2 hours. The Utleys did not return for 5 hours so Beth had to baby-sit overtime. What fraction of the agreed time did she baby-sit?

——————— 14. Stephen's regular working time is 160 hours per month. If he worked 208 hours last month, what fraction of his regular time did Stephen work?

——————— 15. Tom purchased a second-hand car for $750, did some work on it, and then sold it for $1100. For what fraction of the purchase price did he sell it?

_____ 16. Samantha regularly earns $6 an hour. For each overtime hour she is paid $9. What fraction of her regular hourly wage does Samantha earn when she works overtime?

_____ 17. Tracy bought a box of 52 used light bulbs at a yard sale. She found that 39 of them were good. What fractional part of the box of bulbs was good?

_____ 18. Earlene's math class met for 40 hours during the semester. If Earlene cut class 12 hours during the semester, what fractional part of the class time did she miss?

_____ 19. A survey of 175 families indicated that 105 had more than 1 television. What fraction of those surveyed had more than 1 television?

_____ 20. Elvis has a job washing windows at the National Bank building. If the building has 92 windows, what fraction of the job is completed when he has washed 44 of the windows?

☆ Stretching the Topics _____

_____ 1. On a business trip Lydia averaged 62 miles per hour for 4 hours on interstate highways and 52 miles per hour for 3 hours on other roads. What fractional part of the total miles did Lydia drive on interstate highways?

_____ 2. In Jim's math class there are 28 freshmen, 15 sophomores, 7 juniors, and 8 seniors. What fractional part of the class is not classified as freshmen?

_____ 3. Mr. Garcia's will left $210,000 to his wife and the remainder of his $735,000 estate to his children. Use a fraction to compare his children's inheritance to his wife's inheritance.

Check your answers in the back of your book.

If you can complete **Checkup 3.3,** then you are ready to go on to the **Review Exercises** for Chapter 3.

✓ CHECKUP 3.3

_____ 1. There are 525 seats in Cowan Theater and only 210 tickets were sold for the first performance of a recent production. What fractional part of the total seating was sold?

_____ 2. The distance from Chicago to Dallas by way of St. Louis is 888 miles. If the distance from Chicago to St. Louis is 288 miles, what fractional part of the total distance from Chicago to Dallas has been traveled when one reaches St. Louis?

_____ 3. Raymond planted 84 gladiola bulbs this spring. If 60 of them bloomed, what fraction of the bulbs produced no blooms?

_____ 4. Miguel bought a 16-ounce soft drink and spilled about 6 ounces when opening it. What fractional part of the drink did he have left?

_____ 5. Ayre's Jewelers has a ruby ring on display marked $625. If the jewelry store paid $250 for the ring, what fractional part of the amount paid for the ring is the selling price?

Check your answers in the back of your book.

If You Missed Problems:	You Should Review Examples:
1, 2	1–3
3, 4	4
5	5

Summary

In this chapter we discussed the set of numbers called **fractions** and learned to identify the parts of every fractional number.

$$\text{fraction bar} \rightarrow \frac{3 \leftarrow \text{numerator}}{4 \leftarrow \text{denominator}}$$

A fraction whose numerator is *smaller* than its denominator is called a **proper fraction.** A fraction whose numerator is the *same* as its denominator (except $\frac{0}{0}$) is equal to 1. A fraction whose numerator is *not smaller* than its denominator is called an **improper fraction.** We agreed that every improper fraction may be rewritten either as a whole number or as a **mixed number.**

Fraction	Type	Rewritten Form
$\frac{1}{3}$	proper	—
$\frac{8}{5}$	improper	$1\frac{3}{5}$ (mixed number)
$\frac{10}{10}$	improper	1 (whole number)
$\frac{33}{11}$	improper	3 (whole number)

We then learned to **reduce** a fraction to lowest terms by dividing its numerator and denominator by common factors.

Fraction	Factored Form	Reduced Form
$\frac{15}{35}$	$\frac{3 \times 5}{7 \times 5}$	$\frac{3}{7}$
$\frac{188}{24}$	$\frac{2 \times 2 \times 47}{2 \times 2 \times 2 \times 3}$	$\frac{47}{6}$

Finally we learned to switch from word expressions to fractional numbers, watching for key phrases such as "out of" that would tell us how to write our fraction in the form $\frac{\text{part}}{\text{whole}}$.

❑ Speaking the Language of Mathematics

Complete each sentence with the appropriate word or phrase.

1. In a fractional number, the number above the fraction bar is called the _____ , and the number below the fraction bar is called the _____ .

2. If the numerator of a fraction is smaller than its denominator, we call it a(n) _____ fraction.

3. If the numerator of a fraction is larger than its denominator, we call it a(n) _____ fraction.

4. Numbers such as $3\frac{1}{2}$ and $5\frac{3}{8}$ are called _____ _____ .

5. To reduce a fraction, we divide its numerator and denominator by all _____
   _____ .

6. If a class of 39 students contains 15 males, then the fractional part of the class that is male
   is _____ and the fractional part that is female is _____ .

△ Writing About Mathematics

Write your response to each question in complete sentences.

1. Explain the difference between a proper fraction and an improper fraction. Describe the process used to rewrite an improper fraction as a mixed number.

2. Show and explain the steps you would use to reduce $\frac{84}{90}$ to a fraction in its simplest form.

3. Describe three everyday situations in which you would be required to use fractions.

REVIEW EXERCISES for Chapter 3

_____ **1.** Use a fraction to describe the shaded part of the figure.

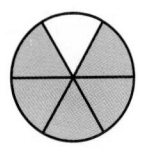

_____ **2.** A baseball player had 139 hits in 378 times at bat. In what fractional part of the times he was at bat did he get a hit?

_____ **3.** Of the 112 runners who started the 10-K run, only 59 finished. What fractional part of the runners beginning the race actually finished?

_____ **4.** In a small company, 38 out of 57 workers voted to form a union. What fractional part of the workers wanted a union?

Find the missing part in each statement.

_____ **5.** $\dfrac{?}{9} = 6$ _____ **6.** $\dfrac{28}{7} = ?$

_____ **7.** $\dfrac{48}{?} = 8$ _____ **8.** $\dfrac{?}{9} = 45$

Represent each of the following using an improper fraction and then a mixed number.

_____ **9.** If 1 carton holds 4 petunias, how many cartons will be needed for 97 petunia plants?

_____ **10.** If 1 box will hold 20 candy bars, how many boxes will be needed for 223 candy bars?

_____ **11.** If 1 bus has a seating capacity of 25, how many buses will be needed to transport 136 students?

Change each improper fraction to a mixed number.

_____ **12.** $\dfrac{25}{4}$ _____ **13.** $\dfrac{79}{8}$

Change each mixed number to an improper fraction.

_____ **14.** $7\dfrac{3}{5}$ _____ **15.** $5\dfrac{2}{3}$

Reduce each fraction to lowest terms.

_____ 16. $\dfrac{35}{95}$

_____ 17. $\dfrac{21}{56}$

_____ 18. $\dfrac{24}{378}$

_____ 19. $\dfrac{66}{220}$

_____ 20. $\dfrac{78}{104}$

_____ 21. $\dfrac{120}{450}$

_____ 22. $\dfrac{70}{100}$

_____ 23. $\dfrac{12}{35}$

_____ 24. A survey showed that 196 out of 324 persons preferred Tasty toothpaste to Klean toothpaste. What fractional part of those surveyed preferred Tasty?

_____ 25. If Rowman's monthly salary is $1525 and he pays $425 a month for rent, what fractional part of his salary is used for rent?

_____ 26. On a Friday, only 21 students out of 45 attended history class. What fractional part of the class did not attend?

_____ 27. Johnny bought a box of books at a yard sale. Of the 35 books in the box, 15 were fiction. What fractional part of the books were non-fiction?

_____ 28. Terri was paid $4 an hour for working in the math lab. For overtime she was paid $7 an hour. What fraction of her regular hourly wage was she paid for working overtime?

Check your answers in the back of your book.

If You Missed Exercises:	You Should Review Examples:	
1	Section 3.1	1
2–4		2, 3
5–8		5, 6
9–11		8
12, 13		9, 10
14, 15		11
16, 17	Section 3.2	1
18–20		2
21, 22		3
23		4
24, 25	Section 3.3	1–3
26, 27		4
28		5

If you have completed the **Review Exercises** and corrected your answers, you are ready to take the **Practice Test** for Chapter 3.

Name _____ **Date** _____

PRACTICE TEST for Chapter 3

		SECTION	EXAMPLE

_____ 1. If a basketball player made 286 out of 335 free-throw tries, what fractional part of his attempts at free throws did he make?   3.1   2

_____ 2. Lovell Memorial Theater's seating capacity is 515. If only 336 persons attended a play, what fractional part of the theater's seats were filled?   3.1   2

Find the missing part of each statement.

_____ 3. $\dfrac{?}{15} = 9$ 3.1   5

_____ 4. $\dfrac{168}{7} = ?$ 3.1   6

_____ 5. Each rack at the Fashion House will display 15 dresses. How many racks will be needed to display 88 dresses? Write your answer as an improper fraction and then as a mixed number.   3.1   8

Change each improper fraction to a mixed number.

_____ 6. $\dfrac{51}{8}$ 3.1   9, 10

_____ 7. $\dfrac{19}{5}$ 3.1   9, 10

Change each mixed number to an improper fraction.

_____ 8. $9\dfrac{7}{8}$ 3.1   11

_____ 9. $23\dfrac{2}{3}$ 3.1   11

Reduce each fraction to lowest terms.

_____ 10. $\dfrac{35}{42}$ 3.2   1

_____ 11. $\dfrac{90}{117}$ 3.2   1

		SECTION	EXAMPLE

_____ 12. $\dfrac{36}{45}$ 3.2   2

_____ 13. $\dfrac{456}{612}$ 3.2   2

_____ 14. $\dfrac{500}{1300}$ 3.2   3

_____ 15. $\dfrac{80}{180}$ 3.2   4

_____ 16. $\dfrac{18}{55}$ 3.2   4

_____ 17. When 3000 athletes were tested for drug usage, 450 were found to   3.3   1–3
have used steroids. What fractional part of those tested had used
steroids?

_____ 18. José and Rick assembled 36 carburetor parts in an hour. If Rick   3.3   1–3
assembled 20 parts, what fractional part of the total did he assemble?

_____ 19. If 430 of 645 homes surveyed had a microwave oven, what fractional   3.3   4
part of those homes did not have a microwave?

_____ 20. Ted can throw the discus 60 meters, but Wilson can only throw it 54   3.3   5
meters. What fractional part of Ted's distance is Wilson's distance?

Check your answers in the back of your book.

SHARPENING YOUR SKILLS after Chapters 1–3

SECTION

_____ 1. Write the word form for 703,450.   1.1

_____ 2. Classify 756 as prime or composite. If the number is composite, write it as a product   2.3
of prime numbers.

_____ 3. Change $\frac{58}{7}$ to a mixed number.   3.1

_____ 4. Change $14\frac{5}{8}$ to an improper fraction.   3.2

Find the sums.

_____ 5. 683 + 2075 + 10,039   1.2

_____ 6.   85,678   1.2
                 29,724
               $+38,005$

Subtract and check.

_____ 7.   401   1.3
               -386

_____ 8.   35,671   1.3
               $-17,893$

_____ 9.   10,000   1.3
               $-\ 3,549$

Multiply.

_____ 10. 782 × 403   2.1

_____ 11. 12 × 3 × 9 × 7   2.1

Divide and check.

_____ 12. 6984 ÷ 8   2.2

_____ 13. 23,115 ÷ 23   2.2

_____ 14. 1029 ÷ 12   2.2

Reduce each fraction to lowest terms.

_____ 15. $\dfrac{168}{280}$ 3.2

SECTION

———— 16. $\dfrac{51}{57}$ 3.2

———— 17. During the month of May the Health Department issued 27 permits for the construction   1.4
of backyard swimming pools. The next month the number of permits issued increased
by 19. How many permits were issued in June?

———— 18. The New York Yankees won 92 out of 154 games. What fractional part of their games   3.1
did they win?

———— 19. Find the area of a school playground that is 600 feet long and 250 feet wide.   2.4

———— 20. In an election for associated student government president, John received 216 votes,   1.4
Ahsan 171 votes, and Bettina 63 votes. If 593 votes were cast, how many votes did
the other candidates receive?

Check your answers in the back of your book.

Multiplying and Dividing Fractions

A flag maker wishes to make 4 signal flags in the shape of right triangles. If each triangle has a height of $8\frac{1}{3}$ inches and a base of $9\frac{3}{5}$ inches, find the number of square inches of fabric needed to make the 4 flags.

Now that we understand the meaning of fractional numbers, we must learn to perform the four basic operations of arithmetic using fractions. In this chapter we concentrate on multiplication and division, as we learn to

1. Multiply fractions.
2. Divide fractions.
3. Simplify complex fractions.
4. Switch from word statements to fractional number statements.

The methods for using a calculator to solve problems of the types encountered in Chapters 3–7 are discussed in Sections 16.1 and 16.2 of Chapter 16.

4.1 Multiplying Fractions

Multiplying Fractions by Whole Numbers

If a recipe calls for $\frac{1}{3}$ cup of flour and we wish to double the recipe, we know we must *multiply* the amount of each ingredient by 2. To figure the amount of flour needed, we must find a way to multiply 2 times $\frac{1}{3}$. Perhaps a drawing will help us again.

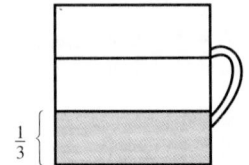   

Do you agree that the shaded part of the second cup is double the shaded part of the first cup? If each portion represents $\frac{1}{3}$ of the cup, then the shaded part in the second cup must represent $\frac{2}{3}$ of the cup. In other words,

$$2 \times \frac{1}{3} = \frac{2}{3}$$

This example can help us decide upon a method for multiplying a whole number times a fraction.

Multiplying a Whole Number Times a Fraction

1. Write the whole number as a fraction with a denominator of 1.
2. Multiply the numerators.
3. Multiply the denominators.
4. Write a new fraction that is the product of the numerators over the product of the denominators.
5. Reduce the fraction if possible.

Let's see how this method works for our flour example.

$$2 \times \frac{1}{3} = \frac{2}{1} \times \frac{1}{3}$$

$$= \frac{2 \times 1}{1 \times 3}$$

$$= \frac{2}{3}$$

Example 1. Ricky grew $\frac{5}{8}$ inches last year. His friend Todd grew 4 times as much. How many inches did Todd grow?

Solution. First we multiply.

$$4 \times \frac{5}{8} = \frac{4}{1} \times \frac{5}{8}$$

$$= \frac{4 \times 5}{1 \times 8}$$

$$= \frac{20}{8}$$

We must remember to reduce if possible.

$$\frac{20}{8} = \frac{5 \times 4}{2 \times 4}$$

$$= \frac{5 \times \overset{1}{\cancel{4}}}{2 \times \underset{1}{\cancel{4}}}$$

$$= \frac{5}{2} \quad \text{or} \quad 2\frac{1}{2} \text{ inches}$$

Todd grew $2\frac{1}{2}$ inches.

Example 2. Multiply $\frac{2}{15} \times 6$.

Solution

$$\frac{2}{15} \times 6 = \frac{2}{15} \times \frac{6}{1}$$

$$= \frac{2 \times 6}{15 \times 1}$$

$$= \frac{12}{15}$$

$$= \frac{4 \times \overset{1}{\cancel{3}}}{5 \times \underset{1}{\cancel{3}}}$$

$$\frac{2}{15} \times 6 = \frac{4}{5}$$

Notice that we remembered to reduce.

You try Example 3.

Example 3. Multiply $\frac{1}{3} \times 12$.

Solution

$$\frac{1}{3} \times 12 = \frac{1}{3} \times \frac{12}{1}$$

$$= \frac{\times}{\times}$$

$$=$$

$$\frac{1}{3} \times 12 = \underline{\qquad}$$

Check your work on page 148. ▶

Perhaps you are wondering about the *meaning* of a multiplication such as $\frac{1}{3} \times 12$. One interpretation that works well is to think that

$$\boxed{\frac{1}{3} \times 12 \quad \text{means} \quad \frac{1}{3} \text{ of } 12}$$

You should agree that if 12 were separated into thirds, then each third would contain 4. So to say that "$\frac{1}{3}$ of 12 is 4" should not seem surprising to you.

⇒ Trial Run

Multiply.

———— 1. $18 \times \dfrac{5}{6}$　　　　　　　　———— 2. $\dfrac{3}{14} \times 35$

———— 3. $\dfrac{3}{16} \times 4$　　　　　　　　———— 4. $48 \times \dfrac{1}{8}$

———— 5. Find $\dfrac{3}{4}$ of 15.　　　　　———— 6. Find $\dfrac{2}{3}$ of 45.

Answers are on page 149.

Multiplying Fractions by Fractions

You will be glad to know that the method for multiplying two fractions is the same method already described for multiplying whole numbers by fractions.

> **Multiplying Fractions**
>
> 1. Multiply the numerators.
> 2. Multiply the denominators.
> 3. Reduce if possible.

For instance

$$\frac{3}{8} \times \frac{5}{2} = \frac{3 \times 5}{8 \times 2}$$

$$= \frac{15}{16}$$

Example 4. Find $\frac{2}{9} \times \frac{27}{40}$.

Solution

$$\frac{2}{9} \times \frac{27}{40} = \frac{2 \times 27}{9 \times 40}$$

$$= \frac{54}{360}$$

$$= \frac{6 \times \overset{1}{\cancel{9}}}{40 \times \underset{1}{\cancel{9}}}$$

$$= \frac{6}{40}$$

$$= \frac{3 \times \overset{1}{\cancel{2}}}{20 \times \underset{1}{\cancel{2}}}$$

$$\frac{2}{9} \times \frac{27}{40} = \frac{3}{20}$$

You complete Example 5.

Example 5. Find $\frac{1}{2}$ of $\frac{2}{3}$.

Solution. Remember that $\frac{1}{2}$ of $\frac{2}{3}$ means $\frac{1}{2} \times \frac{2}{3}$.

$$\frac{1}{2} \times \frac{2}{3} = \frac{1 \times 2}{2 \times 3}$$

$$= \frac{2}{6}$$

$$= \frac{1 \times \overset{1}{\cancel{2}}}{3 \times \underset{1}{\cancel{2}}}$$

$$\frac{1}{2} \times \frac{2}{3} = \underline{\qquad}$$

Check your work on page 148. ▶

⟫ Trial Run

Multiply.

_____ **1.** $\frac{1}{2} \times \frac{1}{2}$

_____ **2.** $\frac{2}{3} \times \frac{6}{7}$

_____ **3.** $\frac{9}{10} \times \frac{5}{9}$

_____ **4.** $\frac{2}{5} \times \frac{3}{4}$

_____ **5.** Find $\frac{7}{9}$ of $\frac{3}{7}$.

_____ **6.** Find $\frac{11}{15}$ of $\frac{6}{77}$.

Answers are on page 149.

In Example 4 we were able to reduce our fraction after multiplying, but it almost seems like double work to multiply the original numerators and the original denominators and then look for common factors for reducing. Instead, perhaps we should have looked for common factors *before* we multiplied. Let's try that approach.

$$\frac{2}{9} \times \frac{27}{40} = \frac{2 \times 27}{9 \times 40}$$ Rule for multiplication.

$$= \frac{2 \times 3 \times 9}{9 \times 2 \times 20}$$ Factor numerator and denominator.

$$= \frac{\overset{1}{\cancel{2}} \times 3 \times \overset{1}{\cancel{9}}}{\underset{1}{\cancel{9}} \times \underset{1}{\cancel{2}} \times 20}$$ Divide out common factors.

$$\frac{2}{9} \times \frac{27}{40} = \frac{3}{20}$$ Multiply remaining factors.

Notice that this method cuts down considerably on the number of steps in a multiplication problem.

Example 6. Multiply $\frac{12}{25} \times \frac{35}{66}$.

Solution

$$\frac{12}{25} \times \frac{35}{66} = \frac{12 \times 35}{25 \times 66}$$

$$= \frac{2 \times \overset{1}{\cancel{6}} \times \overset{1}{\cancel{5}} \times 7}{\underset{1}{\cancel{5}} \times 5 \times \underset{1}{\cancel{6}} \times 11}$$

$$= \frac{2 \times 1 \times 1 \times 7}{1 \times 5 \times 1 \times 11}$$

$$\frac{12}{25} \times \frac{35}{66} = \frac{14}{55}$$

You try Example 7.

Example 7. Multiply $\frac{6}{5} \times \frac{20}{3}$.

Solution

$$\frac{6}{5} \times \frac{20}{3} = \frac{6 \times 20}{5 \times 3}$$

$$= \frac{2 \times 3 \times 4 \times 5}{5 \times 3}$$

$$=$$

Check your work on page 148. ▶

Some students prefer an even shorter method for reducing before multiplying fractions. In using such a method, we divide any one numerator and any one denominator by any common factor *before* writing the product as one fraction. Let's rework Example 6 by this method.

$$\frac{12}{25} \times \frac{35}{66} = \frac{2 \times \overset{1}{\cancel{6}}}{\underset{1}{\cancel{5}} \times 5} \times \frac{\overset{1}{\cancel{5}} \times 7}{\underset{1}{\cancel{6}} \times 11}$$

$$= \frac{2 \times 7}{5 \times 11}$$

$$= \frac{14}{55}$$

After a bit of practice with this method, you may sometimes be able to spot the common factors without writing the numerators and denominators in factored form. Then your work might look like this:

$$\frac{\overset{2}{\cancel{12}}}{\underset{5}{\cancel{25}}} \times \frac{\overset{7}{\cancel{35}}}{\underset{11}{\cancel{66}}} = \frac{2 \times 7}{5 \times 11}$$

$$= \frac{14}{55}$$

Notice that we are still doing the same thing that we did before. We divided 12 and 66 by 6 (their common factor), and we divided 25 and 35 by 5 (their common factor).

Example 8. Multiply $\frac{14}{27} \times \frac{30}{49}$.

Solution

$$\frac{14}{27} \times \frac{30}{49}$$

$$= \frac{\overset{2}{\cancel{14}}}{\underset{9}{\cancel{27}}} \times \frac{\overset{10}{\cancel{30}}}{\underset{7}{\cancel{49}}} \left\{ \begin{array}{l} \text{Divide 14 and 49 by 7} \\ \text{Divide 27 and 30 by 3} \end{array} \right.$$

$$= \frac{20}{63}$$

You try Example 9.

Example 9. Multiply $\frac{16}{33} \times 11$.

Solution

$$\frac{16}{33} \times 11 = \frac{16}{33} \times \frac{11}{1}$$

$$=$$

Check your work on page 148. ▶

IIII➡ Trial Run

Multiply.

_____ 1. $\frac{7}{10} \times \frac{12}{21}$

_____ 2. $\frac{3}{8} \times \frac{5}{6}$

_____ 3. $\frac{3}{4} \times \frac{4}{9}$

_____ 4. $\frac{54}{40} \times \frac{84}{81}$

_____ 5. Find $\frac{21}{25}$ of $\frac{45}{49}$.

_____ 6. Find $\frac{7}{8}$ of $\frac{5}{9}$.

Answers are on page 149.

Multiplying with Mixed Numbers

If a recipe calls for $2\frac{1}{2}$ cups of flour and we wish to make $\frac{1}{3}$ of the recipe, then we must find $\frac{1}{3}$ of $2\frac{1}{2}$. We know that

$$\frac{1}{3} \ \text{of} \ 2\frac{1}{2} \quad \text{means} \quad \frac{1}{3} \times 2\frac{1}{2}$$

but how do we find such a product? First we must change the mixed number $2\frac{1}{2}$ to an improper fraction. We know that $2\frac{1}{2} = \frac{5}{2}$, so our multiplication becomes

$$\frac{1}{3} \times 2\frac{1}{2} = \frac{1}{3} \times \frac{5}{2}$$
$$= \frac{5}{6}$$

We must use $\frac{5}{6}$ cup of flour.

Multiplying with Mixed Numbers

1. Change all mixed numbers to improper fractions.
2. Multiply numerators and multiply denominators, reducing if possible.

Example 10. Multiply $\frac{2}{7} \times 4\frac{1}{2}$.

Solution

$$\frac{2}{7} \times 4\frac{1}{2} = \frac{2}{7} \times \frac{9}{2}$$
$$= \frac{\overset{1}{\cancel{2}}}{7} \times \frac{9}{\underset{1}{\cancel{2}}}$$
$$= \frac{9}{7}$$
$$\frac{2}{7} \times 4\frac{1}{2} = 1\frac{2}{7}$$

You try Example 11.

Example 11. Multiply $\frac{5}{9} \times 1\frac{4}{5}$.

Solution

$$\frac{5}{9} \times 1\frac{4}{5} = \frac{5}{9} \times \frac{9}{5}$$
$$=$$

Check your work on page 149. ▶

Example 12. Multiply $4\frac{1}{8} \times 7\frac{1}{3}$.

Solution

$$4\frac{1}{8} \times 7\frac{1}{3} = \frac{33}{8} \times \frac{22}{3} \qquad \text{Rewrite mixed numbers as improper fractions.}$$

$$= \frac{\overset{11}{\cancel{33}}}{\underset{4}{\cancel{8}}} \times \frac{\overset{11}{\cancel{22}}}{\underset{1}{\cancel{3}}} \qquad \left\{ \begin{array}{l} \text{Divide 33 and 3 by 3.} \\ \text{Divide 8 and 22 by 2.} \end{array} \right.$$

$$= \frac{121}{4} \qquad \text{Multiply remaining factors.}$$

$$4\frac{1}{8} \times 7\frac{1}{3} = 30\frac{1}{4} \qquad \text{Rewrite answer as mixed number.}$$

Notice that we have changed answers that were improper fractions into mixed numbers. Actually, improper fractions are perfectly acceptable numbers. In fact, they are sometimes preferred to mixed numbers (especially in algebra). Perhaps we can agree to *write answers in the same form as the original numbers in the problem.*

⫸ Trial Run

Multiply.

_____ 1. $2\frac{3}{4} \times \frac{1}{3}$ _____ 2. $\frac{3}{7} \times 4\frac{1}{5}$

_____ 3. $4\frac{1}{3} \times 3\frac{3}{4}$ _____ 4. $7\frac{1}{3} \times \frac{1}{11}$

_____ 5. $3\frac{3}{5} \times \frac{5}{18}$ _____ 6. $15\frac{1}{5} \times 32\frac{1}{2}$

Answers are on page 149.

Let's review the method used for multiplying whole numbers, fractions, and mixed numbers.

Multiplying with Fractions

1. Change any whole numbers and mixed numbers to improper fractions.
2. Multiply numerators and multiply denominators, reducing where possible.
3. Write the answer in the same form as the numbers in the original problem.

These same rules apply no matter how many fractional numbers are to be multiplied.

Example 13. Find $\dfrac{5}{6} \times \dfrac{7}{10} \times \dfrac{9}{14}$.

Solution

$$\frac{5}{6} \times \frac{7}{10} \times \frac{9}{14} = \frac{\overset{1}{\cancel{5}}}{\underset{2}{\cancel{6}}} \times \frac{\overset{1}{\cancel{7}}}{\underset{2}{\cancel{10}}} \times \frac{\overset{3}{\cancel{9}}}{\underset{2}{\cancel{14}}}$$

$$= \frac{1 \times 1 \times 3}{2 \times 2 \times 2}$$

$$= \frac{3}{8}$$

You try Example 14.

Example 14. Multiply $3 \times 1\dfrac{5}{9} \times \dfrac{3}{7}$.

Solution

$$3 \times 1\frac{5}{9} \times \frac{3}{7} = \frac{3}{1} \times \frac{14}{9} \times \frac{3}{7}$$

$$=$$

$$=$$

Check your work on page 149. ▶

▶ Examples You Completed

Example 3. Multiply $\dfrac{1}{3} \times 12$.

Solution

$$\frac{1}{3} \times 12 = \frac{1}{3} \times \frac{12}{1}$$

$$= \frac{1 \times 12}{3 \times 1}$$

$$= \frac{12}{3}$$

$$\frac{1}{3} \times 12 = 4$$

Example 5. Find $\dfrac{1}{2}$ of $\dfrac{2}{3}$.

Solution. Remember that $\dfrac{1}{2}$ of $\dfrac{2}{3}$ means $\dfrac{1}{2} \times \dfrac{2}{3}$.

$$\frac{1}{2} \times \frac{2}{3} = \frac{1 \times 2}{2 \times 3}$$

$$= \frac{2}{6}$$

$$= \frac{1 \times \overset{1}{\cancel{2}}}{3 \times \underset{1}{\cancel{2}}}$$

$$\frac{1}{2} \times \frac{2}{3} = \frac{1}{3}$$

Example 7. Multiply $\dfrac{6}{5} \times \dfrac{20}{3}$.

Solution

$$\frac{6}{5} \times \frac{20}{3} = \frac{6 \times 20}{5 \times 3}$$

$$= \frac{2 \times \overset{1}{\cancel{3}} \times 4 \times \overset{1}{\cancel{5}}}{\underset{1}{\cancel{5}} \times \underset{1}{\cancel{3}}}$$

$$= \frac{2 \times 4}{1}$$

$$\frac{6}{5} \times \frac{20}{3} = 8$$

Example 9. Multiply $\dfrac{16}{33} \times 11$.

Solution

$$\frac{16}{33} \times 11 = \frac{16}{\underset{3}{\cancel{33}}} \times \frac{\overset{1}{\cancel{11}}}{1}$$

$$= \frac{16}{3} \text{ or } 5\frac{1}{3}$$

Example 11. Multiply $\dfrac{5}{9} \times 1\dfrac{4}{5}$.

Solution

$$\dfrac{5}{9} \times 1\dfrac{4}{5} = \dfrac{5}{9} \times \dfrac{9}{5}$$

$$= \dfrac{\overset{1}{\cancel{5}}}{\underset{1}{\cancel{9}}} \times \dfrac{\overset{1}{\cancel{9}}}{\underset{1}{\cancel{5}}}$$

$$= 1$$

Example 14. Multiply $3 \times 1\dfrac{5}{9} \times \dfrac{3}{7}$.

Solution

$$3 \times 1\dfrac{5}{9} \times \dfrac{3}{7} = \dfrac{3}{1} \times \dfrac{14}{9} \times \dfrac{3}{7}$$

$$= \dfrac{\overset{1}{\cancel{3}}}{1} \times \dfrac{\overset{2}{\cancel{14}}}{\underset{\underset{1}{\cancel{3}}}{\cancel{9}}} \times \dfrac{\overset{1}{\cancel{3}}}{\underset{1}{\cancel{7}}}$$

$$= 2$$

Answers to Trial Runs

page 142   **1.** 15   **2.** $\dfrac{15}{2}$ or $7\dfrac{1}{2}$ **3.** $\dfrac{3}{4}$ **4.** 6   **5.** $\dfrac{45}{4}$ or $11\dfrac{1}{4}$ **6.** 30

page 143   **1.** $\dfrac{1}{4}$ **2.** $\dfrac{4}{7}$ **3.** $\dfrac{1}{2}$ **4.** $\dfrac{3}{10}$ **5.** $\dfrac{1}{3}$ **6.** $\dfrac{2}{35}$

page 145   **1.** $\dfrac{2}{5}$ **2.** $\dfrac{5}{16}$ **3.** $\dfrac{1}{3}$ **4.** $\dfrac{7}{5}$ **5.** $\dfrac{27}{35}$ **6.** $\dfrac{35}{72}$

page 147   **1.** $\dfrac{11}{12}$ **2.** $\dfrac{9}{5}$ or $1\dfrac{4}{5}$ **3.** $\dfrac{65}{4}$ or $16\dfrac{1}{4}$ **4.** $\dfrac{2}{3}$ **5.** 1   **6.** 494

EXERCISE SET 4.1

Perform the indicated operation.

_____ 1. $24 \times \frac{3}{8}$

_____ 2. $28 \times \frac{5}{7}$

_____ 3. $\frac{3}{16} \times 40$

_____ 4. $\frac{7}{20} \times 45$

_____ 5. $\frac{8}{35} \times 7$

_____ 6. $\frac{9}{56} \times 8$

_____ 7. $63 \times \frac{1}{9}$

_____ 8. $72 \times \frac{1}{4}$

_____ 9. $\frac{5}{8}$ of 25

_____ 10. $\frac{2}{3}$ of 64

_____ 11. $\frac{4}{5}$ of 60

_____ 12. $\frac{7}{8}$ of 72

_____ 13. $\frac{1}{8} \times \frac{1}{8}$

_____ 14. $\frac{1}{7} \times \frac{1}{7}$

_____ 15. $\frac{4}{5} \times \frac{10}{13}$

_____ 16. $\frac{5}{6} \times \frac{12}{17}$

_____ 17. $\frac{8}{25} \times \frac{5}{8}$

_____ 18. $\frac{13}{45} \times \frac{9}{13}$

_____ 19. $\frac{7}{9} \times \frac{16}{21}$

_____ 20. $\frac{5}{6} \times \frac{7}{25}$

_____ 21. $\frac{11}{12}$ of $\frac{4}{11}$

_____ 22. $\frac{13}{15}$ of $\frac{5}{13}$

_____ 23. $\frac{10}{33}$ of $\frac{3}{50}$

_____ 24. $\frac{12}{55}$ of $\frac{11}{72}$

_____ 25. $\frac{8}{9}$ of $\frac{27}{64}$

_____ 26. $\frac{12}{25}$ of $\frac{75}{84}$

_____ 27. $\frac{8}{15} \times \frac{45}{80}$

_____ 28. $\frac{27}{50} \times \frac{70}{81}$

_____ 29. $\frac{5}{7} \times \frac{3}{20}$

_____ 30. $\frac{4}{5} \times \frac{7}{36}$

_____ 31. $\frac{9}{13} \times \frac{13}{45}$

_____ 32. $\frac{8}{21} \times \frac{21}{40}$

_____ 33. $\frac{72}{81} \times \frac{18}{40}$

_____ 34. $\frac{40}{45} \times \frac{27}{56}$

_____ 35. $\frac{16}{17}$ of $\frac{34}{48}$

_____ 36. $\frac{15}{16}$ of $\frac{80}{91}$

_____ 37. $\frac{2}{3}$ of $\frac{5}{8}$

_____ 38. $\frac{2}{5}$ of $\frac{7}{18}$

_____ 39. $\frac{1}{2}$ of $\frac{3}{4}$

_____ 40. $\frac{1}{3}$ of $\frac{4}{5}$

_____ 41. $\frac{2}{3} \times 4$

_____ 42. $\frac{1}{2} \times 5$

_____ 43. $5 \times \frac{3}{4}$

_____ 44. $8 \times \frac{2}{3}$

_____ 45. $2\frac{1}{2} \times 4$

_____ 46. $4\frac{1}{2} \times \frac{1}{3}$

_____ 47. $3 \times 4\frac{1}{9}$

_____ 48. $5 \times 7\frac{16}{25}$

_____ 49. $5\dfrac{1}{4} \times \dfrac{1}{3}$

_____ 50. $3\dfrac{1}{3} \times \dfrac{1}{5}$

_____ 51. $\dfrac{1}{11}$ of $7\dfrac{1}{3}$

_____ 52. $\dfrac{2}{5}$ of $8\dfrac{1}{3}$

_____ 53. $1\dfrac{1}{5} \times 2\dfrac{2}{3}$

_____ 54. $4\dfrac{1}{3} \times 3\dfrac{3}{4}$

_____ 55. $2\dfrac{3}{4} \times 2\dfrac{2}{3}$

_____ 56. $5\dfrac{1}{3} \times 5\dfrac{3}{4}$

_____ 57. $7\dfrac{1}{2} \times 1\dfrac{1}{15}$

_____ 58. $4\dfrac{2}{3} \times 1\dfrac{13}{14}$

_____ 59. $7\dfrac{1}{8} \times \dfrac{8}{57}$

_____ 60. $9\dfrac{2}{3} \times \dfrac{3}{29}$

_____ 61. $\dfrac{3}{8} \times \dfrac{16}{21} \times \dfrac{7}{8}$

_____ 62. $\dfrac{11}{14} \times \dfrac{7}{10} \times \dfrac{15}{22}$

_____ 63. $\dfrac{4}{5} \times \dfrac{7}{36} \times \dfrac{5}{21}$

_____ 64. $\dfrac{14}{39} \times \dfrac{5}{7} \times \dfrac{3}{20}$

_____ 65. $1\dfrac{1}{5} \times \dfrac{15}{16} \times 2\dfrac{2}{3}$

_____ 66. $3\dfrac{1}{2} \times \dfrac{6}{7} \times 1\dfrac{5}{9}$

☆ Stretching the Topics

Multiply.

_____ 1. $1\dfrac{32}{45} \times 1\dfrac{113}{175} \times 1\dfrac{37}{88}$

_____ 2. $1\dfrac{1}{4} \times 1\dfrac{1}{7} \times 1\dfrac{24}{25} \times \dfrac{5}{14}$

_____ 3. $\dfrac{3}{2} \times \dfrac{5}{9} \times \dfrac{4}{15} \times \dfrac{1}{2} \times 3\dfrac{3}{5}$

Check your answers in the back of your book.

If you can find the products in **Checkup 4.1,** you are ready to go on to Section 4.2.

✔ **CHECKUP 4.1**

Perform the indicated operation.

_____ 1. $56 \times \dfrac{7}{8}$

_____ 2. $\dfrac{1}{8}$ of 72

_____ 3. $\dfrac{5}{6} \times \dfrac{21}{25}$

_____ 4. $\dfrac{36}{75} \times \dfrac{40}{84}$

_____ 5. $\dfrac{35}{49}$ of $\dfrac{21}{25}$

_____ 6. $3\dfrac{1}{2} \times \dfrac{5}{21}$

_____ 7. $6 \times 5\dfrac{1}{12}$

_____ 8. $\dfrac{3}{8}$ of $6\dfrac{2}{3}$

_____ 9. $7\dfrac{3}{4} \times \dfrac{4}{31}$

_____ 10. $8\dfrac{2}{3} \times 2\dfrac{7}{13} \times \dfrac{1}{2}$

Check your answers in the back of your book.

If You Missed Problems:	You Should Review Examples:
1, 2	1–3
3–5	4–9
6–9	10–12
10	13, 14

4.2 Dividing Fractions

Before we learn to divide with fractions, we must take a look at the **reciprocal** of a fractional number. If the product of two fractional numbers is **1**, then we say that the two numbers are reciprocals of each other. We find the reciprocal of a fractional number by interchanging its numerator and denominator. For instance, the reciprocal of $\frac{5}{8}$ is $\frac{8}{5}$ because $\frac{5}{8} \times \frac{8}{5} = 1$.

Number	Fractional Form	Reciprocal	Reason
$\frac{4}{3}$	$\frac{4}{3}$	$\frac{3}{4}$	$\frac{4}{3} \times \frac{3}{4} = 1$
$\frac{1}{7}$	$\frac{1}{7}$	$\frac{7}{1}$	$\frac{1}{7} \times \frac{7}{1} = 1$
2	$\frac{2}{1}$	$\frac{1}{2}$	$\frac{2}{1} \times \frac{1}{2} = 1$
$5\frac{1}{3}$	$\frac{16}{3}$	$\frac{3}{16}$	$\frac{16}{3} \times \frac{3}{16} = 1$

The process of interchanging the numerator and denominator of a fraction is called **inverting** the fraction (or turning it upside down).

The reciprocal of a fraction is found by inverting the fraction.

You complete Example 1.

Example 1. Find the reciprocals of $\frac{5}{7}$, 9, $\frac{1}{3}$, and $2\frac{3}{4}$.

Solution

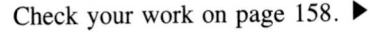

Number	Fractional Form	Reciprocal
$\frac{5}{7}$	$\frac{5}{7}$	
9	$\frac{9}{1}$	
$\frac{1}{3}$	$\frac{1}{3}$	
$2\frac{3}{4}$	$\frac{11}{4}$	

Check your work on page 158. ▶

We note that

Every number *except 0* has a reciprocal.

Dividing Fractional Numbers

To learn how to divide with fractions, we start by using common sense. If you are working 30 hours a week and your number of hours is divided by 2, how many hours will you be working? We can find the answer by computing $30 \div 2 = 15$. Another interpretation for this problem can also be found if we realize that dividing a number by 2 has the same meaning as finding $\frac{1}{2}$ of that number. But finding $\frac{1}{2}$ of a number has the same meaning as multiplying that number by $\frac{1}{2}$.

Therefore we can say here that

$$30 \div 2 = 30 \times \frac{1}{2}$$

$$= \frac{30}{1} \times \frac{1}{2}$$

$$= 15$$

This example gives us a hint about a method for dividing fractional numbers. We found that *division by 2* could also be accomplished using *multiplication by the reciprocal of 2* (or $\frac{1}{2}$).

Dividing Fractional Numbers. To divide two fractions, we multiply the dividend (the first fraction) by the reciprocal of the divisor (the second fraction).

Because the dividend and/or the divisor in our problems might be whole numbers or mixed numbers, there are several steps to be followed in performing division.

Steps in Dividing Fractional Numbers

1. Write the dividend and divisor as fractional numbers.
2. Invert the divisor.
3. Multiply the fractions, reducing if possible.

Example 2. Find $\dfrac{3}{7} \div \dfrac{3}{4}$.

Solution

$$\frac{3}{7} \div \frac{3}{4} = \frac{3}{7} \times \frac{4}{3}$$

$$= \frac{\cancel{3}^{1}}{7} \times \frac{4}{\cancel{3}_{1}}$$

$$= \frac{4}{7}$$

You complete Example 3.

Example 3. Find $\dfrac{13}{9} \div \dfrac{1}{3}$.

Solution

$$\frac{13}{9} \div \frac{1}{3} = \frac{13}{9} \times \frac{3}{1}$$

$$=$$

Check your work on page 159. ▶

Example 4. Find $\dfrac{2}{3} \div 4$.

Solution

$$\frac{2}{3} \div 4 = \frac{2}{3} \div \frac{4}{1}$$

$$= \frac{\cancel{2}^{1}}{3} \times \frac{1}{\cancel{4}_{2}}$$

$$= \frac{1}{6}$$

Example 5. Find $4 \div 7$.

Solution

$$4 \div 7 = \frac{4}{1} \div \frac{7}{1}$$

$$= \frac{4}{1} \cdot \frac{1}{7}$$

$$= \frac{4}{7}$$

Example 6. Find $3 \div 1\dfrac{2}{7}$.

Solution

$$3 \div 1\frac{2}{7} = \frac{3}{1} \div \frac{9}{7}$$

$$= \frac{3}{1} \times \frac{7}{9}$$

$$= \frac{\cancel{3}^{1}}{1} \times \frac{7}{\cancel{9}_{3}}$$

$$= \frac{7}{3}$$

$$= 2\frac{1}{3}$$

You complete Example 7.

Example 7. Find $3\dfrac{1}{7} \div 11$.

Solution

$$3\frac{1}{7} \div 11 = \frac{22}{7} \div \frac{11}{1}$$

$$= \frac{22}{7} \times \frac{1}{11}$$

$$=$$

Check your work on page 159. ▶

Example 8. Find $5\frac{3}{11} \div 2\frac{3}{4}$.

Solution

$$5\frac{3}{11} \div 2\frac{3}{4} = \frac{58}{11} \div \frac{11}{4} \qquad \text{Change mixed numbers to improper fractions.}$$

$$= \frac{58}{11} \times \frac{4}{11} \qquad \text{Invert the divisor and multiply.}$$

$$= \frac{232}{121} \qquad \text{Find the product.}$$

$$= 1\frac{111}{121} \qquad \text{Change improper fraction to mixed number.}$$

▶ Trial Run

Divide.

———— 1. $\dfrac{1}{3} \div \dfrac{3}{8}$

———— 2. $\dfrac{3}{5} \div \dfrac{3}{10}$

———— 3. $\dfrac{25}{7} \div \dfrac{5}{12}$

———— 4. $\dfrac{4}{8} \div \dfrac{5}{6}$

———— 5. $18 \div \dfrac{1}{6}$

———— 6. $5\dfrac{3}{7} \div 19$

———— 7. $3\dfrac{2}{5} \div \dfrac{3}{5}$

———— 8. $4\dfrac{2}{5} \div 8\dfrac{1}{15}$

Answers are on page 159.

Simplifying Complex Fractions

We have already learned that a fraction bar can be used to show division. For instance, we know that $\dfrac{15}{5}$ means $15 \div 5$. Using similar reasoning, we should agree that

$$\dfrac{\dfrac{2}{3}}{\dfrac{1}{2}} \qquad \text{means} \qquad \dfrac{2}{3} \div \dfrac{1}{2}$$

Such a fraction is called a **complex fraction** because it contains a fraction in the numerator and a fraction in the denominator.

> A **complex fraction** is a fraction containing a fraction in the numerator or the denominator or both.

To simplify a complex fraction, we must rewrite it as a division problem and then perform the division by the usual method. For instance,

$$\frac{\frac{2}{3}}{\frac{1}{2}} = \frac{2}{3} \div \frac{1}{2} \qquad \text{Rewrite complex fraction as division.}$$

$$= \frac{2}{3} \times \frac{2}{1} \qquad \text{Invert the divisor and multiply.}$$

$$= \frac{4}{3} \qquad \text{Find the product.}$$

Example 9. Simplify $\dfrac{\frac{5}{8}}{\frac{3}{4}}$.

Solution

$$\frac{\frac{5}{8}}{\frac{3}{4}} = \frac{5}{8} \div \frac{3}{4}$$

$$= \frac{5}{8} \times \frac{4}{3}$$

$$= \frac{5}{\overset{}{\underset{2}{8}}} \times \frac{\overset{1}{4}}{3}$$

$$= \frac{5}{6}$$

You complete Example 10.

Example 10. Simplify $\dfrac{6}{\frac{1}{7}}$.

Solution

$$\frac{6}{\frac{1}{7}} = 6 \div \frac{1}{7}$$

$$= \frac{6}{1} \div \frac{1}{7}$$

$$= \qquad \times$$

$$=$$

Check your work on page 159. ▶

Example 11. Simplify $\dfrac{3\frac{4}{9}}{5}$.

Solution

$$\frac{3\frac{4}{9}}{5} = 3\frac{4}{9} \div 5 \qquad \text{Rewrite complex fraction as division.}$$

$$= \frac{31}{9} \div \frac{5}{1} \qquad \text{Rewrite mixed and whole numbers as fractions.}$$

$$= \frac{31}{9} \times \frac{1}{5} \qquad \text{Invert the divisor and multiply.}$$

$$= \frac{31}{45} \qquad \text{Find the product.}$$

You complete Example 12.

Example 12. Simplify $\dfrac{6\frac{1}{8}}{3\frac{1}{2}}$.

Solution

$$\frac{6\frac{1}{8}}{3\frac{1}{2}} = 6\frac{1}{8} \div 3\frac{1}{2}$$

$$= \frac{49}{8} \div \frac{7}{2}$$

$$= \frac{49}{8} \times \frac{2}{7}$$

$$=$$

Check your work on page 159. ▶

⟹ Trial Run

Simplify.

_____ 1. $\dfrac{\frac{3}{8}}{\frac{3}{5}}$

_____ 2. $\dfrac{\frac{14}{2}}{7}$

_____ 3. $\dfrac{\frac{7}{8}}{14}$

_____ 4. $\dfrac{\frac{1}{10}}{\frac{1}{2}}$

_____ 5. $\dfrac{6\frac{2}{5}}{16}$

_____ 6. $\dfrac{17\frac{1}{2}}{3\frac{1}{2}}$

Answers are on page 159.

As you can see, your ability to divide with fractions depends only on your ability to multiply with fractions. To divide fractions, you must remember to invert the divisor and multiply.

▶ Examples You Completed

Example 1

Number	Fractional Form	Reciprocal
$\frac{5}{7}$	$\frac{5}{7}$	$\frac{7}{5}$
9	$\frac{9}{1}$	$\frac{1}{9}$
$\frac{1}{3}$	$\frac{1}{3}$	$\frac{3}{1}$
$2\frac{3}{4}$	$\frac{11}{4}$	$\frac{4}{11}$

Example 3. Divide $\dfrac{13}{9} \div \dfrac{1}{3}$.

Solution

$$\dfrac{13}{9} \div \dfrac{1}{3} = \dfrac{13}{9} \times \dfrac{3}{1}$$

$$= \dfrac{13}{\underset{3}{\cancel{9}}} \times \dfrac{\overset{1}{\cancel{3}}}{1}$$

$$= \dfrac{13}{3}$$

Example 7. Find $3\dfrac{1}{7} \div 11$.

Solution

$$3\dfrac{1}{7} \div 11 = \dfrac{22}{7} \div \dfrac{11}{1}$$

$$= \dfrac{22}{7} \times \dfrac{1}{11}$$

$$= \dfrac{\overset{2}{\cancel{22}}}{7} \times \dfrac{1}{\underset{1}{\cancel{11}}}$$

$$= \dfrac{2}{7}$$

Example 10. Simplify $\dfrac{6}{\frac{1}{7}}$.

Solution

$$\dfrac{6}{\frac{1}{7}} = 6 \div \dfrac{1}{7}$$

$$= \dfrac{6}{1} \div \dfrac{1}{7}$$

$$= \dfrac{6}{1} \times \dfrac{7}{1}$$

$$= 42$$

Example 12. Simplify $\dfrac{6\frac{1}{8}}{3\frac{1}{2}}$.

Solution

$$\dfrac{6\frac{1}{8}}{3\frac{1}{2}} = 6\dfrac{1}{8} \div 3\dfrac{1}{2}$$

$$= \dfrac{49}{8} \div \dfrac{7}{2}$$

$$= \dfrac{49}{8} \times \dfrac{2}{7}$$

$$= \dfrac{\overset{7}{\cancel{49}}}{\underset{4}{\cancel{8}}} \times \dfrac{\overset{1}{\cancel{2}}}{\underset{1}{\cancel{7}}}$$

$$= \dfrac{7}{4}$$

$$= 1\dfrac{3}{4}$$

Answers to Trial Runs

page 156   **1.** $\dfrac{8}{9}$ **2.** 2   **3.** $\dfrac{60}{7}$ or $8\dfrac{4}{7}$ **4.** $\dfrac{3}{5}$ **5.** 108   **6.** $\dfrac{2}{7}$ **7.** $\dfrac{17}{3}$ or $5\dfrac{2}{3}$ **8.** $\dfrac{6}{11}$

page 158   **1.** $\dfrac{5}{8}$ **2.** 49   **3.** $\dfrac{1}{16}$ **4.** $\dfrac{1}{5}$ **5.** $\dfrac{2}{5}$ **6.** 5

EXERCISE SET 4.2

Use multiplication to find each quotient.

_____ 1. $164 \div 4$ _____ 2. $120 \div 5$ _____ 3. $21 \div 35$

_____ 4. $36 \div 63$ _____ 5. $7 \div 9$ _____ 6. $8 \div 15$

_____ 7. $\dfrac{9}{13} \div 27$ _____ 8. $\dfrac{7}{8} \div 49$ _____ 9. $\dfrac{54}{5} \div 18$

_____ 10. $\dfrac{72}{35} \div 16$ _____ 11. $3\dfrac{2}{5} \div 17$ _____ 12. $9\dfrac{2}{3} \div 29$

Divide.

_____ 13. $\dfrac{1}{5} \div \dfrac{5}{9}$ _____ 14. $\dfrac{1}{7} \div \dfrac{7}{8}$ _____ 15. $\dfrac{4}{9} \div \dfrac{28}{45}$

_____ 16. $\dfrac{5}{7} \div \dfrac{20}{63}$ _____ 17. $\dfrac{35}{9} \div \dfrac{7}{8}$ _____ 18. $\dfrac{72}{5} \div \dfrac{16}{17}$

_____ 19. $\dfrac{3}{5} \div \dfrac{6}{25}$ _____ 20. $\dfrac{4}{9} \div \dfrac{44}{45}$ _____ 21. $63 \div \dfrac{1}{3}$

_____ 22. $84 \div \dfrac{1}{2}$ _____ 23. $9\dfrac{1}{3} \div 28$ _____ 24. $7\dfrac{2}{3} \div 23$

_____ 25. $7\dfrac{2}{3} \div \dfrac{2}{3}$ _____ 26. $9\dfrac{3}{4} \div \dfrac{3}{4}$ _____ 27. $12\dfrac{3}{8} \div 7\dfrac{1}{2}$

_____ 28. $28\dfrac{1}{3} \div 5\dfrac{5}{6}$ _____ 29. $2\dfrac{2}{3} \div 2\dfrac{1}{2}$ _____ 30. $2\dfrac{1}{3} \div 1\dfrac{4}{5}$

_____ 31. $\dfrac{3}{4} \div 3\dfrac{1}{2}$ _____ 32. $\dfrac{1}{2} \div 3\dfrac{1}{4}$ _____ 33. $1\dfrac{3}{4} \div \dfrac{2}{3}$

_____ 34. $4\dfrac{1}{2} \div \dfrac{3}{5}$ _____ 35. $\dfrac{\frac{3}{7}}{\frac{3}{5}}$ _____ 36. $\dfrac{\frac{5}{12}}{\frac{5}{11}}$

_____ 37. $\dfrac{\frac{4}{15}}{\frac{8}{35}}$ _____ 38. $\dfrac{\frac{9}{20}}{\frac{3}{10}}$ _____ 39. $\dfrac{\frac{35}{7}}{2}$

_____ 40. $\dfrac{\frac{28}{4}}{3}$ _____ 41. $\dfrac{\frac{5}{12}}{15}$ _____ 42. $\dfrac{\frac{7}{8}}{14}$

_____ 43. $\dfrac{\dfrac{1}{9}}{\dfrac{1}{3}}$

_____ 44. $\dfrac{\dfrac{1}{15}}{\dfrac{1}{5}}$

_____ 45. $\dfrac{4}{3\dfrac{1}{5}}$

_____ 46. $\dfrac{2}{1\dfrac{3}{13}}$

_____ 47. $\dfrac{7\dfrac{1}{5}}{9}$

_____ 48. $\dfrac{8\dfrac{2}{3}}{13}$

_____ 49. $\dfrac{5\dfrac{1}{5}}{1\dfrac{3}{10}}$

_____ 50. $\dfrac{3\dfrac{2}{3}}{1\dfrac{5}{6}}$

☆ Stretching the Topics

Divide.

_____ 1. $12\dfrac{3}{56} \div 1\dfrac{37}{98}$

_____ 2. $\dfrac{1\dfrac{1}{5} \times 2\dfrac{2}{3}}{4 \times 3\dfrac{3}{5}}$

_____ 3. $\dfrac{1\dfrac{1}{2} \div 2\dfrac{2}{3}}{\dfrac{2}{5} \div 1\dfrac{1}{4}}$

Check your answers in the back of your book.

If you can find the quotients in **Checkup 4.2,** you are ready to go on to Section 4.3.

✓ CHECKUP 4.2

Divide.

_____ 1. $\dfrac{1}{4} \div \dfrac{1}{3}$ 　　　　　 _____ 2. $\dfrac{3}{5} \div \dfrac{7}{11}$

_____ 3. $\dfrac{3}{14} \div \dfrac{6}{7}$ 　　　　　 _____ 4. $\dfrac{8}{15} \div 16$

_____ 5. $28 \div 49$ 　　　　　 _____ 6. $15 \div \dfrac{2}{3}$

_____ 7. $2\dfrac{1}{3} \div 14$ 　　　　　 _____ 8. $2\dfrac{2}{7} \div 2\dfrac{1}{3}$

_____ 9. $\dfrac{\frac{7}{10}}{\frac{1}{4}}$ 　　　　　 _____ 10. $\dfrac{1\frac{1}{2}}{2\frac{2}{3}}$

Check your answers in the back of your book.

If You Missed Problems:	You Should Review Examples:
1–3	2, 3
4–7	4–7
8	8
9	9
10	12

4.3 Switching from Words to Fractional Numbers

We have already discussed some word phrases as they apply to multiplying and dividing with fractional numbers. Let's consider a few of those key phrases.

Word Phrase	Fractional Number Phrase
$\frac{2}{3}$ of 7	$\frac{2}{3} \times \frac{7}{1}$
$\frac{1}{2}$ as much as 60	$\frac{1}{2} \times \frac{60}{1}$
$3\frac{1}{3}$ times as many as 8	$3\frac{1}{3} \times 8$
$\frac{1}{6}$ divided by $\frac{2}{3}$	$\frac{1}{6} \div \frac{2}{3}$
$\frac{1}{4}$ as large as 12	$\frac{1}{4} \times \frac{12}{1}$

Now let's see how we can use these phrases to work some problems stated in words.

Example 1. If $\frac{3}{8}$ of the 120 students in Marty's astronomy class are female, how many females are in the class?

Solution. The key word here is "of." It tells us that we must *multiply*.

$$\frac{3}{8} \times 120 = \frac{3}{8} \times \frac{120}{1}$$

$$= \frac{3}{\underset{1}{\cancel{8}}} \times \frac{\overset{15}{\cancel{120}}}{1}$$

$$= 45$$

So we conclude that there are 45 females in Marty's astronomy class.

Example 2. Heather wishes to make $2\frac{1}{2}$ times as many meatballs for her party as she made last week. If she made 36 meatballs last week, how many will she make for her party?

Solution. The phrase "$2\frac{1}{2}$ times as many" tells us we should *multiply*.

$$2\frac{1}{2} \times 36 = \frac{5}{2} \times \frac{36}{1} \qquad \text{Rewrite mixed and whole numbers as improper fractions.}$$

$$= \frac{5}{\underset{1}{\cancel{2}}} \times \frac{\overset{18}{\cancel{36}}}{1} \qquad \text{Reduce.}$$

$$= 90 \qquad \text{Multiply remaining factors.}$$

She must make 90 meatballs for her party.

Example 3. Henry ran the 440 meter race in $63\frac{1}{2}$ seconds. Eddie took $\frac{9}{10}$ as long to run the same race. What was Eddie's time for the race?

Solution. The phrase "$\frac{9}{10}$ as long" tells us to *multiply*.

$$\frac{9}{10} \times 63\frac{1}{2} = \frac{9}{10} \times \frac{127}{2} \qquad \text{Rewrite mixed number as improper fraction.}$$

$$= \frac{1143}{20} \qquad \text{Find the product.}$$

$$= 57\frac{3}{20} \qquad \text{Rewrite improper fraction as mixed number.}$$

Eddie ran the 440 meter race in $57\frac{3}{20}$ seconds.

Example 4. If a piece of wood 36 feet long is divided into 17 equal pieces, how long is each piece?

Solution. The key phrase "is divided" tells us that we must *divide*.

$$36 \div 17 = \frac{36}{1} \div \frac{17}{1} \qquad \text{Rewrite whole numbers as improper fractions.}$$

$$= \frac{36}{1} \times \frac{1}{17} \qquad \text{Invert the divisor and multiply.}$$

$$= \frac{36}{17} \qquad \text{Find the product.}$$

$$= 2\frac{2}{17} \qquad \text{Rewrite improper fraction as mixed number.}$$

Each piece will be $2\frac{2}{17}$ inches long.

Example 5. If 48 pounds of hamburger are divided into packages containing $1\frac{1}{2}$ pounds each, how many packages will there be?

Solution. Once again we must *divide*.

$$48 \div 1\frac{1}{2} = \frac{48}{1} \div \frac{3}{2}$$

$$= \frac{48}{1} \times \frac{2}{3}$$

$$= \frac{\overset{16}{\cancel{48}}}{1} \times \frac{2}{\underset{1}{\cancel{3}}}$$

$$= 32$$

There will be 32 packages.

In Chapter 2, we learned to use multiplication to find areas of rectangles, squares, and triangles. Recall the formulas for the area of each of those figures.

Square:	Area = side × side
Rectangle:	Area = length × width
Triangle:	Area = (base × height) ÷ 2

Example 6. If Yvonne's rectangular room measures $10\frac{1}{2}$ feet wide by $13\frac{1}{2}$ feet long, find the area of the floor.

Solution

$$\text{Area} = \text{length} \times \text{width}$$

$$= 13\frac{1}{2} \times 10\frac{1}{2} \qquad \text{Use formula for area of rectangle.}$$

$$= \frac{27}{2} \times \frac{21}{2} \qquad \text{Rewrite mixed numbers as improper fractions.}$$

$$= \frac{567}{4} \qquad \text{Find the product.}$$

$$\text{Area} = 141\frac{3}{4} \text{ sq ft} \qquad \text{Rewrite improper fraction as mixed number.}$$

The floor has an area of $141\frac{3}{4}$ square feet.

Now let's solve the problem stated at the beginning of this chapter.

Example 7. A flag maker wishes to make 4 signal flags in the shape of right triangles. If each triangle has a height of $8\frac{1}{3}$ inches and a base of $9\frac{3}{5}$ inches, find the number of square inches of fabric needed to make the 4 flags.

Solution. First we must find the area of *one* triangular flag.

$$\text{Area} = (\text{base} \times \text{height}) \div 2$$

$$= \left(9\frac{3}{5} \times 8\frac{1}{3}\right) \div 2 \qquad \text{Use formula for area of triangle.}$$

$$= \left(\frac{48}{5} \times \frac{25}{3}\right) \div \frac{2}{1} \qquad \text{Change mixed and whole numbers to improper fractions.}$$

$$= \left(\frac{\overset{16}{\cancel{48}}}{\underset{1}{\cancel{5}}} \times \frac{\overset{5}{\cancel{25}}}{\underset{1}{\cancel{3}}}\right) \div \frac{2}{1} \qquad \text{Reduce within parentheses.}$$

$$= \frac{80}{1} \div \frac{2}{1} \qquad \text{Multiply within parentheses.}$$

$$= \frac{80}{1} \times \frac{1}{2} \qquad \text{Invert the divisor and multiply.}$$

$$\text{Area} = 40 \text{ sq in.} \qquad \text{Find the product.}$$

Now we find the area of 4 flags by multiplying 4 times the area of 1 flag.

$$\text{Total area} = 4 \times 40$$

$$= 160$$

To make 4 flags, 160 square inches of fabric will be needed.

EXERCISE SET 4.3

_____ 1. If $\frac{3}{5}$ of 225 students surveyed had a job to help pay college expenses, how many students had a job?

_____ 2. Beverly needs to make $3\frac{1}{2}$ times as many cookies as her recipe makes. If her recipe makes 48 cookies, how many cookies does Beverly need?

_____ 3. Blair can jog 3 miles in 40 minutes. It takes Barney $1\frac{1}{2}$ times as long to jog the same distance. How long does it take Barney to jog the same distance?

_____ 4. If a ribbon 36 inches long is to be used to make bookmarks $2\frac{1}{4}$ inches long, how many bookmarks can be made from the ribbon?

_____ 5. Mr. Bittinger is putting a string of Christmas lights on his house. If the cord is 20 feet long and has a bulb every $\frac{1}{3}$ of a foot, how many bulbs will he need?

_____ 6. Robert's recipe for barbecue sauce uses $1\frac{3}{4}$ cups of vinegar. If he wishes to make only half the recipe, how many cups of vinegar should he use?

_____ 7. Odel bought an 8-pound case of powdered milk. If the powdered milk was in $\frac{1}{2}$-pound boxes, how many boxes were in the case?

_____ 8. Lorenzo has a $119\frac{1}{4}$-acre farm. If $\frac{2}{3}$ of the farm is cropland, how many acres of cropland does the farm have?

_____ 9. No more than $\frac{1}{4}$ of a family's income should be spent on rent. What is the maximum rent a family should pay if its combined monthly income is $1800?

_____ 10. Joe walks $1\frac{2}{5}$ miles to work each day. If Maria walks only $\frac{1}{3}$ as far, how far does Maria walk?

_____ 11. Ms. Garcia has a 12-pound, fully cooked, boneless ham. If she allows $\frac{3}{8}$ of a pound for each serving, how many servings can she get from the ham?

_____ 12. If one allows $\frac{7}{8}$ pound of rib roast per person, how many pounds will be needed to serve 5 people?

_____ 13. Laura knows that it takes $1\frac{3}{4}$ yards of material to make 1 tablecloth. If she is planning a party and will use 5 tables, how many yards of material will she need to make matching tablecloths for all the tables?

_____ 14. Daphne uses $1\frac{1}{2}$ cups of popcorn for each popcorn ball. How many popcorn balls can she make from 15 cups of popcorn?

_____ 15. If $10\frac{1}{2}$ ounces of tomato puree costs 35¢, how much is tomato puree per ounce?

_____ 16. If a $15\frac{3}{4}$-ounce can of green beans costs 49¢, how much do green beans cost per ounce?

_____ 17. If Estelle earned $300 one week for $37\frac{1}{2}$ hours of work, what was she earning per hour?

_____ 18. A recipe for cupcakes calls for $1\frac{2}{3}$ cups of milk. If Ron wishes to make $2\frac{1}{2}$ times as many cupcakes as the recipe makes, how many cups of milk will he use?

_____ 19. Rick has a recipe that calls for $2\frac{1}{4}$ cups of flour, $1\frac{1}{2}$ cups of milk, and $\frac{2}{3}$ cup of cooking oil. How much of each ingredient will he need if he plans to make only half the recipe?

_____ 20. Eugenia is making the costumes for a local theater production. If she needs 5 triangular shawls that have a height $15\frac{3}{4}$ inches and a base of $30\frac{2}{3}$ inches, find how many square inches of material she needs.

_____ 21. Find the area of a square dance floor that measures $14\frac{2}{3}$ feet on each side.

_____ 22. If the area of a rectangular door is 1810 square inches and its height is $60\frac{1}{3}$ inches, what is its width?

_____ 23. After 4 laps around a neighborhood park, Antonio had run $5\frac{1}{3}$ miles. Find the distance he ran in 1 lap.

_____ 24. During a half-hour TV show, there are $6\frac{2}{3}$ minutes of commercials. If each commercial is $1\frac{1}{9}$ minutes long, how many commercials appear during the show?

_____ 25. A set of encyclopedias takes up a shelf that is 44 inches long. If each volume takes up $1\frac{3}{8}$ inches on the shelf, how many volumes are in the set?

☆ Stretching the Topics

_____ 1. Calvin is cooking an $8\frac{3}{4}$-pound roast for his dinner party. The roast loses 2 pounds during cooking. If there will be 9 persons for dinner, what size serving is he allowing for each person?

_____ 2. Todd has a sheet of plywood that has an area of 840 square inches. How many plaques that measure $9\frac{1}{3}$ inches by $7\frac{1}{2}$ inches can he cut from the plywood?

Check your answers in the back of your book.

If you can complete **Checkup 4.3,** then you are ready to do the **Review Exercises** for Chapter 4.

✓ **CHECKUP 4.3**

_____ **1.** Johnny uses $\frac{2}{3}$ cup of fertilizer for each shrub in his yard. If he has 20 shrubs, how many cups of fertilizer does he need?

_____ **2.** If each book is approximately $1\frac{3}{4}$ inches thick, how many books can you put on a shelf that is 63 inches wide?

_____ **3.** A drawing in an advertisement shows the height of a doll to be $2\frac{1}{4}$ inches. The actual doll is $5\frac{1}{2}$ times that tall. How tall is the doll?

_____ **4.** If 15 pounds of candy are to be put into $2\frac{1}{2}$-pound boxes, how many boxes of candy will there be?

_____ **5.** If Kevin has a rectangular garden that measures $24\frac{3}{4}$ feet by $20\frac{1}{3}$ feet, find the area of the garden.

Check your answers in the back of your book.

If You Missed Problems:	You Should Review Examples:
1	1
2	2
3	3
4	5
5	6

Summary

In this chapter we learned to multiply and divide fractional numbers and to simplify complex fractions.

In Order to	We Must	Examples
Multiply fractions	Multiply numerators and multiply denominators, reducing if possible.	$\dfrac{5}{8} \times \dfrac{3}{2} = \dfrac{15}{16}$ $\dfrac{\overset{1}{\cancel{3}}}{\underset{1}{\cancel{4}}} \times \dfrac{\overset{2}{\cancel{8}}}{\underset{3}{\cancel{9}}} = \dfrac{2}{3}$
Divide fractions	Invert the divisor and multiply.	$\dfrac{14}{15} \div \dfrac{2}{3} = \dfrac{\overset{7}{\cancel{14}}}{\underset{5}{\cancel{15}}} \times \dfrac{\overset{1}{\cancel{3}}}{\underset{1}{\cancel{2}}} = \dfrac{7}{5}$ $\dfrac{1}{3} \div 5 = \dfrac{1}{3} \div \dfrac{5}{1}$ $= \dfrac{1}{3} \times \dfrac{1}{5} = \dfrac{1}{15}$
Simplify complex fractions	Rewrite complex fraction as division problem and perform the division.	$\dfrac{\frac{2}{3}}{\frac{4}{5}} = \dfrac{2}{3} \div \dfrac{4}{5}$ $= \dfrac{\overset{1}{\cancel{2}}}{3} \times \dfrac{5}{\underset{2}{\cancel{4}}} = \dfrac{5}{6}$

In our work with division we introduced the **reciprocal** of a fractional number and discovered that every number (except 0) has a reciprocal.

Fractional Number	Reciprocal	Reason
$\dfrac{2}{9}$	$\dfrac{9}{2}$	$\dfrac{2}{9} \times \dfrac{9}{2} = 1$
$\dfrac{1}{8}$	$\dfrac{8}{1}$	$\dfrac{1}{8} \times \dfrac{8}{1} = 1$
$13 \left(\text{or } \dfrac{13}{1} \right)$	$\dfrac{1}{13}$	$\dfrac{13}{1} \times \dfrac{1}{13} = 1$

Finally, we practiced switching from word statements to fractional-number statements in which we used the operations of multiplication and/or division.

❏ Speaking the Language of Mathematics

Complete each statement with the appropriate word or phrase.

1. To multiply fractional numbers, we multiply the _____ and multiply the

   _____ .

2. The product of every number and its _____ is always 1.

3. The reciprocal of $\frac{17}{11}$ is _____ because _____ .

4. To divide fractional numbers we _____ the divisor and then _____ .

5. A fraction whose numerator and/or denominator contains fractions is called a _____

   _____ .

6. To simplify a complex fraction, we can rewrite it as a _____ problem.

△ Writing About Mathematics

Write your response to each question in complete sentences.

1. For each of the following fractional number phrases, write a *word* phrase that has the same meaning.

   (a) $\frac{2}{3} \times 15$

   (b) $4\frac{1}{2} \times \$120$

   (c) $\frac{1}{8} \times 36$ ft

2. Show and explain the steps you would use to simplify the complex fraction $\dfrac{5\frac{1}{3}}{2\frac{5}{6}}$.

3. If your punch recipe will serve 50 people and you wish to serve only 20 people, explain how you would use fractions to decide the correct amount of each ingredient to use.

REVIEW EXERCISES for Chapter 4

Perform the indicated operation.

_____ 1. $16 \times \dfrac{7}{8}$ _____ 2. $\dfrac{3}{10} \times 35$ _____ 3. $\dfrac{5}{8} \times \dfrac{14}{25}$

_____ 4. $\dfrac{1}{5}$ of $\dfrac{10}{21}$ _____ 5. $\dfrac{4}{9} \times \dfrac{3}{5}$ _____ 6. $\dfrac{11}{15} \times \dfrac{3}{22}$

_____ 7. $\dfrac{15}{42} \times \dfrac{21}{55}$ _____ 8. $\dfrac{29}{48} \times 24$ _____ 9. $\dfrac{3}{8} \times 5\dfrac{1}{3}$

_____ 10. $4\dfrac{1}{3} \times 3\dfrac{3}{4}$ _____ 11. $2\dfrac{1}{12} \times 4\dfrac{1}{5}$ _____ 12. $\dfrac{2}{3} \times \dfrac{4}{5} \times 1\dfrac{7}{8}$

_____ 13. $\dfrac{5}{9} \div \dfrac{3}{10}$ _____ 14. $\dfrac{22}{15} \div \dfrac{4}{25}$ _____ 15. $\dfrac{9}{13} \div 6$

_____ 16. $5 \div 8$ _____ 17. $14 \div 1\dfrac{2}{5}$ _____ 18. $9\dfrac{3}{8} \div 15$

_____ 19. $4\dfrac{7}{8} \div 1\dfrac{4}{9}$ _____ 20. $15\dfrac{5}{6} \div 2\dfrac{4}{19}$ _____ 21. $\dfrac{\frac{7}{6}}{\frac{2}{3}}$

_____ 22. $\dfrac{\frac{9}{1}}{\frac{1}{5}}$ _____ 23. $\dfrac{5\frac{2}{3}}{4}$ _____ 24. $\dfrac{7\frac{4}{15}}{2\frac{2}{5}}$

_____ 25. Rachel knows that it takes $1\frac{3}{4}$ yards of material for each costume she is sewing for a dance recital. If there are 12 dancers in the recital, how many yards of material will she need for the costumes?

_____ 26. At a sale after Christmas, a department store advertises all its merchandise at $\frac{3}{4}$ of the original price. If a sweater regularly sells for \$48, what will be the sale price?

_____ 27. Caywood has a $1\frac{1}{2}$-acre garden plot. If he decides to plant $\frac{1}{3}$ of the garden in tomatoes, how much land will be planted in tomatoes?

_____ 28. Katie bought 6 yards of material to make napkins. If each napkin requires $\frac{3}{8}$ of a yard, how many napkins can she make?

_____ 29. A land developer has a $4\frac{1}{2}$-acre lot which is to be divided into $\frac{3}{4}$-acre building sites. How many building sites can he advertise for sale?

_____ 30. Kelvin has a rectangular patio that measures $17\frac{1}{2}$ feet by $19\frac{3}{5}$ feet. Find the area of the patio.

Check your answers in the back of your book.

If You Missed Exercises:	You Should Review Examples:	
1, 2	Section 4.1	1–3
3–7		4–8
8		9
9–11		10–12
12		13, 14
13, 14	Section 4.2	2, 3
15–18		4, 7
19, 20		8
21–24		9–12
25, 26	Section 4.3	1, 2
27		3
28, 29		5
30		6

If you have completed the **Review Exercises** and corrected your answers, you are ready to take the **Practice Test** for Chapter 4.

PRACTICE TEST for Chapter 4

Perform the indicated operation.

		SECTION	EXAMPLE
_____	1. $\dfrac{5}{16} \times 24$	4.1	1–3
_____	2. $\dfrac{3}{7} \times \dfrac{35}{27}$	4.1	4, 5
_____	3. $\dfrac{9}{8} \times \dfrac{40}{63}$	4.1	7
_____	4. $\dfrac{16}{27} \times \dfrac{33}{40}$	4.1	8
_____	5. $\dfrac{19}{39} \times 26$	4.1	9
_____	6. $\dfrac{3}{5} \times 7\dfrac{1}{2}$	4.1	10
_____	7. $11\dfrac{2}{3} \times 12\dfrac{4}{7}$	4.1	12
_____	8. $\dfrac{2}{3} \times 4\dfrac{1}{2} \times \dfrac{5}{6}$	4.1	13, 14
_____	9. $\dfrac{9}{13} \div \dfrac{15}{26}$	4.2	2
_____	10. $\dfrac{72}{81} \div \dfrac{40}{18}$	4.2	3
_____	11. $8 \div 1\dfrac{1}{2}$	4.2	6
_____	12. $4\dfrac{1}{2} \div 3$	4.2	7
_____	13. $8\dfrac{1}{3} \div 1\dfrac{1}{4}$	4.2	8
_____	14. $6\dfrac{2}{5} \div 5\dfrac{1}{3}$	4.2	8

	SECTION	EXAMPLE

_____ 15. $\dfrac{\frac{8}{25}}{\frac{4}{5}}$ ⟶ 4.2 · 9

_____ 16. $\dfrac{10\frac{1}{2}}{1\frac{2}{5}}$ ⟶ 4.2 · 12

_____ 17. Hubert wishes to spend no more than $\frac{1}{4}$ of his monthly salary for rent. If his salary is \$832 a month, what is the most he can pay for rent? ⟶ 4.3 · 1, 2

_____ 18. Find the area of a rectangle that is $7\frac{1}{2}$ inches wide and $9\frac{3}{4}$ inches long. ⟶ 4.3 · 3, 6

_____ 19. If a piece of wire 48 feet long is cut into pieces that are $\frac{3}{4}$ foot long, how many pieces will there be? ⟶ 4.3 · 4, 5

_____ 20. During one week, Wanda drives a total of $16\frac{1}{4}$ miles going to *and* from her job. If she makes the round-trip 5 days a week, how far does Wanda live from her workplace? ⟶ 4.3 · 4, 5

SHARPENING YOUR SKILLS after Chapters 1-4

SECTION

_____ 1. In the whole number 6728, name the hundreds digit.   1.1

_____ 2. Classify 379 as prime or composite. If it is composite, write it as a product of prime   2.3
numbers.

_____ 3. Change $\frac{83}{5}$ to a mixed number.   3.1

_____ 4. Find the sum of 7342 and 3509.   1.2

_____ 5. Fill in the missing number: _____ − 9 = 14 because _____ = 14 + 9.   1.3

_____ 6. Subtract and check: 800 − 379   1.3

_____ 7. Reduce $\frac{78}{126}$ to lowest terms.   3.2

Multiply.

_____ 8. 723 × 502   2.1

_____ 9. 370 × 1500   2.1

_____ 10. 15 × 9 × 27   2.1

_____ 11. $\frac{8}{15} \times 75$ 4.1

_____ 12. $\frac{21}{39} \times \frac{26}{49}$ 4.1

_____ 13. $8\frac{4}{5} \times 3\frac{2}{11}$ 4.1

Divide and check.

_____ 14. 3948 ÷ 12   2.2

_____ 15. 40,002 ÷ 105   2.2

_____ 16. $7\frac{1}{2} \div 3$ 4.2

_____ 17. $9\frac{3}{4} \div 1\frac{1}{2}$ 4.2

_____ 18. $\dfrac{\frac{9}{35}}{\frac{15}{28}}$ 4.2

_____ 19. Find the area of a triangle if the base is 120 feet and the height is 85 feet.   2.4

_____ 20. If a piece of lumber 72 inches long is cut into pieces that are $2\frac{2}{3}$ inches long, how   4.3
many pieces will there be?

Check your answers in the back of your book.

Adding and Subtracting
Fractions

Scott wishes to install a new baseboard around his kitchen. The kitchen measures $12\frac{7}{10}$ feet by $14\frac{1}{2}$ feet, but the two doorways will need no baseboard. One door is $3\frac{1}{8}$ feet wide and the other is $2\frac{9}{10}$ feet wide. How many feet of baseboard will Scott need?

Now that we have learned how to perform the operations of multiplication and division with fractional numbers, we must spend some time with the operations of addition and subtraction. In this chapter we learn how to

1. Add and subtract fractional numbers with the same denominator.
2. Build fractions.
3. Add and subtract fractional numbers with different denominators.
4. Compare fractional numbers.
5. Switch from word statements to fractional number statements.

The methods for using a calculator to solve problems of the types encountered in Chapters 3–7 are discussed in Sections 16.1 and 16.2 of Chapter 16.

5.1 Adding and Subtracting Fractions with the Same Denominator

Suppose Johnny and Sarah share a pizza cut into 8 slices. If Johnny eats $\frac{5}{8}$ of the pizza and Sarah eats $\frac{2}{8}$ of the pizza, what fraction represents the part of the pizza that has been eaten? To solve this problem, we must be able to *add* fractional numbers.

Adding Proper and Improper Fractions

To solve the pizza problem, we must find the sum

$$\frac{5}{8} + \frac{2}{8}$$

Perhaps an illustration will help us see the answer.

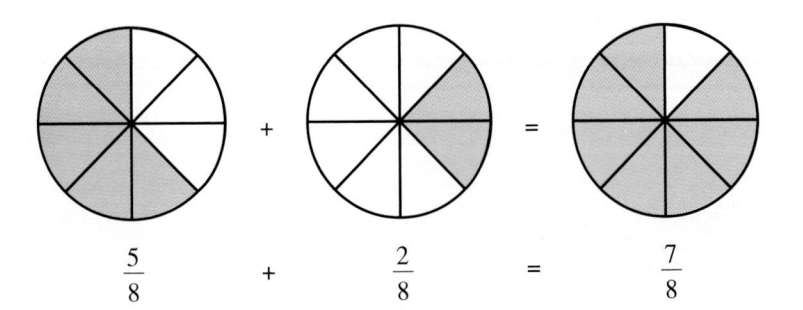

$$\frac{5}{8} \qquad + \qquad \frac{2}{8} \qquad = \qquad \frac{7}{8}$$

Notice that the numerator in our answer is the sum of the numerators in the fractions being added. Notice that the denominator in the answer is the denominator that the two fractions shared. Let's state this rule for adding fractions having the same denominator.

Adding Fractions with the Same Denominator

1. Add the numerators.
2. Keep the same denominator.
3. Reduce the final fraction if possible.

Example 1. Find $\dfrac{3}{16} + \dfrac{5}{16}$.

Solution

$$\dfrac{3}{16} + \dfrac{5}{16} = \dfrac{3 + 5}{16}$$

$$= \dfrac{8}{16}$$

$$= \dfrac{\cancel{8}^{\,1}}{\cancel{16}_{\,2}}$$

$$= \dfrac{1}{2}$$

You complete Example 2.

Example 2. Find $\dfrac{2}{3} + \dfrac{2}{3}$.

Solution

$$\dfrac{2}{3} + \dfrac{2}{3} = \dfrac{\;+\;}{\rule{2em}{0.4pt}}$$

$$= \rule{2em}{0.4pt} \quad \text{or} \quad \rule{2em}{0.4pt}$$

Check your work on page 188. ▶

Adding improper fractions with the same denominator requires no new rules. We continue to add numerators, keep the same denominator, and reduce the answer if possible.

Example 3. Find $\dfrac{19}{17} + \dfrac{21}{17}$.

Solution

$$\dfrac{19}{17} + \dfrac{21}{17} = \dfrac{19 + 21}{17}$$

$$= \dfrac{40}{17} \quad \text{or} \quad 2\dfrac{6}{17}$$

You try Example 4.

Example 4. Find $\dfrac{10}{3} + \dfrac{14}{3}$.

Solution

$$\dfrac{10}{3} + \dfrac{14}{3} = \dfrac{\;+\;}{\rule{2em}{0.4pt}}$$

$$= \dfrac{\rule{2em}{0.4pt}}{3}$$

$$= \rule{2em}{0.4pt}$$

Check your work on page 188. ▶

To add more than two fractions, we continue to use the same process.

Example 5. Find $\dfrac{5}{12} + \dfrac{7}{12} + \dfrac{1}{12}$.

Solution

$$\dfrac{5}{12} + \dfrac{7}{12} + \dfrac{1}{12} = \dfrac{5 + 7 + 1}{12}$$

$$= \dfrac{13}{12} \quad \text{or} \quad 1\dfrac{1}{12}$$

Example 6. Find $\dfrac{17}{15} + \dfrac{19}{15} + \dfrac{29}{15}$.

Solution

$$\dfrac{17}{15} + \dfrac{19}{15} + \dfrac{29}{15} = \dfrac{17 + 19 + 29}{15}$$

$$= \dfrac{65}{15}$$

$$= \dfrac{\cancel{65}^{\,13}}{\cancel{15}_{\,3}}$$

$$= \dfrac{13}{3} \quad \text{or} \quad 4\dfrac{1}{3}$$

⟫ Trial Run

Find the sums.

_____ 1. $\dfrac{5}{21} + \dfrac{2}{21}$

_____ 2. $\dfrac{4}{5} + \dfrac{3}{5}$

_____ 3. $\dfrac{3}{8} + \dfrac{5}{8} + \dfrac{7}{8}$

_____ 4. $\dfrac{23}{15} + \dfrac{7}{15}$

_____ 5. $\dfrac{20}{13} + \dfrac{7}{13}$

_____ 6. $\dfrac{5}{18} + \dfrac{11}{18} + \dfrac{17}{18}$

Answers are on page 189.

Subtracting Proper and Improper Fractions

If a measuring cup contains $\frac{7}{8}$ cup of water and $\frac{3}{8}$ cup of water is poured out, how much water is left in the cup? To solve this problem we must *subtract* and find

$$\frac{7}{8} - \frac{3}{8}$$

To illustrate our problem we have

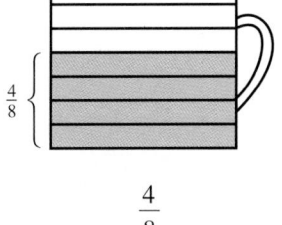

$$\frac{7}{8} \qquad - \qquad \frac{3}{8} \qquad = \qquad \frac{4}{8}$$

If we agree that

$$\frac{7}{8} - \frac{3}{8} = \frac{4}{8} \qquad \text{or} \qquad \frac{1}{2}$$

then we are ready to state the rule for subtracting fractions with the same denominator.

Subtracting Fractions with the Same Denominator

1. Subtract the numerators.
2. Keep the same denominator.
3. Reduce the final fraction if possible.

Example 7. Find $\dfrac{8}{9} - \dfrac{2}{9}$.

Solution

$$\dfrac{8}{9} - \dfrac{2}{9} = \dfrac{8 - 2}{9} \qquad \text{Subtract numerators; keep denominator.}$$

$$= \dfrac{6}{9} \qquad \text{Find the difference.}$$

$$= \dfrac{\cancel{6}^{2}}{\cancel{9}_{3}} \qquad \text{Reduce.}$$

$$= \dfrac{2}{3}$$

You complete Example 8.

Example 8. Find $\dfrac{17}{29} - \dfrac{11}{29}$.

Solution.

$$\dfrac{17}{29} - \dfrac{11}{29} = \dfrac{-}{\rule{1cm}{0.4pt}}$$

$$= \rule{1cm}{0.4pt}$$

Check your work on page 188. ▶

We subtract improper fractions in exactly the same way. We subtract the numerators, keep the same denominator, and reduce the final fraction if possible.

Example 9. Find $\dfrac{19}{11} - \dfrac{8}{11}$.

Solution

$$\dfrac{19}{11} - \dfrac{8}{11} = \dfrac{19 - 8}{11}$$

$$= \dfrac{11}{11}$$

$$= 1$$

You complete Example 10.

Example 10. Find $\dfrac{43}{16} - \dfrac{21}{16}$.

Solution

$$\dfrac{43}{16} - \dfrac{21}{16} = \dfrac{-}{16}$$

$$= \dfrac{-}{16}$$

$$=$$

Check your work on page 188. ▶

⫸ Trial Run

Find each difference.

_____ 1. $\dfrac{7}{8} - \dfrac{3}{8}$

_____ 2. $\dfrac{15}{17} - \dfrac{12}{17}$

_____ 3. $\dfrac{23}{10} - \dfrac{13}{10}$

_____ 4. $\dfrac{38}{15} - \dfrac{11}{15}$

_____ 5. $\dfrac{35}{12} - \dfrac{11}{12}$

_____ 6. $\dfrac{47}{9} - \dfrac{23}{9}$

Answers are on page 189.

To add *and* subtract within the same problem, we must *proceed from left to right* and deal with the fractions two at a time. For instance, let's find $\frac{6}{7} + \frac{3}{7} - \frac{1}{7}$.

$$\frac{6}{7} + \frac{3}{7} - \frac{1}{7} = \frac{6+3}{7} - \frac{1}{7}$$ Perform the addition.

$$= \frac{9}{7} - \frac{1}{7}$$ Find the sum.

$$= \frac{9-1}{7}$$ Perform the subtraction.

$$= \frac{8}{7} \quad \text{or} \quad 1\frac{1}{7}$$ Find the difference.

You try Example 11.

Example 11. Find $\dfrac{6}{25} + \dfrac{27}{25} - \dfrac{13}{25}$.

Solution

$$\frac{6}{25} + \frac{27}{25} - \frac{13}{25} = \frac{6+27}{25} - \frac{13}{25}$$

$$= \frac{\quad}{25} - \frac{\quad}{25}$$

$$=$$

$$=$$

Check your work on page 189. ▶

Example 12. Find $\dfrac{19}{18} - \dfrac{3}{18} - \dfrac{1}{18}$.

Solution

$$\frac{19}{18} - \frac{3}{18} - \frac{1}{18} = \frac{19-3}{18} - \frac{1}{18}$$ Perform first subtraction.

$$= \frac{16}{18} - \frac{1}{18}$$ Find the difference.

$$= \frac{16-1}{18}$$ Perform second subtraction.

$$= \frac{15}{18}$$ Find the difference.

$$= \frac{\overset{5}{\cancel{15}}}{\underset{6}{\cancel{18}}}$$ Reduce.

$$= \frac{5}{6}$$

Example 13. Find $\dfrac{10}{21} - \dfrac{2}{21} - \dfrac{8}{21}$.

Solution

$$\dfrac{10}{21} - \dfrac{2}{21} - \dfrac{8}{21} = \dfrac{10-2}{21} - \dfrac{8}{21} \qquad \text{Perform first subtraction.}$$

$$= \dfrac{8}{21} - \dfrac{8}{21} \qquad \text{Find the difference.}$$

$$= \dfrac{8-8}{21} \qquad \text{Perform second subtraction.}$$

$$= \dfrac{0}{21} \qquad \text{Find the difference.}$$

$$= 0 \qquad \text{Simplify.}$$

Remember that 0 divided by any nonzero number is always 0.

⟶ Trial Run

Perform the indicated operation and reduce.

_____ 1. $\dfrac{8}{9} + \dfrac{7}{9} - \dfrac{5}{9}$ _____ 2. $\dfrac{13}{20} + \dfrac{3}{20} - \dfrac{1}{20}$

_____ 3. $\dfrac{25}{16} - \dfrac{3}{16} - \dfrac{7}{16}$ _____ 4. $\dfrac{22}{27} - \dfrac{5}{27} - \dfrac{17}{27}$

_____ 5. $\dfrac{12}{35} - \dfrac{9}{35} + \dfrac{2}{35}$ _____ 6. $\dfrac{27}{50} - \dfrac{3}{50} + \dfrac{11}{50}$

Answers are on page 189.

Adding and Subtracting Mixed Numbers

If Marci buys 2 packages of hamburger weighing $2\frac{7}{10}$ pounds and $3\frac{1}{10}$ pounds, how many pounds does she have altogether? To compute her total amount, we must *add*. To add mixed numbers, we may add the whole number parts and add the fractional parts. Marci's problem becomes

$$2\frac{7}{10} \quad = \quad 2 + \frac{7}{10}$$

$$+3\frac{1}{10} \quad = \quad 3 + \frac{1}{10}$$

$$\overline{\phantom{+3\frac{1}{10}}} \qquad \overline{5 + \frac{8}{10}}$$

$$\text{But} \quad 5 + \frac{8}{10} \quad = \quad 5 + \frac{4}{5}$$

$$= \quad 5\frac{4}{5}$$

Marci has a total of $5\frac{4}{5}$ pounds. Notice that we reduced the fractional part of the answer.

Example 14. Add $1\frac{2}{3} + 5\frac{2}{3} + 3\frac{1}{3}$.

Solution

$$1\frac{2}{3} = 1 + \frac{2}{3}$$

$$5\frac{2}{3} = 5 + \frac{2}{3}$$

$$+3\frac{1}{3} = 3 + \frac{1}{3}$$

$$9 + \frac{5}{3}$$

But $9 + \frac{5}{3} = 9 + 1\frac{2}{3}$

$$= 10\frac{2}{3}$$

Note that the improper fractional part ($\frac{5}{3}$) was changed to a mixed number ($1\frac{2}{3}$) and added to the whole number part (9).

You complete Example 15.

Example 15. Add $2\frac{9}{11} + 9\frac{4}{11}$.

Solution

$$2\frac{9}{11} = 2 + \frac{9}{11}$$

$$+9\frac{4}{11} = 9 + \frac{4}{11}$$

$$11 + \frac{13}{11}$$

But $11 + \frac{13}{11} = 11 + \underline{}$

$$= \underline{}$$

Check your work on page 189. ▶

Subtraction of mixed numbers is accomplished in a similar way. We subtract the whole number parts and subtract the fractional parts.

Example 16. Find $4\frac{7}{9} - 2\frac{4}{9}$.

Solution

$$4\frac{7}{9} = 4 + \frac{7}{9}$$

$$-2\frac{4}{9} = 2 + \frac{4}{9}$$

$$2 + \frac{3}{9} \qquad \text{Subtract whole numbers; subtract fractions.}$$

But $2 + \frac{3}{9} = 2 + \frac{1}{3}$ Reduce the fraction.

$$= 2\frac{1}{3} \qquad \text{Write answer as mixed number.}$$

To subtract mixed numbers when the fractional part being subtracted is larger than the fractional part being subtracted from, we must do some rewriting. Let's find $7\frac{1}{4} - 5\frac{3}{4}$.

Note that $\frac{3}{4}$ is larger than $\frac{1}{4}$.   Rewrite 7 as $6 + 1$.   Rewrite 1 as $\frac{4}{4}$.   Rewrite $\frac{4}{4} + \frac{1}{4}$ as $\frac{5}{4}$.

$$7\frac{1}{4} = 7 + \frac{1}{4} = 6 + 1 + \frac{1}{4} = 6 + \frac{4}{4} + \frac{1}{4} = 6 + \frac{5}{4}$$

$$-5\frac{3}{4} = 5 + \frac{3}{4} = 5 \qquad + \frac{3}{4} = 5 \qquad + \frac{3}{4} = 5 + \frac{3}{4}$$

$$1 + \frac{2}{4} \qquad \text{SUBTRACT.}$$

But $1 + \frac{2}{4} = 1 + \frac{1}{2}$ Reduce the fraction.

$$= 1\frac{1}{2}$$ Write answer as mixed number.

You complete Example 17.

Example 17. Find $10\frac{1}{8} - 4\frac{7}{8}$.

Solution

Note that $\frac{7}{8}$ is larger than $\frac{1}{8}$.   Rewrite 10 as $9 + 1$.   Rewrite 1 as $\frac{8}{8}$.   Rewrite $\frac{8}{8} + \frac{1}{8}$ as $\frac{9}{8}$.

$$10\frac{1}{8} = 10 + \frac{1}{8} = 9 + 1 + \frac{1}{8} = 9 + \frac{8}{8} + \frac{1}{8} = 9 + \frac{9}{8}$$

$$-4\frac{7}{8} = 4 + \frac{7}{8} = 4 \qquad + \frac{7}{8} = 4 \qquad + \frac{7}{8} = 4 + \frac{7}{8}$$

$$5 + \underline{\qquad} \quad \text{SUBTRACT.}$$

But $5 + \underline{\qquad} = 5 + \underline{\qquad}$

$$= \underline{\qquad}$$

Check your work on page 189. ▶

We mention here that we can also add or subtract mixed numbers by changing them first to *improper fractions*. For instance,

$$2\frac{7}{10} + 3\frac{1}{10} = \frac{27}{10} + \frac{31}{10}$$

$$= \frac{27 + 31}{10}$$

$$= \frac{58}{10}$$

$$= \frac{29}{5}$$

$$= 5\frac{4}{5}$$

$$10\frac{1}{8} - 4\frac{7}{8} = \frac{81}{8} - \frac{39}{8}$$

$$= \frac{81 - 39}{8}$$

$$= \frac{42}{8}$$

$$= \frac{21}{4}$$

$$= 5\frac{1}{4}$$

This is the same sum that we found before in Marci's hamburger problem.

This is the same difference that we found before in Example 17.

⮕ Trial Run

Perform the indicated operation and reduce.

———— 1. $3\frac{5}{8} + 2\frac{7}{8}$

———— 2. $3\frac{1}{4} + 5\frac{3}{4} + 2\frac{1}{4}$

———— 3. $7\frac{3}{8} + 9\frac{7}{8}$

———— 4. $9\frac{7}{12} - 5\frac{5}{12}$

———— 5. $2\frac{3}{16} - 1\frac{7}{16}$

———— 6. $8\frac{1}{6} - 4\frac{5}{6}$

Answers are on page 189.

Let's conclude this section with some examples involving addition and subtraction of mixed numbers, improper fractions, and proper fractions. Remember to proceed from left to right, dealing with the numbers two at a time (especially when subtraction is involved).

Example 18. Find $7\dfrac{2}{3} + \dfrac{2}{3} - \dfrac{5}{3}$.

Solution. Since one of the numbers is an improper fraction, let's use the second method to add and subtract.

$$7\frac{2}{3} + \frac{2}{3} - \frac{5}{3} = \frac{23}{3} + \frac{2}{3} - \frac{5}{3} \qquad \text{Change } 7\tfrac{2}{3} \text{ to the improper fraction } \tfrac{23}{3}.$$

$$= \frac{23 + 2}{3} - \frac{5}{3} \qquad \text{Add the first two fractions.}$$

$$= \frac{25}{3} - \frac{5}{3} \qquad \text{Find the sum.}$$

$$= \frac{25 - 5}{3} \qquad \text{Subtract.}$$

$$= \frac{20}{3} \quad \text{or} \quad 6\frac{2}{3} \qquad \text{Find the difference.}$$

Example 19. Find $4\dfrac{1}{22} - 3\dfrac{7}{22} - \dfrac{5}{22}$.

Solution. First we find $4\dfrac{1}{22} - 3\dfrac{7}{22}$.

$$
\begin{array}{ccccccc}
4 + \dfrac{1}{22} & = & 3 + 1 + \dfrac{1}{22} & = & 3 + \dfrac{22}{22} + \dfrac{1}{22} & = & 3 + \dfrac{23}{22} \\
3 + \dfrac{7}{22} & = & 3 \qquad + \dfrac{7}{22} & = & 3 \qquad + \dfrac{7}{22} & = & 3 + \dfrac{7}{22} \\
\hline
& & & & & & \dfrac{16}{22} \quad \text{SUBTRACT.}
\end{array}
$$

Now we find
$$\frac{16}{22} - \frac{5}{22} = \frac{16 - 5}{22}$$

$$= \frac{11}{22}$$

$$= \frac{1}{2}$$

⇒ Trial Run

Perform the indicated operation and reduce.

_____ 1. $\dfrac{5}{9} + 2\dfrac{1}{9} + \dfrac{16}{9}$

_____ 2. $8\dfrac{3}{5} + \dfrac{4}{5} - \dfrac{12}{5}$

_____ 3. $5\dfrac{2}{3} - 4\dfrac{1}{3} - \dfrac{4}{3}$

_____ 4. $5\dfrac{3}{8} + 2\dfrac{7}{8} - 6\dfrac{1}{8}$

_____ 5. $\dfrac{14}{15} + 2\dfrac{8}{15} - \dfrac{17}{15}$

_____ 6. $3\dfrac{7}{12} - \dfrac{13}{12} - 2\dfrac{5}{12}$

Answers are on page 189.

▶ Examples You Completed

Example 2. Find $\dfrac{2}{3} + \dfrac{2}{3}$.

Solution

$$\dfrac{2}{3} + \dfrac{2}{3} = \dfrac{2 + 2}{3}$$

$$= \dfrac{4}{3} \quad \text{or} \quad 1\dfrac{1}{3}$$

Example 4. Find $\dfrac{10}{3} + \dfrac{14}{3}$.

Solution

$$\dfrac{10}{3} + \dfrac{14}{3} = \dfrac{10 + 14}{3}$$

$$= \dfrac{24}{3}$$

$$= 8$$

Example 8. Find $\dfrac{17}{29} - \dfrac{11}{29}$.

Solution

$$\dfrac{17}{29} - \dfrac{11}{29} = \dfrac{17 - 11}{29}$$

$$= \dfrac{6}{29}$$

Example 10. Find $\dfrac{43}{16} - \dfrac{21}{16}$.

Solution

$$\dfrac{43}{16} - \dfrac{21}{16} = \dfrac{43 - 21}{16}$$

$$= \dfrac{22}{16}$$

$$= \dfrac{\overset{11}{\cancel{22}}}{\underset{8}{\cancel{16}}}$$

$$= \dfrac{11}{8} \quad \text{or} \quad 1\dfrac{3}{8}$$

Example 11. Find $\dfrac{6}{25} + \dfrac{27}{25} - \dfrac{13}{25}$.

Solution

$$\dfrac{6}{25} + \dfrac{27}{25} - \dfrac{13}{25} = \dfrac{6 + 27}{25} - \dfrac{13}{25}$$

$$= \dfrac{33}{25} - \dfrac{13}{25}$$

$$= \dfrac{20}{25}$$

$$= \dfrac{\overset{4}{\cancel{20}}}{\underset{5}{\cancel{25}}}$$

$$= \dfrac{4}{5}$$

Example 15. Find $2\dfrac{9}{11} + 9\dfrac{4}{11}$.

Solution

$$2\dfrac{9}{11} = 2 + \dfrac{9}{11}$$

$$+9\dfrac{4}{11} = 9 + \dfrac{4}{11}$$

$$\overline{\qquad\qquad 11 + \dfrac{13}{11}}$$

But $\quad 11 + \dfrac{13}{11} = 11 + 1\dfrac{2}{11}$

$$= 12\dfrac{2}{11}$$

Example 17. Find $10\dfrac{1}{8} - 4\dfrac{7}{8}$.

Solution

| Note that $\frac{7}{8}$ is larger than $\frac{1}{8}$. | Rewrite 10 as 9 + 1. | Rewrite 1 as $\frac{8}{8}$. | Rewrite $\frac{8}{8} + \frac{1}{8}$ as $\frac{9}{8}$. |

$$10\dfrac{1}{8} = \quad 10 + \dfrac{1}{8} \quad = 9 + 1 + \dfrac{1}{8} = 9 + \dfrac{8}{8} + \dfrac{1}{8} = \quad 9 + \dfrac{9}{8}$$

$$-4\dfrac{7}{8} = \quad 4 + \dfrac{7}{8} \quad = 4 + \dfrac{7}{8} = 4 \quad + \dfrac{7}{8} = \quad 4 + \dfrac{7}{8}$$

$$\overline{\qquad\qquad\qquad\qquad\qquad\qquad\qquad\qquad 5 + \dfrac{2}{8}}\quad \text{SUBTRACT.}$$

But $\quad 5 + \dfrac{2}{8} = 5 + \dfrac{1}{4}$

$$= 5\dfrac{1}{4}$$

Answers to Trial Runs

page 180 1. $\dfrac{1}{3}$ 2. $\dfrac{7}{5}$ or $1\dfrac{2}{5}$ 3. $\dfrac{15}{8}$ or $1\dfrac{7}{8}$ 4. 2 5. $\dfrac{27}{13}$ or $2\dfrac{1}{13}$ 6. $\dfrac{11}{6}$ or $1\dfrac{5}{6}$

page 181 1. $\dfrac{1}{2}$ 2. $\dfrac{3}{17}$ 3. 1 4. $\dfrac{9}{5}$ or $1\dfrac{4}{5}$ 5. 2 6. $\dfrac{8}{3}$ or $2\dfrac{2}{3}$

page 183 1. $\dfrac{10}{9}$ or $1\dfrac{1}{9}$ 2. $\dfrac{3}{4}$ 3. $\dfrac{15}{16}$ 4. 0 5. $\dfrac{1}{7}$ 6. $\dfrac{7}{10}$

page 186 1. $6\dfrac{1}{2}$ 2. $11\dfrac{1}{4}$ 3. $17\dfrac{1}{4}$ 4. $4\dfrac{1}{6}$ 5. $\dfrac{3}{4}$ 6. $3\dfrac{1}{3}$

page 188 1. $\dfrac{40}{9}$ or $4\dfrac{4}{9}$ 2. 7 3. 0 4. $\dfrac{17}{8}$ or $2\dfrac{1}{8}$ 5. $\dfrac{7}{3}$ or $2\dfrac{1}{3}$ 6. $\dfrac{1}{12}$

EXERCISE SET 5.1

Perform the indicated operations and reduce.

_____ 1. $\dfrac{5}{24} + \dfrac{7}{24}$

_____ 2. $\dfrac{2}{9} + \dfrac{1}{9}$

_____ 3. $\dfrac{5}{8} + \dfrac{7}{8}$

_____ 4. $\dfrac{5}{7} + \dfrac{3}{7}$

_____ 5. $\dfrac{2}{15} + \dfrac{4}{15} + \dfrac{7}{15}$

_____ 6. $\dfrac{3}{8} + \dfrac{7}{8} + \dfrac{1}{8}$

_____ 7. $\dfrac{31}{12} + \dfrac{5}{12}$

_____ 8. $\dfrac{13}{24} + \dfrac{11}{24}$

_____ 9. $\dfrac{16}{9} + \dfrac{8}{9}$

_____ 10. $\dfrac{31}{14} + \dfrac{9}{14}$

_____ 11. $\dfrac{4}{21} + \dfrac{13}{21} + \dfrac{16}{21}$

_____ 12. $\dfrac{13}{15} + \dfrac{7}{15} + \dfrac{1}{15}$

_____ 13. $\dfrac{7}{9} - \dfrac{4}{9}$

_____ 14. $\dfrac{11}{15} - \dfrac{8}{15}$

_____ 15. $\dfrac{8}{13} - \dfrac{2}{13}$

_____ 16. $\dfrac{17}{23} - \dfrac{7}{23}$

_____ 17. $\dfrac{35}{12} - \dfrac{11}{12}$

_____ 18. $\dfrac{19}{16} - \dfrac{3}{16}$

_____ 19. $\dfrac{31}{10} - \dfrac{13}{10}$

_____ 20. $\dfrac{21}{8} - \dfrac{3}{8}$

_____ 21. $\dfrac{43}{15} - \dfrac{13}{15}$

_____ 22. $\dfrac{29}{12} - \dfrac{5}{12}$

_____ 23. $\dfrac{52}{27} - \dfrac{4}{27}$

_____ 24. $\dfrac{28}{25} - \dfrac{8}{25}$

_____ 25. $\dfrac{3}{10} + \dfrac{13}{10} - \dfrac{1}{10}$

_____ 26. $\dfrac{5}{12} + \dfrac{23}{12} - \dfrac{7}{12}$

_____ 27. $\dfrac{15}{16} + \dfrac{9}{16} - \dfrac{7}{16}$

_____ 28. $\dfrac{19}{21} + \dfrac{18}{21} - \dfrac{4}{21}$

_____ 29. $\dfrac{12}{25} + \dfrac{17}{25} - \dfrac{4}{25}$

_____ 30. $\dfrac{16}{27} + \dfrac{28}{27} - \dfrac{17}{27}$

_____ 31. $\dfrac{15}{23} - \dfrac{7}{23} - \dfrac{8}{23}$

_____ 32. $\dfrac{17}{29} - \dfrac{5}{29} - \dfrac{12}{29}$

_____ 33. $\dfrac{9}{10} - \dfrac{3}{10} - \dfrac{1}{10}$

_____ 34. $\dfrac{11}{12} - \dfrac{1}{12} - \dfrac{7}{12}$

_____ 35. $\dfrac{35}{13} - \dfrac{7}{13} - \dfrac{2}{13}$

_____ 36. $\dfrac{42}{17} - \dfrac{5}{17} - \dfrac{3}{17}$

_____ 37. $2\dfrac{5}{7} + 3\dfrac{1}{7}$

_____ 38. $5\dfrac{4}{5} + 7\dfrac{3}{5}$

_____ 39. $5\dfrac{3}{8} + 2\dfrac{1}{8} + 1\dfrac{7}{8}$

_____ 40. $6\dfrac{2}{7} + 3\dfrac{1}{7} + 1\dfrac{6}{7}$

_____ 41. $4\dfrac{3}{14} + 2\dfrac{5}{14}$

_____ 42. $5\dfrac{5}{12} + 9\dfrac{1}{12}$

_____ 43. $12\dfrac{5}{8} - 7\dfrac{3}{8}$

_____ 44. $15\dfrac{7}{9} - 8\dfrac{4}{9}$

_____ 45. $2\dfrac{3}{8} - 1\dfrac{7}{8}$

_____ 46. $6\dfrac{1}{5} - 4\dfrac{3}{5}$

_____ 47. $15\dfrac{3}{10} - 8\dfrac{7}{10}$

_____ 48. $19\dfrac{5}{12} - 15\dfrac{7}{12}$

_____ 49. $\dfrac{3}{8} + 2\dfrac{1}{8} + \dfrac{17}{8}$

_____ 50. $\dfrac{5}{6} + 1\dfrac{1}{6} + \dfrac{19}{6}$

_____ 51. $5\dfrac{3}{5} + \dfrac{4}{5} - \dfrac{7}{5}$

_____ 52. $1\dfrac{11}{15} + \dfrac{8}{15} - \dfrac{4}{15}$

_____ 53. $7\dfrac{1}{7} - 5\dfrac{5}{7} - \dfrac{10}{7}$

_____ 54. $6\dfrac{5}{9} - 4\dfrac{7}{9} - \dfrac{13}{9}$

_____ 55. $1\dfrac{3}{8} + 5\dfrac{5}{8} - 4\dfrac{1}{8}$

_____ 56. $5\dfrac{2}{5} + 4\dfrac{1}{5} - 6\dfrac{4}{5}$

_____ 57. $\dfrac{7}{13} + 3\dfrac{2}{13} - \dfrac{15}{13}$

_____ 58. $\dfrac{4}{9} + 3\dfrac{5}{9} - \dfrac{10}{9}$

_____ 59. $7\dfrac{3}{4} - \dfrac{15}{4} - 3\dfrac{1}{4}$

_____ 60. $9\dfrac{1}{3} - \dfrac{13}{3} - 4\dfrac{2}{3}$

☆ Stretching the Topics _____

Perform the indicated operations and reduce.

_____ 1. $3\dfrac{1}{5} + 7\dfrac{4}{5} + 15\dfrac{3}{5} + 9\dfrac{2}{5}$

_____ 2. $21\dfrac{5}{9} - 17\dfrac{4}{9} + 12\dfrac{7}{9} - 8\dfrac{1}{9}$

_____ 3. $121\dfrac{39}{50} + 242\dfrac{17}{50} + 389\dfrac{43}{50} + 126\dfrac{19}{50} + 72\dfrac{31}{50}$

Check your answers in the back of your book.

If you can add and subtract the fractions in **Checkup 5.1,** you are ready to go on to Section 5.2.

✔ # CHECKUP 5.1

Perform the indicated operations and reduce.

_____ 1. $\dfrac{11}{28} + \dfrac{3}{28}$ _____ 2. $\dfrac{15}{16} + \dfrac{19}{16}$

_____ 3. $\dfrac{35}{12} - \dfrac{11}{12}$ _____ 4. $\dfrac{52}{9} - \dfrac{13}{9}$

_____ 5. $\dfrac{19}{18} + \dfrac{3}{18} - \dfrac{1}{18}$ _____ 6. $\dfrac{22}{27} - \dfrac{5}{27} - \dfrac{17}{27}$

_____ 7. $7\dfrac{1}{5} - \dfrac{17}{5} - \dfrac{4}{5}$ _____ 8. $7\dfrac{5}{8} + 9\dfrac{7}{8}$

_____ 9. $4\dfrac{7}{9} - 2\dfrac{5}{9}$ _____ 10. $7\dfrac{1}{6} - 3\dfrac{5}{6}$

Check your work in the back of your book.

If You Missed Problems:	You Should Review Examples:
1, 2	1–6
3, 4	7–10
5–7	11–13
8–10	14–17

5.2 Building Fractions

We have learned that adding and subtracting fractions is not difficult if the fractions have the same denominator. Before we can consider adding and subtracting fractions with *different* denominators we must spend some time learning to **build fractions.**

Building Proper and Improper Fractions

In Chapter 3, we learned to **reduce** a fraction by *dividing* the numerator and denominator by any common factor. To build a fraction, we reverse this procedure and *multiply* the numerator and denominator by the same number.

To build a fraction to an equivalent fraction, we multiply the numerator and denominator by the same number (except 0).

Let's see if this procedure seems reasonable. Consider the simple fraction $\frac{1}{2}$ and multiply the numerator and denominator by the same number, for example, 5.

$$\frac{1 \times 5}{2 \times 5} = \frac{5}{10} \qquad \text{Remember,} \quad \frac{5}{5} = 1$$

Do you agree that the new fraction, $\frac{5}{10}$, is equivalent to the old fraction, $\frac{1}{2}$? Certainly $\frac{1}{2} = \frac{5}{10}$.

When we multiply the numerator and denominator of a fraction by the same nonzero number, we are just multiplying the fraction by **1.** Therefore, multiplying the numerator and denominator of a fraction by the same number will always produce an equivalent fraction.

Let's practice changing some old fractions to new fractions by multiplying the numerator and denominator by the same number.

Old Fraction	Multiply Numerator and Denominator by	New Fraction
$\frac{2}{3}$	5	$\frac{2 \times 5}{3 \times 5} = \frac{10}{15}$
$\frac{1}{5}$	3	$\frac{1 \times 3}{5 \times 3} = \frac{3}{15}$
$\frac{7}{3}$	4	$\frac{7 \times 4}{3 \times 4} = \frac{28}{12}$
$\frac{1}{6}$	2	$\frac{1 \times 2}{6 \times 2} = \frac{2}{12}$
$\frac{11}{14}$	3	$\frac{11 \times 3}{14 \times 3} = \frac{33}{42}$
$\frac{7}{6}$	7	$\frac{7 \times 7}{6 \times 7} = \frac{49}{42}$

In situations where we will be required to build fractions, we will often be given the *new* denominator for a fraction. It will then be our task to decide by what number the old denominator was multiplied to build it to the new denominator. Then we shall be asked to decide what the new numerator must be.

For instance, suppose we would like to change the fraction $\frac{2}{3}$ to a new fraction with a denominator of 24. Our problem is to find a number to complete the statement

$$\frac{2}{3} = \frac{?}{24}$$ By what number was the old denominator (3) multiplied to arrive at the new denominator (24)?

$$\frac{2}{3} = \frac{?}{3 \times 8}$$ The old denominator of 3 was multiplied by 8 to arrive at the new denominator of 24.

$$\frac{2}{3} = \frac{2 \times 8}{3 \times 8}$$ So the old numerator must also be multiplied by 8 to produce an equivalent fraction.

$$\frac{2}{3} = \frac{16}{24}$$ We have found the equivalent fraction.

Example 1. Change $\frac{3}{5}$ to an equivalent fraction with a denominator of 35.

Solution. We must find the missing number in the statement

$$\frac{3}{5} = \frac{?}{35}$$

$$\frac{3}{5} = \frac{?}{5 \times 7}$$

$$\frac{3}{5} = \frac{3 \times 7}{5 \times 7}$$

$$\frac{3}{5} = \frac{21}{35}$$

You try Example 2.

Example 2. Change $\frac{25}{16}$ to an equivalent fraction with a denominator of 48.

Solution. We must find the missing number in the statement

$$\frac{25}{16} = \frac{?}{48}$$

$$\frac{25}{16} = \frac{?}{16 \times 3}$$

$$\frac{25}{16} = \frac{\times}{16 \times 3}$$

$$\frac{25}{16} = \frac{}{48}$$

Check your work on page 197. ▶

If you have trouble finding the number by which the old denominator was multiplied to arrive at the new denominator, you can always *divide* the new denominator by the old denominator. The quotient will tell you the number by which you must multiply to build the fraction.

Example 3. Change $\frac{5}{11}$ to an equivalent fraction with a denominator of 143.

Solution. We must find the missing number in the statement

$$\frac{5}{11} = \frac{?}{143}$$

The number by which 11 must be multiplied to arrive at 143 may not be obvious to you, so let's divide 143 by 11.

$$\begin{array}{r} 13 \\ 11\overline{)143} \\ \underline{11} \\ 33 \\ \underline{33} \\ 0 \end{array}$$

Now we know that

$$\frac{5}{11} = \frac{?}{143}$$

$$\frac{5}{11} = \frac{?}{11 \times 13}$$

$$\frac{5}{11} = \frac{5 \times 13}{11 \times 13}$$

$$\frac{5}{11} = \frac{65}{143}$$

⫸ Trial Run

Change each fraction to an equivalent fraction with the given denominator.

_____ 1. $\dfrac{7}{9} = \dfrac{?}{27}$ _____ 2. $\dfrac{5}{12} = \dfrac{?}{24}$

_____ 3. $\dfrac{4}{15} = \dfrac{?}{60}$ _____ 4. $\dfrac{25}{13} = \dfrac{?}{26}$

_____ 5. $\dfrac{3}{5} = \dfrac{?}{45}$ _____ 6. $\dfrac{27}{20} = \dfrac{?}{100}$

Answers are on page 197.

Building Whole Numbers and Mixed Numbers

We have learned the method for building proper and improper fractions to equivalent fractions having certain denominators. Now we must consider a method for changing whole numbers and mixed numbers to equivalent fractional numbers with certain denominators.

Recall that every whole number can be written as an improper fraction with a denominator of 1. For instance,

$$5 = \frac{5}{1} \quad \text{and} \quad 17 = \frac{17}{1} \quad \text{and} \quad 89 = \frac{89}{1}$$

To change a whole number to a fraction with a certain denominator, we first write the whole number as an improper fraction with a denominator of 1. Then we proceed as before to change the improper fraction to a fraction with the new denominator.

Let's change the whole number 5 to a fraction having a denominator of 7.

$$\frac{5}{1} = \frac{?}{7}$$

$$\frac{5}{1} = \frac{?}{1 \times 7}$$

$$\frac{5}{1} = \frac{5 \times 7}{1 \times 7}$$

$$\frac{5}{1} = \frac{35}{7}$$

You try Example 4.

Example 4. Change 17 to an equivalent fraction with a denominator of 8.

Solution. We write 17 as $\frac{17}{1}$ and consider the statement

$$\frac{17}{1} = \frac{?}{8}$$

$$\frac{17}{1} = \frac{?}{1 \times 8}$$

$$\frac{17}{1} = \frac{\times}{1 \times 8}$$

$$\frac{17}{1} = \frac{}{8}$$

Check your work on page 197. ▶

Example 5. Change 1 to an equivalent fraction with a denominator of 67.

Solution. We write 1 as $\frac{1}{1}$ and consider the statement

$$\frac{1}{1} = \frac{?}{67}$$

$$\frac{1}{1} = \frac{?}{1 \times 67}$$

$$\frac{1}{1} = \frac{1 \times 67}{1 \times 67}$$

$$\frac{1}{1} = \frac{67}{67}$$

As you can see, once we have rewritten a whole number as an improper fraction, the procedure used for building is identical to the procedure used for improper fractions.

▥▶ Trial Run

Change each number to an equivalent fraction with the given denominator.

_____ 1. $13 = \frac{?}{9}$

_____ 2. $1 = \frac{?}{15}$

_____ 3. $\frac{12}{5} = \frac{?}{50}$

_____ 4. $5 = \frac{?}{16}$

Answers are on page 197.

In this section we learned the method for changing a fractional number to an equivalent fraction having a certain denominator. We summarize the necessary steps here.

Building Fractions

1. Write the original fractional number as a proper fraction or an improper fraction.
2. Decide by what number the original denominator must be multiplied to arrive at the new denominator.
3. Multiply the original numerator by that same number.
4. The new fraction will be equivalent to the original fraction.

▶ Examples You Completed

Example 2. Change $\frac{25}{16}$ to an equivalent fraction with a denominator of 48.

Solution. We must find the missing number in the statement

$$\frac{25}{16} = \frac{?}{48}$$

$$\frac{25}{16} = \frac{?}{16 \times 3}$$

$$\frac{25}{16} = \frac{25 \times 3}{16 \times 3}$$

$$\frac{25}{16} = \frac{75}{48}$$

Example 4. Change 17 to an equivalent fraction with a denominator of 8.

Solution. We write 17 as $\frac{17}{1}$ and consider the statement

$$\frac{17}{1} = \frac{?}{8}$$

$$\frac{17}{1} = \frac{?}{1 \times 8}$$

$$\frac{17}{1} = \frac{17 \times 8}{1 \times 8}$$

$$\frac{17}{1} = \frac{136}{8}$$

Answers to Trial Runs

page 195   1. $\frac{21}{27}$ 2. $\frac{10}{24}$ 3. $\frac{16}{60}$ 4. $\frac{50}{26}$ 5. $\frac{27}{45}$ 6. $\frac{135}{100}$

page 196   1. $\frac{117}{9}$ 2. $\frac{15}{15}$ 3. $\frac{120}{50}$ 4. $\frac{80}{16}$

EXERCISE SET 5.2

Change each number to an equivalent fraction with the given denominator.

_____ 1. $\dfrac{2}{3} = \dfrac{?}{12}$ _____ 2. $\dfrac{3}{4} = \dfrac{?}{20}$ _____ 3. $\dfrac{5}{12} = \dfrac{?}{36}$

_____ 4. $\dfrac{7}{16} = \dfrac{?}{32}$ _____ 5. $\dfrac{3}{5} = \dfrac{?}{35}$ _____ 6. $\dfrac{7}{8} = \dfrac{?}{48}$

_____ 7. $\dfrac{4}{7} = \dfrac{?}{49}$ _____ 8. $\dfrac{3}{5} = \dfrac{?}{25}$ _____ 9. $\dfrac{16}{5} = \dfrac{?}{15}$

_____ 10. $\dfrac{14}{3} = \dfrac{?}{12}$ _____ 11. $\dfrac{9}{4} = \dfrac{?}{36}$ _____ 12. $\dfrac{12}{7} = \dfrac{?}{56}$

_____ 13. $\dfrac{9}{10} = \dfrac{?}{100}$ _____ 14. $\dfrac{7}{10} = \dfrac{?}{100}$ _____ 15. $\dfrac{13}{10} = \dfrac{?}{1000}$

_____ 16. $\dfrac{21}{10} = \dfrac{?}{1000}$ _____ 17. $\dfrac{23}{20} = \dfrac{?}{100}$ _____ 18. $\dfrac{17}{20} = \dfrac{?}{100}$

_____ 19. $8 = \dfrac{?}{3}$ _____ 20. $9 = \dfrac{?}{5}$ _____ 21. $1 = \dfrac{?}{16}$

_____ 22. $1 = \dfrac{?}{12}$ _____ 23. $13 = \dfrac{?}{5}$ _____ 24. $10 = \dfrac{?}{3}$

_____ 25. $\dfrac{9}{2} = \dfrac{?}{8}$ _____ 26. $\dfrac{7}{3} = \dfrac{?}{9}$ _____ 27. $\dfrac{13}{12} = \dfrac{?}{36}$

_____ 28. $\dfrac{17}{15} = \dfrac{?}{30}$ _____ 29. $\dfrac{25}{3} = \dfrac{?}{30}$ _____ 30. $\dfrac{21}{4} = \dfrac{?}{40}$

☆ Stretching the Topics _____

Change each number to an equivalent fraction with the given denominator.

_____ 1. $\dfrac{7}{18} = \dfrac{?}{108}$

_____ 2. $\dfrac{38}{45} = \dfrac{?}{675}$

_____ 3. $5\dfrac{7}{66} = \dfrac{?}{396}$

Check your answers in the back of your book.

If you can complete **Checkup 5.2,** you are ready to go on to Section 5.3.

✔ **CHECKUP 5.2**

Change each number to an equivalent fraction with the given denominator.

_____ 1. $\dfrac{3}{4} = \dfrac{?}{12}$ 　　　　　 _____ 2. $\dfrac{9}{15} = \dfrac{?}{60}$

_____ 3. $\dfrac{35}{9} = \dfrac{?}{18}$ 　　　　　 _____ 4. $\dfrac{4}{5} = \dfrac{?}{35}$

_____ 5. $\dfrac{27}{20} = \dfrac{?}{100}$ 　　　　 _____ 6. $\dfrac{19}{2} = \dfrac{?}{6}$

_____ 7. $2 = \dfrac{?}{16}$ 　　　　　 _____ 8. $15 = \dfrac{?}{2}$

Check your answers in the back of your book.

If You Missed Problems:	You Should Review Examples:
1–6	1–3
7, 8	4, 5
9, 10	6

5.3 Adding and Subtracting Fractions with Different Denominators

In Section 5.1 we learned to add fractional numbers having the *same* denominator by adding the numerators and keeping the same denominator.

What happens when we wish to add fractions having *different* denominators? We cannot add fractions unless their denominators are exactly the same. Somehow, *we must make the denominators match.* We must give our fractions a **common denominator.**

Suppose we consider a problem in which the fractions do *not* have a common denominator. Let's try to add $\frac{1}{2} + \frac{1}{4}$. Most of us would agree that $\frac{1}{4}$ cannot be changed in any way to have a denominator of 2. But we would also agree that $\frac{1}{2}$ *can* be changed so that it will have a denominator of 4.

First we must use the techniques of *building fractions.*

$$\frac{1}{2} = \frac{?}{4}$$

$$\frac{1}{2} = \frac{1 \times 2}{2 \times 2}$$

$$\frac{1}{2} = \frac{2}{4}$$

Then our addition problem becomes

$$\frac{1}{2} + \frac{1}{4} = \frac{2}{4} + \frac{1}{4}$$

$$= \frac{2 + 1}{4}$$

$$= \frac{3}{4}$$

Notice that the common denominator was a number into which each of the original denominators (2 and 4) would divide *exactly* (with no remainder). Of course there are other such numbers (8 or 12 or 16 and so on). However, we always try to find the *smallest* number to use as the common denominator. That smallest number is called the **lowest common denominator** (sometimes abbreviated LCD).

For the time being, we will consider a few examples in which one of the original denominators is the lowest common denominator for all the fractions. As you work through these examples, notice the steps that are used.

Adding and Subtracting Fractions with Different Denominators

1. Find the lowest common denominator (LCD) for all the fractions to be added (or subtracted).
2. Change all the fractions to equivalent new fractions having the LCD, using the method for building fractions.
3. Add (or subtract) the new numerators and keep the LCD.
4. Reduce the final fraction if possible.

Example 1. Find $\dfrac{2}{9} + \dfrac{17}{45} - \dfrac{1}{15}$.

Solution. Since 9 and 15 divide exactly into 45, we choose 45 as the LCD.

$$\frac{2}{9} = \frac{?}{45} = \frac{?}{9 \times 5} = \frac{2 \times 5}{9 \times 5} = \frac{10}{45}$$

$$\frac{17}{45} = \frac{17}{45}$$

$$\frac{1}{15} = \frac{?}{45} = \frac{?}{15 \times 3} = \frac{1 \times 3}{15 \times 3} = \frac{3}{45}$$

Our problem becomes

$$\frac{2}{9} + \frac{17}{45} - \frac{1}{15} = \frac{10}{45} + \frac{17}{45} - \frac{3}{45}$$ Rewrite fractions as new fractions with LCD.

$$= \frac{10 + 17}{45} - \frac{3}{45}$$ Perform addition.

$$= \frac{27}{45} - \frac{3}{45}$$ Find the sum.

$$= \frac{24}{45}$$ Perform subtraction.

$$= \frac{\overset{8}{\cancel{24}}}{\underset{15}{\cancel{45}}}$$ Reduce the fraction.

$$= \frac{8}{15}$$

Example 2. Find $7\frac{1}{2} - 3\frac{5}{12}$.

Solution. We subtract mixed numbers by subtracting their whole number parts and their fractional parts, but we must make their fractional parts have the *same denominator*.

$$7\frac{1}{2} \ = \ 7 + \frac{1}{2} \ = \ 7 + \frac{6}{12}$$ Rewrite $\frac{1}{2}$ as $\frac{6}{12}$.

$$\underline{-3\frac{5}{12}} \ = \ \underline{3 + \frac{5}{12}} \ = \ \underline{3 + \frac{5}{12}}$$

$$4 + \frac{1}{12}$$ Subtract.

Therefore, $7\frac{1}{2} - 3\frac{5}{12} = 4\frac{1}{12}$.

⏩ Trial Run

Perform the indicated operation and reduce.

_____ 1. $\frac{2}{3} + \frac{3}{4} + \frac{7}{12}$

_____ 2. $\frac{15}{16} - \frac{7}{8}$

_____ 3. $\frac{5}{8} + \frac{39}{40} - \frac{4}{5}$

_____ 4. $\frac{5}{9} + \frac{1}{4} - \frac{17}{36}$

_____ 5. $8\frac{1}{2} - 5\frac{3}{4}$

_____ 6. $3\frac{7}{10} + 8\frac{9}{20}$

Answers are on page 208.

Suppose we consider an addition problem such as

$$\frac{1}{6} + \frac{3}{4} + \frac{1}{3}$$

None of the original denominators will "work" as the LCD because each of the denominators will *not* divide evenly into 6 or 4 or 3. Any common denominator we choose must be divisible by 6, 4, and 3. Can you think of such a number? The correct choice for the lowest common denominator is 12.

$$\frac{1}{6} = \frac{?}{12} = \frac{?}{6 \times 2} = \frac{1 \times 2}{6 \times 2} = \frac{2}{12}$$

$$\frac{3}{4} = \frac{?}{12} = \frac{?}{4 \times 3} = \frac{3 \times 3}{4 \times 3} = \frac{9}{12}$$

$$\frac{1}{3} = \frac{?}{12} = \frac{?}{3 \times 4} = \frac{1 \times 4}{3 \times 4} = \frac{4}{12}$$

Our problem becomes

$$\frac{1}{6} + \frac{3}{4} + \frac{1}{3} = \frac{2}{12} + \frac{9}{12} + \frac{4}{12} \qquad \text{Rewrite fractions as new fractions with LCD.}$$

$$= \frac{2 + 9 + 4}{12} \qquad \text{Perform addition.}$$

$$= \frac{15}{12} \qquad \text{Find the sum.}$$

$$= \frac{\cancel{15}^{5}}{\cancel{12}_{4}} \qquad \text{Reduce the fraction.}$$

$$= \frac{5}{4}$$

Sometimes, as in this last problem, the LCD will occur to you quickly. In other cases, however, your search for the LCD can become frustrating unless you have an orderly process for finding it. Let's consider developing such an orderly process.

We know that the LCD for a group of fractions must be exactly divisible by each of the original denominators in the group. That means that *the original denominators must be factors of the lowest common denominator*. Therefore, the prime number factors of the original denominators must be included as prime number factors of the LCD. To use the prime number factors to find the LCD, we proceed as follows.

Finding the LCD

1. Write all the original denominators as products of prime number factors.
2. Choose the LCD to be the product of all the *different* prime number factors of the original denominators, with each factor appearing the most number of times that it appears in any *single* denominator.

Let's try to use this method to find the LCD for the fractions $\frac{1}{6}$, $\frac{3}{4}$, and $\frac{2}{3}$. First we write each fraction with its denominator written as a product of prime number factors

$$\frac{1}{6} = \frac{1}{2 \times 3} \qquad \frac{3}{4} = \frac{3}{2 \times 2} \qquad \frac{2}{3} = \frac{2}{3}$$

The prime factors are 2 and 3, so we know the LCD must contain 2 and 3. But how many times does each of these factors appear in any *single* denominator? The factor 2 appears twice in the second denominator. Therefore, 2 must appear *twice* in the LCD. The factor 3 appears just once in the first denominator and just once in the third denominator. Therefore, 3 must appear just *once* in the LCD. We choose

$$\text{LCD} = 2 \times 2 \times 3$$
$$= 12$$

and this is the same LCD we chose earlier for these fractions. Our method appears correct.

Let's practice finding the lowest common denominators for some groups of fractions before we continue with addition and subtraction.

Example 3. Find the LCD for $\frac{6}{7}$, $\frac{3}{14}$, $\frac{9}{10}$ and change each fraction to a new fraction having that LCD.

Solution. First we write each fraction with its denominator in factored form.

$$\frac{6}{7} = \frac{6}{7 \times 1} \qquad \frac{3}{14} = \frac{3}{2 \times 7} \qquad \frac{9}{10} = \frac{9}{2 \times 5}$$

Now we choose the LCD.

$$\text{LCD} = 2 \times 5 \times 7 = 70$$

We change each fraction to a new fraction having the LCD.

$$\frac{6}{7} = \frac{6}{7 \times 1} = \frac{?}{2 \times 5 \times 7} = \frac{6 \times 2 \times 5}{2 \times 5 \times 7} = \frac{60}{70}$$

$$\frac{3}{14} = \frac{3}{2 \times 7} = \frac{?}{2 \times 5 \times 7} = \frac{3 \times 5}{2 \times 5 \times 7} = \frac{15}{70}$$

$$\frac{9}{10} = \frac{9}{2 \times 5} = \frac{?}{2 \times 5 \times 7} = \frac{9 \times 7}{2 \times 5 \times 7} = \frac{63}{70}$$

Example 4. Find the LCD for $\frac{7}{15}$, $\frac{11}{12}$, $\frac{13}{20}$ and change each fraction to a new fraction having that LCD.

Solution. First we write each fraction with its denominator in factored form.

$$\frac{7}{15} = \frac{7}{3 \times 5} \qquad \frac{11}{12} = \frac{11}{2 \times 2 \times 3} \qquad \frac{13}{20} = \frac{13}{2 \times 2 \times 5}$$

Noting that 2 appears *twice* in the second and third denominators, we choose the LCD.

$$\text{LCD} = 2 \times 2 \times 3 \times 5 = 60$$

We change each fraction to a new fraction having the LCD.

$$\frac{7}{15} = \frac{7}{3 \times 5} = \frac{?}{2 \times 2 \times 3 \times 5} = \frac{7 \times 2 \times 2}{2 \times 2 \times 3 \times 5} = \frac{28}{60}$$

$$\frac{11}{12} = \frac{11}{2 \times 2 \times 3} = \frac{?}{2 \times 2 \times 3 \times 5} = \frac{11 \times 5}{2 \times 2 \times 3 \times 5} = \frac{55}{60}$$

$$\frac{13}{20} = \frac{13}{2 \times 2 \times 5} = \frac{?}{2 \times 2 \times 3 \times 5} = \frac{13 \times 3}{2 \times 2 \times 3 \times 5} = \frac{39}{60}$$

Notice that we wrote each original denominator in factored form when we were trying to decide by what number it needed to be multiplied to arrive at the LCD. If you do this, you will find it easy to spot the needed multiplier as the *other* factors of the LCD that are not factors of your original denominator.

Building fractions is a tedious process, but it is not difficult if you proceed in an orderly fashion.

⫸ Trial Run

Find the lowest common denominator for each group of fractions. Change each fraction to a new fraction having that LCD.

———— 1. $\dfrac{3}{5}$, $\dfrac{7}{10}$, $\dfrac{8}{15}$ ———— 2. $\dfrac{1}{2}$, $\dfrac{2}{3}$, $\dfrac{4}{7}$

———— 3. $\dfrac{5}{12}$, $\dfrac{13}{18}$, $\dfrac{1}{6}$ ———— 4. $\dfrac{7}{8}$, $\dfrac{11}{12}$, $\dfrac{4}{15}$

Answers are on page 208.

Remember that fractions cannot be added or subtracted to give a single fraction unless their denominators match. Now that we have learned to rewrite any group of fractions so that they have the same denominator (the LCD), we are ready to add and subtract any fractions we choose. The procedure involves several familiar steps.

> **Adding and Subtracting Fractions with Different Denominators**
>
> 1. Write each of the original denominators in factored form.
> 2. Choose the LCD as the product of all the different factors appearing in the original denominators, with each factor appearing the most number of times that it appears in any single denominator.
> 3. Change each of the original fractions to a new fraction having the LCD by multiplying the numerator and denominator by the necessary factors.
> 4. Multiply the factors in each new numerator and in the LCD.
> 5. Add or subtract the numerators and keep the LCD.
> 6. Reduce the final fraction if possible.

Let's try a few examples using these steps.

Example 5. Find $\dfrac{7}{8} + \dfrac{11}{6} + \dfrac{1}{15}$.

Solution. First we factor the original denominators.

$$\frac{7}{8} = \frac{7}{2 \times 2 \times 2} \qquad \frac{11}{6} = \frac{11}{2 \times 3} \qquad \frac{1}{15} = \frac{1}{3 \times 5}$$

We choose the LCD.

$$\text{LCD} = 2 \times 2 \times 2 \times 3 \times 5 = 120$$

Now we change the original fractions to new fractions having the LCD.

$$\frac{7}{8} = \frac{7}{2 \times 2 \times 2} = \frac{?}{2 \times 2 \times 2 \times 3 \times 5} = \frac{7 \times 3 \times 5}{2 \times 2 \times 2 \times 3 \times 5} = \frac{105}{120}$$

$$\frac{11}{6} = \frac{11}{2 \times 3} = \frac{?}{2 \times 2 \times 2 \times 3 \times 5} = \frac{11 \times 2 \times 2 \times 5}{2 \times 2 \times 2 \times 3 \times 5} = \frac{220}{120}$$

$$\frac{1}{15} = \frac{1}{3 \times 5} = \frac{?}{2 \times 2 \times 2 \times 3 \times 5} = \frac{1 \times 2 \times 2 \times 2}{2 \times 2 \times 2 \times 3 \times 5} = \frac{8}{120}$$

Now we are ready to add the fractions.

$$\frac{7}{8} + \frac{11}{6} + \frac{1}{15} = \frac{105}{120} + \frac{220}{120} + \frac{8}{120} \qquad \text{Rewrite fractions as new fractions with LCD.}$$

$$= \frac{105 + 220 + 8}{120} \qquad \text{Perform addition.}$$

$$= \frac{333}{120} \qquad \text{Find the sum.}$$

$$= \frac{\overset{111}{\cancel{333}}}{\underset{40}{\cancel{120}}} \qquad \text{Divide numerator and denominator by 3.}$$

$$= \frac{111}{40}$$

Example 6. Find $\dfrac{8}{3} + \dfrac{2}{7} - \dfrac{1}{2}$.

Solution. First we factor the original denominators.

$$\frac{8}{3} = \frac{8}{3 \times 1} \qquad \frac{2}{7} = \frac{2}{7 \times 1} \qquad \frac{1}{2} = \frac{1}{2 \times 1}$$

We choose the LCD.

$$\text{LCD} = 2 \times 3 \times 7 = 42$$

Now we change the original fractions to new fractions having the LCD.

$$\frac{8}{3} = \frac{8}{3 \times 1} = \frac{?}{2 \times 3 \times 7} = \frac{8 \times 2 \times 7}{2 \times 3 \times 7} = \frac{112}{42}$$

$$\frac{2}{7} = \frac{2}{7 \times 1} = \frac{?}{2 \times 3 \times 7} = \frac{2 \times 2 \times 3}{2 \times 3 \times 7} = \frac{12}{42}$$

$$\frac{1}{2} = \frac{1}{2 \times 1} = \frac{?}{2 \times 3 \times 7} = \frac{1 \times 3 \times 7}{2 \times 3 \times 7} = \frac{21}{42}$$

We are ready to find the sum.

$$\frac{8}{3} + \frac{2}{7} - \frac{1}{2} = \frac{112}{42} + \frac{12}{42} - \frac{21}{42}$$ Rewrite fractions as new fractions with LCD.

$$= \frac{112 + 12}{42} - \frac{21}{42}$$ Perform addition.

$$= \frac{124}{42} - \frac{21}{42}$$ Find the sum.

$$= \frac{124 - 21}{42}$$ Perform subtraction.

$$= \frac{103}{42}$$ Find the difference.

Will this fraction reduce? Recall that the only factors of the common denominator (42) are 2, 3, and 7. Since none of those factors is a factor of 103, we know this fraction cannot be reduced.

⟫ Trial Run

Perform the indicated operation and reduce.

_____ 1. $\frac{1}{2} + \frac{3}{4} + \frac{3}{5}$ _____ 2. $\frac{7}{9} + \frac{5}{6} - \frac{3}{4}$

_____ 3. $\frac{5}{6} + \frac{9}{5} - \frac{1}{3}$ _____ 4. $\frac{17}{3} - \frac{3}{7} - \frac{1}{2}$

_____ 5. $\frac{11}{12} - \frac{1}{8} - \frac{2}{3}$ _____ 6. $\frac{7}{4} - \frac{5}{13} + \frac{7}{26}$

Answers are on page 208.

From our earlier work with mixed numbers, you should agree that we can find sums (or differences) of such numbers by adding (or subtracting) their whole number parts and fractional parts. Adding or subtracting fractional parts with different denominators requires the method we have just learned.

Example 7. Find $10\frac{1}{4} - 7\frac{11}{15}$.

Solution. We must find the LCD for the fractional parts before we begin to subtract.

$$\frac{1}{4} = \frac{1}{2 \times 2} \qquad \frac{11}{15} = \frac{11}{3 \times 5} \qquad \text{LCD: } 2 \times 2 \times 3 \times 5$$

We change the original fractions to new fractions.

$$\frac{1}{4} = \frac{1}{2 \times 2} = \frac{?}{2 \times 2 \times 3 \times 5} = \frac{1 \times 3 \times 5}{2 \times 2 \times 3 \times 5} = \frac{15}{60}$$

$$\frac{11}{15} = \frac{1}{3 \times 5} = \frac{?}{2 \times 2 \times 3 \times 5} = \frac{11 \times 2 \times 2}{2 \times 2 \times 3 \times 5} = \frac{44}{60}$$

Now we use our usual method for subtracting mixed numbers.

$$10\frac{1}{4} = 10 + \frac{1}{4} = 10 + \frac{15}{60} = 9 + \frac{60}{60} + \frac{15}{60} = 9 + \frac{75}{60}$$

$$- 7\frac{11}{15} = 7 + \frac{11}{15} = 7 + \frac{44}{60} = 7 \qquad + \frac{44}{60} = 7 + \frac{44}{60}$$

$$\overline{\qquad\qquad\qquad\qquad\qquad\qquad\qquad\qquad\qquad\qquad\qquad\qquad 2 + \frac{31}{60}} \quad \text{SUBTRACT.}$$

Therefore, $10\frac{1}{4} - 7\frac{11}{15} = 2\frac{31}{60}$.

⫸ Trial Run

Perform the indicated operation and reduce.

_____ 1. $4\frac{1}{3} + 2\frac{1}{2}$

_____ 2. $10\frac{3}{4} - 6\frac{2}{5}$

_____ 3. $9 + 2\frac{4}{15} - 7\frac{5}{6}$

_____ 4. $4\frac{3}{4} + 7\frac{1}{8} - 11\frac{19}{24}$

Answers are given below.

Answers to Trial Runs

page 202 1. 2 2. $\frac{1}{16}$ 3. $\frac{4}{5}$ 4. $\frac{1}{3}$ 5. $2\frac{3}{4}$ 6. $12\frac{3}{20}$

page 205 1. $\frac{18}{30}, \frac{21}{30}, \frac{16}{30}$ 2. $\frac{21}{42}, \frac{28}{42}, \frac{24}{42}$ 3. $\frac{15}{36}, \frac{26}{36}, \frac{6}{36}$ 4. $\frac{105}{120}, \frac{110}{120}, \frac{32}{120}$

page 207 1. $\frac{37}{20}$ 2. $\frac{31}{36}$ 3. $\frac{23}{10}$ 4. $\frac{199}{42}$ 5. $\frac{1}{8}$ 6. $\frac{85}{52}$

page 208 1. $6\frac{5}{6}$ 2. $4\frac{7}{20}$ 3. $3\frac{13}{30}$ 4. $\frac{1}{12}$

EXERCISE SET 5.3

Perform the indicated operations and reduce.

_____ 1. $\dfrac{3}{4} + \dfrac{2}{3} + \dfrac{1}{12}$

_____ 2. $\dfrac{2}{3} + \dfrac{1}{7} + \dfrac{7}{21}$

_____ 3. $\dfrac{7}{9} + \dfrac{13}{18}$

_____ 4. $\dfrac{4}{5} + \dfrac{9}{10}$

_____ 5. $\dfrac{11}{16} - \dfrac{1}{2}$

_____ 6. $\dfrac{7}{12} - \dfrac{1}{4}$

_____ 7. $\dfrac{19}{6} - \dfrac{5}{3}$

_____ 8. $\dfrac{15}{8} - \dfrac{7}{4}$

_____ 9. $\dfrac{4}{9} + \dfrac{19}{36} - \dfrac{3}{4}$

_____ 10. $\dfrac{3}{5} + \dfrac{11}{15} - \dfrac{2}{3}$

_____ 11. $7\dfrac{5}{8} + 3\dfrac{1}{4}$

_____ 12. $9\dfrac{3}{5} + 2\dfrac{1}{10}$

_____ 13. $9\dfrac{1}{2} - 3\dfrac{7}{12}$

_____ 14. $3\dfrac{1}{3} - 2\dfrac{7}{9}$

_____ 15. $\dfrac{1}{2} + \dfrac{2}{3} + \dfrac{3}{5}$

_____ 16. $\dfrac{2}{7} + \dfrac{1}{3} + \dfrac{1}{2}$

_____ 17. $\dfrac{11}{25} + \dfrac{7}{10} - \dfrac{1}{4}$

_____ 18. $\dfrac{5}{9} + \dfrac{1}{6} - \dfrac{1}{4}$

_____ 19. $\dfrac{7}{15} + \dfrac{5}{7} - \dfrac{3}{5}$

_____ 20. $\dfrac{5}{6} + \dfrac{2}{5} - \dfrac{1}{2}$

_____ 21. $\dfrac{3}{2} + \dfrac{5}{7} + \dfrac{3}{4}$

_____ 22. $\dfrac{8}{3} + \dfrac{3}{5} + \dfrac{4}{9}$

_____ 23. $\dfrac{16}{9} + \dfrac{1}{2} - \dfrac{7}{5}$

_____ 24. $\dfrac{14}{11} + \dfrac{5}{2} - \dfrac{5}{3}$

_____ 25. $\dfrac{7}{10} + \dfrac{5}{12} + \dfrac{7}{15}$

_____ 26. $\dfrac{10}{21} + \dfrac{11}{18} + \dfrac{9}{14}$

_____ 27. $\dfrac{3}{26} + \dfrac{9}{8} + \dfrac{1}{12}$

_____ 28. $\dfrac{5}{33} + \dfrac{10}{9} + \dfrac{2}{15}$

_____ 29. $\dfrac{12}{17} - \dfrac{1}{4} + \dfrac{3}{5}$

_____ 30. $\dfrac{12}{13} - \dfrac{2}{3} + \dfrac{3}{4}$

_____ 31. $6\dfrac{1}{2} + 7\dfrac{2}{3}$

_____ 32. $9\dfrac{1}{2} + 8\dfrac{1}{5}$

_____ 33. $5\dfrac{4}{5} - 2\dfrac{1}{2}$

_____ 34. $10\dfrac{6}{7} - 8\dfrac{1}{3}$

_____ 35. $9\dfrac{2}{7} - 6\dfrac{3}{4}$

_____ 36. $4\dfrac{1}{3} - 1\dfrac{3}{4}$

_____ 37. $9 + 3\dfrac{3}{10} - 8\dfrac{4}{15}$

_____ 38. $7 + 4\dfrac{5}{6} - 5\dfrac{3}{8}$

_____ 39. $1\dfrac{2}{3} + 5\dfrac{3}{4} - 4\dfrac{7}{12}$

_____ 40. $2\dfrac{1}{5} + 4\dfrac{3}{4} - 5\dfrac{19}{20}$

☆ Stretching the Topics _____

Perform the indicated operations and reduce.

_____ 1. $\dfrac{15}{56} + \dfrac{29}{63} + \dfrac{31}{72}$

_____ 2. $\dfrac{3}{5} + \dfrac{5}{7} + \dfrac{2}{3} - \dfrac{1}{2} - \dfrac{2}{11}$

_____ 3. $8\dfrac{4}{5} + 6\dfrac{9}{10} - 3\dfrac{3}{4} - 7\dfrac{24}{25}$

Check your answers in the back of your book.

If you can do the problems in **Checkup 5.3,** you are ready to go to Section 5.4.

✔ CHECKUP 5.3

Perform the indicated operations and reduce.

——————— 1. $\dfrac{1}{2} + \dfrac{3}{5} + \dfrac{9}{10}$

——————— 2. $\dfrac{11}{8} - \dfrac{3}{4}$

——————— 3. $5\dfrac{1}{2} - 3\dfrac{5}{6}$

——————— 4. $\dfrac{3}{5} + \dfrac{7}{10} + \dfrac{8}{15}$

——————— 5. $\dfrac{7}{8} + \dfrac{11}{12} - \dfrac{4}{15}$

——————— 6. $7\dfrac{5}{12} + 3\dfrac{1}{9}$

——————— 7. $9\dfrac{1}{10} - 4\dfrac{7}{8}$

——————— 8. $3\dfrac{1}{4} + 2\dfrac{5}{6} - 1\dfrac{7}{8}$

Check your work in the back of your book.

If You Missed Problems:	You Should Review Examples:
1–3	1, 2
4, 5	5, 6
6–8	7

5.4 Comparing Fractional Numbers

In Chapter 1, we learned to compare whole numbers by noting their positions on the number line. We wrote statements such as

$$3 < 5 \qquad \text{meaning} \qquad \text{``3 is less than 5''}$$

$$7 > 2 \qquad \text{meaning} \qquad \text{``7 is greater than 2''}$$

$$9 = 5 + 4 \qquad \text{meaning} \qquad \text{``9 is equal to 5 + 4''}$$

We would like to be able to use similar statements to compare fractional numbers. Unfortunately, we are not always sure where two fractions lie on the number line, so we may not know immediately which fraction is larger.

When two fractions have the *same* denominator, comparing them is not so difficult. For instance, to compare $\frac{7}{8}$ and $\frac{5}{8}$ we can visualize each fraction as a shaded part of a square divided into 8 equal portions.

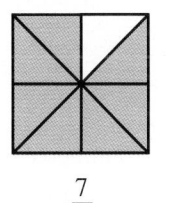

$$\frac{7}{8} \qquad\qquad\qquad \frac{5}{8}$$

It is clear from the drawings that $\frac{7}{8} > \frac{5}{8}$.

This was probably your guess even before looking at the drawing because you reasoned that the numerator 7 was larger than the numerator 5 for these fractions with the same denominator. Indeed, this reasoning is perfectly sound.

Comparing Fractions with the Same Denominator. If two fractions have the same denominator, then the fraction with the larger numerator is the larger fraction.

Example 1. Use $<$, $>$, or $=$ to compare $\dfrac{31}{17}$ and $\dfrac{34}{17}$.

Solution. Since $31 < 34$ and both fractions have the same denominator, we know

$$\frac{31}{17} < \frac{34}{17}$$

Example 2. Use $<$, $>$, or $=$ to compare $\dfrac{1}{8} + \dfrac{1}{8}$ and $\dfrac{7}{8} - \dfrac{5}{8}$.

Solution. First find $\dfrac{1}{8} + \dfrac{1}{8}$.

$$\frac{1}{8} + \frac{1}{8} = \frac{1+1}{8} = \frac{2}{8}$$

Now find $\dfrac{7}{8} - \dfrac{5}{8}$.

$$\frac{7}{8} - \frac{5}{8} = \frac{7-5}{8} = \frac{2}{8}$$

Since $\dfrac{2}{8} = \dfrac{2}{8}$ we conclude that

$$\frac{1}{8} + \frac{1}{8} = \frac{7}{8} - \frac{5}{8}$$

What happens when we try to compare fractions with *different* denominators? As before, we must make the denominators match. We do this by choosing a lowest common denominator and changing each of the fractions to a new fraction having that LCD. Then we simply compare the new numerators.

For instance, to compare $\frac{7}{8}$ and $\frac{29}{31}$ we must find their LCD.

$$\frac{7}{8} = \frac{7}{2 \times 2 \times 2} \qquad \frac{29}{31} = \frac{29}{31 \times 1}$$

The LCD is $2 \times 2 \times 2 \times 31$.

$$\frac{7}{8} = \frac{7}{2 \times 2 \times 2} = \frac{?}{2 \times 2 \times 2 \times 31} = \frac{7 \times 31}{2 \times 2 \times 2 \times 31} = \frac{217}{248}$$

$$\frac{29}{31} = \frac{29}{1 \times 31} = \frac{?}{2 \times 2 \times 2 \times 31} = \frac{29 \times 2 \times 2 \times 2}{2 \times 2 \times 2 \times 31} = \frac{232}{248}$$

Now we can compare $\frac{7}{8}$ and $\frac{29}{31}$ using $\frac{217}{248}$ and $\frac{232}{248}$. We see that

$$\frac{217}{248} < \frac{232}{248}, \qquad \text{so} \qquad \frac{7}{8} < \frac{29}{31}$$

Example 3. On her first test Sara correctly answered 23 out of 25 questions. On the second test she correctly answered 27 out of 30. On which test did she have a larger fractional number of correct answers?

Solution. On the first test, the fractional part correct was $\frac{23}{25}$. On the second test, the fractional part correct was $\frac{27}{30}$. We can compare $\frac{23}{25}$ and $\frac{27}{30}$ by finding their LCD.

$$\frac{23}{25} = \frac{23}{5 \times 5} \qquad \frac{27}{30} = \frac{9}{10} = \frac{9}{2 \times 5}$$

Notice that we reduced the second fraction before seeking the LCD to avoid more factors than necessary. Now we choose the LCD as $2 \times 5 \times 5$.

$$\frac{23}{25} = \frac{23}{5 \times 5} = \frac{?}{2 \times 5 \times 5} = \frac{23 \times 2}{2 \times 5 \times 5} = \frac{46}{50}$$

$$\frac{27}{30} = \frac{9}{10} = \frac{9}{2 \times 5} = \frac{?}{2 \times 5 \times 5} = \frac{9 \times 5}{2 \times 5 \times 5} = \frac{45}{50}$$

Since $\qquad \dfrac{46}{50} > \dfrac{45}{50}$

we know $\qquad \dfrac{23}{25} > \dfrac{27}{30}$

so Sara did better on her first test than she did on her second test.

Example 4. Compare $\dfrac{43}{16}$ and $2\dfrac{1}{3}$.

Solution. We write both numbers as improper fractions and find the LCD.

$$\frac{43}{16} = \frac{43}{2 \times 2 \times 2 \times 2} \qquad 2\frac{1}{3} = \frac{7}{3} = \frac{7}{1 \times 3}$$

The LCD is $2 \times 2 \times 2 \times 2 \times 3$.

$$\frac{43}{16} = \frac{43}{2 \times 2 \times 2 \times 2} = \frac{?}{2 \times 2 \times 2 \times 2 \times 3} = \frac{43 \times 3}{2 \times 2 \times 2 \times 2 \times 3} = \frac{129}{48}$$

$$\frac{7}{3} = \frac{7}{1 \times 3} = \frac{?}{2 \times 2 \times 2 \times 2 \times 3} = \frac{7 \times 2 \times 2 \times 2 \times 2}{2 \times 2 \times 2 \times 2 \times 3} = \frac{112}{48}$$

Since $\dfrac{129}{48} > \dfrac{112}{48}$

we know $\dfrac{43}{16} > 2\dfrac{1}{3}$

⇒ Trial Run

Use $<$, $>$, or $=$ to compare the following pairs of numbers.

_____ 1. $\dfrac{7}{8}, \dfrac{3}{8}$ _____ 2. $\dfrac{8}{9}, \dfrac{17}{18}$

_____ 3. $\dfrac{3}{4}, \dfrac{4}{5}$ _____ 4. $\dfrac{13}{15}, \dfrac{19}{24}$

_____ 5. $\dfrac{59}{8}, 7\dfrac{3}{8}$ _____ 6. $3\dfrac{5}{8}, \dfrac{15}{4}$

Answers are below.

Answers to Trial Run

page 215   1. $\dfrac{7}{8} > \dfrac{3}{8}$ 2. $\dfrac{8}{9} < \dfrac{17}{18}$ 3. $\dfrac{3}{4} < \dfrac{4}{5}$ 4. $\dfrac{13}{15} > \dfrac{19}{24}$ 5. $\dfrac{59}{8} = 7\dfrac{3}{8}$ 6. $3\dfrac{5}{8} < \dfrac{15}{4}$

Name _____ Date _____

EXERCISE SET 5.4

Use $<$, $>$, or $=$ to compare the following pairs of numbers.

_____ 1. $\dfrac{9}{11}$, $\dfrac{5}{11}$ _____ 2. $\dfrac{11}{13}$, $\dfrac{7}{13}$ _____ 3. $\dfrac{3}{5}$, $\dfrac{7}{10}$

_____ 4. $\dfrac{5}{6}$, $\dfrac{7}{12}$ _____ 5. $\dfrac{2}{3}$, $\dfrac{3}{4}$ _____ 6. $\dfrac{5}{8}$, $\dfrac{2}{3}$

_____ 7. $\dfrac{21}{25}$, $\dfrac{7}{9}$ _____ 8. $\dfrac{3}{4}$, $\dfrac{40}{49}$ _____ 9. $\dfrac{7}{15}$, $\dfrac{5}{24}$

_____ 10. $\dfrac{10}{21}$, $\dfrac{13}{28}$ _____ 11. $\dfrac{15}{3}$, $3\dfrac{1}{2}$ _____ 12. $\dfrac{17}{4}$, $4\dfrac{2}{3}$

_____ 13. $\dfrac{19}{3}$, $6\dfrac{1}{3}$ _____ 14. $\dfrac{25}{7}$, $3\dfrac{4}{7}$ _____ 15. $5\dfrac{3}{5}$, $\dfrac{57}{10}$

_____ 16. $4\dfrac{1}{4}$, $\dfrac{67}{16}$ _____ 17. $3\dfrac{5}{14}$, $3\dfrac{8}{21}$ _____ 18. $2\dfrac{4}{15}$, $2\dfrac{16}{35}$

_____ 19. $\dfrac{53}{30}$, $\dfrac{29}{20}$ _____ 20. $\dfrac{61}{28}$, $\dfrac{59}{21}$

_____ 21. If Sarah is $60\frac{1}{3}$ inches tall and her classmate Rita is $60\frac{2}{7}$ inches tall, which student is taller?

_____ 22. James lives $2\frac{2}{3}$ miles from campus and Tanika lives $2\frac{7}{10}$ miles from campus. Who lives closer to campus?

_____ 23. In 1986, Dominique Wilkins led the NBA in scoring with $30\frac{3}{10}$ points per game. In 1984, Adrian Dantley led with $30\frac{3}{5}$ points per game. Whose scoring average was better?

_____ 24. Which figure has a larger area: a square measuring $11\frac{1}{2}$ feet on each side or a rectangle that is $10\frac{1}{2}$ feet wide and $12\frac{1}{2}$ feet long?

☆ Stretching the Topics _____

_____ 1. Arrange the fractions $\frac{5}{9}$, $\frac{4}{7}$, and $\frac{3}{8}$ in order from smallest to largest.

_____ 2. Compare $5\frac{750}{1000}$ and $\frac{23}{4}$.

_____ 3. Arrange the numbers $3\frac{1}{8}$, $\frac{15}{4}$, $3\frac{3}{16}$, and $\frac{7}{2}$ in order from smallest to largest.

Check your answers in the back of your book.

If you can complete **Checkup 5.4,** then you are ready to go to Section 5.5.

 **CHECKUP 5.4**

Use <, >, or = to compare the following pairs of numbers.

_____ 1. $\dfrac{3}{4}$, $\dfrac{5}{8}$ _____ 2. $\dfrac{5}{7}$, $\dfrac{7}{9}$

_____ 3. $\dfrac{13}{15}$, $\dfrac{17}{18}$ _____ 4. $\dfrac{19}{14}$, $1\dfrac{3}{10}$

_____ 5. $\dfrac{27}{20}$, $1\dfrac{7}{20}$ _____ 6. $2\dfrac{7}{8}$, $2\dfrac{5}{6}$

Check your work in the back of your book.

If You Missed Problems:	You Should Review Examples:
1–3	1–3
4–6	4

5.5 Switching from Words to Fractional Numbers

Having learned how to add and subtract proper fractions, improper fractions, whole numbers, and mixed numbers, we can put those skills to use in solving some problems stated in words. As in earlier sections where we switched from words to numbers, we must always decide what operation (addition or subtraction) is to be performed.

Example 1. Last year, Mimi was $61\frac{3}{8}$ inches tall. This year she has grown $1\frac{7}{8}$ inches. What is her height?

Solution. Since Mimi's height has *increased*, we know we must *add*.

$$61\frac{3}{8} = 61 + \frac{3}{8}$$

$$\frac{+\ 1\frac{7}{8} = \quad 1 + \frac{7}{8}}{62 + \frac{10}{8}} \qquad \text{ADD.}$$

$$62 + \frac{10}{8} = 62 + \frac{5}{4} \qquad \text{Reduce the fraction.}$$

$$= 62 + 1\frac{1}{4} \qquad \text{Rewrite } \frac{5}{4} \text{ as } 1\frac{1}{4}.$$

$$= 63\frac{1}{4} \qquad \text{Add the whole number parts.}$$

Mimi is now $63\frac{1}{4}$ inches tall.

Let's solve the problem stated at the beginning of this chapter.

Example 2. Scott wishes to install a new baseboard around his kitchen. The kitchen measures $12\frac{7}{10}$ feet by $14\frac{1}{2}$ feet, but the two doorways will need no baseboard. One door is $3\frac{1}{8}$ feet wide and the other is $2\frac{9}{10}$ feet wide. How many feet of baseboard will Scott need?

Solution. To find the distance around a rectangular room, we need its perimeter. Perhaps a drawing of the kitchen will help.

To find the total perimeter, we must *add*

$$12\frac{7}{10} + 12\frac{7}{10} + 14\frac{1}{2} + 14\frac{1}{2}$$

But to leave out the doorways, we must *subtract* them from the total perimeter. The amount of baseboard needed is

$$12\frac{7}{10} + 12\frac{7}{10} + 14\frac{1}{2} + 14\frac{1}{2} - 3\frac{1}{8} - 2\frac{9}{10}$$

The LCD for the fractions is 40. Our problem becomes

$$12\,\frac{28}{40} + 12\,\frac{28}{40} + 14\,\frac{20}{40} + 14\,\frac{20}{40} - 3\,\frac{5}{40} - 2\,\frac{36}{40}$$ Rewrite fractional parts.

$$= \quad\quad 52\,\frac{96}{40} \quad\quad - 3\,\frac{5}{40} - 2\,\frac{36}{40}$$ Perform additions.

$$= \quad\quad 49\,\frac{91}{40} \quad\quad\quad - 2\,\frac{36}{40}$$ Perform first subtraction.

$$= 47\,\frac{55}{40}$$ Perform second subtraction.

$$= 47\,\frac{11}{8}$$ Reduce fraction.

$$= 47 + 1\,\frac{3}{8}$$ Rewrite improper fraction.

$$= 48\,\frac{3}{8}$$ Add whole number parts.

Scott needs $48\frac{3}{8}$ feet of baseboard.

Example 3. Helena spends $\frac{1}{4}$ of her monthly paycheck for rent. She spends $\frac{1}{7}$ of her paycheck for groceries. What part of her paycheck is left after she pays her rent and buys groceries?

Solution. We can represent Helena's whole paycheck with the number 1. To find the fractional part left, we must *subtract* the fractional parts she has spent.

$$1 - \frac{1}{4} - \frac{1}{7} \quad\quad \text{LCD: } 28$$

$$= \frac{28}{28} - \frac{7}{28} - \frac{4}{28}$$

$$= \frac{28 - 7}{28} - \frac{4}{28}$$

$$= \frac{21}{28} - \frac{4}{28}$$

$$= \frac{21 - 4}{28}$$

$$= \frac{17}{28}$$

Helena has $\frac{17}{28}$ of her paycheck left.

Example 4. Ms. Portland borrowed a cap from a colleague to wear in the university's graduation exercises. Ms. Portland wears a size $7\frac{3}{8}$ hat, and her colleague wears a size $7\frac{1}{4}$. Is the hat she borrowed too large or too small?

Solution. This is a problem in *comparing* fractional numbers. We change the mixed numbers to improper fractions, and then write each one with the same denominator.

$$\text{Ms. Portland:} \quad 7\frac{3}{8} = \frac{59}{8}$$

$$\text{Colleague:} \quad 7\frac{1}{4} = \frac{29}{4} = \frac{29 \times 2}{4 \times 2} = \frac{58}{8}$$

$$\frac{59}{8} > \frac{58}{8}$$

Therefore her colleague's hat is too small.

EXERCISE SET 5.5

_____ 1. Jo Ann's new baby girl weighed $7\frac{3}{4}$ pounds at birth. At her first visit to the doctor's office today, the nurse told Jo Ann her baby had gained $2\frac{1}{3}$ pounds. What was the baby's weight at the first visit?

_____ 2. If $\frac{5}{8}$ of a cake has been eaten, how much of the cake is left?

_____ 3. Lew spends $\frac{1}{10}$ of his salary each month for gasoline. His car payment each month is $\frac{1}{5}$ of his salary. What fractional part of his monthly salary is spent on his car?

_____ 4. A painting crew at Western University completed painting $\frac{1}{9}$ of the classrooms during the first week of semester break. They painted $\frac{3}{8}$ of the rooms the second week and $\frac{1}{6}$ of the rooms the third week. If summer school started after 3 weeks, what fraction of the rooms had been painted before classes began?

_____ 5. Paul's recipe for cupcakes calls for $\frac{3}{4}$ cup of sugar for the batter and $2\frac{2}{3}$ cups of sugar for the icing. How much sugar does he need?

_____ 6. Lamar bought $5\frac{1}{2}$ pounds of steak, a $3\frac{3}{4}$-pound chuck roast, and $8\frac{2}{3}$ pounds of hamburger. How many pounds of beef did he buy?

_____ 7. The river near Felecia's home has risen to $37\frac{7}{8}$ feet. If "flood stage" is $43\frac{5}{6}$ feet, how much farther can the river rise before it begins to flood?

_____ 8. Vanessa is buying paper to wrap her Christmas presents. There are $140\frac{2}{3}$ square inches in one roll of Ready Wrap and $135\frac{2}{5}$ square inches in one roll of Pretty Paper. How many more square inches does Vanessa get if she buys Ready Wrap?

_____ 9. Earl is fencing his yard which measures $150\frac{3}{4}$ feet by $250\frac{2}{3}$ feet, but there will be two gates where no fencing will be needed. One gate is $3\frac{1}{3}$ feet wide and the other is $6\frac{1}{2}$ feet wide. How many feet of fencing will Earl need?

_____ 10. At the Union Bank, savings certificates earn $8\frac{3}{4}$ percent interest. National Bank's interest rate is $9\frac{4}{5}$ percent. What is the difference in the interest rates?

_____ 11. LaRhonda is making punch for her sister's wedding reception. She needs 64 ounces of pineapple juice. She finds 2 cans of juice in the pantry. One of them contains $15\frac{2}{3}$ ounces and the other contains $22\frac{1}{2}$ ounces. How many more ounces of juice does LaRhonda need?

_____ 12. Merideth kept a record of how much gas she used on a trip to Birmingham to visit her mother. She bought $9\frac{3}{10}$ gallons, $11\frac{1}{2}$ gallons, and $9\frac{3}{4}$ gallons. How much gasoline did she buy?

_____ 13. One day recently the stock market reported that at the beginning of the day Zerox stock shares sold for $102\frac{1}{8}$ dollars and at the close of the day a share sold for $101\frac{3}{4}$ dollars. Find by how much the price changed during the day.

_____ 14. Aaron jogged $2\frac{1}{2}$ miles on Friday, $3\frac{1}{5}$ miles on Saturday, and $1\frac{5}{6}$ miles on Sunday. How many miles did Aaron jog during the 3 days?

_____ 15. Camilla is buying material to build a bookcase. The clerk at the hardware store asked her whether she wanted $\frac{7}{8}$-inch nails or $\frac{11}{12}$-inch nails. She said she wanted the shorter nails. Which size nail did she buy?

_____ 16. Tim jogged $3\frac{2}{5}$ miles on Monday and $3\frac{7}{10}$ miles on Tuesday. On which day did he jog the longer distance?

_____ 17. The Edmonds bought 2 plots of land. One plot was $3\frac{3}{4}$ acres and the other was $4\frac{5}{6}$ acres. How many acres of land did they buy?

_____ 18. In her will Aunt Vina left $\frac{1}{4}$ of her estate to her nephew, $\frac{2}{5}$ of her estate to her niece, and the remainder to her brother. What fractional part of the estate did the brother receive?

_____ 19. A rectangular piece of plywood is $6\frac{3}{8}$ feet wide by $10\frac{1}{3}$ feet long. How many feet longer is the piece of plywood than it is wide?

_____ 20. Samantha uses $\frac{1}{5}$ of her salary for rent, $\frac{1}{8}$ for food, and $\frac{3}{10}$ for utilities. What fractional part of her salary is left for other things?

☆ Stretching the Topics _____

_____ 1. Martin spends $\frac{1}{10}$ of his salary each month for gasoline. His car payment is $\frac{1}{5}$ of his salary. If his monthly salary is \$1250, how much does he spend each month on other things?

_____ 2. Simon jogged $2\frac{1}{2}$ miles on Monday, $3\frac{1}{5}$ miles on Tuesday, $1\frac{5}{6}$ miles on Wednesday, and $2\frac{2}{3}$ miles on Thursday. What is his average distance for the 4 days? If Simon has a goal of jogging a total of 12 miles every 5 days, how far must he jog on Friday?

_____ 3. Find the missing lengths in the drawing at the right. (Assume all sides meet in right angles.)

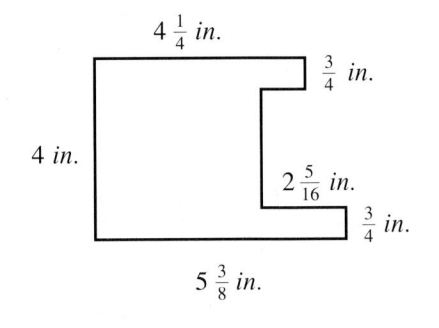

Check your work in the back of your book.

If you can do the problems in **Checkup 5.5,** you are ready to do the **Review Exercises** for Chapter 5.

✓ **CHECKUP 5.5**

_____ 1. Last weekend, Maranda worked $3\frac{2}{3}$ hours baby-sitting and $5\frac{1}{4}$ hours life-guarding at the pool. How many hours did she work last weekend?

_____ 2. The three sides of a triangle are $\frac{2}{3}$ feet, $\frac{7}{8}$ feet, and $\frac{5}{6}$ feet. Find the distance around the triangle.

_____ 3. A corner field is a rectangle that measures $1\frac{3}{16}$ miles along Bunzer Road and $1\frac{5}{48}$ miles along Shiloh Road. Which road borders the longer measurement?

_____ 4. Kris needs 8 cups of sugar for his candy recipes. He finds $3\frac{2}{3}$ cups in the cannister and he borrows $2\frac{1}{4}$ cups from his neighbor. How much more sugar does he need?

_____ 5. Each spring Ms. Green plants $\frac{1}{4}$ of her garden in green beans, $\frac{2}{5}$ in peas, and the rest is set out in tomato plants. What fractional part of her garden is used to raise tomatoes?

Check your answers in the back of your book.

If You Missed Problems:	You Should Review Examples:
1, 2	1
3	4
4, 5	2, 3

Summary

In this chapter, we learned first how to add, subtract, and compare fractions with the *same denominator*.

In Order to	We Must	Examples
Add fractions	Add numerators and keep same denominator.	$\dfrac{3}{7} + \dfrac{2}{7} = \dfrac{5}{7}$
Subtract fractions	Subtract numerators and keep same denominator.	$\dfrac{17}{11} - \dfrac{3}{11} = \dfrac{14}{11}$
Compare fractions	Compare numerators	$\dfrac{5}{23} < \dfrac{7}{23}$ because $5 < 7$

When fractions did *not* have the same denominator, we learned to use the technique of **building fractions** to rewrite the original fractions as equivalent new fractions having the same denominator or **lowest common denominator** (LCD). Then we added, subtracted, or compared the new fractions by the methods just described.

Original Problem	LCD	Building Process	New Problem
$\dfrac{1}{6} + \dfrac{7}{10}$ $= \dfrac{1}{2 \times 3} + \dfrac{7}{2 \times 5}$	$2 \times 3 \times 5$	$\dfrac{1}{6} = \dfrac{1}{2 \times 3} = \dfrac{1 \times 5}{2 \times 3 \times 5} = \dfrac{5}{30}$ $\dfrac{7}{10} = \dfrac{7}{2 \times 5} = \dfrac{7 \times 3}{2 \times 5 \times 3} = \dfrac{21}{30}$	$\dfrac{5}{30} + \dfrac{21}{30}$ $= \dfrac{26}{30} = \dfrac{13}{15}$
$4 - \dfrac{3}{11}$ $= \dfrac{4}{1} - \dfrac{3}{11}$	1×11	$\dfrac{4}{1} = \dfrac{4 \times 11}{1 \times 11} = \dfrac{44}{11}$ $\dfrac{3}{11} = \dfrac{3}{11}$	$\dfrac{44}{11} - \dfrac{3}{11}$ $= \dfrac{41}{11}$ or $3\dfrac{8}{11}$
Compare $1\dfrac{2}{7}$ and $1\dfrac{1}{4}$ $\dfrac{9}{7}$ and $\dfrac{5}{4}$	7×4	$1\dfrac{2}{7} = \dfrac{9}{7} = \dfrac{9 \times 4}{7 \times 4} = \dfrac{36}{28}$ $1\dfrac{1}{4} = \dfrac{5}{4} = \dfrac{5 \times 7}{4 \times 7} = \dfrac{35}{28}$	$\dfrac{36}{28} > \dfrac{35}{28}$ $1\dfrac{2}{7} > 1\dfrac{1}{4}$

The same techniques are used to add, subtract, and compare proper fractions, improper fractions, whole numbers, and mixed numbers.

We concluded the chapter with word problems that required us to add, subtract, or compare fractional numbers.

❑ Speaking the Language of Mathematics ──────────────

Complete each statement with the appropriate word or phrase.

1. To add or subtract fractions with the same denominator, we add or subtract their _____ , keeping the same _____ .

2. To compare two fractions with the same denominator, we compare their _____ .

3. To add or subtract fractions with different denominators, we must find the _____ _____ _____ for all the original fractions.

4. The process of multiplying the numerator and denominator of a fraction by the same number is called _____ fractions.

5. To add (or subtract) mixed numbers, we may add (or subtract) their _____ _____ parts and their _____ parts.

△ Writing About Mathematics ──────────────

Write your response to each question in complete sentences.

1. If two fractions have different denominators, which operation is easier to perform with those fractions: addition or multiplication? Why?

2. How do you compare two fractions that have different denominators?

3. If you know what fractional part of a farm has been planted in each of several crops, describe how you would find the fractional part that is *not* planted.

REVIEW EXERCISES for Chapter 5

Perform the indicated operation and reduce.

_____ 1. $\dfrac{5}{24} + \dfrac{7}{24}$

_____ 2. $\dfrac{5}{13} + \dfrac{12}{13}$

_____ 3. $\dfrac{19}{27} - \dfrac{7}{27}$

_____ 4. $\dfrac{12}{25} + \dfrac{19}{25} - \dfrac{2}{25}$

_____ 5. $\dfrac{43}{16} - \dfrac{7}{16} - \dfrac{5}{16}$

_____ 6. $2\dfrac{3}{8} + 4\dfrac{7}{8} - \dfrac{15}{8}$

Change each number to an equivalent fraction with the given denominator.

_____ 7. $\dfrac{4}{5} = \dfrac{?}{25}$

_____ 8. $\dfrac{12}{7} = \dfrac{?}{42}$

_____ 9. $12 = \dfrac{?}{9}$

_____ 10. $1 = \dfrac{?}{48}$

Perform the indicated operations and reduce.

_____ 11. $\dfrac{2}{3} + \dfrac{4}{5} + \dfrac{7}{15}$

_____ 12. $8\dfrac{2}{3} - 4\dfrac{5}{12}$

_____ 13. $\dfrac{5}{7} + \dfrac{3}{14} - \dfrac{9}{10}$

_____ 14. $\dfrac{11}{15} + \dfrac{1}{12} + \dfrac{17}{20}$

_____ 15. $12\dfrac{6}{7} + 5\dfrac{1}{3}$

_____ 16. $7\dfrac{5}{12} + 3\dfrac{8}{9}$

_____ 17. $12\dfrac{5}{8} - 7\dfrac{9}{16}$

_____ 18. $18\dfrac{2}{3} - 12\dfrac{3}{4}$

_____ 19. $13\dfrac{1}{3} + 2\dfrac{1}{4} - 6\dfrac{7}{8}$

_____ 20. $11\dfrac{3}{8} - 2\dfrac{1}{2} - 5\dfrac{1}{6}$

Use $<$, $>$, or $=$ to compare the following pairs of numbers.

_____ 21. $\dfrac{5}{6}, \ \dfrac{2}{3}$

_____ 22. $\dfrac{3}{8}, \ \dfrac{4}{7}$

_____ 23. $\dfrac{13}{10}, \ \dfrac{5}{4}$

_____ 24. $2\dfrac{4}{5}, \ 2\dfrac{11}{12}$

_____ 25. Darren and Warren are hanging pictures in their apartment. If Warren needs $3\frac{3}{4}$ feet of wire and Darren needs $2\frac{1}{2}$ feet, how many feet of wire should they buy?

_____ 26. A rectangular plot of land is $135\frac{2}{5}$ feet wide and $142\frac{5}{6}$ feet long. How much longer is the plot than it is wide?

_____ 27. A company's stock increased from $24\frac{3}{4}$ points to $31\frac{5}{8}$ points. Find by how much the stock's value increased.

_____ 28. A truck with a full load of cargo weighs $2\frac{3}{10}$ tons. If the cargo weighs $\frac{4}{5}$ of a ton, what is the weight of the truck?

_____ 29. Carol wishes to put a ceiling border around her newly wallpapered den. The den measures $15\frac{3}{4}$ feet by $18\frac{1}{2}$ feet, but she will need no border over the bookcases. One bookcase is $2\frac{1}{8}$ feet wide and the other is $3\frac{3}{10}$ feet wide. How many feet of border will Carol need?

_____ 30. Bucky is on a diet. During the past 3 weeks he has lost $2\frac{1}{2}$ pounds, $1\frac{3}{4}$ pounds, and $2\frac{1}{8}$ pounds. If Bucky weighed $186\frac{1}{2}$ pounds at the beginning of his diet, how much does he weigh now?

Check your answers in the back of your book.

If You Missed Exercises:	You Should Review Examples:	
1, 2	Section 5.1	1–4
3		7–10
4, 5		11–13
6		14–19
7, 8	Section 5.2	1–3
9, 10		4, 5
11	Section 5.3	1
12		2
13, 14		5, 6
15–20		7
21, 22	Section 5.4	1–3
23, 24		4
25–28	Section 5.5	1, 4
29		2
30		3

If you have completed the **Review Exercises** and corrected your errors, you are ready to take the **Practice Test** for Chapter 5.

Name _____   **Date** _____

PRACTICE TEST for Chapter 5

Perform the indicated operations and reduce.

		SECTION	EXAMPLE
_____	**1.** $\dfrac{9}{16} + \dfrac{17}{16} + \dfrac{11}{16}$	5.1	6
_____	**2.** $\dfrac{18}{25} - \dfrac{3}{25}$	5.1	7, 8
_____	**3.** $\dfrac{39}{45} - \dfrac{16}{45} - \dfrac{4}{45}$	5.1	12, 13
_____	**4.** $9\dfrac{1}{6} - 3\dfrac{5}{6}$	5.1	16, 17

Change each number to an equivalent fraction with the given denominator.

_____	**5.** $\dfrac{5}{9} = \dfrac{?}{63}$	5.2	1–3
_____	**6.** $4 = \dfrac{?}{24}$	5.2	4

Perform the indicated operations and reduce.

_____	**7.** $\dfrac{3}{4} + \dfrac{11}{12} - \dfrac{2}{3}$	5.3	1
_____	**8.** $8\dfrac{1}{2} - 4\dfrac{7}{18}$	5.3	2
_____	**9.** $9\dfrac{2}{3} + 4\dfrac{5}{6}$	5.3	2
_____	**10.** $\dfrac{3}{5} + \dfrac{9}{20} + \dfrac{5}{6}$	5.3	5, 6
_____	**11.** $\dfrac{7}{8} + \dfrac{9}{12} - \dfrac{7}{9}$	5.3	5, 6
_____	**12.** $\dfrac{8}{7} + \dfrac{3}{4} + \dfrac{7}{15}$	5.3	6
_____	**13.** $12 - 9\dfrac{3}{5}$	5.3	7
_____	**14.** $2\dfrac{7}{8} + 9\dfrac{1}{3}$	5.3	7

	SECTION	EXAMPLE

———— 15. $9\dfrac{7}{20} - 6\dfrac{4}{5}$ 5.3   7

Use $<$, $>$, or $=$ to compare the following pairs of numbers.

———— 16. $\dfrac{11}{12}$, $\dfrac{4}{5}$ 5.4   3

———— 17. $\dfrac{48}{15}$, $3\dfrac{1}{12}$ 5.4   4

———— 18. Juan needs $3\frac{2}{3}$ gallons of paint for the trim on his house and $4\frac{2}{5}$ gallons for his garage. How many gallons of paint should he buy?   5.5   1

———— 19. For her Independence Day barbecue, Hattie will need $18\frac{3}{4}$ pounds of ribs. She has a package of ribs in her freezer that weighs $7\frac{3}{10}$ pounds. How many pounds will she need to buy?   5.5   4

———— 20. A paving crew resurfaced $\frac{3}{5}$ of a highway 1 week and $\frac{1}{3}$ of the highway the next week. If they plan to finish in 3 weeks, how much of the highway must they resurface during the third week?   5.5   3

Check your answers in the back of your book.

SHARPENING YOUR SKILLS after Chapters 1–5

SECTION

_____ 1. Write the word form for 20,309.   1.1

_____ 2. Classify 397 as prime or composite. If the number is composite, write it as a product   2.3
of prime numbers.

_____ 3. Change $31\frac{5}{8}$ to an improper fraction.   3.1

_____ 4. Reduce $\frac{42}{63}$ to lowest terms.   3.2

_____ 5. Change $\frac{3}{8}$ to an equivalent fraction with a denominator of 48.   5.2

_____ 6. Use $<$, $>$, or $=$ to compare $\frac{9}{5}$ and $1\frac{1}{3}$.   5.4

Perform the indicated operations and reduce if possible.

_____ 7. $89 + 204 + 5632$ 1.2

_____ 8.   73,296   1.2
         14,314
      $+\,20,005$

_____ 9. $\dfrac{8}{15} + \dfrac{7}{15} + \dfrac{23}{15}$ 5.1

_____ 10. $\dfrac{3}{4} + \dfrac{9}{20} + \dfrac{7}{10}$ 5.3

_____ 11. $3\dfrac{7}{8} + 12\dfrac{2}{3}$ 5.3

_____ 12.   35,504   1.3
      $-\,18,736$

_____ 13. $700 - 329$ 1.3

_____ 14. $\dfrac{29}{35} - \dfrac{8}{35}$ 5.1

_____ 15. $7\dfrac{3}{4} - 3\dfrac{7}{8}$ 5.3

_____ 16. $34 - 12\dfrac{3}{5}$ 5.3

_____ 17. $15 \times 7 \times 19$ 2.1

_____ 18. 6005×307 2.1

SECTION

_____ 19. $\dfrac{16}{45} \times 27$

4.1

_____ 20. $\dfrac{4}{5} \times 3\dfrac{7}{8} \times \dfrac{15}{2}$

4.1

_____ 21. $2528 \div 79$

2.2

_____ 22. $10\dfrac{2}{3} \div 1\dfrac{1}{15}$

4.2

_____ 23. $\dfrac{12\frac{2}{3}}{2\frac{1}{9}}$

4.2

_____ 24. In one week Nick jogs 33 miles. He covers the same route each day. If he jogs every day except Saturday, how long is the route he follows?

4.3

_____ 25. The paving crew for Warren County planned to resurface 620 miles of highway this summer. By the end of July, they had paved 372 miles. What fractional part of the job remained?

3.3

Check your answers in the back of your book.

Working with Decimal Numbers

Martie has recorded her checking account transactions in her check register but she has not done the necessary subtractions. Do them for her and find the balance in her account.

CHECK NO.	DATE	CHECKS DRAWN OR DEPOSITS MADE	BALANCE FORWARD ▷	√	339	17
329	6/3	TO Casual Clothier	DEDUCT CHECK − / ADD DEPOSIT +		− 33	60
			BALANCE ▷			
330	6/4	TO Cash	DEDUCT CHECK − / ADD DEPOSIT +		− 20	00
			BALANCE ▷			
331	6/5	TO Hometown Gas Co.	DEDUCT CHECK − / ADD DEPOSIT +		− 59	28
			BALANCE ▷			
332	6/5	TO Enrico's Restaurant	DEDUCT CHECK − / ADD DEPOSIT +		− 8	57
			BALANCE ▷			
333	6/6	TO Global Insurance Co.	DEDUCT CHECK − / ADD DEPOSIT +		−139	00
			BALANCE ▷			
334	6/7	TO Shoe Shack	DEDUCT CHECK − / ADD DEPOSIT +		− 40	80
			BALANCE ▷			

When you receive a check for $73.28 or when you read that unemployment has reached 10.2 percent, you are being asked to deal with **decimal numbers.** Let us turn our attention to the study of decimal numbers and the operations of arithmetic.

In this chapter we learn how to

1. Understand decimal numbers.
2. Compare decimal numbers.
3. Round decimal numbers.
4. Add and subtract decimal numbers.
5. Switch from word statements to decimal number statements.

The methods for using a calculator to solve problems of the types encountered in Chapters 3–7 are discussed in Sections 16.1 and 16.2 of Chapter 16.

6.1 Understanding Decimal Numbers

Every decimal number can be thought of as having two parts separated by the decimal point. The part to the left of the decimal point is the **whole number** part, and the part to the right of the decimal point is the **fractional part.** For instance, we can label the parts of 73.28 as follows

$$
\begin{array}{c}
\text{fractional part} \\
\downarrow \\
73.28 \\
\end{array}
$$

decimal point

whole number part

Understanding Place Value in Decimal Numbers

From Chapter 1 we recall that each digit in a whole number has a certain **place value.** We learned to give meaning to a whole number according to the digits in the ones place, the tens place, the hundreds place, the thousands place, and so on. We remember that 2789 means 2 *thousands, 7 hundreds, 8 tens,* and 9 *ones.* The value of each place has *10 times* the value of the place immediately to its *right.* In other words, each place has *one-tenth* the value of the place immediately to its *left.* Using this idea, we will be able to give value to the places that lie to the right of the decimal point in a decimal number.

Since the right-most place in a *whole number* is the ones place, the first place to the right of the decimal point should have one-tenth the value of the ones place. The value of the ones place is 1, so we must find

$$ \frac{1}{10} \ \text{ of } \ 1 = \frac{1}{10} \times 1 = \frac{1}{10} $$

Thus the first place to the right of the decimal point is the **tenths place.** The next place should have one-tenth the value of the tenths place. Since the value of the tenths place is $\frac{1}{10}$, we must find:

$$ \frac{1}{10} \ \text{ of } \ \frac{1}{10} = \frac{1}{10} \times \frac{1}{10} = \frac{1}{100} $$

The next place must be the **hundredths place.** Using similar reasoning for each place thereafter, we can label the places to the right of the decimal point as

. 2 3 1 5 7

tenths | hundredths | thousandths | ten thousandths | hundred thousandths

Each digit to the right of the decimal point occupies a position with a certain *fractional* place value. To read the fractional part of a decimal number, we note the position where the last digit appears. The place value of that position tells us whether we are dealing with tenths or hundredths or thousandths and so on. The digits themselves tell us how many tenths or hundredths or thousandths or whatever we have. Let's practice reading some decimal numbers and then writing them as equivalent fractional numbers.

Decimal Form	Word Form	Fractional Form
0.3	3 tenths	$\dfrac{3}{10}$
0.19	19 hundredths	$\dfrac{19}{100}$
0.257	257 thousandths	$\dfrac{257}{1000}$
73.28	73 and 28 hundredths	$73\dfrac{28}{100} = \dfrac{7328}{100}$

Example 1. Write 0.017 as a fraction.

Solution. $0.017 = \dfrac{17}{1000}$

Note that a 0 is sometimes placed to the left of the decimal point when there is no whole number part. This is done simply to draw attention to the location of the decimal point and is the accepted international notation.

Example 2. Write $\dfrac{37}{10,000}$ as a decimal number.

Solution. $\dfrac{37}{10,000}$ is 37 ten thousandths, so we know the last digit appears in the ten thousandths position.

$$\frac{37}{10,000} = 0.0037$$

Example 3. Write 2.54 in words.

Solution. 2.54 means 2 and 54 hundredths.

Notice that in reading a decimal number we say "and" when we reach the decimal point. This signals that we have finished the whole number part and are moving on to read the fractional part.

In the fractional part of a decimal number, zeros may be attached after the last nonzero digit without changing the value of the number. Thus we may write

$$0.5 = 0.50 = 0.500$$

Do you see why this is true? Recall from our work with building fractions that

$$\frac{5}{10} = \frac{50}{100} = \frac{500}{1000}$$

Example 4. Rewrite 1.2 as a decimal number that ends in the thousandths place.

Solution. Remembering that we may attach zeros after the last nonzero digit in the fractional part of a decimal number, we write

$$1.2 = 1.200$$

You try Example 5.

Example 5. Rewrite $3 as a decimal number that ends in the hundredths place.

Solution. $3 = _____

Check your work on page 239. ▶

As you can see from Example 5, attaching zeros in the fractional part of a decimal number allows us to express a whole number of dollars in "dollars and cents" form. This will be very useful to us later on in the chapter.

⇒ Trial Run

_____ 1. Write 0.025 as a fraction. _____ 2. Write 0.25 as a fraction.

_____ 3. Write $\frac{213}{1000}$ as a decimal number. _____ 4. Write $\frac{423}{10,000}$ as a decimal number.

_____ 5. Write 3.052 in words. _____ 6. Write 1.35 in words.

_____ 7. Write ten and thirteen thousandths as a decimal number. _____ 8. Rewrite 4.52 as a decimal number that ends in the thousandths place.

_____ 9. Rewrite 2.4 as a decimal number that ends in the hundredths place.

Answers are on page 239.

Comparing Decimal Numbers

Perhaps our work with comparing fractional numbers can provide us with a clue for comparing decimal numbers. Recall that we compared two fractions by writing them as fractions with the same denominator. Then we saw that the fraction with the larger numerator represented the larger fraction. For instance,

$$\frac{83}{100} > \frac{81}{100} \quad \text{because} \quad 83 > 81$$

Similar reasoning will help us compare decimal numbers such as 0.11 and 0.12. We can reason that

$$0.11 = \frac{11}{100} \quad \text{and} \quad 0.12 = \frac{12}{100}$$

and comparing the fractional numbers, we see that

$$\frac{11}{100} < \frac{12}{100}$$

so $0.11 < 0.12$

Suppose we wish to compare two decimal numbers that do not contain the same number of places to the right of the decimal point.

Example 6. Compare 0.314 and 0.3137.

Solution. We know that we may attach zeros after the last nonzero digit in the fractional part of any decimal number. Let's attach zeros where needed to make both numbers contain the same number of places to the right of the decimal point.

$$0.314 = 0.3140$$
$$0.3137 = 0.3137$$

Now we see that we are comparing

$$\frac{3140}{10,000} \quad \text{and} \quad \frac{3137}{10,000}$$

Our conclusion?

$$0.3140 > 0.3137$$
$$0.314 > 0.3137$$

Perhaps you have thought of a way to compare decimal numbers without rewriting them as fractional numbers. Indeed this is possible, if you will follow these steps.

Comparing Decimal Numbers

1. Write both decimal numbers with the same number of places to the right of the decimal point, attaching zeros where needed after the last nonzero digit in the fractional part.
2. Ignoring the decimal point, compare the numbers as though they were whole numbers.

Let's practice this method as we compare 6.3 and 6.299. First we attach zeros to give both decimal numbers the same number of places to the right of the decimal point.

$$6.3 \quad = 6.300$$
$$6.299 = 6.299$$

Now we must compare 6.300 and 6.299. Ignoring the decimal point, we see that 6300 > 6299, so we conclude that

$$6.300 > 6.299$$
$$6.3 \quad > 6.299$$

Example 7. Write in order, from smallest to largest: 0.034, 0.03359, and 0.0343.

Solution. We attach the needed zeros.

$$0.034 \quad = 0.03400$$
$$0.03359 = 0.03359$$
$$0.0343 \quad = 0.03430$$

Now we ignore the decimal points, and from smallest to largest, we write

$$3359, \ 3400, \ 3430$$

We conclude that the decimal numbers belong in the order

$$0.03359, \ 0.034, \ 0.0343$$

⫸ Trial Run ━━━━━━━━━━━━━━━━━━━━━━━━━━━━━━━━━━━━━

Compare the following numbers using < or >.

_____ **1.** 0.035 and 0.037   _____ **2.** 15.5 and 15.07

_____ **3.** 0.042 and 0.0423   _____ **4.** 0.025 and 0.13

_____ **5.** 5.1 and 5.087   _____ **6.** 0.0395 and 0.004

Write the numbers in order from smallest to largest.

_____ **7.** 0.11, 0.02, 0.0154   _____ **8.** 1.5, 1.053, 1.0054

Answers are on page 239.

Rounding Decimal Numbers

Sometimes it is desirable to **round** decimals to some particular place. As was the case in rounding whole numbers (Chapter 1), to round a decimal number to a certain place, we look at the place immediately to the *right*. If we wish to round a decimal number to the tenths place, we look at the hundredths place. If we wish to round a decimal number to the hundredths place, we look at the thousandths place, and so on.

In general we agree to use the following procedure in rounding decimal numbers to a certain place.

Rounding Decimal Numbers

1. Look at the digit in the place immediately to the right of the place to which you wish to round the number.
2. If the digit in the place to the right is less than 5, leave the digit in the rounding place as it is.
3. If the digit in the place to the right is 5 or greater than 5, add 1 to the digit in the rounding place.
4. Drop all digits to the right of the rounding place.
5. State that your original decimal number is approximately equal to ($\doteq$) your rounded decimal number.

Example 8. Round 17.28 to the tenths place.

Solution. Since the digit in the hundredths place (8) is greater than 5,

$$17.28 \doteq 17.3$$

You complete Example 9.

Example 9. Round 0.01354 to the thousandths place.

Solution. Since the digit in the ten thousandths place (_____) is _____ ,

$$0.01354 \doteq \underline{\quad\quad}$$

Check your work on page 239. ▶

Example 10. Round 0.01354 to the ten thousandths place.

Solution. Since the digit in the hundred thousandths place (4) is less than 5,

$$0.01354 \doteq 0.0135$$

⇒ Trial Run

Round each number to the tenths place.

_____ **1.** 5.087   _____ **2.** 0.548

Round each number to the thousandths place.

_____ **3.** 0.02357   _____ **4.** 3.1343

Round each number to the hundredths place.

_____ **5.** 1.195   _____ **6.** 0.0729

Round each number to the ten thousandths place.

_____ **7.** 0.23486   _____ **8.** 2.003235

Answers are below.

▶ Examples You Completed

Example 5. Rewrite $3 as a decimal number that ends in the hundredths place.

Solution. $3 = $3.00

Example 9. Round 0.01354 to the thousandths place.

Solution. Since the digit in the ten thousandths place (5) is 5,

$$0.01354 \doteq 0.014$$

Answers to Trial Runs

page 236 **1.** $\dfrac{25}{1000}$ **2.** $\dfrac{25}{100}$ **3.** 0.213 **4.** 0.0423 **5.** Three and fifty-two thousandths
6. One and thirty-five hundredths **7.** 10.013 **8.** 4.520 **9.** 2.40

page 238 **1.** 0.035 < 0.037 **2.** 15.5 > 15.07 **3.** 0.042 < 0.0423 **4.** 0.025 < 0.13
5. 5.1 > 5.087 **6.** 0.0395 > 0.004 **7.** 0.0154, 0.02, 0.11 **8.** 1.0054, 1.053, 1.5

page 239 **1.** 5.1 **2.** 0.5 **3.** 0.024 **4.** 3.134 **5.** 1.20 **6.** 0.07 **7.** 0.2349 **8.** 2.0032

EXERCISE SET 6.1

Write each decimal number as a fraction.

_____ 1. 0.03   _____ 2. 0.127

_____ 3. 0.0073   _____ 4. 0.02439

Write each fraction as a decimal number.

_____ 5. $\dfrac{345}{1000}$ _____ 6. $\dfrac{5}{100}$

_____ 7. $\dfrac{56}{100,000}$ _____ 8. $\dfrac{7835}{10,000}$

Write each decimal number in words.

_____ 9. 5.038   _____ 10. 2.75

_____ 11. 0.0075   _____ 12. 0.974

_____ 13. Write six and nine hundredths as a decimal number.

_____ 14. Write two hundred four and eighteen thousandths as a decimal number.

_____ 15. Rewrite 8.7 as a decimal number that ends in the thousandths place.

_____ 16. Rewrite 0.6 as a decimal number that ends in the hundredths place.

_____ 17. Rewrite 0.3 as a decimal number that ends in the ten thousandths place.

_____ 18. Rewrite 1.37 as a decimal number that ends in the thousandths place.

Compare the following numbers using < or >.

_____ 19. 0.45 and 0.049   _____ 20. 0.56 and 0.058

_____ 21. 13.0092 and 13.09   _____ 22. 28.0085 and 28.08

_____ 23. 0.038 and 0.04   _____ 24. 0.793 and 0.80

_____ 25. 6.1 and 6.054   _____ 26. 9.3 and 9.087

_____ 27. 12.03 and 12.2   _____ 28. 65.09 and 65.3

Write the numbers in order from smallest to largest.

_____ 29. 0.351, 0.052, 0.0367

_____ 30. 0.861, 0.092, 0.125

_____ 31. 2.36, 2.072, 2.3

_____ 32. 4.5, 4.23, 4.079

Round each number to the tenths place.

_____ 33. 8.362

_____ 34. 12.742

_____ 35. 15.034

_____ 36. 13.072

_____ 37. 0.872

_____ 38. 0.943

_____ 39. 2.95

_____ 40. 3.96

Round each number to the thousandths place.

_____ 41. 52.75348

_____ 42. 83.25447

_____ 43. 6.0395

_____ 44. 7.0498

_____ 45. 0.0428

_____ 46. 0.0673

_____ 47. 200.243862

_____ 48. 500.234675

Round each number to the hundredths place.

_____ 49. 7.034

_____ 50. 9.056

_____ 51. 25.325

_____ 52. 35.432

_____ 53. 0.86723

_____ 54. 0.65437

_____ 55. 153.995

_____ 56. 721.996

Round each number to the ten thousandths place.

_____ 57. 0.73845

_____ 58. 0.93762

_____ 59. 12.043949

_____ 60. 15.057349

_____ 61. 0.003724

_____ 62. 0.00738

_____ 63. 1356.72948

_____ 64. 1275.83546

☆ Stretching the Topics

_____ **1.** Compare 0.04783 and 0.04779 using $<$ or $>$.

_____ **2.** Write in order from smallest to largest: 1.0598, 1.0173, 1.103, 1.101, 1.1001, 1.1102.

_____ **3.** Round 3.8605967 to the hundred thousandths place.

Check your answers in the back of your book.

If you can complete **Checkup 6.1,** you are ready to go to Section 6.2.

✓ **CHECKUP 6.1**

_____ 1. Write 0.039 as a fraction.

_____ 2. Write $\dfrac{8}{10,000}$ as a decimal number.

_____ 3. Write 0.865 in words.

_____ 4. Rewrite 0.38 as a decimal number that ends in the thousandths place.

_____ 5. Compare 0.73 and 0.0839.

_____ 6. Write in order from smallest to largest: 3.76, 3.0934, 3.672.

_____ 7. Round 9.349 to the tenths place.

_____ 8. Round 0.04952 to the thousandths place.

_____ 9. Round 2.3676 to the hundredths place.

_____ 10. Round 0.0003524 to the ten thousandths place.

Check your answers in the back of your book.

If You Missed Problems:	You Should Review Examples:
1	1
2, 3	2, 3
4	4, 5
5, 6	6, 7
7–10	8–10

6.2 Adding and Subtracting Decimal Numbers

The methods used for adding and subtracting fractional numbers will help us do the same for decimal numbers. However, we will discover that addition and subtraction of decimal numbers is easier than addition and subtraction of fractional numbers.

Adding Decimal Numbers

In adding fractional numbers, we learned that we could only add fractions that were alike. That is, we could only add fractions having the same denominator.

In adding decimal numbers, a similar condition must be met. The decimal numbers being added must contain the same number of places in their fractional parts. If the decimal numbers meet this condition, then we add them by lining up the decimal points beneath each other and adding the digits in each column. For instance,

$$\begin{array}{r} 0.2 \\ +0.3 \\ \hline 0.5 \end{array} \quad \text{and} \quad \begin{array}{r} 0.07 \\ +0.09 \\ \hline 0.16 \end{array}$$

Notice that the decimal point in the answer appears directly beneath the decimal points in the numbers being added. Notice also that the rules for "carrying" are the same rules for carrying used in adding whole numbers.

Addition of decimal numbers can be performed horizontally. However, to avoid careless mistakes, you may choose to write the problems vertically.

Example 1. Find $17.4 + 5.3$.

Solution

$$\begin{array}{r} 17.4 \\ +5.3 \\ \hline 22.7 \end{array}$$

You try Example 2.

Example 2. Find $0.62 + 3.11 + 0.37$.

Solution

$$\begin{array}{r} 0.62 \\ 3.11 \\ +0.37 \\ \hline \end{array}$$

The sum is _____ .

Check your work on page 249. ▶

Notice from Example 2 that we may drop any zeros appearing after the last nonzero digit in the fractional part of a decimal number.

⫸ Trial Run

Find the sums.

_____ 1. $0.09 + 0.08$

_____ 2. $0.7 + 0.8$

_____ 3. $13.5 + 7.5$

_____ 4. $0.73 + 4.22 + 2.46$

_____ 5. $3.72 + 0.49 + 0.08$

_____ 6. $12.35 + 4.06 + 0.04$

Answers are on page 250.

In finding the sum of these decimal numbers

$$1.3 + 0.269 + 14.56$$

we notice that their fractional parts do *not* contain the same number of places. Remembering that we may attach zeros to the right of the last nonzero digit in the fractional part of a decimal number, we may easily make each of the shorter numbers as long as the longest one. Then we add by the usual method.

$$\begin{aligned} &1.3 \ \ + 0.269 + 14.56 \\ =\ &1.300 + 0.269 + 14.560 \end{aligned}$$

$$\begin{array}{r} 1.300 \\ 0.269 \\ +\,14.560 \\ \hline 16.129 \end{array}$$

We note that attaching trailing zeros makes our decimal numbers "alike" in the same way that writing fractions with a common denominator made the fractions "alike".

Example 3. Find $6 + 2.03 + 0.1$.

Solution

$$\begin{array}{r} 6.00 \\ 2.03 \\ +\,0.10 \\ \hline 8.13 \end{array}$$

You try Example 4.

Example 4. Find

$$\$118 + \$235.43 + \$10.57$$

Solution

$$\begin{array}{r} \$118.00 \\ 235.43 \\ +\ \ \ 10.57 \\ \hline \$ \end{array}$$

The sum is $ _____ .

Check your work on page 249. ▶

▥▶ Trial Run

Find the sums.

_____ **1.** $2.4 + 3.095$ _____ **2.** $0.783 + 1.34$

_____ **3.** $5.3 + 0.368 + 15.72$ _____ **4.** $7.6 + 1.085 + 9.35$

_____ **5.** $12 + 3.0429 + 0.5$ _____ **6.** $\$125 + \$3.85 + \$75.68$

Answers are on page 250.

Subtracting Decimal Numbers

Subtraction of decimal numbers is similar to addition in that we must line up like places beneath each other before we subtract in the usual way. For instance, to find the difference $0.8 - 0.2$, we write

$$\begin{array}{r} 0.8 \\ -0.2 \\ \hline 0.6 \end{array}$$

Notice that the decimal point in the answer appears directly beneath the decimal points in the numbers being subtracted.

Example 5. Find $6.295 - 2.173$.

Solution.
$$\begin{array}{r} 6.295 \\ -2.173 \\ \hline 4.122 \end{array}$$

We *check* our subtraction by addition.

$$\begin{array}{r} 2.173 \\ +4.122 \\ \hline 6.295 \end{array} \checkmark$$

You try Example 6.

Example 6. Find $\$93.87 - \$20.05$, and check.

Solution.
$$\begin{array}{r} \$93.87 \\ -\ 20.05 \\ \hline \$ \end{array}$$

CHECK:
$$\begin{array}{r} \$20.05 \\ +\ \underline{} \\ \$ \end{array} \checkmark$$

Check your work on page 249. ▶

If "borrowing" is necessary in subtracting decimal numbers, we do so in exactly the same way that we borrowed in subtracting whole numbers.

Example 7. Find $0.8234 - 0.5764$, and check.

Solution.
$$\begin{array}{r} \overset{\overset{11}{7\ \cancel{1}\ 13}}{0.8\cancel{2}\cancel{3}4} \\ -0.5764 \\ \hline 0.2470 \end{array}$$

CHECK:
$$\begin{array}{r} 0.5764 \\ +0.2470 \\ \hline 0.8234 \end{array} \checkmark$$

The difference is 0.247.

Example 8. Find $\$101.05 - \$77.19$, and check.

Solution.
$$\begin{array}{r} \overset{\overset{10\ \ 9}{9\ \cancel{0}\ \cancel{10}\ 15}}{\$1\cancel{0}\cancel{1}.\cancel{0}\cancel{5}} \\ -\ \ 77.19 \\ \hline \$\ 23.86 \end{array}$$

CHECK:
$$\begin{array}{r} \$\ 77.19 \\ +\ \ 23.86 \\ \hline \$101.05 \end{array} \checkmark$$

The difference is $23.86.

⦀➡ Trial Run

Subtract and check.

———— 1. $0.86 - 0.34$ ———— 2. $8.527 - 3.423$

———— 3. $0.5439 - 0.1846$ ———— 4. $13.095 - 11.986$

———— 5. $\$203.15 - \$89.79$ ———— 6. $\$300.50 - \$95.89$

Answers are on page 250.

If the decimal numbers being subtracted do not contain the same number of places in their fractional parts, we use our old trick of attaching zeros after the last nonzero digit.

Example 9. Find $16.39 - 11.5$, and check.

Solution.

$$
\begin{array}{r}
16.39 \\
-11.50 \\
\hline
4.89
\end{array}
$$

Attach 1 zero.

Subtract.

CHF :

$$
\begin{array}{r}
11.50 \\
+\ 4.89 \\
\hline
16.39 \quad \sqrt{}
\end{array}
$$

Example 10. Find $2.6 - 1.572$, and check.

Solution.

$$
\begin{array}{r}
2.600 \\
-1.572 \\
\hline
1.028
\end{array}
$$

Attach 2 zeros.

Subtract.

CHECK:

$$
\begin{array}{r}
1.572 \\
+1.028 \\
\hline
2.600 \quad \sqrt{}
\end{array}
$$

You try Example 11.

Example 11. Find $\$83 - \$79.63$, and check.

Solution.

$$
\begin{array}{r}
\$83.00 \\
-\ 79.63 \\
\hline
\$
\end{array}
$$

CHECK:

$$
\begin{array}{r}
\$79.63 \\
+\ \underline{} \\
\$ \quad \sqrt{}
\end{array}
$$

Check your work on page 250. ▶

Example 12. Find $3 - 1.6321$, and check.

Solution.

$$
\begin{array}{r}
3.0000 \\
-1.6321 \\
\hline
1.3679
\end{array}
$$

Locate the decimal point and attach 3 zeros.

Subtract.

CHECK:

$$
\begin{array}{r}
1.6321 \\
+1.3679 \\
\hline
3.0000 \quad \sqrt{}
\end{array}
$$

▌▌▶ **Trial Run**

Subtract and check.

_____ 1. $4.53 - 3.2$　　　　　_____ 2. $18.78 - 12.9$

_____ 3. $13.2 - 9.685$　　　　_____ 4. $9.15 - 0.038$

_____ 5. $\$95 - \$84.87$　　　　_____ 6. $9 - 3.1376$

Answers are on page 250.

If a problem involves more than one operation, we perform the additions or subtractions in order from left to right, working with just two numbers at a time. This is the same approach we used with whole numbers and with fractional numbers.

If Martina receives $21.50 and $13 for baby-sitting and spends $8.29, how much does she have left? Here we must add the amounts Martina received and then subtract the amount she has spent. The problem can be written as $\$21.50 + \$13 - \$8.29$ but we work it in two steps.

FIRST WE ADD:	THEN WE SUBTRACT:
$\begin{array}{r} \$21.50 \\ +\ 13.00 \\ \hline \$34.50 \end{array}$	$\begin{array}{r} \$34.50 \\ -\ 8.29 \\ \hline \$26.21 \end{array}$

Martina has $26.21 left.

You try Example 13.

Example 13. Find 7.132 + 3.8 − 2.999.

Solution. First we   Then we
add: subtract:

$$
\begin{array}{r} 7.132 \\ +3.800 \\ \hline \end{array}
\qquad
\begin{array}{r} -2.999 \\ \hline \end{array}
$$

So 7.132 + 3.8 − 2.999 = _____ .

Check your work on page 250. ▶

Example 14. Find 43.2 − 16.98 − 8.631.

Solution. We perform the first subtraction: Then the second subtraction:

$$
\begin{array}{r} 43.20 \\ -16.98 \\ \hline 26.22 \end{array}
\qquad\qquad
\begin{array}{r} 26.220 \\ -\;\;8.631 \\ \hline 17.589 \end{array}
$$

So 43.2 − 16.98 − 8.631 = 17.589.

⫸ Trial Run

Perform the indicated operations.

_____ **1.** 9.342 + 1.03 − 4.008   _____ **2.** $10 + $3.75 − $8.56

_____ **3.** 5.32 − 0.0351 + 1.06   _____ **4.** $75 − $32.86 − $25.98

Answers are on page 250.

▶ Examples You Completed

Example 2. Find 0.62 + 3.11 + 0.37.

Solution

$$
\begin{array}{r} 0.62 \\ 3.11 \\ +0.37 \\ \hline 4.10 \end{array}
$$

The sum is 4.1.

Example 4. Find $118 + $235.43 + $10.57.

Solution

$$
\begin{array}{r} \$118.00 \\ 235.43 \\ +\;\;\;10.57 \\ \hline \$364.00 \end{array}
$$

The sum is $364.

Example 6. Find $93.87 − $20.05, and check.

Solution.
$$
\begin{array}{r} \$93.87 \\ -\;20.05 \\ \hline \$73.82 \end{array}
\qquad
\text{CHECK:}
\quad
\begin{array}{r} \$20.05 \\ +\;73.82 \\ \hline \$93.87 \end{array}\;\checkmark
$$

Example 11. Find $83 − $79.63, and check.

Solution.
$$
\begin{array}{r} \$83.00 \\ -\;79.63 \\ \hline \$\;3.37 \end{array}
\qquad
\text{CHECK:}
\quad
\begin{array}{r} \$79.63 \\ +\;\;\;3.37 \\ \hline \$83.00 \end{array}\;\checkmark
$$

Example 13. Find $7.132 + 3.8 - 2.999$.

Solution. First we   Then we subtract:
add:

$$
\begin{array}{r}
7.132 \\
+\,3.800 \\
\hline
10.932
\end{array}
\qquad
\begin{array}{r}
10.932 \\
-\,2.999 \\
\hline
7.933
\end{array}
$$

So $7.132 + 3.8 - 2.999 = 7.933$.

Answers to Trial Runs

page 245 **1.** 0.17 **2.** 1.5 **3.** 21.0 **4.** 7.41 **5.** 4.29 **6.** 16.45

page 246 **1.** 5.495 **2.** 2.123 **3.** 21.388 **4.** 18.035 **5.** 15.5429 **6.** $204.53

page 247 **1.** 0.52 **2.** 5.104 **3.** 0.3593 **4.** 1.109 **5.** $113.36 **6.** $204.61

page 248 **1.** 1.33 **2.** 5.88 **3.** 3.515 **4.** 9.112 **5.** $10.13 **6.** 5.8624

page 249 **1.** 6.364 **2.** $5.19 **3.** 6.3449 **4.** $16.16

EXERCISE SET 6.2

Perform the indicated operations.

_____ 1. 0.07 + 0.08

_____ 2. 0.06 + 0.05

_____ 3. 0.9 + 0.3

_____ 4. 0.7 + 0.5

_____ 5. 15.5 + 8.7

_____ 6. 29.3 + 7.5

_____ 7. 0.21 + 1.74 + 2.39

_____ 8. 0.72 + 3.45 + 6.29

_____ 9. 6.39 + 0.76 + 0.09

_____ 10. 8.24 + 0.39 + 0.07

_____ 11. 13.76 + 5.07 + 0.08

_____ 12. 12.72 + 3.09 + 0.08

_____ 13. 8.6 + 4.026

_____ 14. 13.5 + 7.068

_____ 15. 0.398 + 2.72

_____ 16. 0.935 + 3.26

_____ 17. 8.2 + 0.298 + 13.52

_____ 18. 9.4 + 0.372 + 16.67

_____ 19. 8.3 + 2.072 + 8.27

_____ 20. 9.3 + 3.089 + 7.15

_____ 21. 13 + 4.0765 + 0.9

_____ 22. 17 + 3.0824 + 0.7

_____ 23. $235 + $9.75 + $29.98

_____ 24. $176 + $12.13 + $8.09

_____ 25. 0.89 − 0.47

_____ 26. 0.95 − 0.63

_____ 27. 9.872 − 6.451

_____ 28. 8.786 − 4.325

_____ 29. 0.3986 − 0.2742

_____ 30. 0.6835 − 0.2846

_____ 31. 13.072 − 9.835

_____ 32. 16.089 − 8.376

_____ 33. 9.0072 − 8.3684

_____ 34. 7.0035 − 3.8276

_____ 35. $365.24 − $78.79

_____ 36. $276.35 − $98.86

_____ 37. $500.78 − $98.89

_____ 38. $400.36 − $87.59

_____ 39. 8.76 − 3.9

_____ 40. 9.24 − 4.6

_____ 41. 18.7 − 13.23

_____ 42. 19.6 − 12.42

_____ 43. 71.62 − 8.0024

_____ 44. 83.53 − 9.0031

_____ 45. 8.72 − 0.364

_____ 46. 9.65 − 0.238

_____ 47. $85 − $19.79

_____ 48. $76 − $29.95

_____ 49. 12 − 3.762

_____ 50. 19 − 7.398

_____ 51. 8.763 + 2.07 − 5.036   _____ 52. 9.685 + 3.01 − 9.035

_____ 53. $12 + $5.95 − $8.56   _____ 54. $19 + $7.38 − $13.29

_____ 55. 4 − 2.03 + 2.5   _____ 56. 9 − 6.07 + 6.3

_____ 57. $100 − $13.98 − $27.45   _____ 58. $200 − $86.42 − $22.87

_____ 59. 5 + 0.7 − 3.025   _____ 60. 9 + 0.6 − 5.045

☆ Stretching the Topics

Perform the indicated operations.

_____ 1. 0.871 + 1.06 + 0.37 + 0.0035 + 3

_____ 2. 7.5 + 2 + 0.085 + 103 + 12.71 + 0.0666

_____ 3. 15.9 + 181.44 − 19.5 − 12.8

_____ 4. (2019.82 + 7065.88) − (228.73 + 3585.69)

Check your answers in the back of your book.

If you can complete **Checkup 6.2,** you are ready to go to Section 6.3.

✓ CHECKUP 6.2

Perform the indicated operations.

_____ **1.** 0.83 + 0.9

_____ **2.** 49.6 + 3.002

_____ **3.** 8 + 3.07 + 12.094

_____ **4.** $9.75 + $13 + $3.05

_____ **5.** 5.39 − 2.7

_____ **6.** 9.005 − 3.72

_____ **7.** 13 − 4.139

_____ **8.** $20 − $15.78

_____ **9.** 3.089 + 0.0845 − 1.32

_____ **10.** $50 − $7.85 − $19.46

Check your answers in the back of your book.

If You Missed Problems:	You Should Review Examples:
1, 2	1, 2
3, 4	3, 4
5–8	5–12
9, 10	13, 14

6.3 Switching from Words to Decimal Numbers

Many situations in our daily lives require that we be able to compare, add, or subtract decimal numbers.

Working with Dollars and Cents

One of the main everyday uses for decimal numbers occurs in dealing with money. Dollars and cents can always be expressed using decimal numbers.

Suppose you visit the local grocery store and purchase a package of hamburger for $1.79, rolls for 89 cents, frozen french fries for 69 cents, and a candy stick for 8 cents. To compute your total bill you must find the *sum* of all your purchases. But one of the amounts is expressed in dollars, while the others are expressed in cents. Before adding, we must change all the amounts to dollars or to cents. Since the purchases will obviously sum to more than 1 dollar, we would probably be wise to change the amounts to dollars.

$$\begin{array}{r} \$1.79 \\ 0.89 \\ 0.69 \\ +\,0.08 \\ \hline \$3.45 \end{array}$$

Your total bill is $3.45.

Suppose you give the grocery clerk a $10 bill. How much change should you receive? To compute the amount of change, you must find the *difference* between $10 and $3.45. Remember to attach zeros where necessary.

$$\begin{array}{r} \$10.00 \\ -3.45 \\ \hline \$6.55 \end{array}$$

You should receive $6.55 in change.

Example 1. If Heather buys a shirt for $14.98, pants for $17.99, a belt for $4, and socks for 99 cents, what is her total bill?

Solution. We notice that one price is expressed in cents, so we first change that amount to dollars. Then we find the sum.

$$\begin{array}{r} \$14.98 \\ 17.99 \\ 4.00 \\ +\,0.99 \\ \hline \$37.96 \end{array}$$

Heather's total bill is $37.96.

Many people pay their bills by writing **checks** on the money they keep in a checking account. The correct form for writing a check is illustrated on the following page.

```
┌─────────────────────────────────────────────────────────────────┐
│  Heather Copeland                                        2358     │
│  1234 Vine Street                                                 │
│  Hometown, N.J. 08800          March 12          19 90            │
│                                                                   │
│                                                                   │
│  Pay to ____Casual Clothier_____  $ 37.96 _____        │
│                                                                   │
│  Thirty-seven and 96/100 ————————————————————————— dollars        │
│                                                                   │
│                                    Heather Copeland                │
│  HOMETOWN NATIONAL BANK                                           │
│  Hometown, N.J. 08800                                             │
└─────────────────────────────────────────────────────────────────┘
```

Notice that the amount appears twice on a check. First it is written using a decimal number. Then it is repeated using words, with the cents expressed in fractional form. By having the check-writer repeat the amount in words, the bank hopes to prevent anyone from changing the amount once the check has been written.

You try Example 2.

Example 2. Write a check to Appliances Unlimited in the amount of $256.59.

```
┌─────────────────────────────────────────────────────────────────┐
│  Martin Ortiz                                            1309     │
│  803 Centenary Dr.                                                │
│  Anytown, KY 42100            _____  19 ____           │
│                                                                   │
│                                                                   │
│  Pay to _____  $ _____       │
│                                                                   │
│  _____ dollars        │
│                                                                   │
│                                                                   │
│  ANYTOWN BANK & TRUST CO.         _____        │
│  Anytown, KY 42100                                                │
└─────────────────────────────────────────────────────────────────┘
```

Check your work on page 261. ▶

If you have a checking account, you must keep a careful record of the checks you write and the deposits that you make to your account. The amount remaining in your account at any time is called your **balance.** To be sure that you do not spend more money than you have in your account, you should *subtract* the amount of each check as you write it and *add* the amount of each deposit as you make it.

Your record of checks and deposits is kept in your **check register.** Part of a typical check register is shown here.

CHECK NO.	DATE	CHECKS DRAWN OR DEPOSITS MADE	BALANCE FORWARD	√	91	23
827	3/10	TO *Super Drugstore*	DEDUCT CHECK − ADD DEPOSIT +		− 7	39
			BALANCE ▷		83	84
	3/11	TO *deposit paycheck*	DEDUCT CHECK − ADD DEPOSIT +		+800	00
			BALANCE ▷		883	84
828	3/13	TO *Ma Bell Telephone Co.*	DEDUCT CHECK − ADD DEPOSIT +		− 37	81
			BALANCE ▷		846	03
829	3/13	TO *Estate Farm Insurance*	DEDUCT CHECK − ADD DEPOSIT +		− 79	28
			BALANCE ▷		766	75
830	3/15	TO *Mall Apartments, Inc.*	DEDUCT CHECK − ADD DEPOSIT +		−275	00
			BALANCE ▷		491	75
831	3/15	TO *Foodtown*	DEDUCT CHECK − ADD DEPOSIT +		− 88	90
			BALANCE ▷		402	85

Notice how each deposit is added to the balance and each check is subtracted. A check register that is always kept up to date allows you to keep careful track of your expenditures. By glancing at the previous check register, we can answer such questions as

What was the balance in the account on March 12?
The balance was $883.84

On what date was the insurance premium paid?
The date on the check to Estate Farm was March 13.

How much was spent at the grocery store on March 15?
The amount of the check to Foodtown was $88.90.

With what check was the phone bill paid?
The phone bill was paid with check number 828.

Example 3. Use the following check register to record these transactions in Sid's checking account. (His next check is number 2359.)

Nov. 1: deposit $1237 paycheck
Nov. 1: mortgage payment of $489.20 to Liberty Bank
Nov. 2: payment of $30 to John Holley, M.D.
Nov. 2: car payment of $197.92 to Credit Union
Nov. 3: check for $150 cash at bank
Nov. 3: payment of $50 on Sears charge account

Solution

CHECK NO.	DATE	CHECKS DRAWN OR DEPOSITS MADE	BALANCE FORWARD ▷	√	107	86
	11/1	TO *deposit paycheck*	DEDUCT CHECK − / ADD DEPOSIT +		+ 1237	00
			BALANCE ▷		1344	86
2359	11/1	TO *Liberty Bank* *mortgage*	DEDUCT CHECK − / ADD DEPOSIT +		− 489	20
			BALANCE ▷		855	66
2360	11/2	TO *John Holley, MD*	DEDUCT CHECK − / ADD DEPOSIT +		− 30	00
			BALANCE ▷		825	66
2361	11/2	TO *Credit Union* *car payment*	DEDUCT CHECK − / ADD DEPOSIT +		− 197	92
			BALANCE ▷		627	74
2362	11/3	TO *Cash*	DEDUCT CHECK − / ADD DEPOSIT +		− 150	00
			BALANCE ▷		477	74
2363	11/3	TO *Sears* *Charge account*	DEDUCT CHECK − / ADD DEPOSIT +		− 50	00
			BALANCE ▷		427	74

Now complete Example 4, which is the problem stated at the beginning of this chapter.

Example 4. Martie has recorded her checking account transactions in her check register but she has not done the necessary subtractions. Do them for her and find the balance in her account.

CHECK NO.	DATE	CHECKS DRAWN OR DEPOSITS MADE	BALANCE FORWARD ▷	√	339	17
329	6/3	TO *Casual Clothiers* *dress*	DEDUCT CHECK − / ADD DEPOSIT +		− 33	60
			BALANCE ▷			
330	6/4	TO *Cash*	DEDUCT CHECK − / ADD DEPOSIT +		− 20	00
			BALANCE ▷			
331	6/5	TO *Hometown Gas Co.* *gas bill*	DEDUCT CHECK − / ADD DEPOSIT +		− 59	28
			BALANCE ▷			
332	6/5	TO *Enrico's* *dinner*	DEDUCT CHECK − / ADD DEPOSIT +		− 8	57
			BALANCE ▷			
333	6/6	TO *Global Insurance Co.* *life insurance*	DEDUCT CHECK − / ADD DEPOSIT +		− 139	00
			BALANCE ▷			
334	6/7	TO *Shoe Shack* *shoes*	DEDUCT CHECK − / ADD DEPOSIT +		− 40	80
			BALANCE ▷			

Check your work on page 261. ▶

⇒ Trial Run

1. Write the amount $309.95 in words as it would appear on a check.
   _____ dollars

2. Write the amount three thousand twenty-five and 15/100 dollars in numbers as it would appear on a check. $_____

_____ 3. Bud made purchases of $32.98, $23.15, and $7.83 at Pzitz Department Store. If he decides to use a check for the total amount, for how much should he write it?

_____ 4. Edwina earns $253.85 per week but deductions of $25.39, $15.32, and $6 are made before her check is written. What is the amount of her check?

_____ 5. On July 2, Sandy had a balance of $84.30. During the next week, the following transactions were recorded in her check register.
   July 3:   check for $23.87 to Meacham's Hams
   July 4:   check for $18.73 to Farmer's Market
   July 6:   deposit of $135.75
   July 8:   check for $62.60 to Leoni Insurance Co.
   What was Sandy's balance on July 9?

Answers are on page 261.

Working with Other Word Statements

Decimal numbers appear in everyday situations other than those involving money. Let's consider a few word problems in which we must add, subtract, or compare decimal numbers.

Example 5. At the start of his vacation trip, Barry's car odometer read 37823.7 miles. At the end of the trip, the reading was 39010.2 miles. How far did Barry drive?

Solution. We must find the *difference* between the two readings, so we subtract.

$$
\begin{array}{r}
39010.2 \\
-37823.7 \\
\hline
1186.5
\end{array}
$$

Barry traveled 1186.5 miles.

Example 6. Marie is making chili for her party and the recipe calls for 6 pounds of ground beef. In her freezer she discovers 4 packages of ground beef weighing 1.37 lb, 1.98 lb, 2.07 lb, and 1.01 lb. Does she have enough ground beef?

Solution. First we must find out how many pounds of ground beef she has by *adding* all the package weights.

$$
\begin{array}{r}
1.37 \\
1.98 \\
2.07 \\
+1.01 \\
\hline
6.43
\end{array}
$$

Now we must compare 6.43 pounds to the 6 pounds she needs. We agree that

$$6.43 > 6.00$$

so Marie has enough ground beef.

Example 7. Ms. Patton wishes to buy fringe to sew around the edge of a rectangular tablecloth. If the tablecloth is 2.1 meters long and 1.3 meters wide, how much fringe should she buy?

Solution. Let's illustrate the tablecloth.

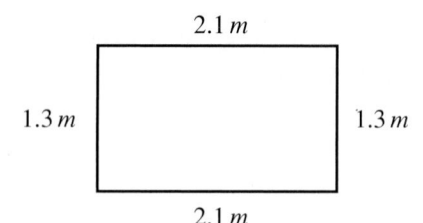

2.1 m

1.3 m   1.3 m

2.1 m

Since the fringe will go around the edges of the tablecloth, we must find the total distance around the outside of the rectangle, or its *perimeter*. We find the perimeter by *adding* the lengths of the 4 sides.

$$
\begin{array}{r}
2.1 \\
1.3 \\
2.1 \\
+\,1.3 \\
\hline
6.8
\end{array}
$$

Ms. Patton should buy 6.8 meters of fringe.

You complete Example 8.

Example 8. Beth and her Dad jogged the following distances over the past 5 days.

Beth: 2.8 mi, 3.1 mi, 2.25 mi, 3 mi, 2.9 mi
Dad: 3.4 mi, 2 mi, 2.7 mi, 3.3 mi, 2.9 mi

Who jogged farther? How much farther?

Solution. First we must find the total distance for each jogger.

	Beth	*Dad*
	2.80	3.4
	3.10	2.0
	2.25	2.7
	3.00	3.3
	+2.90	+2.9

Now we must compare _____ miles and _____ miles. We see that _____ < _____ , so we agree that _____ jogged farther. To find out *how much* farther, we must _____ :

Therefore, _____ jogged _____ miles farther than _____ .

Check your work on page 261. ▶

▶ **Examples You Completed** ⎯⎯⎯⎯⎯⎯⎯⎯⎯⎯⎯⎯⎯⎯⎯⎯⎯

Example 2

Martin Ortiz	1309
803 Centenary Dr.	
Anytown, KY 42110	*October 25* 19 *85*

Pay to *Appliances Unlimited* $ *256.59*

Two hundred fifty-six and 59/100⎯⎯⎯⎯ dollars

ANYTOWN BANK & TRUST CO.   *Martin Ortiz*
Anytown, KY 42100

Example 4

CHECK NO.	DATE	CHECKS DRAWN OR DEPOSITS MADE		BALANCE FORWARD ▷	✓	339	17
329	6/3	TO *Casual Clothiers*	*dress*	DEDUCT CHECK − / ADD DEPOSIT +		− 33	60
				BALANCE ▷		305	57
330	6/4	TO *Cash*		DEDUCT CHECK − / ADD DEPOSIT +		− 20	00
				BALANCE ▷		285	57
331	6/5	TO *Hometown Gas Co.*	*gas bill*	DEDUCT CHECK − / ADD DEPOSIT +		− 59	28
				BALANCE ▷		226	29
332	6/5	TO *Enrico's*	*dinner*	DEDUCT CHECK − / ADD DEPOSIT +		− 8	57
				BALANCE ▷		217	72
333	6/6	TO *Global Insurance Co.*	*life insurance*	DEDUCT CHECK − / ADD DEPOSIT +		−139	00
				BALANCE ▷		78	72
334	6/7	TO *Shoe Shack*	*shoes*	DEDUCT CHECK − / ADD DEPOSIT +		− 40	80
				BALANCE ▷		37	92

Martie has a balance of $37.92 in her checking account.

Example 8 (*Solution*). First we must find the total distance for each jogger.

Beth	*Dad*
2.80	3.4
3.10	2.0
2.25	2.7
3.00	3.3
+2.90	+2.9
14.05 miles	14.3 miles

Now we must compare 14.05 miles and 14.3 miles. We see that 14.05 < 14.3, so we agree that Dad jogged farther. To find out *how much* farther, we must *subtract*: 14.30 − 14.05 = 0.25. Therefore, Dad jogged 0.25 miles farther than Beth.

Answers to Trial Run ⎯⎯⎯⎯⎯⎯⎯⎯⎯⎯⎯⎯⎯⎯⎯⎯⎯⎯⎯⎯⎯

page 259   **1.** Three hundred nine and $\frac{95}{100}$ dollars.   **2.** $3025.15   **3.** $63.96   **4.** $207.14
5. $114.85

EXERCISE SET 6.3

Write the following amounts in words as they would appear on a check.

_____ dollars   **1.** $423.78

_____ dollars   **2.** $586.83

_____ dollars   **3.** $2036

_____ dollars   **4.** $5009

_____ dollars   **5.** $15.50

_____ dollars   **6.** $18.25

Write the following amounts in decimal numbers as they would appear on a check.

$_____   **7.** Fifty-nine and $\frac{75}{100}$ dollars.

$_____   **8.** Sixty-eight and $\frac{98}{100}$ dollars.

$_____   **9.** Three thousand nine dollars.

$_____   **10.** Seven thousand eleven dollars.

$_____   **11.** Three hundred twenty and $\frac{1}{100}$ dollars.

$_____   **12.** Five hundred thirty and $\frac{2}{100}$ dollars.

13. Use the following check register to record these transactions in Aaron's checking account and find his final balance. His balance on April 30 is $45.87 and his next check number is 215.

   May 1: deposit $786.03 paycheck
   May 1: rent payment of $265
   May 2: payment of $73.58 to Municipal Utilities
   May 3: car payment of $168.93 to Union Bank
   May 3: payment of $25 to Visa
   May 5: payment to Saveway Market of $43.88
   May 5: deposit $28.96 travel expense check

CHECK NO.	DATE	CHECKS DRAWN OR DEPOSITS MADE	BALANCE FORWARD ⟹	√		
		TO	DEDUCT CHECK − ADD DEPOSIT +			
			BALANCE ⟹			
		TO	DEDUCT CHECK − ADD DEPOSIT +			
			BALANCE ⟹			
		TO	DEDUCT CHECK − ADD DEPOSIT +			
			BALANCE ⟹			
		TO	DEDUCT CHECK − ADD DEPOSIT +			
			BALANCE ⟹			
		TO	DEDUCT CHECK − ADD DEPOSIT +			
			BALANCE ⟹			
		TO	DEDUCT CHECK − ADD DEPOSIT +			
			BALANCE ⟹			
		TO	DEDUCT CHECK − ADD DEPOSIT +			
			BALANCE ⟹			

14. Use the next check register to record the following transactions in Lydia Vanderbilt's checking account and find her final balance. Her balance on February 28 was $396.82 and her next check number is 2736.

   March 1: deposit alimony check, $1500
   March 2: payment to Andre's Boutique, $123.78
   March 2: payment to Pippin's Hair Salon, $25.00
   March 3: deposit stock dividend check, $1115.88
   March 4: payment to American Express, $250.00
   March 5: payment to Theater Tickets, Inc., $86.93
   March 5: payment to Cedar Hall Country Club, $575.00

CHECK NO.	DATE	CHECKS DRAWN OR DEPOSITS MADE	BALANCE FORWARD ▷	√		
		TO	DEDUCT CHECK − / ADD DEPOSIT +			
			BALANCE ▷			
		TO	DEDUCT CHECK − / ADD DEPOSIT +			
			BALANCE ▷			
		TO	DEDUCT CHECK − / ADD DEPOSIT +			
			BALANCE ▷			
		TO	DEDUCT CHECK − / ADD DEPOSIT +			
			BALANCE ▷			
		TO	DEDUCT CHECK − / ADD DEPOSIT +			
			BALANCE ▷			
		TO	DEDUCT CHECK − / ADD DEPOSIT +			
			BALANCE ▷			
		TO	DEDUCT CHECK − / ADD DEPOSIT +			
			BALANCE ▷			

_____ 15. On a shopping trip to Nashville, Maxine wrote checks for $75.98, $32.79, $128.93, and $15.34. If her checking balance before she left was $584.17, how much did she have in her account after shopping?

_____ 16. When Linda sat down to write checks for her monthly bills she had $1378.15 in her checking account. After writing checks for $287.19, $75.83, $186.29, and $16.97, what was her balance?

_____ 17. The amount of precipitation (in inches) for Newburg during the first 6 months of the year was as follows: January, 2.6; February, 2.8; March, 3.3; April, 3.5; May, 3.2; June 2.9. Find the total inches of rainfall for the first 6 months of the year.

_____ 18. For lunch Olivia had a super burger for $1.85, salad bar for $1.69, and a carton of milk for $0.35. The sales tax was $0.19. Find the total cost of lunch. How much change did she receive from a ten-dollar bill?

_____ 19. When Mario had mononucleosis his temperature at the doctor's office was 102.7 degrees. By the next morning, after he had begun medication, his temperature was 99.3 degrees. By how many degrees had his temperature fallen over night?

_____ 20. In 1982, Johncock won the Indianapolis 500 with an average speed of 162.026 miles per hour. The year before Unser won, averaging 139.085 miles per hour. How much faster was Johncock's car than Unser's?

_____ 21. Judson ordered several items from Penney's catalog with the following weights (in pounds): 2.3, 7.9, 1.3, 0.5, and 4.6. Find the total number of pounds on which he must pay shipping fees.

_____ 22. The maximum speed of a rabbit over a quarter-mile distance is 39.35 miles per hour. A quarterhorse can cover the same distance 8.15 miles per hour faster. Find the speed of a quarterhorse over the quarter-mile.

_____ 23. Walter can throw the discus 58.7 meters but Ted can throw it 60.26 meters. How much farther can Ted throw the discus than Walter?

_____ 24. In 4 days of driving, Lana traveled 200.9, 158.7, 369.4, and 276 miles. Find the total distance driven in 4 days.

_____ 25. Last semester, Audrey's grade point average dropped from 3.82 to 2.74. By how many points did her average drop?

☆ Stretching the Topics ────────────────────

_____
_____
_____
_____ 1. Anita spent the following amounts on items for lunch: $1.89, $0.74, $1.03, and $0.97. Her brother Carlos spent $2.08, $0.87, $1.15, and $1.10 for lunch. Find the total Anita spent for lunch. Find the amount Carlos spent. Who spent more for lunch and how much more? If they had a ten-dollar bill to pay for both lunches, how much change did they receive?

_____ 2. On his first 4 dives, Greg scored 7.9, 8.0, 6.3, and 8.4. What must he score on his fifth dive to have an average of at least 8.0?

_____
_____
_____
_____
_____ 3. The barometer readings for 5 days were Monday, 30.9; Tuesday, 31.2; Wednesday, 30.5; Thursday, 29.4; Friday, 29.7. Which day had the highest reading? Which day had the lowest reading? Between which 2 consecutive days did the greatest change in the barometer readings occur? How much was that change? Find the difference between the highest and lowest readings.

Check your answers in the back of your book.

If you can complete **Checkup 6.3,** you are ready to do the **Review Exercises** for Chapter 6.

✔ CHECKUP 6.3

_____ dollars   **1.** Write $73.85 as it would appear in words on a check.

$_____   **2.** Write the amount five hundred six dollars and three cents as it would appear in decimal numbers on a check.

_____   **3.** At the beginning of the month Camiela had a balance of $343.83. Her total deposits for the month were $984.36 and her total withdrawals were $1123.14. A service charge of $4.50 was also deducted. Find Cam's balance at the end of the month.

_____   **4.** For the top of the table he is building Wilburn glues together 2 pieces of wood which are 1.34 inches and 0.752 inch thick and a piece of formica which is 0.25 inch thick. Find the thickness of the tabletop.

_____   **5.** The record qualifying speed for ''Thunder on the Ohio V'' Hydroplane Races was set by Chip Hanover at 133.050 miles per hour. The slowest qualifying speed was 95.119 miles per hour. Find the difference between the slowest and fastest qualifying speeds.

Check your answers in the back of your book.

If You Missed Problems:	You Should Review Examples:
1, 2	1, 2
3	3, 4
4	6, 7
5	8

Summary

In this chapter we learned to identify the parts of a decimal number.

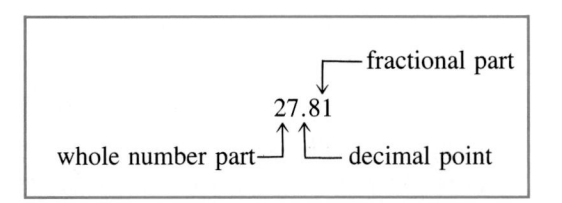

Then we discussed the place values for the fractional part of decimal numbers and learned to write them in word form and fractional form.

Decimal Form	Word Form	Fractional Form
0.9	9 tenths	$\dfrac{9}{10}$
0.07	7 hundredths	$\dfrac{7}{100}$
0.123	123 thousandths	$\dfrac{123}{1000}$
6.47	6 and 47 hundredths	$6\dfrac{47}{100} = \dfrac{647}{100}$

We discovered that we could attach zeros after the last nonzero digit in the fractional part of any decimal number without changing the value of the number. We found this fact very useful in adding, subtracting, and comparing decimal numbers.

We then learned to **round** decimal numbers to a particular place by looking at the digit in the place immediately to the right of the rounding place.

Rounded to the	The number 6.2457 becomes
ones place	6
tenths place	6.2
hundredths place	6.25
thousandths place	6.246

We learned to add and subtract decimal numbers by lining up their decimal points directly beneath each other and performing the addition and subtraction in the usual way. We attached zeros when necessary and used the techniques of ''carrying'' and ''borrowing'' as learned earlier with whole numbers.

To find	$3.7 + 8.59$	$\$15.10 - \$7.87$
We write	$\begin{array}{r} 3.70 \\ +8.59 \\ \hline 12.29 \end{array}$	$\begin{array}{r} \$15.10 \\ -7.87 \\ \hline \$7.23 \end{array}$

Finally, we practiced switching from problems stated in words to problems involving decimal numbers. We paid particular attention to changing cents to dollars, making use of the fact that there are 100 cents in 1 dollar.

❏ Speaking the Language of Mathematics

Complete each statement with the appropriate word or phrase.

1. The part to the right of the decimal point in a decimal number is called its _____ part.

2. In the decimal number 13.57, the 5 occupies the _____ place and the 7 occupies the _____ place.

3. To add or subtract decimal numbers, we must line up their _____ _____ .

4. To round a decimal number to the hundredths place, we first look at the digit in the _____ place.

5. To change cents to dollars, we remember that _____ cents equals _____ dollar.

△ Writing About Mathematics

Write your response to each question in complete sentences.

1. Explain how you would round the decimal number 2.7863 to the hundredths place.

2. Describe how you would subtract a decimal number, with 3 places in its fractional part, from a whole number.

3. Why do you think it is important to keep an up-to-date check register?

REVIEW EXERCISES for Chapter 6

Write each decimal number as a fraction.

_____ 1. 0.045   _____ 2. 0.31

Write each fraction as a decimal number.

_____ 3. $\dfrac{28}{1000}$ _____ 4. $\dfrac{9}{10,000}$

Write each decimal number in words.

_____ 5. 0.0783   _____ 6. 7.802

_____ 7. Rewrite 9.3 as a decimal number that ends in the thousandths place.

_____ 8. Rewrite 1.2 as a decimal number that ends in the hundredths place.

Compare the numbers using < or >.

_____ 9. 6.02 and 6.0195   _____ 10. 0.00486 and 0.0386

_____ 11. 0.397 and 0.4   _____ 12. 4.098 and 4.1

_____ 13. Write the numbers in order from smallest to largest: 3.26, 3.09, 3.029.

_____ 14. Round 16.072 to the tenths place.

_____ 15. Round 0.65432 to the thousandths place.

_____ 16. Round 189.6845 to the hundredths place.

_____ 17. Round 0.0467092 to the ten thousandths place.

Perform the indicated operations.

_____ 18. 19.23 + 8.07   _____ 19. 14.2 + 0.039

_____ 20. 9.63 + 0.87 + 5.32   _____ 21. 3.6 + 0.294 + 31.25

_____ 22. 3.07 + 0.892 + 16   _____ 23. 17.65 + 0.739 + 22.3

_____ 24. $723 + $51.38 + $9.26   _____ 25. $815.13 + $2342.50 + $78.09

_____ 26. 0.97 − 0.63   _____ 27. 8.769 − 6.342

_____ 28. 8.0093 − 6.2476   _____ 29. $300.23 − $196.89

_____ 30. 6.32 − 4.9   _____ 31. 73.26 − 4.0082

_____ 32. 9.37 − 0.8936   _____ 33. $96 − $45.16

_____ **34.** 9.832 + 1.05 − 6.072   _____ **35.** $15 + $12.83 + $13.97

_____ **36.** 8 − 3.02 + 2.6   _____ **37.** $500 − $123.55 − $38.05

Write the following amounts in words as they would appear on a check.

_____ dollars **38.** $329.07

_____ dollars **39.** $7,006.25

Write the following amounts in decimal numbers as they would appear on a check.

$_____ **40.** Four hundred eight and $\frac{6}{100}$ dollars.

$_____ **41.** Seven thousand three hundred sixty-eight and $\frac{65}{100}$ dollars.

_____ **42.** Carlita had $1578.13 in her checking balance before paying her monthly bills. After writing checks for $300, $13.25, $73.80, $26.87, and $178.38, what was her balance?

_____ **43.** A football field is 109.68 meters long and 48.72 meters wide. Find the distance around (perimeter) the football field.

_____ **44.** The odometer on Tony's car reads 39,386.7 miles. If he plans to have his car serviced when the odometer reads 41,500 miles, how much farther can he drive before servicing his car?

_____ **45.** The cost of a certain mail order item is $19.95. The postage is $1.84 and the handling charge is $2. Find the total cost including postage and handling.

Check your answers in the back of your book.

If You Missed Exercises:	You Should Review Examples:	
1–4	Section 6.1	1, 2
5–8		3–5
9–13		6, 7
14–17		8–10
18–25	Section 6.2	1–4
26–33		5–12
34–37		13, 14
38–42	Section 6.3	1–4
43–45		5–8

If you have completed the **Review Exercises** and corrected your errors, you are ready to take the **Practice Test** for Chapter 6.

PRACTICE TEST for Chapter 6

		SECTION	EXAMPLE
_____	1. Write 0.0397 as a fraction.	6.1	1
_____	2. Write $\frac{8}{100}$ as a decimal number.	6.1	2
_____	3. Write 502.06 in words.	6.1	3
_____	4. Rewrite 8.23 as a decimal number that ends in the ten thousandths place.	6.1	4, 5
_____	5. Compare 87.034 and 87.304 using < or >.	6.1	6
_____	6. Round 8.349 to the tenths place.	6.1	8
_____	7. Round 0.57042 to the thousandths place.	6.1	9

Perform the indicated operations.

		SECTION	EXAMPLE
_____	8. 19.6 + 8.2	6.2	1
_____	9. 0.72 + 5.61 + 0.45	6.2	2
_____	10. 9 + 3.07 + 0.2	6.2	3
_____	11. 35 + 0.623 + 10.3	6.2	3
_____	12. $225 − $89.73 + $8.07	6.2	4
_____	13. 8.953 − 6.702	6.2	5
_____	14. $586.45 − $496.59	6.2	8
_____	15. 12.396 − 6.95	6.2	9
_____	16. 27.3 − 16.934	6.2	10
_____	17. $75 − $38.92	6.2	11
_____	18. 19.362 + 8.5 − 3.988	6.2	13
_____	19. 42.4 − 18.3 − 9.745	6.2	14
_____	20. $300 − $192.35 − $68	6.2	13, 14

		SECTION	EXAMPLE

_____ 21. Write $98.89 in words as it would appear on a check.   6.3   1, 2

_____ 22. Oliver would like to be able to throw the discus 65 meters. On the   6.3   5
first try he throws the discus 61.26 meters. How much farther must
Oliver throw if he is to reach his goal on the second try?

_____ 23. In 3 weeks of driving, Kalani used 18.6, 15.5, and 21.7 gallons of   6.3   6
gasoline. If her budget only allows her to use 70 gallons per month,
how many gallons can she use during the fourth week?

_____ 24. Gene is purchasing framing for his new rectangular window that is   6.3   7
1.8 meters long and 1.2 meters wide. How much framing should
he buy?

_____ 25. During the first 4 months of the year, the following inches of rainfall   6.3   8
_____   were recorded in 2 cities: Grove City recorded 3.6, 4.5, 2.7, and 4.6
inches, and Waverly recorded 3.7, 3.9, 2.1, and 3.8 inches. Which
city had more rainfall? How much more rainfall did that city have?

SHARPENING YOUR SKILLS after Chapters 1–6

SECTION

_____ 1. Write the word form for 507,209.   1.1

_____ 2. Classify 252 as prime or composite. If the number is composite, write it as a product of prime numbers.   2.3

_____ 3. Change $24\frac{2}{3}$ to an improper fraction.   3.1

_____ 4. Reduce $\frac{456}{612}$ to lowest terms.   3.2

_____ 5. Use $<$, $>$, or $=$ to compare $7\frac{1}{9}$ and $\frac{23}{3}$.   5.4

_____ 6. Write 0.0423 as a fraction.   6.1

_____ 7. Round 7.096 to the hundredths place.   6.1

Perform the indicated operations.

_____ 8. $3728 + 50{,}007 + 15{,}372$ 1.2

_____ 9. $5000 - 3786$ 1.3

_____ 10. $15 - 11\frac{2}{3}$ 5.3

_____ 11. $\frac{7}{8} + \frac{3}{5} - \frac{5}{6}$ 5.3

_____ 12. $7\frac{1}{2} - 5\frac{3}{4} + 8\frac{2}{3}$ 5.3

_____ 13. $0.72 + 9 + 3.4$ 6.2

_____ 14. $\$115 - \$78.98$ 6.2

_____ 15. $29.03 - 5.6 - 18.75$ 6.2

_____ 16. $15 \times 25 \times 200$ 2.1

_____ 17. $12\frac{2}{3} \times 13\frac{1}{2}$ 4.1

_____ 18. $415{,}761 \div 407$ 2.2

_____ 19. $15 \div \frac{2}{3}$ 4.2

_____ 20. $\dfrac{8\frac{3}{4}}{1\frac{3}{7}}$ 4.2

_____ 21. Irene's bank balance is $872. She writes checks for $39, $125, and $78. What is her   1.4
new balance?

_____ 22. If Olivia drove 65 miles per hour for 2 hours and 55 miles per hour for 3 hours, how   2.4
far did she drive?

_____ 23. Travis lives 24 blocks from work. If he walks 4 blocks to catch a bus, for what   3.3
fractional part of the total distance does he ride the bus?

_____ 24. Find the area of a square patio that measures $15\frac{2}{3}$ feet on each side.   4.3

_____ 25. Richard uses $\frac{1}{4}$ of his salary for rent, $\frac{1}{6}$ for food, and $\frac{1}{5}$ for utilities. What fractional part   5.5
of his salary is left for other things?

Check your answers in the back of your book.

CHAPTER

Multiplying and Dividing Decimal Numbers

If Scott Spann can swim the 200-yard breaststroke in 2.02 minutes and Graham Smith can swim the same race in 2 minutes 2 seconds, who is the faster swimmer?

Now that we understand the meaning of decimal numbers and have learned how to compare, add, and subtract decimal numbers, we must tackle the two remaining operations of arithmetic: multiplication and division.

Our work with multiplication and division of fractional numbers will be very helpful as we learn how to

1. Multiply decimal numbers.
2. Divide decimal numbers.
3. Change fractional numbers to decimal numbers.
4. Switch from word statements to decimal number statements.

The methods for using a calculator to solve problems of the types encountered in Chapters 3–7 are discussed in Sections 16.1 and 16.2 of Chapter 16.

7.1 Multiplying Decimal Numbers

Perhaps if we recall that decimal numbers can be written as fractional numbers, we will be able to arrive at a method for multiplying decimal numbers. For instance,

$$0.3 = \frac{3}{10} \qquad 0.17 = \frac{17}{100}$$

$$0.2 = \frac{2}{10} \qquad 0.039 = \frac{39}{1000}$$

Learning to Multiply with Decimal Numbers

To multiply decimal numbers such as 0.3×0.2, we may rewrite the decimal numbers as fractions, and find the product by the rules for multiplying fractions.

$$\frac{3}{10} \times \frac{2}{10} = \frac{6}{100}$$

So $0.3 \times 0.2 = 0.06$.

Let's try another multiplication before we decide upon a rule for multiplying decimal numbers. Let's find this product:

$$0.5 \times 0.07 = \frac{5}{10} \times \frac{7}{100}$$

$$= \frac{35}{1000}$$

So $0.5 \times 0.07 = 0.035$.

We have found that

$$0.2 \times 0.3 = 0.06$$

$$0.5 \times 0.07 = 0.035$$

In each product, we can see that the digits in the answer are found by multiplying the digits in the original numbers. But how do we decide upon the *place* where those digits belong? How do we decide upon the location of the decimal point?

Notice that in each product, the number of places to the right of the decimal point is equal to the *sum* of the number of places to the right of the decimal point in the factors being multiplied. For instance, in the product $0.3 \times 0.2 = 0.06$,

0.3   contains *1* place to the right of the decimal point

0.2   contains *1* place to the right of the decimal point

0.06 contains *2* places to the right of the decimal point

Likewise, in the product $0.5 \times 0.07 = 0.035$,

0.5   contains *1* place to the right of the decimal point

0.07   contains *2* places to the right of the decimal point

0.035 contains *3* places to the right of the decimal point

To Multiply Decimal Numbers

1. Multiply the digits (ignoring the decimal points).
2. Count the total number of places to the right of the decimal point in all of the numbers being multiplied. This total will tell how many places there will be to the right of the decimal point in the answer.
3. Locate the decimal point in the answer by "counting off" the places from right to left, starting with the rightmost digit in the answer.

Let's use this method to multiply 0.2×0.8. Perhaps we should find the product vertically.

$$\begin{array}{r} 0.8 \quad \text{(1 place)} \\ \times 0.2 \quad \text{(1 place)} \\ \hline .16 \quad \text{(2 places)} \end{array}$$

So $0.2 \times 0.8 = 0.16$.

If necessary, we may attach zeros in front of our product if they are needed to reach the correct location of the decimal point. Consider this example.

$$\begin{array}{r} 0.182 \quad \text{(3 places)} \\ \times \ 0.05 \quad \text{(2 places)} \\ \hline .00910 \quad \text{(5 places)} \end{array}$$

Therefore, $0.182 \times 0.05 = 0.00910$.

To multiply decimal numbers in which the multiplication requires more than one step, we proceed as we did with whole numbers. We do not worry about the location of the decimal point until we have found the product of the digits.

Example 1. Find 0.036×0.14.

Solution

$$\begin{array}{r} 0.036 \quad \text{(3 places)} \\ \times \ 0.14 \quad \text{(2 places)} \\ \hline 144 \\ 36 \quad \ \ \\ \hline .00504 \quad \text{(5 places)} \end{array}$$

So $0.036 \times 0.14 = 0.00504$.

You complete Example 2.

Example 2. Find 53.25×4.2.

Solution

$$\begin{array}{r} 53.25 \quad \text{(2 places)} \\ \times \ 4.2 \quad \text{(1 place)} \\ \hline \\ \hline \text{(3 places)} \end{array}$$

So $53.25 \times 4.2 = $ _____.

Check your work on page 281. ▶

Example 3. Find 2.00203×0.0115.

Solution

$$\begin{array}{r} 2.00203 \quad \text{(5 places)} \\ 0.0115 \quad \text{(4 places)} \\ \hline 1001015 \\ 200203 \quad \ \\ 200203 \quad \quad \ \\ \hline .023023345 \quad \text{(9 places)} \end{array}$$

So $2.00203 \times 0.0115 = 0.023023345$.

🏃➡ **Trial Run** ━━

Find each product.

_____ 1. 0.024×0.37 _____ 2. 8.93×0.018

_____ 3. 35.6×3.21 _____ 4. 57.03×0.006

_____ 5. 4.235×0.0046 _____ 6. 272.5×0.08

Answers are on page 281.

Using Shortcuts in Multiplying Decimal Numbers

In Chapter 2, we learned that we could omit some unnecessary steps in multiplying whole numbers containing the digit 0. We may use the same technique in multiplying a decimal number by a number with zeros at the end.

$$
\begin{array}{rl}
7.56 & \text{(2 places)} \\
\underline{\times\ \ 21\ 00} & \text{(0 places)} \\
756\ 00 & \\
\underline{1512\quad\ } & \\
15876.00 & \text{(2 places)}
\end{array}
$$

So $7.56 \times 2100 = 15{,}876$.

Notice that we located the decimal point in our answer by the usual method of counting places. Notice also that we were careful to line up the rightmost digit of each product beneath the digit we were multiplying by.

You try Example 4.

Example 4. Use the shortcut to find 0.0739×500.

Solution

$$
\begin{array}{rl}
0.0739 & \text{(4 places)} \\
\underline{\times\qquad 500} & \text{(0 places)} \\
 & \text{(4 places)}
\end{array}
$$

So $0.0739 \times 500 =$ _____

Check your work on page 281. ▶

Example 5. Use the shortcut to find 5.3209×1000.

Solution

$$
\begin{array}{rl}
5.3209 & \text{(4 places)} \\
\underline{\times\qquad 1000} & \text{(0 places)} \\
5320.9000 & \text{(4 places)}
\end{array}
$$

So $5.3209 \times 1000 = 5320.9$.

The product found in Example 5 is especially interesting because the digits in the answer are the same digits appearing in one of the factors. The other factor contained the digit 1 followed by 3 zeros. Naturally, when we multiplied the digit 1 times the digits in the other factor, those digits remained the same. But the decimal point did *not* stay in the same location. In fact, it moved 3 places to the right. It is not just a coincidence that when we multiplied 5.3209 by 1 followed by *3* zeros the decimal point moved *3* places to the right. This will *always* occur.

To multiply a decimal number by a whole number that is 1 followed by a certain number of zeros, we may simply move the decimal point in the decimal number that certain number of places to the *right*.

Perhaps a chart will help make this handy shortcut clearer.

To Multiply a Decimal Number by	Move the Decimal Point Right	Examples
10	1 place	$8.32 \times 10 = 83.2$
100	2 places	$0.73 \times 100 = 73.$
1000	3 places	$0.0061 \times 1000 = 006.1 = 6.1$
10,000	4 places	$0.237 \times 10,000 = 2370.$

We could continue this chart on and on, but the message is clear. Count the number of zeros following the 1 to decide how many places to move the decimal point to the right.

Example 6. Use the shortcut to find $0.07 \times 100,000$.

Solution. We count 5 zeros following the 1, so

$$0.07 \times 100,000 = 07000.$$
$$= 7000$$

Notice that we may need to attach zeros at the end of our number to reach the correct location of the decimal point.

You complete Example 7.

Example 7. Use the shortcut to find 0.00295×100.

Solution. We count 2 zeros following the 1, so

$$0.00295 \times 100 =$$
$$= \_\_\_\_$$

Check your work on page 281. ▶

⟫ Trial Run

Use shortcuts to find each product.

_____ 1. 0.0873×600

_____ 2. 9.7642×1000

_____ 3. $0.04 \times 10,000$

_____ 4. 0.00357×100

_____ 5. 45.37×1000

_____ 6. 0.03876×10

Answers are on page 281.

We have discussed shortcuts to use when multiplying a decimal number by a whole number ending in zeros. Now let's recall the shortcut used to multiply whole numbers containing 0 as a digit in some other place. We may use the same method to multiply decimal numbers containing zeros as digits. We continue to line up each product carefully, and we locate the decimal point in the answer by the usual rule.

Example 8. Use shortcuts to find 13.009 × 0.071.

Solution

$$
\begin{array}{rl}
0.071 & \text{(3 places)} \\
\times 13.009 & \text{(3 places)} \\
\hline
639 & \\
213 & \\
71 & \\
\hline
.923\ 639 & \text{(6 places)}
\end{array}
$$

So 13.009 × 0.071 = 0.923639.

Example 9. Use shortcuts to find 7.32 × 100900.

Solution

$$
\begin{array}{rl}
7.32 & \text{(2 places)} \\
\times 1009\ 00 & \text{(0 places)} \\
\hline
6588\ 00 & \\
732 & \\
\hline
738588.00 & \text{(2 places)}
\end{array}
$$

So 7.32 × 100,900 = 738,588.

Example 10. Use shortcuts to multiply 11.86 × 2.003 × 1000.

Solution. We know that we may multiply 3 numbers in any order we choose. Looking at these numbers, we see that we can multiply the last 2 numbers easily, using a shortcut.

$$11.86 \times \underbrace{2.003 \times 1000}$$

$$= 11.86 \times 2003$$

Notice that this choice reduced the number of decimal places to keep track of. Now we finish the multiplication.

$$
\begin{array}{rl}
11.86 & \text{(2 places)} \\
\times 20\ 03 & \text{(0 places)} \\
\hline
35\ 58 & \\
2372 & \\
\hline
23755.58 & \text{(2 places)}
\end{array}
$$

Therefore, 11.86 × 2.003 × 1000 = 23,755.58.

⮞ Trial Run

Find each product.

_____ 1. 7.36 × 2.07

_____ 2. 0.085 × 15.006

_____ 3. 6.24 × 20,030

_____ 4. 34.29 × 5070

_____ 5. 5.72 × 3.005 × 1000

_____ 6. 0.472 × 0.503 × 100

Answers are on page 281.

▶ **Examples You Completed** _____

Example 2. Find 53.25×4.2.

Solution

$$
\begin{array}{r}
53.25 \quad \text{(2 places)} \\
\times \quad 4.2 \quad \text{(1 place)} \\
\hline
10\ 650 \\
213\ 00 \\
\hline
223.650 \quad \text{(3 places)}
\end{array}
$$

So $53.25 \times 4.2 = 223.65$.

Example 4. Use the shortcut to find 0.0739×500.

Solution

$$
\begin{array}{r}
0.0739 \quad \text{(4 places)} \\
\times \quad 500 \quad \text{(0 places)} \\
\hline
36.9500 \quad \text{(4 places)}
\end{array}
$$

So $0.0739 \times 500 = 36.95$.

Example 7. Use the shortcut to find 0.00295×100.

Solution. We count 2 zeros following the 1, so

$$0.00295 \times 100 = 00.295$$
$$= 0.295$$

Answers to Trial Runs _____

page 278   1. 0.00888   2. 0.16074   3. 114.276   4. 0.34218   5. 0.019481   6. 21.8

page 279   1. 52.38   2. 9764.2   3. 400   4. 0.357   5. 45,370   6. 0.3876

page 280   1. 15.2352   2. 1.27551   3. 124,987.2   4. 173,850.3   5. 17,188.6   6. 23.7416

EXERCISE SET 7.1

Find the products.

_____ **1.** 0.034 × 0.72

_____ **2.** 0.047 × 0.39

_____ **3.** 9.36 × 0.063

_____ **4.** 7.86 × 0.054

_____ **5.** 45.9 × 6.32

_____ **6.** 75.4 × 5.43

_____ **7.** 89.03 × 0.009

_____ **8.** 63.04 × 0.007

_____ **9.** 1.348 × 0.0023

_____ **10.** 4.147 × 0.0037

_____ **11.** 342.6 × 0.04

_____ **12.** 652.3 × 0.07

_____ **13.** 236 × 0.0025

_____ **14.** 435 × 0.0034

_____ **15.** 35.2 × 12.5

_____ **16.** 45.5 × 13.2

_____ **17.** 0.842 × 24

_____ **18.** 0.936 × 32

_____ **19.** 0.097 × 0.04

_____ **20.** 0.086 × 0.03

_____ **21.** 0.0739 × 500

_____ **22.** 0.0834 × 300

_____ **23.** 8.625 × 1000

_____ **24.** 3.971 × 1000

_____ **25.** 0.023 × 10,000

_____ **26.** 0.045 × 10,000

_____ **27.** 0.00487 × 100

_____ **28.** 0.00379 × 100

_____ **29.** 38.72 × 1000

_____ **30.** 46.83 × 1000

_____ **31.** 0.072 × 10

_____ **32.** 0.083 × 10

_____ **33.** 885.36 × 10,000

_____ **34.** 776.29 × 10,000

_____ **35.** 0.0035 × 2000

_____ **36.** 0.0074 × 3000

_____ **37.** 7.49 × 500

_____ **38.** 3.26 × 700

_____ **39.** 8.62 × 1.03

_____ **40.** 2.74 × 2.03

_____ **41.** 0.032 × 12.004

_____ **42.** 0.057 × 12.003

_____ **43.** 4.62 × 5.002

_____ **44.** 5.73 × 2.003

_____ **45.** 13.26 × 3070

_____ **46.** 12.71 × 2050

_____ **47.** 3.2 × 4.005 × 1000

_____ **48.** 4.3 × 7.001 × 1000

_____ **49.** 0.731 × 0.203 × 100

_____ **50.** 0.345 × 0.105 × 100

☆ Stretching the Topics

Perform the indicated operations.

_____ **1.** 256 × 1.59 × 0.15 × 18 × 12.3

_____ **2.** 0.5028 × 0.03 × 5000 × 201

_____ **3.** (206 × 3.43) − (12.903 × 5.08)

Check your answers in the back of your book.

If you can complete **Checkup 7.1,** you are ready to go on to Section 7.2.

✓ CHECKUP 7.1

Multiply.

_____ 1. 0.027×0.78

_____ 2. 56.9×3.24

_____ 3. 3.27×0.008

_____ 4. 0.0974×1000

_____ 5. 15.2×100

_____ 6. 0.238×10

_____ 7. 0.295×3.02

_____ 8. 7.85×2.06

_____ 9. 87.39×0.205

_____ 10. $3.86 \times 10,000 \times 0.505$

Check your answers in the back of your book.

If You Missed Problems:	You Should Review Examples:
1–3	1–3
4–6	4–7
7–9	8, 9
10	10

7.2 Dividing Decimal Numbers

Suppose that we wish to find a quotient of decimal numbers such as $0.6 \div 0.3$. Perhaps we can use fractions again to find the answer to this division problem.

$$0.6 \div 0.3 = \frac{6}{10} \div \frac{3}{10}$$

$$= \frac{6}{10} \times \frac{10}{3}$$

$$= \frac{\overset{2}{\cancel{6}}}{\underset{1}{\cancel{10}}} \times \frac{\overset{1}{\cancel{10}}}{\underset{1}{\cancel{3}}}$$

$$= 2$$

Therefore, $0.6 \div 0.3 = 2$.

Learning to Divide Decimal Numbers

Let's try a different approach to finding this same quotient. Recall that we can also state a division problem using a fraction bar.

$$0.6 \div 0.3 = \frac{0.6}{0.3}$$

(dividend = 0.6, divisor = 0.3)

Because we are more familiar with division by whole numbers, let's try to change the divisor (0.3) to a whole number by moving the decimal point *one place to the right*. From our work in the last section, we know we can accomplish this by *multiplying the denominator* (0.3) *by 10*. But if we multiply the denominator of a fraction by 10, we know we must also multiply the numerator by 10. Let's do that.

$$0.6 \div 0.3 = \frac{0.6}{0.3}$$

$$= \frac{0.6 \times 10}{0.3 \times 10}$$

$$= \frac{6}{3}$$

$$0.6 \div 0.3 = 2$$

Notice that this answer agrees with the answer we found before.

The handy fact to remember in dividing decimal numbers is

> We always want the divisor to be a whole number.

To change a decimal number divisor to a whole number, we simply move the decimal point to the right as many places as necessary to change it to a whole number. At the same time, we must move the decimal point in the dividend the *same* number of places to the right.

Using the familiar long division form to write our old problem, we have

$$0.3\overline{)0.6}$$

$$0.3_\wedge\overline{)0.6_\wedge}$$ Move the decimal point to make the divisor a whole number. Move the decimal point in the dividend the *same* number of places.

$$0.3_\wedge\overline{)0.6_\wedge}^{\,\cdot}$$ Locate the decimal point in the quotient *directly above* the new decimal point in the dividend.

$$0.3_\wedge\overline{)0.6_\wedge}^{\,2.}$$ Ignore the decimal points in the divisor and dividend, and *divide*.

Again, our answer is 2.

Let's summarize the steps we have used to divide decimal numbers.

Dividing Decimal Numbers

1. Move the decimal point in the divisor to the right as many places as necessary to make the divisor a whole number.
2. Move the decimal point in the dividend to the right the same number of places, attaching zeros to the end of the dividend if necessary.
3. Locate the decimal point in the quotient directly above the new location of the decimal point in the dividend.
4. Ignore the decimal points in the divisor and dividend and divide, using long division.
5. Check your quotient by multiplication.

Example 1. Find 0.035 ÷ 0.05 and check.

Solution

$$
\begin{array}{r}
.7 \\
0.05_\wedge\overline{)0.03_\wedge5} \\
\underline{3\ 5} \\
0
\end{array}
$$

CHECK:

$$
\begin{array}{r}
0.05 \quad \text{(2 places)} \\
\times\ 0.7 \quad \text{(1 place)} \\
\hline
.035 \quad \text{(3 places)}
\end{array}
$$

Therefore, 0.035 ÷ 0.05 = 0.7.

You complete Example 2.

Example 2. Find 1.3939 ÷ 5.3 and check.

Solution

$$5.3_\wedge\overline{)1.3_\wedge939}^{\,\cdot}$$

CHECK:

So 1.3939 ÷ 5.3 = _____ .

Check your work on page 291. ▶

Example 3. Find $0.464 \div 0.000116$.

Solution

$$
\begin{array}{r}
4000. \\
0.000116\,\overline{)0.464000}_{\wedge} \\
\underline{464} \\
00 \\
\underline{0} \\
00 \\
\underline{0} \\
00 \\
\underline{0}
\end{array}
$$

Notice that we attached zeros in the dividend so that we could move the decimal point 6 places to the right.

So $0.464 \div 0.000116 = 4000$.

Trial Run

Divide and check.

_____ 1. $0.072 \div 0.06$

_____ 2. $48 \div 0.32$

_____ 3. $1.1607 \div 7.3$

_____ 4. $1.065 \div 0.00213$

_____ 5. $2.697 \div 4.65$

_____ 6. $0.1764 \div 0.063$

Answers are on page 291.

Working with Repeating Decimals

In the quotients we have found so far, we eventually arrived at a point where our subtractions yielded 0. At that point, we stopped the division process. Quotients such as these, that *end* after a certain number of places, are called **terminating decimal numbers.**

Let's try a few division problems that do not turn out so "nicely". Consider the quotient $2.8 \div 0.03$.

$$
\begin{array}{r}
93. \\
0.03\,\overline{)2.80}_{\wedge} \\
\underline{2\,7} \\
10 \\
\underline{9} \\
1
\end{array}
\qquad\qquad
\begin{array}{r}
93.333 \\
0.03\,\overline{)2.80\,_{\wedge}000} \\
\underline{2\,7} \\
10 \\
\underline{9} \\
1\,0 \\
\underline{9} \\
10 \\
\underline{9} \\
10 \\
\underline{9} \\
1
\end{array}
$$

Let's attach some more zeros in the dividend. This is perfectly all right since we know we may attach as many zeros as we like to the right of the last digit in the fractional part of a decimal number.

It looks as though we could continue this same process forever without getting a 0 in our subtractions. If we attach more zeros in the dividend, we will continue to come up with 3's in the quotient. The quotient 93.333 . . . is an example of a **repeating decimal number.** Here the digit 3 is repeated forever in the fractional part of the number.

There is another way to write such a decimal number with repeating digits in its fractional part. We observe the digit or digits that are repeated and use a **bar** to write

$$93.333 \ldots = 93.\overline{3}$$

The bar over the 3 tells the reader that the digit 3 is repeated forever in the fractional part of this decimal number.

Example 4. Find $0.241 \div 0.06$.

Solution

```
              4.0166
    0.06 )0.24ˌ1000
             24
             0 1
               0
              10
               6
              40
              36
              40
              36
               4
```

We notice that the digit 6 will continue to repeat in the quotient. Therefore,

$$0.241 \div 0.06 = 4.01\overline{6}$$

You try Example 5.

Example 5. Find $21.4 \div 9.9$.

Solution

```
    9.9 )21.4ˌ0000
```

Therefore, $21.4 \div 9.9 =$ _____ .

Check your work on page 291. ▶

Example 6. Find $0.8238 \div 19.98$.

Solution

```
                    .04123123
    19.98 )0.82ˌ38000000
             79 92
              2 460
              1 998
               4620
               3996
               6240
               5994
               2460
               1998
               4620
               3996
               6240
               5994
                246
```

We see that the digits 123 will continue to be repeated in the quotient. Therefore, $0.8238 \div 19.98 = 0.04\overline{123}$.

⫸ Trial Run

Divide.

———— **1.** $2.3 \div 0.045$ ⏐ ———— **2.** $0.019 \div 0.003$

———— **3.** $9.24 \div 0.09$ ⏐ ———— **4.** $80.9 \div 0.99$

———— **5.** $388.2 \div 11.1$ ⏐ ———— **6.** $0.4766 \div 0.066$

Answers are on page 291.

Rounding Decimal Numbers in Division

In Chapter 6, we learned to round a decimal number to a certain place by looking at the place immediately to the right. For instance, consider rounding 0.34715 to different places.

Rounded to the	Rounded Number Becomes
tenths place	$0.34715 \doteq 0.3$
hundredths place	$0.34715 \doteq 0.35$
thousandths place	$0.34715 \doteq 0.347$
ten thousandths place	$0.34715 \doteq 0.3472$

Now we turn our attention to rounding some quotients. If we are told the rounding place *before* we perform the division, we need only continue the division process *one place beyond* that rounding place. Then we will know what the digit in the rounding place should be. Suppose we wish to find $0.895 \div 3.2$, *rounded to the tenths place*.

$$
\begin{array}{r}
.27 \\
3.2\overline{)0.8\wedge95} \\
6\ 4 \\
\hline
2\ 55 \\
2\ 24 \\
\hline
31
\end{array}
$$

We do not need to go any further because we have reached the hundredths place in our quotient. To round to the tenths place, we look at the hundredths place and decide that 0.27 rounds to 0.3.

Therefore, $0.895 \div 3.2 \doteq 0.3$.

Example 7. Find $8.512 \div 0.336$ and round to the thousandths place.

Solution. We must carry our quotient out to the *ten thousandths* place.

$$
\begin{array}{r}
25.3333 \\
0.336\overline{)8.512\,0000} \\
6\ 72 \\
\overline{1\ 792} \\
1\ 680 \\
\overline{112\ 0} \\
100\ 8 \\
\overline{11\ 20} \\
10\ 08 \\
\overline{1\ 120} \\
1\ 008 \\
\overline{1120} \\
1008 \\
\overline{112}
\end{array}
$$

To round our quotient to the thousandths place, we look at the ten thousandths place and decide that 25.3333 rounds to 25.333. Therefore, $8.512 \div 0.336 \doteq 25.333$.

Example 8. Find $69.3 \div 6.5$ and round to the ones place.

Solution. We must carry our quotient only to the tenths place.

$$
\begin{array}{r}
1\,0.6 \\
6.5\overline{)69.3\,0} \\
65 \\
\overline{4\ 3} \\
0 \\
\overline{4\ 3\ 0} \\
3\ 9\ 0 \\
\overline{4\ 0}
\end{array}
$$

To round our quotient to the ones place, we look at the tenths place and decide that 10.6 rounds to 11.
Therefore, $69.3 \div 6.5 \doteq 11$.

⁣⟫ Trial Run

_____ 1. Find $6.354 \div 0.07$ and round to the thousandths place.

_____ 2. Find $83.7 \div 5.06$ and round to the ones place.

_____ 3. Find $0.0375 \div 0.022$ and round to the hundredths place.

_____ 4. Find $19.955 \div 2.5$ and round to the tenths place.

_____ 5. Find $0.639 \div 0.007$ and round to the hundredths place.

_____ 6. Find $13 \div 0.3$ and round to the ten thousandths place.

Answers are on page 291.

▶ Examples You Completed

Example 2. Find 1.3939 ÷ 5.3 and check.

Solution

```
          .263          CHECK:
5.3,)1.3,939           0.263   (3 places)
    1 06            ×   5.3    (1 place)
    3 33               789
    3 18              1 315
     159             1.3939    (4 places)
     159
       0
```

So 1.3939 ÷ 5.3 = 0.263.

Example 5. Find 21.4 ÷ 9.9.

Solution

```
            2.1616
9.9,)21.4,0000
     19 8
      1 60
        99
        6 10
        5 94
          160
           99
          610
          594
           16
```

Therefore, 21.4 ÷ 9.9 = 2.$\overline{16}$.

Answers to Trial Runs

page 287   **1.** 1.2   **2.** 150   **3.** 0.159   **4.** 500   **5.** 0.58   **6.** 2.8

page 289   **1.** 51.$\overline{1}$ **2.** 6.$\overline{3}$ **3.** 102.$\overline{6}$ **4.** 81.$\overline{71}$ **5.** 34.$\overline{972}$ **6.** 7.2$\overline{21}$

page 290   **1.** 90.771   **2.** 17   **3.** 1.70   **4.** 8.0   **5.** 91.29   **6.** 43.3333

EXERCISE SET 7.2

Divide.

_____ **1.** $0.078 \div 0.06$ _____ **2.** $0.091 \div 0.07$ _____ **3.** $18 \div 0.15$

_____ **4.** $37.8 \div 0.27$ _____ **5.** $0.5254 \div 3.7$ _____ **6.** $0.4004 \div 2.8$

_____ **7.** $0.642 \div 0.00214$ _____ **8.** $0.648 \div 0.00324$ _____ **9.** $1.4175 \div 5.67$

_____ **10.** $2.6528 \div 8.29$ _____ **11.** $0.1944 \div 0.054$ _____ **12.** $0.2236 \div 0.043$

_____ **13.** $671 \div 0.305$ _____ **14.** $612 \div 0.408$ _____ **15.** $27.744 \div 27.2$

_____ **16.** $38.584 \div 36.4$ _____ **17.** $24.06 \div 2.005$ _____ **18.** $42.056 \div 3.004$

_____ **19.** $9.66 \div 276$ _____ **20.** $20.79 \div 385$ _____ **21.** $8.6 \div 0.03$

_____ **22.** $6.5 \div 0.06$ _____ **23.** $0.21 \div 1.1$ _____ **24.** $0.189 \div 3.3$

_____ **25.** $72.763 \div 0.9$ _____ **26.** $53.426 \div 0.6$ _____ **27.** $36.42 \div 2.7$

_____ **28.** $45.1 \div 3.6$ _____ **29.** $0.0016 \div 0.24$ _____ **30.** $0.0014 \div 0.21$

_____ **31.** $700 \div 0.063$ _____ **32.** $400 \div 0.036$ _____ **33.** $53.1 \div 2.22$

_____ **34.** $75.2 \div 1.32$

Divide and round to the tenths place.

_____ **35.** $4.397 \div 0.08$ _____ **36.** $1.1394 \div 0.03$

_____ **37.** $2.045 \div 3.7$ _____ **38.** $3.12 \div 4.8$

Divide and round to the thousandths place.

_____ **39.** $5.732 \div 0.09$ _____ **40.** $3.257 \div 0.06$

_____ **41.** $57.3 \div 2.45$ _____ **42.** $87.4 \div 3.52$

Divide and round to the ones place.

_____ **43.** $273.6 \div 42.7$ _____ **44.** $476.3 \div 81.2$

_____ **45.** $32.47 \div 0.05$ _____ **46.** $83.42 \div 0.07$

Divide and round to the hundredths place.

_____ **47.** $0.764 \div 0.007$ _____ **48.** $0.349 \div 0.006$

_____ **49.** $17 \div 0.3$ _____ **50.** $24 \div 0.7$

☆ Stretching the Topics

Perform the indicated operations.

_____ 1. $80,360.28 \div 8.008$

_____ 2. $(2820.47 - 967.29) \div (149.5 + 228.7)$

_____ 3. $(11,594.52 \div 128.4) \div 0.003$

Check your answers in the back of your book.

If you can do the problems in **Checkup 7.2,** you are ready to do Section 7.3.

✓ CHECKUP 7.2

Divide.

_____ 1. 0.273 ÷ 0.07   _____ 2. 1.6849 ÷ 8.3

_____ 3. 2.58 ÷ 0.0043   _____ 4. 305.2 ÷ 8.72

_____ 5. 3.5 ÷ 0.27   _____ 6. 8.21 ÷ 0.003

_____ 7. 526 ÷ 4.5

_____ 8. Find 0.75576 ÷ 0.6 and round to the hundredths place.

_____ 9. Find 51.319 ÷ 7.3 and round to the tenths place.

_____ 10. Find 0.0635 ÷ 0.008 and round to the ones place.

Check your answers in the back of your book.

If You Missed Problems:	You Should Review Examples:
1–4	1–3
5–7	4–6
8–10	7, 8

7.3 Changing Fractional Numbers to Decimal Numbers

There are many situations in which it is necessary to take a number written in fractional form and change it to decimal form. Most calculators, for instance, perform all their calculations with decimal numbers. Let's spend some time developing a method for changing fractional numbers to decimal numbers.

Changing Fractional Numbers with "Nice" Denominators

Actually, we have already learned how to change fractional numbers with "nice" denominators (10 or 100 or 1000 or 10,000 and so on) into decimal numbers. In fact, we have made such changes frequently in our work with decimal numbers. We know, for instance, that

$$\frac{2}{10} = 0.2 \qquad \frac{9}{1000} = 0.009$$

$$\frac{19}{100} = 0.19 \qquad \frac{27}{10,000} = 0.0027$$

To make these changes merely required that we read the fractional number aloud and then interpret it as a decimal number with that name.

You try Example 1.

Example 1. Change each fractional number to a decimal number.

Fractional Number		Decimal Number
$\dfrac{293}{1000}$	=	_____
$\dfrac{7}{100}$	=	_____
$\dfrac{4}{10,000}$	=	_____

Check your work on page 300. ▶

We also used this method for changing mixed numbers and improper fractions (with "nice" denominators) into decimal numbers. In changing a mixed number to a decimal number, we know that the whole number part of the mixed number will become the part to the left of the decimal point in the decimal number. The fractional part of the mixed number will become the part to the right of the decimal point in the decimal number. For instance,

$$\text{whole number part}$$
$$2\frac{7}{10} = 2.7$$
$$\text{fractional part}$$

You try Example 2.

Example 2. Change $50\frac{19}{100}$ to a decimal number.

Solution

$$50\frac{19}{100} = \underline{\hspace{3cm}}$$

Check your work on page 300. ▶

Example 3. Change $496\frac{1}{1000}$ to a decimal number.

Solution

$$496\frac{1}{1000} = 496.001$$

To write an improper fraction with a "nice" denominator as a decimal number we can first change it to a mixed number and then proceed as before. For instance, if we wish to write $\frac{289}{100}$ as a decimal number, we first write it as a mixed number.

$$\frac{289}{100} = 2\frac{89}{100}$$
$$= 2.89$$

You try Example 4.

Example 4. Change $\frac{17}{10}$ to a decimal number.

Solution

$$\frac{17}{10} =$$

$$= \underline{\hspace{2cm}}$$

Check your work on page 300. ▶

Example 5. Change $\frac{6005}{1000}$ to a decimal number.

Solution

$$\frac{6005}{1000} = 6\frac{5}{1000}$$
$$= 6.005$$

▐▌▶ Trial Run

Change each of the following to a decimal number.

_____ 1. $\frac{735}{1000}$

_____ 2. $\frac{7}{100}$

_____ 3. $\frac{45}{10,000}$

_____ 4. $8\frac{37}{100}$

_____ 5. $27\frac{5}{10,000}$

_____ 6. $\frac{24}{10}$

_____ 7. $\frac{278}{100}$

_____ 8. $\frac{23,721}{1000}$

Answers are on page 301.

From Examples 1, 4, and 5 you may have noticed a shortcut to use when dividing by 10, 100, 1000, 10000, and so on. Can you spot a pattern in those divisions?

$$\frac{293}{1000} = 0.293 \qquad \frac{17}{10} = 1.7$$

$$\frac{7}{100} = 0.07 \qquad \frac{6005}{1000} = 6.005$$

$$\frac{4}{10,000} = 0.0004$$

Notice that the digits in each quotient are exactly the same digits appearing in each dividend. But what about the decimal point's location in the quotient? Although we did not write the decimal points in the dividends, we know that they belong to the right of the last digit. Notice how the number of zeros in each divisor compares to the location of the decimal point in each quotient.

To divide a number by a whole number divisor that is 1 followed by a certain number of zeros, we simply move the decimal point in the dividend that certain number of places to the *left* to arrive at the quotient.

Perhaps we can use a chart to illustrate this new shortcut.

To Divide a Number by	Move the Decimal Point Left	Examples
10	1 place	$\frac{29}{10} = 2.9$
100	2 places	$\frac{17}{100} = 0.17$
1000	3 places	$\frac{3}{1000} = 0.003$
10,000	4 places	$\frac{49}{10,000} = 0.0049$

Example 6. Use a shortcut to find $39 \div 1000$.

Solution. Counting the zeros following 1 in the divisor, we see that we must move the decimal point in the dividend 3 places to the left.

$$39 \div 1000 = \frac{39}{1000}$$

$$= 0.039$$

You try Example 7.

Example 7. Use a shortcut to find $\frac{7911}{100}$.

Solution. Counting the zeros following 1 in the divisor, we see that we must move the decimal point in the dividend _____ places to the left.

$$\frac{7911}{100} = \underline{}$$

Check your work on page 300. ▶

We see that this shortcut works whether the fractional number is proper or improper. In fact, it also works when the dividend is not a whole number. For instance, to find $34.7 \div 100$ we simply move the decimal point in the dividend 2 places to the left.

$$34.7 \div 100 = \frac{34.7}{100}$$

$$= 0.347$$

Of course, any of these quotients can be found using long division, but our shortcut certainly cuts down on the steps required.

Example 8. Use a shortcut to find $0.32 \div 10,000$.

Solution. Counting 4 zeros following the 1 in the divisor, we see that we must move the decimal point in the dividend 4 places to the left.

$$0.32 \div 10,000 = \frac{0.32}{10,000}$$

$$= 0.000032$$

You try Example 9.

Example 9. Use a shortcut to find $687.5 \div 100,000$.

Solution

$$687.5 \div 100,000 = \frac{687.5}{100,000}$$

$$= \underline{}$$

Check your work on page 300. ▶

⫸ **Trial Run**

Use a shortcut to find each quotient.

_____ 1. $72 \div 1000$

_____ 2. $386 \div 10$

_____ 3. $83.5 \div 10,000$

_____ 4. $0.035 \div 1000$

_____ 5. $0.0058 \div 10$

_____ 6. $435.072 \div 100$

_____ 7. $0.725 \div 100$

_____ 8. $376.5 \div 100,000$

Answers are on page 301.

Changing Other Fractional Numbers to Decimal Numbers

Suppose the denominator of a fractional number is *not* 10 or 100 or 1000 or 10,000 and so on. How can we change fractional numbers such as

$$\frac{1}{6} \qquad \frac{22}{7} \qquad \frac{84}{93}$$

to decimal numbers? Unfortunately, there is no nice shortcut for making such changes. The only method available is to treat the fraction as a division problem and find the quotient by long division.

Example 10. Change $\frac{1}{6}$ to a decimal number.

Solution. We must divide.

```
      .166
   6)1.000
      6
      40
      36
      40
      36
       4
```

We see that our quotient is a *repeating decimal* and we write

$$\frac{1}{6} = 0.1\overline{6}$$

You try Example 11.

Example 11. Change $\frac{22}{7}$ to a decimal number and round to the thousandths place.

Solution. We must carry our quotient out to the ten thousandths place.

```
        .
   7)22.0000
```

$$\frac{22}{7} \doteq \underline{}$$

Check your work on page 301. ▶

Example 12. Change $\frac{84}{93}$ to a decimal number and round to the hundredths place.

Solution. We carry our quotient out to the thousandths place.

```
         .903
   93)84.000
      83 7
         30
          0
        300
        279
         21
```

Now rounding to the hundredths place, we decide that

$$\frac{84}{93} \doteq 0.90$$

Notice that our answer to Example 10 is *exact,* but our answers to Examples 11 and 12 are *approximations.*

⫸ Trial Run

——— 1. Change $\dfrac{38}{7}$ to a decimal number and round to the thousandths place.

——— 2. Change $\dfrac{5}{6}$ to a decimal number and round to the hundredths place.

——— 3. Change $\dfrac{56}{73}$ to a decimal number and round to the tenths place.

——— 4. Change $\dfrac{369}{115}$ to a decimal number and round to the hundred thousandths place.

——— 5. Change $\dfrac{270}{11}$ to a decimal number and round to the tenths place.

——— 6. Change $\dfrac{8754}{1000}$ to a decimal number and round to the hundredths place.

Answers are on page 301.

▶ Examples You Completed

Example 1.

$$\frac{293}{1000} = 0.293$$

$$\frac{7}{100} = 0.07$$

$$\frac{4}{10,000} = 0.0004$$

Example 2. Change $50\dfrac{19}{100}$ to a decimal number.

Solution

$$50\frac{19}{100} = 50.19$$

Example 4. Change $\frac{17}{10}$ to a decimal number.

Solution

$$\frac{17}{10} = 1\frac{7}{10}$$
$$= 1.7$$

Example 7. Use a shortcut to find $\frac{7911}{100}$.

Solution. Counting the zeros following 1 in the divisor, we see that we must move the decimal point in the dividend 2 places to the left.

$$\frac{7911}{100} = 79.11$$

Example 9. Use a shortcut to find $687.5 \div 100,000$.

Solution

$$687.5 \div 100,000 = \frac{687.5}{100,000}$$
$$= 0.006875$$

Example 11. Change $\frac{22}{7}$ to a decimal number and round to the thousandths place.

Solution. We must carry our quotient out to the ten thousandths place.

$$
\begin{array}{r}
3.1428 \\
7\overline{)22.0000} \\
\underline{21} \\
1\,0 \\
\underline{7} \\
30 \\
\underline{28} \\
20 \\
\underline{14} \\
60 \\
\underline{56} \\
4
\end{array}
$$

Now rounding to the thousandths place, we decide

$$\frac{22}{7} \doteq 3.143$$

Answers to Trial Runs

page 296 **1.** 0.735 **2.** 0.07 **3.** 0.0045 **4.** 8.37 **5.** 27.0005 **6.** 2.4 **7.** 2.78
8. 23.721

page 298 **1.** 0.072 **2.** 38.6 **3.** 0.00835 **4.** 0.000035 **5.** 0.00058 **6.** 4.35072 **7.** 0.00725
8. 0.003765

page 300 **1.** 5.429 **2.** 0.83 **3.** 0.8 **4.** 3.20870 **5.** 24.5 **6.** 8.75

EXERCISE SET 7.3

Change each of the following to a decimal number.

_____ 1. $\dfrac{879}{1000}$ _____ 2. $\dfrac{935}{1000}$

_____ 3. $\dfrac{72}{10,000}$ _____ 4. $\dfrac{43}{10,000}$

_____ 5. $\dfrac{9}{100}$ _____ 6. $\dfrac{7}{100}$

_____ 7. $\dfrac{378}{100}$ _____ 8. $\dfrac{576}{100}$

_____ 9. $23 \dfrac{74}{100,000}$ _____ 10. $81 \dfrac{24}{100,000}$

Use a shortcut to find each quotient.

_____ 11. $76 \div 1000$ _____ 12. $37 \div 1000$

_____ 13. $39.7 \div 10$ _____ 14. $42.5 \div 10$

_____ 15. $0.035 \div 100$ _____ 16. $0.042 \div 100$

_____ 17. $429.6 \div 100,000$ _____ 18. $269.3 \div 100,000$

_____ 19. $0.39 \div 10,000$ _____ 20. $0.25 \div 10,000$

Change to a decimal number and round to the tenths place.

_____ 21. $\dfrac{7395}{100}$ _____ 22. $\dfrac{2497}{100}$

_____ 23. $\dfrac{360}{11}$ _____ 24. $\dfrac{440}{9}$

Change to a decimal number and round to the thousandths place.

_____ 25. $\dfrac{74}{3}$ _____ 26. $\dfrac{85}{6}$

_____ 27. $\dfrac{83}{100,000}$ _____ 28. $\dfrac{72}{100,000}$

Change to a decimal number and round to the hundredths place.

_____ 29. $\dfrac{9375}{1000}$ _____ 30. $\dfrac{8763}{1000}$

_____ 31. $\dfrac{842}{7}$ _____ 32. $\dfrac{732}{7}$

Change to a decimal number and round to the ten thousandths place.

_____ 33. $\dfrac{35}{100,000}$ _____ 34. $\dfrac{46}{100,000}$

_____ 35. $\dfrac{1}{6}$ _____ 36. $\dfrac{2}{3}$

Change to a decimal number and round to the ones place.

_____ 37. $\dfrac{8975}{100}$ _____ 38. $\dfrac{7986}{100}$

_____ 39. $\dfrac{9236}{15}$ _____ 40. $\dfrac{726}{23}$

☆ Stretching the Topics _____

_____ 1. Change $35\frac{12}{21}$ to a decimal number and round to the hundredths place.

_____ 2. Change $\frac{486}{39}$ to a decimal number and round to the thousandths place.

Check your answers in the back of your book.

If you can do the problems in **Checkup 7.3,** you are ready to go on to Section 7.4.

✓ **CHECKUP 7.3**

Change each of the following to a decimal number.

_____ 1. $\dfrac{48}{100,000}$

_____ 2. $3\dfrac{23}{10,000}$

_____ 3. $\dfrac{8765}{100}$

_____ 4. $\dfrac{129}{1000}$

Use a shortcut to find each quotient.

_____ 5. $8 \div 1000$

_____ 6. $365 \div 10$

_____ 7. $0.0456 \div 100$

_____ 8. $356.3 \div 10,000$

_____ 9. Change $\dfrac{35}{10,000}$ to a decimal number and round to the thousandths place.

_____ 10. Change $\dfrac{729}{13}$ to a decimal number and round to the tenths place.

Check your answers in the back of your book.

If You Missed Problems:	You Should Review Examples:
1–4	1–5
5–8	6, 7
9, 10	10–12

7.4 Switching from Words to Decimal Numbers

Now we can put our skills with decimal numbers to use in solving some problems stated in words. As usual, we must be on the lookout for the key phrases which tell us what operation to perform.

Example 1. If a restaurant bill of $83.04 is to be split equally among 3 couples, how much will each couple pay?

Solution. The phrase "split equally" tells us we must *divide*.

$$
\begin{array}{r}
27.68 \\
3\overline{)83.04} \\
\underline{6} \\
23 \\
\underline{21} \\
2\,0 \\
\underline{1\,8} \\
24 \\
\underline{24} \\
0
\end{array}
$$

Each couple must pay $27.68.

Example 2. Michelle works in the mathematics department office for $3.35 per hour. During the last pay period, she worked 29.5 hours. What was her pay?

Solution. Since Michelle's pay rate is $3.35 for 1 hour of work, we must *multiply* that rate times the number of hours she worked.

$$
\begin{array}{r}
3.3\,5 \quad \text{(2 places)} \\
\times 2\,9.5 \quad \text{(1 place)} \\
\hline
1\,6\,7\,5 \\
30\,1\,5 \\
67\,0 \\
\hline
98.8\,2\,5 \quad \text{(3 places)}
\end{array}
$$

Michelle's pay was $98.83.

Example 3. Find the area of a rectangular postage stamp that is 2.2 centimeters wide and 1.7 centimeters long.

Solution. To find the area of a rectangle we multiply its length times its width.

$$
\begin{array}{r}
2.2 \quad \text{(1 place)} \\
\times 1.7 \quad \text{(1 place)} \\
\hline
1\,54 \\
2\,2 \\
\hline
3.74 \quad \text{(2 places)}
\end{array}
$$

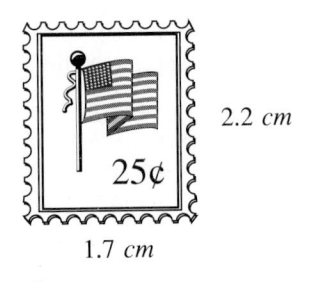

2.2 *cm*

1.7 *cm*

The area is 3.74 square centimeters.

Example 4. The center section of a quilt is made up of a square of checked fabric surrounded by 4 triangles of solid fabric. If the dimensions are as illustrated, find the amount of checked fabric and the amount of solid fabric needed. Round to the *tenths* place.

Solution. To find the amount of checked fabric we must find the area of the square by multiplying side times side.

Area of square = 2.26 × 2.26

```
    2.26   (2 places)
  × 2.26   (2 places)
    1356
     452
   4 52
  5.1076   (4 places)
```

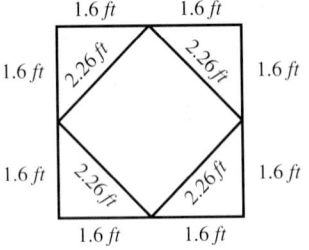

We need 5.1076 square feet. Rounding here to the tenths place, we must be careful. If we round to 5.1 by our usual rules, we will not have enough fabric. Instead, we must round "up" and buy 5.2 square feet of checked fabric.

To find the amount of solid fabric, we must find the area of each triangle by multiplying base times height and dividing by 2.

Area of triangle = (1.6 × 1.6) ÷ 2

```
     1.6          1.28
   × 1.6      2)2.56
      96          2
     1 6          0 5
    2.56            4
                   16
                   16
                    0
```

Area of triangle = 1.28 square feet

Since there are 4 triangles, we must multiply this area by 4.

```
     1.28
   ×    4
     5.12
```

The area of the 4 triangles is 5.12 square feet, but we must round *up* again and buy 5.2 square feet of solid fabric.

Running times for races are usually expressed using decimal numbers. Let's see if we can change some times stated in hours or minutes or seconds into times stated as decimal numbers.

Suppose we wish to express 37 minutes as some decimal number of *hours*. Remembering that 1 hour contains 60 minutes, we first write 37 minutes as a *fractional number* of hours.

$$\frac{\text{part}}{\text{whole}} = \frac{37}{60} \text{ hours}$$

Now we can change this fractional number to a decimal number using long division.

```
        .6166
  60)37.0000
     36 0
      1 00
        60
       400
       360
       400
       360
        40
```

Since the quotient is a repeating decimal, we write

$$37 \text{ min} = 0.61\overline{6} \text{ hr}$$

Example 5. If Secretariat ran the 1973 Belmont Stakes in 2 minutes 15 seconds, write the horse's running time as a decimal number of minutes.

Solution. We know that there are 60 seconds in 1 minute, so 15 seconds is

$$\frac{15}{60} \text{ minute} = \frac{1}{4} \text{ minute}$$

and the running time is

$$2\frac{1}{4} \text{ minutes}$$

We write this mixed number as an improper fraction

$$2\frac{1}{4} = \frac{9}{4}$$

```
        2.25
    4)9.00
      8
      ‾‾‾
      1 0
        8
        ‾‾
        20
        20
        ‾‾
         0
```

and divide to find the decimal number. Secretariat's running time was 2.25 minutes.

Now let's solve the problem stated at the beginning of this chapter.

Example 6. If Scott Spann can swim the 200 yard breaststroke in 2.02 minutes and Graham Smith can swim the same race in 2 minutes 2 seconds, who is the faster swimmer?

Solution. To compare these times we can change both to decimal numbers.

Spann: 2.02 min

Smith: $2 \text{ min } 2 \text{ sec} = 2\frac{2}{60} \text{ min}$

$$= 2\frac{1}{30} \text{ min}$$

$$= \frac{61}{30} \text{ min}$$

$$= 2.0\overline{3} \text{ min}$$

```
          2.033
    30)61.000
       60
       ‾‾‾
        1 0
          0
        ‾‾‾
        1 00
          90
        ‾‾‾‾
         100
          90
         ‾‾‾
          10
```

Since Smith's time is greater than Spann's time, we conclude that Spann is the faster swimmer.

EXERCISE SET 7.4

_____ 1. The tips at Mel's Diner for one evening totaled $63.75. If the tips were split equally among the 3 waitresses, how much did each waitress get in tips that evening?

_____ 2. Carol and 3 of her friends share equally the rent on a 2-bedroom apartment. If the rent is $295 per month, how much does each person pay per month?

_____ 3. Billy Joe delivers pizzas for Uno's Pizza Parlor for $3.65 an hour. Last week he worked 32.5 hours. What was his pay?

_____ 4. Rhonda lifeguards at the city pool parttime for $3.75 an hour. If she works 15.5 hours per week, what is her weekly salary?

_____ 5. Carlos bought 15.6 gallons of gasoline at $1.069 per gallon. Find the cost of the gasoline.

_____ 6. If Michelle's car averages 19.75 miles per gallon of gasoline, how far can she travel on a full 16-gallon tank?

_____ 7. The tax on property is computed by multiplying its assessed value by the tax rate. If the tax rate for Perry County is 0.01235, find the tax bill on property that is evaluated at $65,000.

_____ 8. If the Social Security tax rate is 0.0765, find how much social security tax was deducted from a monthly salary of $1832.

_____ 9. The first 30 kilowatt-hours of electricity cost $4.89. For each kilowatt-hour above 30 the cost is $0.0545. Find the total electric bill for a consumer using 1225 kilowatt-hours.

_____ 10. The yearly premium per $1000 of life insurance for a 40-year-old female is $18.24. If Jane is 40 years old, find her premium on a $50,000 life insurance policy.

_____ 11. Mr. Collins sold 2667.6 bushels of corn for $7869.42. How much per bushel was he paid for his corn?

_____ 12. If Renaldo sells 87.5 shares of stock for $1860.25, find the value of each share of his stock.

_____ 13. An 8-bottle carton of Dewdrop was on sale for $1.59. What was the cost per bottle to the nearest cent?

_____ 14. At Hereford Heaven's sale on USDA choice beef, 150 pounds is advertised at $41.25 per payment for 3 payments. How much per pound is the beef at Hereford Heaven?

_____ 15. In 1980, Buddy Baker averaged 177.602 miles per hour over the 500-mile track at Daytona. Find how long it took him to finish the race. (Round answer to nearest hundredth of an hour.)

_____ 16. If Mary can run 880 yards in 1 minute 58 seconds and Rosalyn can run the same distance in 1.08 minutes, who is the faster runner?

_____ 17. One year Joe Montana completed 311 passes for 3565 yards. What was his average number of yards per pass? (Round answer to nearest tenth of a yard.)

_____ 18. A football field is 109.68 meters long and 48.72 meters wide. Find the area of the football field. (Round answer to nearest hundredth of a square meter.)

_____ 19. Agnes has a triangular macramé wall hanging that has a base of 22.5 inches and a height of 14.75 inches. How many square inches of wall space will the hanging cover? (Round answer to the nearest square inch.)

_____ 20. For her political fund-raising dinner, Maryon is making square tableclothes of red, white, and blue fabric. The center is a red square bordered by 2 white triangles and 2 blue triangles. If the dimensions are as shown, find the amounts of red, white, and blue fabric needed for 1 tablecloth. (Round answer to tenths place.)

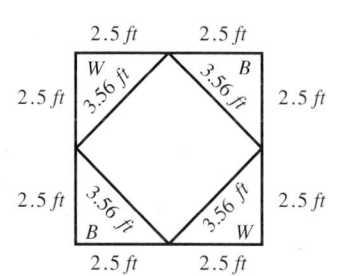

☆ Stretching the Topics

_____ 1. The formula $C = \pi \times d$ gives the circumference (C) of a circle when the diameter (d) and π (use 3.14) are known. Find the circumference of a circle that has a diameter 0.27 miles. (Round answer to hundredths place.)

_____ 2. Bruce's car averages 21.5 miles per gallon on the interstate and 16.5 miles per gallon on two-lane roads. On a trip, he drives 129 miles on interstate highway and 67 miles on two-lane roads. If gasoline costs $1.059 per gallon, how much will the trip cost? (Round answer to the nearest cent.)

_____ 3. Rates for electrical services are as follows: first 40 kilowatt-hours, $3.85; next 260 kilowatt-hours, $0.0635 per kwh; next 1500 kilowatt-hours, $0.0424 per kwh; any additional kilowatt-hours, $0.0315 per kwh. Find the cost of using 2550 kilowatt-hours of electricity.

Check your answers in the back of your book.

If you can complete **Checkup 7.4** correctly, then you are ready to do the **Review Exercises** for Chapter 7.

✓ **CHECKUP 7.4**

_____ 1. Frances bought a package of ground beef weighing 3.24 pounds. If ground beef sells for $1.39 per pound, find the cost of the package to the nearest cent.

_____ 2. If Carl's car averages 15.25 miles per gallon of gasoline, how many gallons will he need for a trip of 1220 miles?

_____ 3. A case of 24 cans of green beans costs $12.25. What is the cost per can to the nearest cent?

_____ 4. How many square feet of carpet is needed to cover a living room floor that measures 15.5 feet by 24.25 feet? (Round answer to nearest square foot.)

_____ 5. The record speed for the Indianapolis Speedway is 163.465 miles per hour. If the record-holding car finished the race in 3.06 hours, how many miles was the race? (Round to the nearest mile.)

Check your answers in the back of your book.

If You Missed Problems:	You Should Review Examples:
1	2
2, 3	1
4	3
5	1

Summary

In this chapter we used our knowledge of fractional numbers to help us understand multiplication and division of decimal numbers.

To locate the decimal point in the product of decimal numbers, we learned to count the total number of places to the right of the decimal point in all of the factors being multiplied. That total told us how many places would be to the right of the decimal point in the answer.

As with whole number multiplication, shortcuts could be used in multiplying decimal numbers containing zero digits. It continued to be important to line up the rightmost digit of each product beneath the digit by which we were multiplying.

To find: 0.123×0.42

We write: $\begin{array}{r} 0.123 \quad \text{(3 places)} \\ \times\ \ 0.42 \quad \text{(2 places)} \\ \hline 246 \quad\quad\quad \\ 492 \quad\quad\quad\quad \\ \hline 0.05166 \quad \text{(5 places)} \end{array}$

Product: $0.123 \times 0.42 = 0.05166$

To find: 1.05×200

We write: $\begin{array}{r} 1.05 \quad \text{(2 places)} \\ \times\ \ \ 200 \quad \text{(0 places)} \\ \hline 210.00 \quad \text{(2 places)} \end{array}$

Product: $1.05 \times 200 = 210$

To divide decimal numbers, we learned to move the decimal points in the divisor *and* the dividend to make the divisor a *whole number*. After locating the decimal point in the quotient directly above the new decimal point in the dividend, we then divide by the method of long division.

In dividing decimal numbers we discovered that the quotient might be a **terminating** decimal number or a **repeating** decimal number. We also learned to round a quotient to a specified place.

To find: $0.8 \div 0.25$

We write: $\begin{array}{r} 3.2 \\ 0.25\overline{)0.80,0} \\ 75 \\ \hline 5\,0 \\ 5\,0 \\ \hline 0 \end{array}$

Quotient: $0.8 \div 0.25$
$= 3.2$

To find: $0.7 \div 0.3$

We write: $\begin{array}{r} 2.33 \\ 0.3\overline{)0.7,00} \\ 6 \\ \hline 1\,0 \\ 9 \\ \hline 10 \\ 9 \\ \hline 1 \end{array}$

Quotient: $0.7 \div 0.3$
$= 2.\overline{3}$

We noted shortcuts for multiplying and dividing decimal numbers by a whole number that is 1 followed by a certain number of zeros.

In Order to	Move Decimal Point	Examples
Multiply by 10	1 place right	$7.64 \times 10 = 76.4$
Multiply by 100	2 places right	$7.64 \times 100 = 764$
Multiply by 1000	3 places right	$7.64 \times 1000 = 7640$
Divide by 10	1 place left	$7.64 \div 10 = 0.764$
Divide by 100	2 places left	$7.64 \div 100 = 0.0764$
Divide by 1000	3 places left	$7.64 \div 1000 = 0.00764$

To change a fractional number to an equivalent decimal number, we learned to divide the denominator into the numerator, using the method for dividing decimal numbers.

Finally we used multiplication and division to change problems stated in words to problems involving decimal numbers. We paid particular attention to area problems and also learned to change hours and minutes to decimal numbers of hours.

☐ Speaking the Language of Mathematics

Complete each statement with the appropriate word or phrase.

1. If we multiply 2.76×3.2, there will be _____ places to the right of the decimal point in the product.

2. In dividing decimal numbers, we begin by moving the decimal point in the divisor to make it a _____ _____ .

3. To multiply a decimal number by 1000, we must move its decimal point _____ places to the _____ .

4. To divide a decimal number by 100, we must move its decimal point _____ places to the _____ .

5. A decimal number such as $3.4\overline{13}$ is called a _____ decimal number.

6. To round a quotient to the tenths place, we must carry out the division to the _____ place.

7. To change a fractional number to a decimal number, we divide the fraction's _____ into its _____ .

8. To change times to decimal numbers, we must remember that there are _____ seconds in 1 minute and there are _____ minutes in 1 hour.

△ Writing About Mathematics

Write your response to each question in complete sentences.

1. Show and explain the steps you would use to change the fractional number $\frac{5}{8}$ to a decimal number.

2. Explain how to locate the decimal point in the product when you are multiplying two decimal numbers. Explain how to locate the decimal point in the quotient when you are dividing two decimal numbers.

3. Can you think of at least two situations in everyday life in which you would be required to round the decimal number 1.514 ''up'' to 1.52?

REVIEW EXERCISES for Chapter 7

Perform the indicated operations.

_____ 1. 0.056×0.37

_____ 2. 28.9×8.54

_____ 3. 322×0.018

_____ 4. 0.086×0.03

_____ 5. 9.342×1000

_____ 6. 0.083×10

_____ 7. 7.36×2.04

_____ 8. 0.017×11.005

_____ 9. 19.62×5030

_____ 10. $4.6 \times 5.007 \times 100$

_____ 11. $4.536 \div 0.021$

_____ 12. $0.1176 \div 5.6$

_____ 13. $264 \div 0.24$

_____ 14. $8.034 \div 0.0013$

_____ 15. $38.61 \div 0.045$

_____ 16. $20.3 \div 0.0028$

_____ 17. Find $100.4 \div 8.4$ and round to the tenths place.

_____ 18. Find $0.718 \div 0.009$ and round to the hundredths place.

_____ 19. Find $23.6 \div 0.12$ and round to the thousandths place.

_____ 20. Find $0.09585 \div 0.045$ and round to the ones place.

_____ 21. Use a shortcut to find $0.089 \div 100$.

_____ 22. Change $5\frac{7}{1000}$ to a decimal number.

_____ 23. Change $\frac{453}{9}$ to a decimal number and round to the tenths place.

_____ 24. Change $\frac{96}{100,000}$ to a decimal number and round to the thousandths place.

_____ 25. Change $\frac{38}{7}$ to a decimal number and round to the hundredths place.

_____ 26. Change $\frac{4}{9}$ to a decimal number and round to the ten thousandths place.

_____ 27. Change $\frac{627}{32}$ to a decimal number and round to the ones place.

_____ 28. Margo owns 85.25 shares of stock. If the stock is valued at $12.83 per share, find the value of her stock.

_____ 29. Joey bought 13.5 gallons of gasoline at $1.059 per gallon. Find the cost of the gasoline. (Round answer to nearest cent.)

_____ 30. Find the area of a rectangle that is 17.8 centimeters long and 9.7 centimeters wide.

_____ 31. The distance for the Indianapolis 500 race is 804.68 kilometers. If each lap is 4.0234 kilometers, how many laps will the winner make during the race?

_____ 32. The receipts from a pantomime show were $2307.25. If tickets cost $2.75 each, find how many persons attended the show.

_____ 33. If 5.25 pounds of steak cost $23.57, find the cost of steak per pound. (Round answer to nearest cent.)

Check your answers in the back of your book.

If You Missed Exercises:	You Should Review Examples:	
1–4	Section 7.1	1–3
5, 6		4–7
7–9		8, 9
10		10
11–16	Section 7.2	1–3
17–20		4–8
21	Section 7.3	1–5
22		6–9
23–27		10–12
28–30	Section 7.4	2–4
31–33		1

If you have completed the **Review Exercises** and corrected your answers, then you are ready to take the **Practice Test** for Chapter 7.

PRACTICE TEST for Chapter 7

Perform the indicated operations.   SECTION   EXAMPLES

_____ 1. 0.093 × 0.28   7.1   1

_____ 2. 84.32 × 5.7   7.1   2

_____ 3. 0.3452 × 600   7.1   4

_____ 4. 21.003 × 0.092   7.1   8

_____ 5. 12.34 × 4.002 × 100   7.1   10

_____ 6. 2.930 ÷ 7.4   7.2   2

_____ 7. 7.176 ÷ 0.00312   7.2   3

_____ 8. 41.2 ÷ 3.3   7.2   5

_____ 9. Find 7.823 ÷ 0.411 and round to hundredths place.   7.2   7

_____ 10. Find 87.3 ÷ 9.5 and round to tenths place.   7.2   7

_____ 11. Change $\frac{2371}{100}$ to a decimal number.   7.3   3

_____ 12. Use a shortcut to find $\frac{7.5}{10,000}$.   7.3   8

_____ 13. Change $\frac{96}{13}$ to a decimal number and round to tenths place.   7.3   11, 12

_____ 14. Change $\frac{389}{57}$ to a decimal number and round to hundredths place.   7.3   11, 12

_____ 15. Change $\frac{35}{49}$ to a decimal number and round to thousandths place.   7.3   11, 12

_____ 16. Ms. Humphrey is buying material for a rectangular tablecloth. If the tablecloth is to be 6.5 feet long and 4.25 feet wide, how many square feet of material should she buy?   7.4   2

_____ 17. Oliver earns $8.78 per hour. If he worked 36.5 hours last week, find how much he earned. (Round to the nearest cent.)   7.4   2

_____ 18. If 64 ounces of laundry detergent cost $4.41, find the cost per ounce of the detergent. (Round to the nearest cent.)   7.4   1

		SECTION	EXAMPLES

_____ 19. Mr. Hunter sold 1542.8 bushels of soybeans for a total of $9025.38.   7.4   1
How much per bushel was he paid for his soybeans?

_____ 20. Mark set a record of 1 minute and 52 seconds for the 200-meter   7.4   5, 6
freestyle. Eric's time was 1.97 minutes. Express Mark's time as a
decimal number of minutes rounded to the hundredths place. Who is
the faster swimmer and by how much?

Check your answers in the back of your book.

SHARPENING YOUR SKILLS after Chapters 1–7

		SECTION

_____ 1. In the whole number 56,087, name the hundreds digit.   1.1

_____ 2. Change $\frac{41}{3}$ to a mixed number.   3.1

_____ 3. Change $\frac{7}{8}$ to an equivalent fraction with a denominator of 56.   5.2

_____ 4. Write 85.365 in words.   6.1

_____ 5. Compare 15.057 and 15.507 using $<$ or $>$.   6.1

_____ 6. Change $\frac{429}{10,000}$ to a decimal number.   7.3

_____ 7. Find the sum of 8739 and 15,789.   1.2

_____ 8. Fill in the missing number: _____ $-$ 29 = 64 because _____ = 29 + 64.   1.3

Perform the indicated operations.

_____ 9. $\dfrac{7}{8} + \dfrac{9}{10} - \dfrac{5}{6}$ 5.3

_____ 10. $3\dfrac{1}{2} + 7\dfrac{5}{8} + 9\dfrac{2}{3}$ 5.3

_____ 11. \$325 $-$ \$83.95 $-$ \$72.38   6.2

_____ 12. 209.3 $-$ 165.025   6.2

_____ 13. 758,910 $\div$ 615   2.2

_____ 14. 23 $\times$ 105 $\times$ 72   2.1

_____ 15. $\dfrac{15}{8} \times \dfrac{40}{63}$ 4.1

_____ 16. $\dfrac{7}{6} \times 9\dfrac{1}{3}$ 4.1

_____ 17. $7\dfrac{1}{2} \div 2\dfrac{1}{4}$ 4.2

_____ 18. 0.073 $\times$ 0.19   7.1

_____ 19. 10.35 $\times$ 3.002 $\times$ 100   7.1

_____ 20. 159.06 $\div$ 72.3   7.2

_____ 21. 21.84 $\div$ 0.00312   7.2

SECTION

_____ 22. If Jackie's new car gets 21 miles per gallon, how far can she expect to travel on a full   2.4
16-gallon tank?

_____ 23. If Phil earned $292 for $36\frac{1}{2}$ hours of work, what was he earning per hour?   4.3

_____ 24. Vincent wishes to install a new baseboard around his family room. The room measures   5.5
$12\frac{3}{4}$ feet by $15\frac{1}{2}$ feet, but the two doorways will need no baseboard. One door is $3\frac{1}{6}$
feet wide and the other is $2\frac{7}{8}$ feet wide. How many feet of baseboard will Vincent
need?

_____ 25. Use the following check register to record the transactions in Mr. Painter's checking   6.3
account and find his final balance. His balance on October 31 was $78.38.
November 1:   deposit $898.73 paycheck
November 1:   rent payment, $310
November 5:   payment of $63.72 to utility company
November 6:   payment to Sureway market, $43.72
November 10:   payment to Union Cable TV, $15.75
November 12:   deposit $75, repayment of loan

CHECK NO.	DATE	CHECKS DRAWN OR DEPOSITS MADE	BALANCE FORWARD ▷	√		
		TO	DEDUCT CHECK − ADD DEPOSIT +			
			BALANCE ▷			
		TO	DEDUCT CHECK − ADD DEPOSIT +			
			BALANCE ▷			
		TO	DEDUCT CHECK − ADD DEPOSIT +			
			BALANCE ▷			
		TO	DEDUCT CHECK − ADD DEPOSIT +			
			BALANCE ▷			
		TO	DEDUCT CHECK − ADD DEPOSIT +			
			BALANCE ▷			
		TO	DEDUCT CHECK − ADD DEPOSIT +			
			BALANCE ▷			

Check your answers in the back of your book.

Working with Ratios and Proportions

The key on a scale drawing of a new shopping center states that $\frac{3}{4}$ inches represents 5 feet. If a building is 100 feet long, how long will it be in the drawing?

Many everyday problems of arithmetic can be solved using the ideas of **ratio** and **proportion.** To figure gas mileage, to cut down or increase a recipe, to compare package prices, to draw scale diagrams, or to fill quotas, we can use ratios and proportions.

In this chapter we shall learn to

1. Write ratios and proportions.
2. Find the missing part of a proportion.
3. Switch from word statements to ratios and proportions.

8.1 Understanding Ratios

If there are 39 males and 23 females in a certain class, then we can say that the *ratio* of males to females is 39 to 23.

If instructions for thinning shellac call for mixing 4 pounds of shellac with 2 quarts of alcohol, then we can say that the *ratio* of shellac to alcohol is 4 pounds to 2 quarts.

Perhaps you can see that

> A **ratio** is a way to compare two numbers that are related to each other.

Writing Ratios of Whole Numbers

There are several ways to write a ratio.

Word Form	Colon Form	Fraction Form
39 to 23	39 : 23	$\dfrac{39}{23}$

In each case, a ratio tells us how many of one thing we should have for every so many of another thing. For instance, if the ratio of shellac to thinner is 4 pounds to 2 quarts, then we know we should use 4 pounds of shellac for every 2 quarts of thinner. We may write this ratio as 4 to 2 or 4 : 2 or $\frac{4}{2}$.

Using the fraction bar to write a ratio allows us to reduce our ratio.

$$\frac{4}{2} = \frac{2}{1}$$

This new ratio tells us that we should use 2 pounds of shellac for every 1 quart of thinner. We have not changed the ratio. We have merely rewritten it in reduced form. We will usually write our ratios in fractional form so that we can reduce when possible.

Let's practice interpreting some ratios by writing them in words.

Ratio	Words
$\dfrac{\text{lemonade mix}}{\text{water}} = \dfrac{1 \text{ can}}{3 \text{ cans}}$	We must use 1 can of lemonade mix for every 3 cans of water.
$\dfrac{\text{marriages}}{\text{divorces}} = \dfrac{27}{10}$	For every 27 marriages, there will be 10 divorces.
$\dfrac{\text{unemployed}}{\text{employed}} = \dfrac{9}{100}$	For every 9 unemployed persons, there are 100 employed persons.
$\dfrac{\text{paint}}{\text{area}} = \dfrac{1 \text{ gal}}{450 \text{ sq ft}}$	We must use 1 gallon of paint to cover every 450 square feet of area.

Notice that first writing a ratio in words helps us keep straight which number belongs in which part of the ratio fraction.

Example 1. At Hometown University, there are 12,600 students and 600 faculty members. Write the ratio of the numbers of students to faculty.

Solution

$$\frac{\text{students}}{\text{faculty}} = \frac{12,600}{600}$$

$$= \frac{126}{6} \qquad \text{Divide 12,600 and 600 by 100.}$$

$$= \frac{21}{1} \qquad \text{Divide 126 and 6 by 6.}$$

There are 21 students for every 1 faculty member.

In Example 1, notice that we reduced our ratio fraction, but we did *not* write

$$\frac{21}{1} = 21$$

because that would destroy the idea of a ratio. *A ratio must consist of two numbers.*

You complete Example 2.

Example 2. In Warren County, there are 15,400 registered Democrats and 4400 registered Republicans. Write the ratio of the numbers of Republicans to Democrats.

Solution

$$\frac{\text{Republicans}}{\text{Democrats}} = \frac{4400}{15,400}$$

$$= \underline{\hspace{1cm}} \qquad \text{Divide 4400 and 15,400 by 100.}$$

$$= \underline{\hspace{1cm}} \qquad \text{Divide 44 and 154 by 2.}$$

$$\frac{\text{Republicans}}{\text{Democrats}} = \underline{\hspace{1cm}} \qquad \text{Divide 22 and 77 by 11.}$$

There are _____ Republicans for every _____ Democrats in Warren County.

Check your work on page 326. ▶

Example 3. An 8-ounce jar of instant coffee sells for $3.28. Find the ratio of cents to ounces.

Solution. First we change $3.28 to 328 cents. Now we can write our ratio.

$$\frac{\text{cents}}{\text{ounces}} = \frac{328}{8}$$

$$= \frac{41}{1} \qquad \text{Divide 328 and 8 by 8.}$$

We must pay 41 cents for every 1 ounce of coffee. Or, in other words, this instant coffee costs 41 cents per ounce.

Example 4. If there are 3500 unemployed people in a city with a work force of 38,500 people, write the ratio of the number of unemployed people to the total work force.

Solution

$$\frac{\text{unemployed}}{\text{work force}} = \frac{3500}{38,500}$$

$$= \frac{35}{385} \qquad \text{Divide 3500 and 38,500 by 100.}$$

$$= \frac{7}{77} \qquad \text{Divide 35 and 385 by 5.}$$

$$\frac{\text{unemployed}}{\text{work force}} = \frac{1}{11} \qquad \text{Divide 7 and 77 by 7.}$$

Of every 11 people in the work force, 1 person is unemployed.

ⅠⅢ➡ **Trial Run**

_____ 1. If Felecia has decided that she will save $300 each month from her $1200 take-home pay, write the ratio of her savings to her take-home pay.

_____ 2. In a math class last semester 40 students passed and 6 failed. Write the ratio of the number of students passing to those failing.

_____ 3. José can assemble 20 parts of a carburetor in an hour, but Rick can assemble only 16 parts in the same time. Write the ratio of the number of parts Rick can assemble to the number José can assemble.

_____ 4. In Marcia's aerobics exercise class, there are 18 married persons and 27 single persons. Write the ratio of the numbers of married persons to single persons.

_____ 5. If Amy consumed 350 calories at lunch and 875 calories at dinner, write the ratio of the number of calories consumed at lunch to those consumed at dinner.

_____ 6. A 10-ounce box of snack crackers costs $1.30. Find the ratio of the numbers of cents to ounces.

Answers are on page 326.

Writing Ratios of Decimal and Fractional Numbers

We have seen that ratios can be expressed as fractions with a whole number in the numerator and a whole number in the denominator. But suppose we wish to find the ratio of two *decimal numbers* or of two *fractional numbers*. Can such ratios be written using whole numbers?

Suppose Marcia travels 212.5 miles on 12.5 gallons of gas. Let's try to write the ratio of the number of miles traveled to the number of gallons of gas.

$$\frac{\text{number of miles}}{\text{number of gallons}} = \frac{212.5}{12.5}$$

This ratio is certainly correct, but we would prefer to have whole numbers in the numerator and denominator. Here, we would like to move the decimal point 1 place to the right in the numerator and in the denominator. From our work with fractions and decimal numbers, we know we can accomplish this by multiplying the numerator and denominator by 10.

$$\frac{\text{miles}}{\text{gallons}} = \frac{212.5}{12.5}$$

$$= \frac{212.5 \times 10}{12.5 \times 10}$$

$$= \frac{2125}{125}$$

$$= \frac{85}{5} \qquad \text{Divide 2125 and 125 by 25.}$$

$$\frac{\text{miles}}{\text{gallons}} = \frac{17}{1} \qquad \text{Divide 85 and 5 by 5.}$$

We conclude that Marcia can travel 17 miles on every 1 gallon of gas in her tank. In other words, her car's miles per gallon rating is 17.

Example 5. If a census study shows that there are 2.3 children per family in the United States, use whole numbers to write the ratio of children to families.

Solution. We recall that "2.3 children per family" means "2.3 children for every 1 family." Our ratio must be

$$\frac{\text{number of children}}{\text{number of families}} = \frac{2.3}{1}$$

To change the numerator to a whole number, we can multiply by 10. Of course, if we multiply the numerator by 10, we must also multiply the denominator by 10.

$$\frac{\text{number of children}}{\text{number of families}} = \frac{2.3 \times 10}{1 \times 10}$$

$$= \frac{23}{10}$$

There are 23 children for every 10 families.

As you can see from Example 5, changing to whole numbers often makes a ratio easier to understand. It is difficult to picture 1 family with 2.3 children! It is not difficult to picture 10 families with 23 children.

Example 6. Mario works as a watchman for $5.25 per hour. He also works as a grounds-keeper for $6.30 per hour. Use a ratio of whole numbers to compare his hourly pay as a groundskeeper to his hourly pay as a watchman.

Solution

$$\frac{\text{groundskeeper pay}}{\text{watchman pay}} = \frac{6.30}{5.25}$$

We must multiply the numerator and denominator by 100 to move the decimal points 2 places to the right.

$$\frac{\text{groundskeeper pay}}{\text{watchman pay}} = \frac{6.30 \times 100}{5.25 \times 100}$$

$$= \frac{630}{525}$$

$$= \frac{126}{105} \qquad \text{Divide 630 and 525 by 5.}$$

$$= \frac{6}{5} \qquad \text{Divide 126 and 105 by 21.}$$

For every $6 earned as a groundskeeper, Mario could earn only $5 as a watchman.

Suppose we consider some ratios in which we must compare *fractional numbers* in the numerator and denominator.

Example 7. A meat loaf recipe calls for $\frac{1}{2}$ cup of bread crumbs for every $1\frac{1}{2}$ pounds of ground beef. Write the ratio of ground beef to bread crumbs.

Solution

$$\frac{\text{pounds of ground beef}}{\text{cups of bread crumbs}} = \frac{1\frac{1}{2}}{\frac{1}{2}}$$

This is a *complex fraction* like those discussed in Chapter 5. To simplify this complex fraction, we write the mixed number as an improper fraction and then treat the entire fraction as a division problem.

$$\frac{\text{pounds of ground beef}}{\text{cups of bread crumbs}} = \frac{\frac{3}{2}}{\frac{1}{2}}$$

$$= \frac{3}{2} \div \frac{1}{2}$$

$$= \frac{3}{2} \times \frac{2}{1}$$

$$\frac{\text{pounds of ground beef}}{\text{cups of bread crumbs}} = \frac{3}{1}$$

For every 3 pounds of ground beef, we must use 1 cup of bread crumbs.

Example 8. On a road map, the distance from Montreal to New York City measures $2\frac{1}{5}$ inches and the actual distance from Montreal to New York is 330 miles. Use a ratio of whole numbers to compare the map distance from Montreal to New York to the actual distance.

Solution

$$\frac{\text{map distance}}{\text{actual distance}} = \frac{2\frac{1}{5}}{330}$$

$$= \frac{\frac{11}{5}}{330}$$

$$= \frac{11}{5} \div \frac{330}{1}$$

$$= \frac{11}{5} \cdot \frac{1}{330}$$

$$= \frac{\overset{1}{\cancel{11}}}{5} \cdot \frac{1}{\underset{30}{\cancel{330}}}$$

$$= \frac{1}{150}$$

Every distance of 1 inch on the map represents an actual distance of 150 miles.

Let's summarize the method we have used to change ratios of decimal numbers or fractional numbers to ratios of whole numbers.

Ratios of Decimal Numbers. Multiply numerator and denominator by the number (10, 100, 1000, 10,000, and so on) needed to change the decimal numbers to whole numbers.

Ratios of Fractional Numbers. Treat the ratio of fractional numbers as a complex fraction and divide.

⫸ Trial Run

_____ 1. If Natalie travels 159.5 miles on 7.25 gallons of gasoline, use whole numbers to write the ratio of miles traveled to gallons of gasoline used.

_____ 2. A recent survey shows that there are 1.7 color televisions per family in the United States. Use whole numbers to write the ratio of the number of color televisions to the number of families.

_____ 3. A recipe for candy calls for $\frac{1}{2}$ cup of corn syrup and $2\frac{1}{2}$ cups of sugar. Use whole numbers to write the ratio of cups of sugar to cups of syrup.

_____ 4. On a scale drawing of a house the length of the living room was $6\frac{1}{4}$ inches and the width was $3\frac{3}{4}$ inches. Use whole numbers to write the ratio of the length to the width.

_____ **5.** A factory manufactured 350 small radio parts in $2\frac{1}{3}$ hours. Use whole numbers to write the ratio of the number of parts to the number of hours.

_____ **6.** A pulley with a diameter of 3.5 inches has a circumference of 11 inches. Use whole numbers to write the ratio of the circumference to the diameter.

Answers are given below.

▶ **Example You Completed** _____

Example 2. In Warren County, there are 15,400 registered Democrats and 4400 registered Republicans. Write the ratio of the numbers of Republicans to Democrats.

Solution

$$\frac{\text{Republicans}}{\text{Democrats}} = \frac{4400}{15,400}$$

$$= \frac{44}{154} \qquad \text{Divide 4400 and 15,400 by 100.}$$

$$= \frac{22}{77} \qquad \text{Divide 44 and 154 by 2.}$$

$$\frac{\text{Republicans}}{\text{Democrats}} = \frac{2}{7} \qquad \text{Divide 22 and 77 by 11.}$$

There are 2 Republicans for every 7 Democrats in Warren County.

Answers to Trial Runs _____

page 322   1. $\frac{1}{4}$ 2. $\frac{20}{3}$ 3. $\frac{4}{5}$ 4. $\frac{2}{3}$ 5. $\frac{2}{5}$ 6. $\frac{13}{1}$

page 326   1. $\frac{22}{1}$ 2. $\frac{17}{10}$ 3. $\frac{5}{1}$ 4. $\frac{5}{3}$ 5. $\frac{150}{1}$ 6. $\frac{22}{7}$

EXERCISE SET 8.1

_____ 1. The PTA sold 524 adult's tickets and 786 children's tickets. Write the ratio of the number of adult's tickets to the number of children's tickets.

_____ 2. The width of a house is 30 feet and the length is 50 feet. Write the ratio of the length to the width.

_____ 3. The recommended length of time for cooking a 5-pound roast in a microwave oven is 36 minutes. Write the ratio of the weight in pounds to the length of cooking time in minutes.

_____ 4. For lunch at the school cafeteria, 2468 brownies were baked for 1234 students. Write the ratio of the number of students to the number of brownies.

_____ 5. In 1982, the St. Louis Cardinals played 162 games and won 92. Write the ratio of games won to games played.

_____ 6. A 12-ounce box of rice cereal costs $1.56. Find the ratio of ounces to cents.

_____ 7. At Central University, 45 athletic scholarships are given to football players, while only 15 scholarships are given to basketball players. Write the ratio of the number of basketball scholarships to the number of football scholarships.

_____ 8. If Bernice has decided that she will save $250 each month from her $1050 take-home pay, write the ratio of her savings to her take-home pay.

_____ 9. Last month the city of Callen had 12 days with precipitation and 18 days with no trace of precipitation. Write the ratio of the number of days with no precipitation to the number of days with precipitation.

_____ 10. If Jeff can travel 352 miles on a 16-gallon tank of gasoline, write the ratio of miles to gallons.

_____ 11. Evelyn Ashford holds the women's indoor track record for running 50 yards in 5.83 seconds. Use whole numbers to write the ratio of yards to seconds.

_____ 12. In 1989, the average price received by farmers for hogs was $43.90 per 100 pounds of weight. Use whole numbers to write the ratio of price to weight.

_____ 13. One hundred grams of lima beans have 18.4 grams of protein and the same amount of spinach has 2.3 grams of protein. Use whole numbers to write the ratio of grams of protein in spinach to grams of protein in lima beans.

_____ 14. For refinishing antique furniture, Mr. Savage uses a mixture of $2\frac{3}{4}$ cups of woodcoat and $1\frac{1}{2}$ cups of lacquer thinner. Use whole numbers to write the ratio of cups of lacquer thinner to cups of woodcoat.

_____ 15. A circle with a radius of 1.4 inches has an area of 6.16 square inches. Use whole numbers to write the ratio of the area to the radius.

_____ **16.** A study of TV viewing habits of children reveals that each child under age ten watches, on the average, 3.4 hours of TV per day. Use whole numbers to write the ratio of the number of hours of TV watched per day to the number of children.

_____ **17.** Theo works as a clerk in the bookstore for $5.60 per hour. He also works as a cook at the Burger Hut for $4.20 per hour. Use a ratio of whole numbers to compare his hourly pay as a clerk to his hourly pay as a cook.

_____ **18.** A punch recipe calls for $\frac{1}{4}$ cup of lemon juice for every $2\frac{1}{2}$ cups of orange juice. Use whole numbers to write the ratio of cups of lemon juice to cups of orange juice.

_____ **19.** On a road map the distance from Indianapolis to Philadelphia measures $6\frac{7}{8}$ inches and the distance from Indianapolis to Atlanta measures $5\frac{1}{4}$ inches. Use a ratio of whole numbers to compare the distance from Indianapolis to Atlanta to the distance from Indianapolis to Philadelphia.

_____ **20.** At the track meet, Ray pole-vaulted a height of 16.75 feet but Hans beat him with a height of 18.3 feet. Use a ratio of whole numbers to compare the height of Ray's pole vault to the height Hans pole vaulted.

_____ **21.** The electrical resistance of 6 feet of wire is 3.75 ohms. Use a ratio of whole numbers to compare the ratio of feet of wire to ohms of resistance.

_____ **22.** In his chemistry class, Lou found by experiment that 21.6 grams of water would yield 2.4 grams of hydrogen. Use a ratio of whole numbers to compare grams of water to grams of hydrogen.

_____ **23.** A 3-ounce serving of liver contains 7.5 milligrams of iron. Use a ratio of whole numbers to compare milligrams of iron to ounces of liver.

_____ **24.** On a street map of San Diego, $\frac{1}{2}$ inch represents $1\frac{1}{4}$ miles. Use a ratio of whole numbers to compare inches to miles.

☆ Stretching the Topics ─────────────────────

_____ **1.** During the 1988 baseball season, the Barons won 150 games and lost 65 games. Write the ratio of games won to games played.

_____ **2.** Dominic's car traveled 131.25 miles on 6.25 gallons of gasoline. Tory's car traveled 198 miles on 11 gallons. Use a ratio of whole numbers to compare miles to gallons for each car. Which car had better gas mileage?

_____ **3.** The perimeter of a rectangle is 36 inches and the width is 7 inches. Find the ratio of the width to the length.

Check your answers in the back of your book.

If you can write the ratios in **Checkup 8.1,** you are ready to go on to Section 8.2.

✓ CHECKUP 8.1

_____ 1. At the Regional Medical Center there are 38 male nurses and 190 female nurses. Write the ratio of male nurses to female nurses.

_____ 2. Each month Kenneth saves $360 from his paycheck and spends $840. Write the ratio of the amount he saves to the amount he spends.

_____ 3. The quilt that Amanda is piecing requires $5\frac{1}{3}$ yards of print fabric and 8 yards of solid fabric. Use whole numbers to write the ratio of the amount of print fabric to the amount of solid fabric.

_____ 4. At the Quarter Horse Show, Buddy's horse "ran the barrels" in a time of 16.26 seconds. Larry's horse ran the same course with a time of 21.68 seconds. Use whole numbers to write the ratio comparing the time of Larry's horse to the time of Buddy's horse.

_____ 5. On a scale drawing of the new stadium, $\frac{3}{4}$ inch represents 10 feet. Use whole numbers to write the ratio of inches in the scale drawing to feet in the constructed stadium.

Check your answers in the back of your book.

If You Missed Problems:	You Should Review Examples:
1, 2	1, 2
3	7
4	5, 6
5	8

8.2 Working with Proportions

We know that a ratio is a means of comparing two numbers. Most often, we have used a fraction to express a ratio. Now we must consider **proportions,** which are statements about ratios.

A **proportion** is a statement that says two ratios are equal.

Recognizing Proportions

Actually, we worked with proportions in Section 8.1. Whenever we *reduced* a ratio, we were stating that the original ratio was equal to the reduced ratio. For instance, we agreed that $\frac{4}{2} = \frac{2}{1}$. Such a statement is a proportion.

Similarly, we wrote a proportion when we said that $\frac{2.3}{1} = \frac{23}{10}$. Here, we multiplied the numerator and denominator of the original ratio by 10 to arrive at the new ratio.

Again, we wrote a proportion when we decided that $\frac{1\frac{1}{2}}{\frac{1}{2}} = \frac{3}{1}$. In this case we simplified the complex fraction on the left-hand side to obtain the ratio on the right-hand side.

Deciding whether two ratios are equal is not a difficult process. Let's consider each of the proportions. Notice what happens when we multiply the numerator of the ratio on the left times the denominator of the ratio on the right and compare that product to the product found by multiplying the denominator of the ratio on the left times the numerator of the ratio on the right.

$$\frac{4}{2} \diagdown \!\!\!\!\diagup \frac{2}{1} \qquad \frac{2.3}{1} \diagdown \!\!\!\!\diagup \frac{23}{10} \qquad \frac{1\frac{1}{2}}{\frac{1}{2}} \diagdown \!\!\!\!\diagup \frac{3}{1}$$

$$4 \times 1 = 2 \times 2 \qquad 2.3 \times 10 = 1 \times 23$$

$$4 = 4 \qquad\qquad 23 = 23$$

$$1\frac{1}{2} \times 1 = 3 \times \frac{1}{2}$$

$$\frac{3}{2} \times \frac{1}{1} = \frac{3}{1} \times \frac{1}{2}$$

$$\frac{3}{2} = \frac{3}{2}$$

In each case, the products are equal.

We conclude that we can decide whether two ratios are equal using the following steps.

Equality of Ratios

1. Multiply the numerator of the ratio on the left times the denominator of the ratio on the right.
2. Multiply the denominator of the ratio on the left times the numerator of the ratio on the right.
3. If those products are the same, then the ratios are equal.

The products found in these steps are sometimes called **cross products.**

Two ratios are equal if their cross products are equal.

Example 1. Decide whether the ratio $\frac{15}{12}$ is equal to the ratio $\frac{10}{8}$.

Solution

$$\frac{15}{12} \overset{?}{=} \frac{10}{8}$$

$$\frac{15}{12} \overset{?}{\times} \frac{10}{8}$$

$$15 \times 8 \overset{?}{=} 12 \times 10$$

$$120 = 120$$

We conclude that $\frac{15}{12} = \frac{10}{8}$.

Example 2. Decide whether the ratio $2\frac{1}{3} : \frac{1}{6}$ is equal to the ratio $14 : 1$.

Solution. We write our ratios as fractions.

$$\frac{2\frac{1}{3}}{\frac{1}{6}} \overset{?}{=} \frac{14}{1}$$

$$\frac{\frac{7}{3}}{\frac{1}{6}} \overset{?}{\times} \frac{14}{1} \qquad \text{Change mixed number to improper fraction.}$$

$$\frac{7}{3} \times 1 \overset{?}{=} \frac{1}{6} \times 14 \qquad \text{Write the cross products.}$$

$$\frac{7}{3} \overset{?}{=} \frac{14}{6} \qquad \text{Find the products.}$$

$$\frac{7}{3} = \frac{7}{3} \qquad \text{Reduce on the right.}$$

We conclude that the ratios are equal.

You complete Example 3.

Example 3. Decide whether the ratio $\dfrac{3.5}{0.2}$ is equal to the ratio $\dfrac{5.25}{0.3}$.

Solution

$$\frac{3.5}{0.2} \overset{?}{=} \frac{5.25}{0.3}$$

$$3.5 \times \underline{} \overset{?}{=} 0.2 \times \underline{}$$

$$\underline{} \overset{?}{=} \underline{}$$

Since the cross products are _____ , we conclude that

$$\frac{3.5}{0.2} = \frac{5.25}{0.3}$$

Check your work on page 337. ▶
Then complete Example 4.

Example 4. Decide whether the ratio 25 : 16 is the same as the ratio 5 : 4.

Solution

$$\frac{25}{16} \overset{?}{=} \frac{5}{4}$$

$$25 \times \underline{} \overset{?}{=} 16 \times \underline{}$$

$$\underline{} \overset{?}{=} \underline{}$$

Since the cross products are _____ , the ratios _____ the same.

Check your work on page 337. ▶

⫸ Trial Run

Use cross products to decide whether the following pairs of ratios are equal.

_____ 1. $\dfrac{35}{14}, \dfrac{5}{2}$

_____ 2. $\dfrac{9}{4}, \dfrac{3}{2}$

_____ 3. $\dfrac{3\frac{2}{3}}{\frac{1}{9}}, \dfrac{36}{1}$

_____ 4. $\dfrac{5\frac{1}{2}}{\frac{1}{4}}, \dfrac{22}{1}$

_____ 5. $\dfrac{1.5}{0.3}, \dfrac{3.5}{0.8}$

_____ 6. $\dfrac{6.4}{9.6}, \dfrac{2.8}{4.2}$

Answers are on page 337.

Solving Proportions

Suppose Terry knows he can travel 180 miles on $10 worth of gas. If he has only $3 to spend on gas, how far can Terry travel? This problem contains 1 known ratio of gas money to number of miles and 1 part of a second equivalent ratio. We must develop a method for finding the missing number in the second ratio. The problem can be written

$$\left.\frac{\text{amount of gas money}}{\text{number of miles}}\right\} \; \frac{10}{180} = \frac{3}{?} \; \left\{\frac{\text{amount of gas money}}{\text{number of miles}}\right.$$

and we would like to find ? in the second ratio.

Finding the missing quantity in statements like this is called **solving a proportion.** In order to solve a proportion, we must make use of the fact that two ratios are equal if their cross products are equal. The procedure we will follow is more easily understood if we use letters of the alphabet instead of ? to represent the missing quantity. Let's rewrite Terry's problem, letting m stand for the number of miles he can travel on $3 worth of gas.

$$\frac{10}{180} = \frac{3}{m}$$

Now we use cross products and see that

$$10 \times m = 180 \times 3$$

$$10 \times m = 540$$

This equation says that 10 times m is equal to 540. We would like m to be by itself on the left-hand side of this equation. But how can we get rid of the 10 that is multiplying m?

To undo multiplication, we must *divide*.

Here, we must divide the left-hand side of the equation by 10. But this is an *equation*. Whatever we do to one side we must also do to the other side.

We may divide both sides of an equation by any number (except 0).

Returning to our problem, we have

$$10 \times m = 540$$

$$\frac{10 \times m}{10} = \frac{540}{10} \qquad \text{Divide both sides by } 10.$$

$$\frac{\overset{1}{\cancel{10}} \times m}{\underset{1}{\cancel{10}}} = \frac{540}{10} \qquad \text{Reduce on both sides.}$$

$$m = 54$$

If Terry can travel 180 miles on $10 worth of gas, then he can travel 54 miles on $3 worth of gas. Let's check our proportion, just to be sure.

$$\left.\frac{\text{amount of gas money}}{\text{number of miles}}\right\} \quad \frac{10}{180} \overset{?}{=} \frac{3}{54}$$

$$10 \times 54 \overset{?}{=} 180 \times 3$$

$$540 = 540 \quad \checkmark$$

Example 5. Find the missing part in the proportion $\dfrac{7}{13} = \dfrac{a}{52}$.

Solution

$$\frac{7}{13} = \frac{a}{52}$$

$7 \times 52 = 13 \times a$ Write the cross products.

$364 = 13 \times a$ Multiply on the left side.

$\dfrac{364}{13} = \dfrac{13 \times a}{13}$ Divide both sides by 13.

$\dfrac{\overset{28}{\cancel{364}}}{\underset{1}{\cancel{13}}} = \dfrac{\overset{1}{\cancel{13}} \times a}{\underset{1}{\cancel{13}}}$ Reduce on both sides.

$28 = a$

Example 6. Find the missing part in the proportion $\dfrac{4}{1\frac{1}{3}} = \dfrac{6}{c}$.

Solution

$\dfrac{4}{\frac{4}{3}} = \dfrac{6}{c}$ Change mixed number to improper fraction.

$4 \times c = \dfrac{4}{3} \times \dfrac{6}{1}$ Write the cross products.

$4 \times c = 8$ Multiply on the right side.

$\dfrac{4 \times c}{4} = \dfrac{8}{4}$ Divide both sides by 4.

$\dfrac{\overset{1}{\cancel{4}} \times c}{\underset{1}{\cancel{4}}} = \dfrac{\overset{2}{\cancel{8}}}{\underset{1}{\cancel{4}}}$ Reduce on both sides.

$c = 2$

Example 7. Find the missing part in the proportion $\dfrac{\$2.50}{6} = \dfrac{d}{9}$.

Solution

$$\frac{\$2.50}{6} = \frac{d}{9}$$

$\$2.50 \times 9 = 6 \times d$ Write the cross products.

$\$22.50 = 6 \times d$ Find the product on the left side.

$$\frac{\$22.50}{6} = \frac{6 \times d}{6}$$ Divide both sides by 6

$$\frac{\$22.50}{6} = \frac{\overset{1}{\cancel{6}} \times d}{\underset{1}{\cancel{6}}}$$ Reduce on the right side.

$$\frac{\$22.50}{6} = d$$

$\$3.75 = d$ Find the quotient on the left side.

⮕ Trial Run

Find the missing part in each proportion.

_____ 1. $\dfrac{8}{15} = \dfrac{a}{45}$

_____ 2. $\dfrac{9}{4} = \dfrac{72}{c}$

_____ 3. $\dfrac{b}{2} = \dfrac{25}{3}$

_____ 4. $\dfrac{7}{2\frac{1}{3}} = \dfrac{5}{c}$

_____ 5. $\dfrac{\$3.28}{12} = \dfrac{d}{6}$

_____ 6. $\dfrac{2.3}{a} = \dfrac{5}{2}$

Answers are on page 337.

▶ Examples You Completed

Example 3. Decide whether the ratio $\dfrac{3.5}{0.2}$ is equal to the ratio $\dfrac{5.25}{0.3}$.

Solution

$$\frac{3.5}{0.2} = \frac{5.25}{0.3}$$

$$3.5 \times 0.3 \stackrel{?}{=} 0.2 \times 5.25$$

$$1.05 = 1.05$$

Since the cross products are equal, we conclude that

$$\frac{3.5}{0.2} = \frac{5.25}{0.3}$$

Example 4. Decide whether the ratio 25:16 is the same as the ratio 5:4.

Solution

$$\frac{25}{16} \stackrel{?}{=} \frac{5}{4}$$

$$25 \times 4 \stackrel{?}{=} 16 \times 5$$

$$100 \neq 80$$

Since the cross products are *not* equal, the ratios are *not* the same.

Answers to Trial Runs

page 333   **1.** yes   **2.** no   **3.** no   **4.** yes   **5.** no   **6.** yes

page 336   **1.** 24   **2.** 32   **3.** $\dfrac{50}{3}$ or $16\dfrac{2}{3}$ **4.** $\dfrac{5}{3}$ or $1\dfrac{2}{3}$ **5.** \$1.64   **6.** 0.92

EXERCISE SET 8.2

Use cross products to decide whether the following pairs of ratios are equal.

_____ 1. $\dfrac{5}{7}, \dfrac{15}{14}$

_____ 2. $\dfrac{7}{8}, \dfrac{14}{24}$

_____ 3. $\dfrac{10}{15}, \dfrac{4}{6}$

_____ 4. $\dfrac{6}{10}, \dfrac{9}{15}$

_____ 5. $\dfrac{3}{5}, \dfrac{9}{25}$

_____ 6. $\dfrac{3}{4}, \dfrac{9}{16}$

_____ 7. $\dfrac{4\frac{1}{2}}{\frac{1}{4}}, \dfrac{9}{8}$

_____ 8. $\dfrac{3\frac{2}{3}}{\frac{1}{5}}, \dfrac{11}{15}$

_____ 9. $\dfrac{2\frac{1}{6}}{\frac{1}{12}}, \dfrac{26}{1}$

_____ 10. $\dfrac{3\frac{2}{5}}{\frac{1}{10}}, \dfrac{34}{1}$

_____ 11. $\dfrac{\frac{5}{7}}{3\frac{1}{2}}, \dfrac{2}{5}$

_____ 12. $\dfrac{\frac{3}{5}}{2\frac{1}{2}}, \dfrac{2}{3}$

_____ 13. $\dfrac{\frac{2}{3}}{\frac{5}{6}}, \dfrac{4}{5}$

_____ 14. $\dfrac{\frac{3}{7}}{\frac{9}{10}}, \dfrac{10}{21}$

_____ 15. $\dfrac{1.6}{0.4}, \dfrac{3.2}{0.8}$

_____ 16. $\dfrac{5.6}{0.7}, \dfrac{6.4}{0.8}$

_____ 17. $\dfrac{3.5}{7}, \dfrac{1.25}{2.5}$

_____ 18. $\dfrac{4.8}{2.8}, \dfrac{6}{3.5}$

_____ 19. $\dfrac{3.2}{9.1}, \dfrac{4.3}{8.2}$

_____ 20. $\dfrac{4.5}{8.3}, \dfrac{5.6}{9.4}$

Find the missing part in each proportion.

_____ 21. $\dfrac{6}{7} = \dfrac{a}{14}$

_____ 22. $\dfrac{5}{12} = \dfrac{a}{36}$

_____ 23. $\dfrac{25}{4} = \dfrac{75}{c}$

_____ 24. $\dfrac{16}{9} = \dfrac{80}{c}$

_____ 25. $\dfrac{n}{21} = \dfrac{3}{7}$

_____ 26. $\dfrac{n}{35} = \dfrac{4}{5}$

_____ 27. $\dfrac{13}{x} = \dfrac{2}{3}$

_____ 28. $\dfrac{15}{x} = \dfrac{2}{5}$

_____ 29. $\dfrac{a}{11} = \dfrac{4}{7}$

_____ 30. $\dfrac{a}{13} = \dfrac{5}{9}$

_____ 31. $\dfrac{9}{3\frac{1}{2}} = \dfrac{4}{c}$

_____ 32. $\dfrac{5}{1\frac{1}{2}} = \dfrac{6}{c}$

_____ 33. $\dfrac{n}{5\frac{1}{3}} = \dfrac{9}{4}$

_____ 34. $\dfrac{n}{6\frac{2}{3}} = \dfrac{3}{20}$

_____ 35. $\dfrac{\$3.60}{15} = \dfrac{d}{7}$

_____ 36. $\dfrac{\$4.05}{9} = \dfrac{d}{6}$ _____ 37. $\dfrac{3.5}{9} = \dfrac{7}{x}$ _____ 38. $\dfrac{4.5}{8} = \dfrac{9}{x}$

_____ 39. $\dfrac{a}{7.2} = \dfrac{2}{5}$ _____ 40. $\dfrac{a}{6.4} = \dfrac{3}{4}$

☆ Stretching the Topics _____

Find the missing part in each proportion.

_____ 1. $\dfrac{1\frac{1}{2}}{4\frac{7}{8}} = \dfrac{c}{3\frac{3}{4}}$

_____ 2. $\dfrac{4.75}{14.25} = \dfrac{1.25}{n}$

_____ 3. $\dfrac{\$1.40}{d} = \dfrac{\$73.78}{\$42.16}$

Check your answers in the back of your book.

If you can do the problems in **Checkup 8.2,** then you are ready to go on to Section 8.3.

✓ CHECKUP 8.2

Use cross products to decide if the following pairs of ratios are equal.

_____ 1. $\dfrac{45}{20}, \dfrac{9}{4}$

_____ 2. $\dfrac{3\frac{1}{2}}{\frac{2}{3}}, \dfrac{21}{5}$

_____ 3. $\dfrac{2.3}{0.5}, \dfrac{11.5}{2.5}$

_____ 4. $\dfrac{7.6}{4}, \dfrac{1.9}{1}$

Find the missing part in each proportion.

_____ 5. $\dfrac{7}{12} = \dfrac{a}{36}$

_____ 6. $\dfrac{15}{4} = \dfrac{45}{n}$

_____ 7. $\dfrac{b}{9} = \dfrac{13}{3}$

_____ 8. $\dfrac{9}{4\frac{1}{2}} = \dfrac{2}{c}$

_____ 9. $\dfrac{\$4.35}{8} = \dfrac{d}{24}$

_____ 10. $\dfrac{3.6}{a} = \dfrac{3}{2}$

Check your answers in the back of your book.

If You Missed Problems:	You Should Review Examples:
1–4	1–4
5–7	5
8	6
9, 10	7

8.3 Switching from Word Statements to Proportions

In Section 8.1, we practiced switching from ratios stated in words to ratios stated with numbers. We were careful to write the words corresponding to each part of the ratio to help us keep the right numbers in the right part of the fraction. As we learn to switch from word statements to proportions, we will again use words to label the parts of the ratios to keep the numbers in the right places.

Example 1. At the University Cafeteria hamburgers must consist of 5 parts pure beef for every 2 parts of "filler." If each hamburger contains 3 ounces of beef, how many ounces of filler can be used in each hamburger?

Solution. We can let f stand for the ounces of filler. Our proportion becomes

$$\left.\begin{array}{l}\text{amount of beef}\\\text{amount of filler}\end{array}\right\} \quad \frac{5}{2} = \frac{3}{f}$$

$$5 \times f = 2 \times 3$$

$$5 \times f = 6$$

$$\frac{5 \times f}{5} = \frac{6}{5}$$

$$f = \frac{6}{5}$$

$$f = 1\frac{1}{5}$$

Each hamburger must contain 3 ounces pure beef and $1\frac{1}{5}$ ounces filler.

Example 2. Each month, Claude's spending and saving are in the ratio of 7 : 3. If he saves $360 this month, how much will he spend?

Solution. We can let d stand for the dollars spent. Our proportion becomes

$$\left.\begin{array}{l}\text{amount spent}\\\text{amount saved}\end{array}\right\} \quad \frac{7}{3} = \frac{d}{360}$$

$$7 \times 360 = 3 \times d$$

$$2520 = 3 \times d$$

$$\frac{2520}{3} = \frac{3 \times d}{3}$$

$$\frac{2520}{3} = d$$

$$840 = d$$

If Claude saves $360, then he will spend $840.

Now let's solve the problem stated at the beginning of the chapter.

Example 3. The key on a scale drawing of a new shopping center states that $\frac{3}{4}$ inches represents 5 feet. If a building is 100 feet long, how long will it be in the drawing?

Solution. We can let l stand for the length of the building in the drawing. Our proportion becomes

$$\left. \frac{\text{length in drawing (in.)}}{\text{actual length (ft)}} \right\} \quad \frac{\frac{3}{4}}{5} = \frac{l}{100}$$

$$\frac{3}{4} \times 100 = 5 \times l$$

$$\frac{3}{\cancel{4}} \times \frac{\cancel{100}^{\,25}}{1} = 5 \times l$$
$$\scriptstyle 1$$

$$75 = 5 \times l$$

$$\frac{75}{5} = \frac{5 \times l}{5}$$

$$15 = l$$

The 100-foot building will be 15 inches long in the scale drawing.

Example 4. At the grocery store, a 12-ounce box of Total cereal costs $1.29. What should be the price of a 16-ounce box of Total to make it a better choice for the customer?

Solution. We can let c stand for the cost of the 16-ounce box. Our proportion becomes

$$\left. \frac{\text{weight}}{\text{cost}} \right\} \quad \frac{12}{1.29} = \frac{16}{c}$$

$$12 \times c = 16 \times 1.29$$

$$12 \times c = 20.64$$

$$\frac{12 \times c}{12} = \frac{20.64}{12}$$

$$c = \$1.72$$

If the 16-ounce box costs *less* than $1.72, it is a better choice. If the 16-ounce box costs *more* than $1.72, the 12-ounce box is a better choice. If the 16-ounce box costs *exactly* $1.72, either size is an equally good choice.

As you can see, ratios and proportions are very handy for solving everyday problems. If you become skillful at recognizing and solving proportions, you will find that they can be very useful in many situations. We will find further use for proportions in dealing with percents in Chapter 9.

⫸ **Trial Run**

————— **1.** The ratio of a baseball player's hits to his times at bat is 9 to 25. If he was at bat 375 times, how many hits did he get?

————— **2.** In a company the ratio of workers voting for a strike to those voting against it was 3 to 2. If 51 workers voted to strike, how many voted against it?

————— **3.** The key on a map shows that 1 inch represents 6 miles. If the distance on the map between two cities is $7\frac{1}{4}$ inches, how many miles apart are they?

_____ 4. André can travel 248 miles in 4 hours. If he continues at the same rate of speed, how many hours will it take him to travel 1085 miles?

_____ 5. The ratio of teachers to students at Parker High School is 3 to 70. If the school presently has 24 teachers, what is the student enrollment?

_____ 6. If 5 pounds of roast beef will serve 16 people, how many pounds would be needed to serve a banquet for 192 persons?

Answers are below.

Answers to Trial Run

page 342   1. 135   2. 34   3. $43\frac{1}{2}$ **4.** $17\frac{1}{2}$ 5. 560 students   6. 60

EXERCISE SET 8.3

_____ 1. For the Math Club picnic, Debbie decided that 1 pound of hamburger would serve 4 people. If 76 people sign up to attend the picnic, how many pounds of hamburger should Debbie buy?

_____ 2. If George's punch recipe calls for 1 can of frozen orange juice for every 12 cups of punch, how many cans of juice will he need to make 84 cups of punch?

_____ 3. The ratio of a quarterback's completed passes to attempted passes is 3 to 5. If 485 passes were attempted, how many were completed?

_____ 4. The ratio of your salary to your friend's salary is 5 to 4. Find his salary if you earn $1500 per month.

_____ 5. In a psychology class the ratio of females to males is 3 to 2. If there are 38 males in the class, how many females are there?

_____ 6. On a road map, 1 inch represents 15 miles. If the distance between Birmingham and Chattanooga is 11 inches on the map, how far apart are the two cities?

_____ 7. In a certain math course the ratio of success to failure was 20 to 3. If 15 students failed the course, how many passed?

_____ 8. Lyle can travel 378 miles in 7 hours. Averaging the same rate of speed, how far can he travel in 10 hours?

_____ 9. The ratio of faculty to administrators at Central University is 6 : 1. If there are 138 faculty members, how many administrators are employed at Central University?

_____ 10. If 2 cans of corn will serve 7 people, how many cans will be needed to serve 56 people?

_____ 11. Tom's spending and saving are in the ratio of 8 : 3. If Tom saves $240 this month, how much will he spend?

_____ 12. On a scale drawing of a bookcase, $\frac{3}{4}$ inch represents 2 feet. If the height of the bookcase is 5 feet, how many inches will represent the height in the drawing?

_____ 13. If 1 gallon of paint will cover 300 square feet, how many gallons would be needed to cover an area of 1425 square feet?

_____ 14. If 2.5 pounds of shrimp cost $7.85, how much would 12 pounds cost?

_____ 15. If 3 out of every 100 blenders produced at a factory are defective, how many defective blenders will there be in a shipment of 10,500 blenders?

_____ 16. If 2 doughnuts sell for $0.25, what is the cost of 3 dozen doughnuts?

_____ 17. A picture that is to be enlarged measures $3\frac{3}{8}$ inches wide and 4 inches long. If the enlargement is to have a length of 16 inches, what will be its width?

_____ 18. A machine can make 250 razor blades in 8 minutes. How long will it take to make 1750 razor blades?

_____ 19. In Cayuga County, the ratio of the registered Democrats to Republicans is 5 : 4. If there are 30,725 registered Democrats, how many residents are registered Republicans?

_____ 20. To make concrete, 5 gallons of water are mixed with 3 bags of cement. How many gallons of water are needed for 48 bags of cement?

_____ 21. A nurse has been instructed to administer 50 milligrams of an intravenous dosage every 3 hours. How many milligrams of this medication would be administered in 24 hours?

_____ 22. A credit card user is charged $1.80 interest each month for every $100 of unpaid balance. How much interest would be charged for a month in which the unpaid balance was $1350?

_____ 23. If 22.5 grams of water yield 2.5 grams of hydrogen, how many grams of hydrogen would 270 grams of water yield?

_____ 24. The ratio of a person's weight on earth to that person's weight on Mars is 5 : 2. If Chang's weight on earth is 185 pounds, how much would he weigh on Mars?

☆ Stretching the Topics _____

_____ 1. The water rate for the city of Spring Grove is $1.25 for each 1000 gallons of water used. How many gallons can the Malones use if their water bill is not to exceed $50?

_____ 2. Sixteen out of every 150 cars usually fail to pass a routine safety inspection. If 2700 cars were inspected, how many would you expect to pass the inspection?

_____ 3. If 1 roll of wallpaper will cover an area of 31 square feet, how many full rolls should be bought to paper the walls of a room that is 20 feet long, 16 feet wide, and 8 feet tall?

Check your answers in the back of your book.

If you can do the problems in **Checkup 8.3,** you are ready to do the **Review Exercises** for Chapter 8.

✓ **CHECKUP 8.3**

_____ 1. If Jeff can travel 352 miles on a full 16-gallon tank of gasoline, how many gallons will he need for a trip of 836 miles?

_____ 2. The ratio of a baseball team's games won to games lost was 5 : 3. If the team lost 72 games, how many did it win?

_____ 3. At Union Bank, the employees voted 5 to 2 in favor of a Christmas party. If 14 employees voted against having a party, how many voted for the party?

_____ 4. In Carrie's recipe for meat loaf the ratio of ground beef to sausage is 3 : 2. If she is using a package of ground beef that weighs 4.2 pounds, how many pounds of sausage should she use?

_____ 5. The key on a map shows that 1 inch represents 16 miles. If the distance on the map between two cities is $9\frac{2}{3}$ inches, how many miles apart are the cities?

Check your answers in the back of your book.

If You Missed Problems:	You Should Review Examples:
1	1
2–4	2
5	3

Summary

In this chapter we learned to use a **ratio** to compare two numbers that are related to each other. Ratios can be stated in several ways.

Word Form	Colon Form	Fraction Form
19 to 7	19:7	$\dfrac{19}{7}$

Fraction form is often preferred because it allows us to reduce a ratio when possible.

We also learned to rewrite ratios of decimal numbers or fractional numbers as ratios of whole numbers, which are often easier to interpret.

Original Ratio	Method Used	Ratio of Whole Numbers
$\dfrac{3.73}{1.6}$	Multiply numerator and denominator by 100.	$\dfrac{3.73 \times 100}{1.6 \times 100} = \dfrac{373}{160}$
$\dfrac{1\frac{1}{3}}{\frac{3}{4}}$	Simplify as a complex fraction.	$\dfrac{1\frac{1}{3}}{\frac{3}{4}} = \dfrac{\frac{4}{3}}{\frac{3}{4}} = \dfrac{4}{3} \div \dfrac{3}{4}$
		$= \dfrac{4}{3} \times \dfrac{4}{3}$
		$= \dfrac{16}{9}$

We found we could decide whether two ratios are the same by checking to see if their **cross products** are equal. For example,

$$\frac{2.4}{3} \overset{?}{=} \frac{3.6}{4.5}$$

$$2.4 \times 4.5 \overset{?}{=} 3 \times 3.6$$

$$10.8 = 10.8$$

These ratios *are* the same.

A statement that says that two ratios are equal is called a **proportion.** We learned to use cross products to write an equation that would help us find the missing part in a proportion. To solve such an equation we discovered that we could use division to undo multiplication. For example,

$$\frac{4}{3} = \frac{a}{10}$$

$4 \times 10 = 3 \times a$ Write cross products.

$40 = 3 \times a$ Multiply on the left side.

$$\frac{40}{3} = \frac{\overset{1}{\cancel{3}} \times a}{\underset{1}{\cancel{3}}}$$ Divide both sides by 3.

$13\frac{1}{3} = a$ Simplify the left side.

Finally, we learned to write proportions to solve some everyday problems stated in words.

☐ Speaking the Language of Mathematics ─────────────

Complete each statement with the appropriate word or phrase.

1. A comparison such as 14 : 23 is called a _____ .

2. To rewrite the ratio $\frac{2.7}{3.1}$ as a ratio of whole numbers, we may multiply the _____ and the _____ by _____ .

3. Two ratios are equal if their _____ _____ are equal.

4. A statement that says that two ratios are equal is called a _____ .

5. If we divide the left side of an equation by 5, we must also divide the _____ _____ by 5.

△ Writing About Mathematics ─────────────────────

Write your response to each question in complete sentences.

1. Write a sentence to describe the following ratio: $\dfrac{\text{number of TVs}}{\text{number of VCRs}} = \dfrac{5}{2}$.

2. How can you tell whether two ratios are equal?

3. Describe the mathematics that you would use in making a scale drawing of a room.

REVIEW EXERCISES for Chapter 8

_____ 1. A coffee maker uses 3 tablespoons of coffee for every 8 ounces of water. Write the ratio of tablespoons of coffee to ounces of water.

_____ 2. If it takes 12 gallons of gasoline to drive a certain car 216 miles, write the ratio of gallons of gasoline to miles driven.

_____ 3. If a 4-ounce hamburger patty contains 22.67 milligrams of fat, use a ratio of whole numbers to compare milligrams of fat to ounces of hamburger.

_____ 4. On a trip Anthony traveled 463.68 miles in 11.2 hours. Use a ratio of whole numbers to compare miles to hours.

_____ 5. On a street map $\frac{1}{2}$ inch represents $1\frac{1}{10}$ miles. Use a ratio of whole numbers to compare inches to miles.

Use cross products to decide whether the following pairs of ratios are equal.

_____ 6. $\dfrac{5}{6}, \dfrac{45}{54}$ _____ 7. $\dfrac{11}{21}, \dfrac{2}{3}$

_____ 8. $\dfrac{4\frac{3}{8}}{5\frac{1}{4}}, \dfrac{5}{6}$ _____ 9. $\dfrac{3\frac{1}{3}}{4\frac{2}{3}}, \dfrac{6}{7}$

_____ 10. $\dfrac{43.7}{2.3}, \dfrac{19}{1}$ _____ 11. $\dfrac{3.6}{1.25}, \dfrac{25.2}{8.5}$

Find the missing part in each proportion.

_____ 12. $\dfrac{n}{4} = \dfrac{2}{3}$ _____ 13. $\dfrac{8}{n} = \dfrac{4}{5}$

_____ 14. $\dfrac{6}{17} = \dfrac{3}{n}$ _____ 15. $\dfrac{5}{9} = \dfrac{n}{4}$

_____ 16. $\dfrac{n}{7\frac{1}{2}} = \dfrac{9}{4}$ _____ 17. $\dfrac{8\frac{1}{2}}{3} = \dfrac{n}{4}$

_____ 18. $\dfrac{5.75}{23} = \dfrac{c}{5}$ _____ 19. $\dfrac{a}{8.6} = \dfrac{5}{2}$

_____ 20. If it takes 12 gallons of gasoline to drive a car 216 miles, how many gallons will it take to drive 414 miles?

_____ 21. An investment of $2500 earned $237.50. How much would have to be invested at the same rate to earn $1425 in the same amount of time?

_____ 22. Running against one opponent, Mayor Sloan won an election by the ratio of 9 : 5. If she received 11,700 votes, how many votes did her opponent receive? How many citizens voted in the mayoral election?

_____ 23. If $\frac{1}{2}$ inch on a map represents 50 miles, how many miles would $6\frac{3}{4}$ inches represent?

_____ 24. If 3 cassettes cost $23.94, how many cassettes can be bought for $119.70?

_____ 25. A photograph measuring 8 inches by 10 inches is to be enlarged so that its length is 25 inches. What will be the width of the enlarged picture?

Check your answers in the back of your book.

If You Missed Exercises:	You Should Review Examples:	
1, 2	Section 8.1	1–4
3, 4		5, 6
5		7, 8
6, 7	Section 8.2	1
8, 9		2
10, 11		3
12–15		5
16, 17		6
18, 19		7
20, 21	Section 8.3	1
22		2
23		3
24		4
25		1–4

If you have completed the **Review Exercises** and corrected your errors, you are ready to take the **Practice Test** for Chapter 8.

PRACTICE TEST for Chapter 8

		SECTION	EXAMPLE

_____ 1. At the Regional Medical Center, there are 153 patients and 18 nurses. Write the ratio of the number of nurses to the number of patients.   **8.1**   **1**

_____ 2. A farmer plants 575 acres of corn and 1300 acres of soybeans. Write the ratio of acres of corn to acres of soybeans.   **8.1**   **2**

_____ 3. A 9.6-ounce box of dry milk sells for $1.44. Find the ratio of cents to ounces.   **8.1**   **3**

_____ 4. Last month Damian earned $860.75 and spent $172.15 for food. Use a ratio of whole numbers to compare his food expenses to his income.   **8.1**   **6**

_____ 5. A dieter lost $5\frac{3}{4}$ pounds in 2 weeks. Use a ratio of whole numbers to compare the number of pounds to the number of weeks.   **8.1**   **7**

Use cross products to decide whether the following pairs of ratios are equal.

_____ 6. $\dfrac{3}{8}, \dfrac{39}{104}$ **8.2**   **1**

_____ 7. $\dfrac{65}{72}, \dfrac{5}{6}$ **8.2**   **1**

_____ 8. $\dfrac{3\frac{1}{4}}{\frac{1}{8}}, \dfrac{26}{1}$ **8.2**   **2**

_____ 9. $\dfrac{4.6}{0.3}, \dfrac{27.6}{1.8}$ **8.2**   **3**

Find the missing part in each proportion.

_____ 10. $\dfrac{3}{8} = \dfrac{a}{72}$ **8.2**   **5**

_____ 11. $\dfrac{65}{d} = \dfrac{13}{15}$ **8.2**   **5**

_____ 12. $\dfrac{2\frac{1}{4}}{3} = \dfrac{n}{12}$ **8.2**   **6**

_____ 13. $\dfrac{c}{65} = \dfrac{5}{4\frac{1}{3}}$ **8.2**   **6**

		SECTION	EXAMPLE

_____ 14. $\dfrac{8.5}{c} = \dfrac{187}{143}$ ⟶ 8.2 — 7

_____ 15. $\dfrac{\$5.35}{5} = \dfrac{\$200}{n}$ ⟶ 8.2 — 7

_____ 16. If a recipe calls for $\frac{2}{3}$ cup of milk for a casserole to serve 4 persons, how much milk is needed to increase the recipe to serve 12? ⟶ 8.3 — 1

_____ 17. Each month Sylvia's spending and saving are in the ratio of 8 : 3. If she saves \$150 this month, how much will she spend? ⟶ 8.3 — 2

_____ 18. If $\frac{1}{4}$ inch on a map represents 25 miles, how many inches are required to represent 275 miles? ⟶ 8.3 — 3

_____ 19. At Key Market, 3 cans of green beans cost \$1.25. How much will 15 cans cost? ⟶ 8.3 — 4

Check your answers in the back of your book.

Name _____   **Date** _____

SHARPENING YOUR SKILLS after Chapters 1–8

		SECTION

_____ 1. Classify 2340 as prime or composite. If the number is composite, write it as a product of prime numbers.   2.3

_____ 2. Reduce $\frac{84}{315}$ to lowest terms.   3.2

_____ 3. Use $<$, $>$, or $=$ to compare $\frac{4}{5}$ and $\frac{52}{65}$.   5.4

_____ 4. Round 0.75024 to the thousandths place.   6.1

_____ 5. Write $57.89 in words as it would appear on a check.   6.3

_____ 6. Change $\frac{387}{59}$ to a decimal number and round to the hundredths place.   7.3

_____ 7. Find $38.7 \div 4.9$ and round your answer to the tenths place.   7.2

_____ 8. Use cross products to decide whether the ratios $\frac{45}{63}$ and $\frac{5}{7}$ are equal.   8.2

Perform the indicated operations.

_____ 9. $10,000 - 3876$ 1.3

_____ 10. 787×329 2.1

_____ 11. $138,567 \div 429$ 2.2

_____ 12. $\frac{7}{8} \times \frac{56}{63}$ 4.1

_____ 13. $\frac{5}{6} \times 90$ 4.1

_____ 14. $3\frac{3}{8} \times 12\frac{2}{3}$ 4.1

_____ 15. $\frac{72}{81} \div \frac{80}{18}$ 4.2

_____ 16. $\dfrac{\frac{9}{16}}{1\frac{1}{3}}$ 4.2

_____ 17. $\frac{5}{8} + \frac{7}{12} - \frac{5}{9}$ 5.3

_____ 18. $12\frac{3}{5} - 4\frac{7}{8}$ 5.3

SECTION

_____ 19. $29.263 + 5.8 - 4.889$ 6.2

_____ 20. $12 + 4.02 + 0.3$ 6.2

_____ 21. 0.2543×800 7.1

_____ 22. $0.2784 \div 0.0032$ 7.2

_____ 23. Find the missing part in the proportion $\dfrac{3}{7} = \dfrac{n}{63}$.   8.2

_____ 24. A survey of 1155 voters indicated that 385 were undecided about the lottery   3.3
              proposition. What fractional part of those surveyed were undecided?

_____ 25. Lenora ordered several items from a catalog with the following weights in pounds: 1.3,   6.3
              6.7, 1.9, and 0.7. Find the total weight on which she must pay postage.

_____ 26. Mr. Norris sold 1524 bushels of soybeans for $12,344.40. How much per bushel was   7.4
              he paid for his soybeans?

Check your answers in the back of your book.

9

Working with Percents

The sales tax in New Jersey is 6 percent of the selling price. If the sales tax on a new car is $583.20, what is the selling price of the car?

It is impossible to read a newspaper or watch TV these days without coming across the word **percent.** It is very important that you be able to understand percents and perform calculations involving percents if you wish to be an informed citizen and consumer.

In this chapter we learn to

1. Understand the meaning of percents as fractions and decimal numbers.

2. Find percents using proportions.

3. Find any missing part of a percent statement using a proportion.

The methods for using a calculator to solve problems of the types encountered in Chapters 9 and 10 are discussed in Section 16.3 of Chapter 16.

9.1 Understanding Percents as Fractions and Decimal Numbers

Our work with fractional and decimal numbers will be very important throughout this chapter. Shortly, we shall discover that

> Every percent can be written as a fraction and as a decimal number.

The word **percent** comes from the Latin phrase *per centum* which means *per hundred*. Once you understand this fact, you are well on your way to mastering the study of percents.

> Percent *means* per hundred.

To represent the word "percent" the symbol % is often used. It carries the same meaning as the word percent. So we can write percents in several ways:

Word Form	Meaning	Symbol Form
5 percent	5 *per* 100	5%
28 percent	28 *per* 100	28%
50 percent	50 *per* 100	50%
$\frac{1}{2}$ percent	$\frac{1}{2}$ *per* 100	$\frac{1}{2}$%
0.7 percent	0.7 *per* 100	0.7%

Changing Percents to Fractions

If we know that 11 percent of the people in the labor force are unemployed, then we may conclude that 11 out of every 100 people are unemployed. From our earlier work with ratios and fractions, we know that we may write *11 out of 100* using the fraction $\frac{11}{100}$. Therefore we see that

$$11\% = \frac{11}{100}$$

and we have discovered how to change a percent to a fraction.

Changing a Percent to a Fraction

1. The numerator of the fraction is the number in front of the percent symbol.
2. The denominator of the fraction is 100.

Example 1. Change 79% to a fraction.

Solution

$$79\% = \frac{79}{100}$$

You complete Example 2.

Example 2. Change 3% to a fraction.

Solution

$$3\% = \underline{\hspace{2cm}}$$

Check your work on page 366. ▶

Example 3. Change 64% to a fraction.

Solution

$$64\% = \frac{64}{100}$$

$$= \frac{16}{25} \qquad \text{Divide 64 and 100 by 4.}$$

Notice that we *reduced* our fraction by the usual method.

The reduced form of a fraction often makes it easier to describe the meaning of a percent. For instance, if we are told that 64 percent of entering college freshmen will eventually graduate, then the reduced form of the fraction says that 16 out of every 25 entering freshmen will graduate.

You complete Example 4.

Example 4. Change 75% to a fraction.

Solution

$$75\% = \frac{75}{100}$$

$$= \qquad \text{Divide 75 and 100 by 25.}$$

The reduced form of this fraction tells us that if 75 percent of the registered voters in Kentucky are Democrats, then _____ out of every _____ registered voters in Kentucky are Democrats.

Check your work on page 366. ▶

What happens when the number in front of the percent symbol is *larger* than 100? The meaning of percent is still the same, so to change 171 percent to a fraction, we write

$$171\% = \frac{171}{100}$$

For instance, if the population of a town today is 171 percent of what it was last year, our fraction tells us that there are 171 people now for every 100 people last year. This fraction can be changed to a mixed number ($\frac{171}{100} = 1\frac{71}{100}$) but we will not usually do this because it destroys the idea of "per." It *is* worth noting, however, that

> If the number in front of the percent symbol is larger than 100, then the percent represents a fraction larger than 1.

Example 5. Change 280% to a fraction.

Solution

$$280\% = \frac{280}{100}$$

$$= \frac{28}{10} \qquad \text{Divide 280 and 100 by 10.}$$

$$= \frac{14}{5} \qquad \text{Divide 28 and 10 by 2.}$$

Example 6. Change 500% to a fraction.

Solution

$$500\% = \frac{500}{100}$$

$$= \frac{5}{1}$$

$$= 5$$

▶ **Trial Run**

Change the percents to fractions.

_____ 1. 83%

_____ 2. 7%

_____ 3. 55%

_____ 4. 137%

_____ 5. 150%

_____ 6. 400%

Answers are on page 367.

Is it possible to change a percent such as $\frac{1}{2}\%$ to a fraction? According to the meaning of a percent, we know that

$$\frac{1}{2}\% = \frac{\frac{1}{2}}{100}$$

But this is a complex fraction, so we must use our methods for simplifying complex fractions.

$$\frac{1}{2}\% = \frac{\frac{1}{2}}{100} \qquad \text{Use the meaning of percent.}$$

$$= \frac{1}{2} \div \frac{100}{1} \qquad \text{Rewrite complex fraction as a division problem.}$$

$$= \frac{1}{2} \times \frac{1}{100} \qquad \text{Invert the divisor and multiply.}$$

$$\frac{1}{2}\% = \frac{1}{200} \qquad \text{Find the product.}$$

We can therefore translate $\frac{1}{2}$ percent to mean 1 out of every 200.

You complete Example 7.

Example 7. Change $\frac{3}{4}\%$ to a fraction.

Solution

$$\frac{3}{4}\% = \frac{\frac{3}{4}}{100}$$

$$= \frac{3}{4} \div \frac{100}{1}$$

$$= \underline{\hspace{1cm}} \times \underline{\hspace{1cm}}$$

$$\frac{3}{4}\% = \underline{\hspace{1cm}}$$

Check your work on page 367. ▶

Example 8. Change $12\frac{1}{2}\%$ to a fraction.

Solution. First we rewrite the mixed number as an improper fraction.

$$12\frac{1}{2}\% = \frac{25}{2}\%$$

$$= \frac{\frac{25}{2}}{100}$$

$$= \frac{25}{2} \div \frac{100}{1}$$

$$= \frac{\overset{1}{\cancel{25}}}{2} \times \frac{1}{\underset{4}{\cancel{100}}}$$

$$12\frac{1}{2}\% = \frac{1}{8}$$

▶ Trial Run

Change the percents to fractions.

_____ 1. $\frac{1}{4}\%$

_____ 2. $\frac{3}{5}\%$

_____ 3. $15\frac{3}{4}\%$

_____ 4. $21\frac{2}{3}\%$

_____ 5. $112\frac{1}{2}\%$

_____ 6. $33\frac{1}{3}\%$

Answers are on page 367.

If the number in front of the percent symbol is a decimal number, how can we change the percent to a fraction? Let's use the meaning of percent to change 0.7 percent to a fraction.

$$0.7\% = \frac{0.7}{100}$$

This is correct, but we are not very happy about writing a fraction with a decimal number in the numerator. We can change that decimal number to a whole number by *multiplying by 10* to move the decimal point 1 place to the right. Of course, if we multiply the numerator of the fraction by 10, we must also multiply the denominator by 10.

$$0.7\% = \frac{0.7}{100}$$

$$= \frac{0.7 \times 10}{100 \times 10}$$

$$0.7\% = \frac{7}{1000}$$

Example 9. Change 1.05% to a fraction.

Solution

$$1.05\% = \frac{1.05}{100}$$ Use the meaning of percent.

$$1.05\% = \frac{1.05 \times 100}{100 \times 100}$$ Multiply numerator and denominator by 100.

$$= \frac{105}{10,000}$$

$$= \frac{21}{2000}$$ Divide 105 and 10,000 by 5.

▶ Trial Run

Change the percents to fractions.

_____ **1.** 5.25%   _____ **2.** 0.75%

_____ **3.** 35.7%   _____ **4.** 55.35%

Answers are on page 367.

Changing Percents to Decimal Numbers

In the last section, we learned how to change percents to fractions. In Chapter 7, we learned how to change fractions to decimal numbers. If we can put these two ideas together, we should be able to change percents to decimal numbers.

We know that we can write 7 percent as the fraction $\frac{7}{100}$. But $\frac{7}{100}$ is 7 hundredths, and 7 hundredths can be written as the decimal number 0.07. Therefore we have

$$7\% = \frac{7}{100} = 0.07$$

and we have changed our percent to a decimal number.

You try Example 10.

Example 10. Change 19% to a decimal number.

Solution

$$19\% = \frac{19}{100}$$

$$19\% = \underline{\quad\quad}$$

Check your work on page 367. ▶

Example 11. Change 37.1% to a decimal number.

Solution

$$37.1\% = \frac{37.1}{100}$$

$$= \frac{37.1 \times 10}{100 \times 10}$$

$$= \frac{371}{1000}$$

$$37.1\% = 0.371$$

Can you see a shortcut that we might use to change a percent to a decimal number? Notice that the digits in the decimal number are the same digits that are in front of the percent symbol. Notice that the decimal point is located *2 places to the left* of its location in the percent.

Changing a Percent to a Decimal Number

1. Locate the decimal point in the percent.
2. Drop the percent (%) symbol.
3. Move the decimal point 2 places to the *left*.

Let's practice changing some percents to decimal numbers without first writing them as fractions.

Example 12. Change 10% to a decimal number.

Solution

$$10\% = 10.\%$$
$$= .10$$
$$10\% = 0.1$$

Example 13. Write 4% as a decimal number.

Solution

$$4\% = 4.\%$$
$$= .04$$
$$4\% = 0.04$$

Notice that we attached a 0 that was needed to locate the decimal point in the correct position.

You complete Example 14.

Example 14. Write 0.02% as a decimal number.

Solution

$$0.02\% = \underline{\qquad}$$

Check your work on page 367. ▶

Example 15. Write 219% as a decimal number.

Solution

$$219\% = 219.\%$$
$$= 2.19$$
$$219\% = 2.19$$

Just as we saw earlier, if the number in front of the percent symbol is larger than 100 (see Example 15), the decimal number will be *larger than 1*.

Sometimes it is necessary to change a fractional or mixed number percent to a decimal number. In such cases we may first write the number in front of the percent symbol as a decimal number. Then we may use the method of this section to change the percent to a decimal number.

Example 16. Change $\frac{1}{2}$% to a decimal number.

Solution

$$\frac{1}{2}\% = 0.5\%$$ Write the fraction $\frac{1}{2}$ as the decimal number 0.5.

$$= .005$$ Drop the percent (%) symbol; move the decimal point 2 places *left*.

$$\frac{1}{2}\% = 0.005$$

Example 17. Change $7\frac{2}{3}\%$ to a decimal number, rounded to the ten thousandths place.

Solution

$$7\frac{2}{3}\% = \frac{23}{3}\%$$ Write the mixed number as an improper fraction.

$$= 7.6\overline{6}\%$$ Divide to write the improper fraction as a decimal number.

$$= .076\overline{6}$$ Drop the percent (%) symbol; move the decimal point 2 places *left*.

$$7\frac{2}{3}\% \doteq 0.0767$$ Round to the ten thousandths place.

Changing a Fractional or Mixed Number Percent to a Decimal Number

1. Write the number in front of the percent symbol as a proper or improper fraction.
2. Rewrite the number in front of the percent as a decimal number (using long division if necessary).
3. Drop the percent (%) symbol.
4. Move the decimal point 2 places to the *left*, and round as indicated.

⫸ Trial Run

Change the percents to decimal numbers.

_____ 1. 15%   _____ 2. 3%

_____ 3. 1.7%   _____ 4. 325%

_____ 5. $\frac{1}{20}\%$ _____ 6. $4\frac{1}{3}\%$ (Round to the thousandths place.)

Answers are on page 367.

Changing Decimal Numbers to Percents

If we can change percents to decimal numbers, surely we can do the opposite and change decimal numbers to percents. Suppose we wish to change the decimal number 0.39 to a percent. Our thinking goes like this

$$0.39 \text{ is } 39 \text{ hundredths}$$

$$0.39 = \frac{39}{100}$$

$$0.39 = 39\%$$

The steps used here are just the reverse of the steps used to change percents to decimal numbers.

> **Changing a Decimal Number to a Percent**
>
> **1.** Locate the decimal point in the decimal number.
> **2.** Move the decimal point 2 places to the *right*.
> **3.** Attach a percent (%) symbol.

Example 18. Change 0.032 to a percent.

Solution

$$0.032 = 003.2\%$$
$$0.032 = 3.2\%$$

Example 19. Change 7.1 to a percent.

Solution

$$7.1 = 710.\%$$
$$7.1 = 710\%$$

⫸ Trial Run

Change the decimal numbers to percents.

_____ **1.** 0.03   _____ **2.** 0.058

_____ **3.** 0.375   _____ **4.** 0.006

_____ **5.** 2.5   _____ **6.** 1.75

Answers are on page 367.

Changing Fractions to Percents

If we are asked to change a fraction with a denominator of 100 to a percent, we know we can use the meaning of percent to make such a change.

Example 20. Change $\dfrac{43}{100}$ to a percent.

Solution

$$\frac{43}{100} = 43\%$$

You complete Example 21.

Example 21. Change $\dfrac{187}{100}$ to a percent.

Solution

$$\frac{187}{100} = \underline{\quad}\%$$

Check your work on page 367. ▶

As long as the denominator of the fraction is 100, it takes just 1 step to change a fraction to a percent. The numerator of the fraction will be the number in front of the % symbol. But suppose the denominator of a fraction is *not* 100. Somehow we must rewrite such a fraction as a new fraction having a denominator of 100. Then the numerator of the new fraction will be the percent.

For instance, to change $\frac{1}{4}$ to a percent, we would like to find ? in the statement

$$\frac{1}{4} = \frac{?}{100}$$

But wait a minute. This looks like a *proportion* with a missing part. From Chapter 8, we know how to solve such a proportion. First we let *p* stand for the missing number, and our proportion becomes

$$\frac{1}{4} = \frac{p}{100}$$

$1 \times 100 = 4 \times p$ Write the cross products.

$100 = 4 \times p$ Find the product on the left.

$$\frac{100}{4} = \frac{4 \times p}{4}$$ Divide both sides by 4.

$25 = p$ Find the quotients.

We conclude that our proportion is

$$\frac{1}{4} = \frac{25}{100}$$

and we can read the percent as the numerator of the fraction with the denominator of 100.

$$\frac{1}{4} = 25\%$$

Changing a Fraction to a Percent

1. Write a proportion with your fraction equal to the fraction $\frac{p}{100}$.

2. Solve the proportion (using cross products) to find p.
3. Round your answer as instructed in the problem.
4. Attach a percent (%) symbol to p.

Example 22. Change $\frac{2}{5}$ to a percent.

Solution. We must solve the proportion

$$\frac{2}{5} = \frac{p}{100}$$

$$2 \times 100 = 5 \times p$$

$$200 = 5 \times p$$

$$\frac{200}{5} = \frac{5 \times p}{5}$$

$$40 = p$$

Therefore $\dfrac{2}{5} = 40\%$

Example 23. Change $\frac{7}{11}$ to a percent, rounded to the tenths place.

Solution. We must solve the proportion

$$\frac{7}{11} = \frac{p}{100}$$

$$7 \times 100 = 11 \times p$$

$$700 = 11 \times p$$

$$\frac{700}{11} = \frac{11 \times p}{11}$$

$$\frac{700}{11} = p$$

$$63.6 \doteq p$$

Therefore $\dfrac{7}{11} \doteq 63.6\%$

You complete Example 24.

Example 24. Change $\frac{5}{4}$ to a percent.

Solution. We must solve the proportion

$$\frac{5}{4} = \frac{p}{100}$$

$$5 \times \underline{\hspace{1cm}} = \underline{\hspace{1cm}} \times p$$

$$500 = 4 \times p$$

$$\frac{500}{4} = \frac{4 \times p}{4}$$

$$\underline{\hspace{1cm}} = p$$

Therefore $\frac{5}{4} = \underline{\hspace{0.5cm}}\%$

Check your work on page 367. ▶

Example 25. Change $\frac{231}{382}$ to a percent, rounded to the tenths place.

Solution. We must solve the proportion

$$\frac{231}{382} = \frac{p}{100}$$

$$231 \times 100 = 382 \times p$$

$$23{,}100 = 382 \times p$$

$$\frac{23{,}100}{382} = \frac{382 \times p}{382}$$

$$60.5 \doteq p$$

Therefore $\frac{231}{382} \doteq 60.5\%$

⇒ Trial Run

Change each fraction to a percent.

_____ 1. $\frac{65}{100}$

_____ 2. $\frac{138}{100}$

_____ 3. $\frac{1}{125}$

_____ 4. $\frac{5}{8}$

_____ 5. $\frac{11}{9}$ (Round to tenths place.)

_____ 6. $\frac{285}{370}$ (Round to tenths place.)

Answers are on page 367.

We have done quite a bit of changing in this section. Perhaps a chart would help us keep our methods straight.

To Change	To	We Must	Examples
a percent	a fraction	Use a denominator of 100.	$83\% = \dfrac{83}{100}$
			$121\% = \dfrac{121}{100}$
a percent	a decimal number	Move the decimal point 2 places to the *left* and drop the % symbol.	$91\% = 0.91$
			$7\% = 0.07$
			$150\% = 1.5$
			$\dfrac{2}{5}\% = 0.4\% = 0.004$
a decimal number	a percent	Move the decimal point 2 places to the *right* and attach a % symbol.	$0.32 = 32\%$
			$0.5 = 50\%$
			$0.013 = 1.3\%$
a fraction	a percent	Use a proportion, letting p stand for the unknown percent. Solve for p and attach a % symbol.	$\dfrac{1}{5} = \dfrac{p}{100}$
			$1 \times 100 = 5 \times p$
			$100 = 5 \times p$
			$\dfrac{100}{5} = \dfrac{5 \times p}{5}$
			$20 = p$
			So $\dfrac{1}{5} = 20\%$

▶ Examples You Completed ─────────────

Example 2. Change 3% to a fraction.

Solution

$$3\% = \frac{3}{100}$$

Example 4. Change 75% to a fraction.

Solution

$$75\% = \frac{75}{100}$$

$$75\% = \frac{3}{4}$$

The reduced form of this fraction tells us that if 75 percent of the registered voters in Kentucky are Democrats, then 3 out of every 4 registered voters in Kentucky are Democrats.

Example 7. Change $\frac{3}{4}\%$ to a fraction.

Solution

$$\frac{3}{4}\% = \frac{\frac{3}{4}}{100}$$

$$= \frac{3}{4} \div \frac{100}{1}$$

$$= \frac{3}{4} \times \frac{1}{100}$$

$$\frac{3}{4}\% = \frac{3}{400}$$

Example 21. Change $\frac{187}{100}$ to a percent.

Solution

$$\frac{187}{100} = 187\%$$

Example 10. Change 19% to a decimal number.

Solution

$$19\% = \frac{19}{100}$$

$$19\% = 0.19$$

Example 14. Write 0.02% as a decimal number.

Solution

$$0.02\% = .0002$$

$$0.02\% = 0.0002$$

Example 24. Change $\frac{5}{4}$ to a percent.

Solution. We must solve the proportion

$$\frac{5}{4} = \frac{p}{100}$$

$$5 \times 100 = 4 \times p$$

$$500 = 4 \times p$$

$$\frac{500}{4} = \frac{4 \times p}{4}$$

$$125 = p$$

Therefore $\qquad \frac{5}{4} = 125\%$

Answers to Trial Runs

page 358   **1.** $\frac{83}{100}$ **2.** $\frac{7}{100}$ **3.** $\frac{11}{20}$ **4.** $\frac{137}{100}$ **5.** $\frac{3}{2}$ **6.** $\frac{4}{1}$

page 359   **1.** $\frac{1}{400}$ **2.** $\frac{3}{500}$ **3.** $\frac{63}{400}$ **4.** $\frac{13}{60}$ **5.** $\frac{9}{8}$ **6.** $\frac{1}{3}$

page 360   **1.** $\frac{21}{400}$ **2.** $\frac{3}{400}$ **3.** $\frac{357}{1000}$ **4.** $\frac{1107}{2000}$

page 362   **1.** 0.15   **2.** 0.03   **3.** 0.017   **4.** 3.25   **5.** 0.0005   **6.** 0.043

page 363   **1.** 3%   **2.** 5.8%   **3.** 37.5%   **4.** 0.6%   **5.** 250%   **6.** 175%

page 365   **1.** 65%   **2.** 138%   **3.** 0.8%   **4.** 62.5%   **5.** 122.2%   **6.** 77.0%

EXERCISE SET 9.1

Change the percents to fractions. Reduce if possible.

_____ 1. 79%   _____ 2. 37%   _____ 3. 8%

_____ 4. 6%   _____ 5. 40%   _____ 6. 75%

_____ 7. 250%   _____ 8. 180%   _____ 9. 500%

_____ 10. 300%   _____ 11. $\frac{1}{2}$%   _____ 12. $\frac{1}{4}$%

_____ 13. $\frac{3}{10}$%   _____ 14. $\frac{1}{5}$%   _____ 15. $8\frac{2}{3}$%

_____ 16. $9\frac{1}{3}$%   _____ 17. $115\frac{1}{2}$%   _____ 18. $120\frac{1}{4}$%

_____ 19. $33\frac{1}{3}$%   _____ 20. $66\frac{2}{3}$%   _____ 21. 4.25%

_____ 22. 9.75%   _____ 23. 43.9%   _____ 24. 59.6%

_____ 25. 20.75%   _____ 26. 80.25%   _____ 27. 0.5%

_____ 28. 0.6%   _____ 29. 170.05%   _____ 30. 150.05%

Change the percents to decimal numbers.

_____ 31. 23%   _____ 32. 35%   _____ 33. 62.3%

_____ 34. 47.9%   _____ 35. 80%   _____ 36. 30%

_____ 37. 6%   _____ 38. 8%   _____ 39. 0.03%

_____ 40. 0.07%   _____ 41. 375%   _____ 42. 550%

_____ 43. 10.23%   _____ 44. 15.82%   _____ 45. $\frac{1}{4}$%

_____ 46. $\frac{1}{5}$%   _____ 47. $\frac{3}{10}$%   _____ 48. $\frac{7}{10}$%

_____ 49. $5\frac{1}{2}$%   _____ 50. $12\frac{3}{4}$%

_____ 51. $10\frac{1}{8}$% (Round to the ten thousandths place.)

_____ 52. $11\frac{5}{6}$% (Round to the ten thousandths place.)

368

Change the decimal numbers to percents.

_____ 53. 0.09   _____ 54. 0.07   _____ 55. 0.036

_____ 56. 0.057   _____ 57. 0.275   _____ 58. 0.315

_____ 59. 0.008   _____ 60. 0.002   _____ 61. 3.8

_____ 62. 9.6   _____ 63. 1.32   _____ 64. 1.76

_____ 65. 2.535   _____ 66. 4.256

Change the fractions to percents. (Round answers to the tenths place.)

_____ 67. $\dfrac{35}{100}$ _____ 68. $\dfrac{54}{100}$ _____ 69. $\dfrac{8}{100}$

_____ 70. $\dfrac{9}{100}$ _____ 71. $\dfrac{145}{100}$ _____ 72. $\dfrac{235}{100}$

_____ 73. $\dfrac{2}{5}$ _____ 74. $\dfrac{3}{4}$ _____ 75. $\dfrac{9}{1000}$

_____ 76. $\dfrac{7}{1000}$ _____ 77. $\dfrac{3}{8}$ _____ 78. $\dfrac{5}{8}$

_____ 79. $\dfrac{5}{16}$ _____ 80. $\dfrac{3}{16}$ _____ 81. $\dfrac{5}{9}$

_____ 82. $\dfrac{7}{11}$ _____ 83. $\dfrac{12}{25}$ _____ 84. $\dfrac{13}{20}$

_____ 85. $\dfrac{15}{8}$ _____ 86. $\dfrac{7}{3}$ _____ 87. $\dfrac{36}{125}$

_____ 88. $\dfrac{27}{40}$

☆ Stretching the Topics _____

_____ 1. Change 0.005% to a fraction.

_____ 2. Change $\frac{179}{52}$ to a percent and round your answer to the hundredths place.

_____ 3. In a recent election, 75% of the state's eligible voters cast ballots in favor of a lottery. Use whole numbers to write a ratio comparing the number of voters favoring the lottery to the total number of eligible voters.

Check your answers in the back of your book.

If you can complete **Checkup 9.1,** you are ready to go on to Section 9.2.

✓ CHECKUP 9.1

Change the percents to fractions.

_____ 1. 85%   _____ 2. $\frac{3}{4}$%   _____ 3. 6.5%

Change the percents to decimal numbers.

_____ 4. 22.5%   _____ 5. $3\frac{1}{2}$%

Change the decimal numbers to percents.

_____ 6. 4.5   _____ 7. 0.623

Change the fractions to percents.

_____ 8. $\frac{83}{100}$ _____ 9. $\frac{4}{5}$ _____ 10. $\frac{2}{3}$ (Round to tenths place.)

Check your answers in the back of your book.

If You Missed Problems:	You Should Review Examples:
1–3	1–9
4, 5	10–17
6, 7	18, 19
8–10	20–25

9.2 Using the Percent Proportion

The important **percent proportion**

$$\frac{\text{numerator}}{\text{denominator}} = \frac{p}{100}$$

will be very useful to us in solving many kinds of problems involving percents. The only part of this proportion that never changes is the 100 in the denominator on the right-hand side.

Finding p in the Percent Proportion

Having learned to change any fraction to a percent using the percent proportion, we are ready to consider solving percent problems stated in words. For instance,

If 15 out of 25 members of the track team are seniors, what *percent* of the track team is made up of seniors?

In this situation we have two numbers that can be compared using a *ratio*. But we know that we can always represent a ratio of two numbers as a *fraction*. Then, using the percent proportion, we can change that fraction to a *percent*. Here, we consider the ratio of the number of seniors (15) to the total number of team members (25).

$$\left.\begin{array}{r}\text{seniors} \\ \hline \text{total}\end{array}\right\}\quad \frac{15}{25} = \frac{3}{5} \qquad \text{Reduce the fraction.}$$

$$\frac{3}{5} = \frac{p}{100} \qquad \text{Write the percent proportion.}$$

$$3 \times 100 = 5 \times p \qquad \text{Write the cross products.}$$

$$300 = 5 \times p$$

$$\frac{300}{5} = \frac{5 \times p}{5} \qquad \text{Divide both sides by } 5$$

$$60 = p$$

We see that 60 percent of the track team is made up of seniors.

Finding What Percent One Number Is of Another Number

1. Write the ratio of the numbers as a fraction (putting the number following the word "of" in the denominator).
2. Reduce the ratio fraction, if possible.
3. Set your ratio fraction equal to $\frac{p}{100}$ and solve for p in that proportion.
4. The p you find will be the percent.

Example 1. If Amy's salary last year was $12,000 and she received a $1200 raise this year, what was her *percentage* raise?

Solution. To find Amy's percentage raise, we must consider the ratio

$$\left.\begin{array}{r}\text{raise} \\ \hline \text{old salary}\end{array}\right\}\quad \frac{1200}{12,000} = \frac{1}{10}$$

Now we change the reduced fraction to a percent by using the percent proportion.

$$\frac{1}{10} = \frac{p}{100}$$

$$1 \times 100 = 10 \times p$$

$$100 = 10 \times p$$

$$\frac{100}{10} = \frac{10 \times p}{10}$$

$$10 = p$$

We conclude that Amy received a 10 percent raise.

Example 2. Last year's enrollment at City College was 8000 students. This year's enrollment is 8560 students. What percent of last year's enrollment is this year's enrollment?

Solution. Since last year's enrollment follows the word ''of'' in the question, our ratio is $\frac{\text{this year's enrollment}}{\text{last year's enrollment}}$. First we reduce the ratio fraction.

$$\frac{8560}{8000} = \frac{856}{800} \qquad \text{Divide 8560 and 8000 by 10.}$$

$$= \frac{107}{100} \qquad \text{Divide 856 and 800 by 8.}$$

Now we solve the percent proportion

$$\frac{107}{100} = \frac{p}{100}$$

Since the denominators of both fractions are 100, we know the numerators must be the same.

$$107 = p$$

Therefore, we conclude that this year's enrollment is 107 percent of last year's enrollment.

Example 3. Of the 3,615,000-square-mile area of the United States, 78,000 square miles are water. What percent of the U.S. area is water? (Round your answer to the tenths place.)

Solution. Since the U.S. area follows the word "of" in the question, our ratio must be

$$\left.\frac{\text{water area}}{\text{U.S. area}}\right\} \frac{78,000}{3,615,000} = \frac{78}{3615}$$

Now we solve the percent proportion

$$\frac{78}{3615} = \frac{p}{100}$$

$$78 \times 100 = 3615 \times p$$

$$7800 = 3615 \times p$$

$$\frac{7800}{3615} = \frac{3615 \times p}{3615}$$

$$\frac{7800}{3615} = p$$

$$2.2 \doteq p$$

We conclude that approximately 2.2 percent of the area of the United States is water.

⫸ Trial Run

_____ 1. What percent of 35 is 7?

_____ 2. What percent of 125 is 40?

_____ 3. There are 45 students enrolled in Albert's psychology class. If there are 27 women in the class, what percent of the class is women?

_____ 4. Mr. Robinson has a 586-acre farm. If he has planted 325 acres of corn, what percent of the farm is planted in corn? (Round to whole percent.)

_____ 5. In the last election, 385 persons voted in the Raleigh precinct. There were 176 Republican votes cast. What percent of the total vote in the precinct was Republican? (Round to whole percent.)

_____ 6. Pat's house payment each month is $243. If his monthly salary is $1093, what percent of his salary is his house payment? (Round to tenths place.)

Answers are on page 383.

Finding the Denominator in the Percent Proportion

Keeping the percent proportion in mind, consider this problem.

Julie knows that 15 percent of all the students in her math class are freshmen. If there are 12 freshmen in the class, what is the total number of students in the class?

To solve Julie's problem we see that the percent proportion we must use is

$$\frac{\text{number of freshman students}}{\text{total number of students}} = \frac{p}{100}$$

We know the number of freshman students is 12, and we know the percent, *p,* is 15, but we do *not* know the total number of students. We can let *d* stand for the missing denominator and solve this proportion using cross products.

$$\frac{12}{d} = \frac{15}{100}$$

$$12 \times 100 = 15 \times d$$

$$1200 = 15 \times d$$

$$\frac{1200}{15} = \frac{15 \times d}{15}$$

$$80 = d$$

Since the denominator, *d,* represents the total number of students, we know there are 80 students in Julie's math class.

Example 4. During one season Bob Lanier made 55 percent of the field goal shots he attempted. If he made 726 field goals, how many shots did Lanier attempt?

Solution. Our ratio is

$$\frac{\text{field goals made}}{\text{field goals attempted}}$$

and we put the known numbers into the percent proportion.

$$\frac{726}{d} = \frac{55}{100}$$

$$726 \times 100 = 55 \times d$$

$$72,600 = 55 \times d$$

$$\frac{72,600}{55} = \frac{55 \times d}{55}$$

$$1320 = d$$

Lanier attempted 1320 field goals.

Now we can solve the problem stated at the beginning of this chapter.

Example 5. The sales tax in New Jersey is 6 percent of the selling price. If the sales tax on a new car is $583.20, what is the selling price of the car?

Solution. Our ratio is

$$\frac{\text{sales tax}}{\text{selling price}}$$

and we put the known numbers into the percent proportion.

$$\frac{583.20}{d} = \frac{6}{100}$$

$$583.20 \times 100 = 6 \times d$$

$$58{,}320 = 6 \times d$$

$$\frac{58{,}320}{6} = \frac{6 \times d}{6}$$

$$9720 = d$$

The selling price of the new car is $9720.

Example 6. 96.512 is 116 percent of what number?

Solution. We see that p is 116, but the number following the word "of" is unknown. Our percent proportion becomes

$$\frac{96.512}{d} = \frac{116}{100}$$

$$96.512 \times 100 = 116 \times d$$

$$9651.2 = 116 \times d$$

$$\frac{9651.2}{116} = \frac{116 \times d}{116}$$

$$\frac{9651.2}{116} = d$$

$$83.2 = d$$

Therefore, 96.512 is 116 percent of 83.2.

⟫ Trial Run

_____ **1.** At midseason the Toronto Bluejays had won 55 percent of their games. If they had won 44 games, how many games had they played?

_____ **2.** Twenty-four percent of what amount is $32.40?

_____ **3.** 46.25 is 125 percent of what number?

_____ **4.** Ellen saves 12 percent of her salary each month. If she deposits $119.20 each month in her savings account, what is her monthly salary?

_____ **5.** The rate per day for a semiprivate room at St. Mary's Medical Center has been increased by 11.5 percent. If the increase is $20.20, what was the rate per day before the increase? (Round answer to cents.)

_____ **6.** If he works on holidays, Calvin's hourly wages are 175 percent of what he regularly earns. On Memorial Day last year he earned $15.75 per hour. What does he regularly earn per hour? (Round answer to cents.)

Answers are on page 383.

Finding the Numerator in the Percent Proportion

In the percent proportion

$$\frac{numerator}{denominator} = \frac{p}{100}$$

let's consider situations in which we know the denominator and we know p, but we must find the numerator.

Helena spends 18 percent of her monthly take-home pay for groceries. If her monthly take-home pay is $900, how much does she spend for groceries?

Here, the percent proportion we must consider is

$$\frac{grocery\ money}{take\text{-}home\ pay} = \frac{p}{100}$$

Putting the known numbers in the proper places and using n for the missing numerator, our proportion becomes

$$\frac{n}{900} = \frac{18}{100}$$

$$n \times 100 = 900 \times 18$$

$$n \times 100 = 16{,}200$$

$$\frac{n \times 100}{100} = \frac{16{,}200}{100}$$

$$n = 162$$

So Helena spends $162 for groceries each month.

Example 7. The baseball team won 70 percent of its games this season. If the team played a total of 30 games, how many games did the team win?

Solution. We know the ratio we need is

$$\frac{\text{games won}}{\text{games played}}$$

and we put the known numbers into the percent proportion.

$$\frac{n}{30} = \frac{70}{100}$$

$$n \times 100 = 30 \times 70$$

$$n \times 100 = 2100$$

$$\frac{n \times 100}{100} = \frac{2100}{100}$$

$$n = 21$$

Therefore, the team won 21 games.

Example 8. If the Tennessee sales tax is 6.5 percent of the selling price, how much sales tax will you pay on a record whose selling price is $7.95? (Round your answer to cents.)

Solution. Our ratio is

$$\frac{\text{sales tax}}{\text{selling price}}$$

and we put the known numbers into the percent proportion.

$$\frac{n}{7.95} = \frac{6.5}{100}$$

$$n \times 100 = 7.95 \times 6.5$$

$$n \times 100 = 51.675$$

$$\frac{n \times 100}{100} = \frac{51.675}{100}$$

$$n = \frac{51.675}{100}$$

Using the shortcut to divide by 100 (by moving the decimal point 2 places to the *left*), we find

$$n = \$0.51675$$

Rounding *up* to the nearest cent, you will pay $0.52 in sales tax.

Example 9. Find 129 percent of 500.

Solution. We know p is 129 and we know the denominator of our ratio is 500 (because 500 follows the word "of"). Our percent proportion becomes

$$\frac{n}{500} = \frac{129}{100}$$

$$n \times 100 = 500 \times 129$$

$$n \times 100 = 64{,}500$$

$$\frac{n \times 100}{100} = \frac{64{,}500}{100}$$

$$n = 645$$

So 129 percent of 500 is 645.

▐▐▶ Trial Run

_____ 1. June sells real estate and receives a 5.5 percent commission on each property she sells. If she sells a house for $64,500, how much is her commission?

_____ 2. A basketball player made 93.75 percent of her free throws during one season. If she shot 304 times, how many free throws did she make?

_____ 3. A couch regularly sells for $738. In the sale catalog the price is advertised as being reduced by 33 percent. How much will Jim save if he buys the couch on sale?

_____ 4. In a recent survey, 8,280,000 households were polled. If 86 percent had color televisions, how many households owned a color TV?

_____ 5. The work force at the Curtis Cap Factory this year is 250 percent of last year's. If 36 persons were employed last year, how many are employed this year?

_____ 6. The cost of Mrs. Day's arthritis prescription increased by 15 percent this month. If her medicine cost $40.80 last month, how much more must she pay this month?

Answers are on page 383.

You may remember from earlier mathematics classes that your teacher talked about "the 3 kinds of percent problems." For each kind of percent problem, you used a different method to find the answer. If you are a typical student, you probably found that it was very easy to get confused about which method to use for which problem.

In Section 9.2 we have actually solved all 3 kinds of percent problems, but we used the *percent proportion* to do them all! The nice thing about the percent proportion is that it can be used to solve *any* kind of percent problem.

The percent proportion

$$\frac{\text{numerator}}{\text{denominator}} = \frac{p}{100}$$

can be used to solve any percent problem.

Once you learn to put the given numbers in the proper places in the percent proportion, you can always solve the proportion for the missing piece of information. To be sure you can do this, let's try one more problem of each type.

Example 10. A drugstore advertises a senior citizen discount of 15 percent of the selling price. What will be the senior citizen discount on a prescription that regularly sells for $6.80? What will be the discounted price?

Solution. The ratio we must use to find the discount is

$$\frac{\text{discount}}{\text{selling price}}$$

and our percent proportion becomes

$$\frac{n}{6.80} = \frac{15}{100}$$

$$n \times 100 = 6.80 \times 15$$

$$n \times 100 = 102$$

$$\frac{n \times 100}{100} = \frac{102}{100}$$

$$n = 1.02$$

The discount will be $1.02 and the discounted price will be

$$\$6.80 - \$1.02 = \$5.78$$

You complete Example 11.

Example 11. In a freshman class of 4100 students, 2337 are receiving financial aid. What percent of the freshman class is receiving aid?

Solution. The ratio we must use is

$$\frac{\text{freshmen receiving aid}}{\text{all freshmen}}$$

and our percent proportion becomes

$$\frac{2337}{4100} = \frac{p}{100}$$

$$2337 \times \underline{\hphantom{xx}} = \underline{\hphantom{xx}} \times p$$

$$\underline{\hphantom{xx}} = 4100 \times p$$

$$\frac{233{,}700}{4100} = \frac{4100 \times p}{4100}$$

$$\underline{\hphantom{xx}} = p$$

Therefore, ____ percent of the freshman class is receiving aid.

Check your work on page 383. ▶

Example 12. In Union County, 13 percent of the total work force is unemployed. If 1664 people are unemployed, how many people are in the total work force in Union County?

Solution. The ratio we must use is

$$\frac{\text{number unemployed}}{\text{total work force}}$$

and our percent proportion becomes

$$\frac{1664}{d} = \frac{13}{100}$$

$$1664 \times 100 = 13 \times d$$

$$166{,}400 = 13 \times d$$

$$\frac{166{,}400}{13} = \frac{13 \times d}{13}$$

$$12{,}800 = d$$

The total work force in Union County is 12,800 people.

||||▶ **Trial Run**

  _____ **1.** At the local canning factory, 38 out of 57 workers voted to accept a new wage agreement. What percent of the workers voted for the new agreement?

  _____ **2.** In the 10-mile ride, 306 bicyclists finished the race. If the number finishing the race was 72 percent of the total number entered, how many were entered in the race?

_____ 3. During a snowy winter day, 35 percent of the students in the Union County school system were absent. If the total enrollment is 3440, how many students were absent?

_____ 4. Last week a department store fired 15 of its 90 sales clerks. What percent of the sales force was let go?

_____ 5. A presidential candidate in the last election decided to reduce his $400,000 a month campaign budget by 22 percent. By how much did he reduce his budget? What was his new monthly budget?

_____ 6. At the local theater 28 percent of those attending an R-rated movie were under 16 years of age. If there were 182 persons under age 16, what was the total attendance for that movie?

Answers are on page 383.

Finding Some Percent of a Number Without the Percent Proportion (Alternate Method)

We can use the percent proportion to solve any kind of percent problem. However, sometimes you will find it easier to use a different method to solve problems in which you are asked to _find a specific percent of some number_.

Finding a Percent of Some Number

1. Change the percent to a decimal number (by moving the decimal point 2 places to the left and dropping the percent symbol).
2. Multiply this decimal number times the other number.

Example 13. Find 6 percent of $97.

Solution. 6% of $97 = 0.06 × $97.

$$
\begin{array}{r}
97 \\
0.06 \\
\hline
5.82
\end{array}
$$

Therefore, 6 percent of $97 is $5.82.

Example 14. Find 5.75 percent of $2800.

Solution. 5.75% of $2800 = 0.0575 × $2800.

$$
\begin{array}{r}
0.0575 \\
2800 \\
\hline
46\ 0000 \\
115\ 0 \quad\ \\
\hline
161.0000
\end{array}
$$

Therefore, 5.75 percent of $2800 is $161.

Example 15. Find 123 percent of 6400.

Solution. 123% of 6400 = 1.23 × 6400.

$$
\begin{array}{r}
1.23 \\
\underline{6400} \\
492\ 00 \\
\underline{738} \\
7872.00
\end{array}
$$

If you would rather continue to use the percent proportion to solve Example 15 you may certainly do so.

$$\frac{n}{6400} = \frac{123}{100}$$

$$n \times 100 = 6400 \times 123$$

$$n \times 100 = 787{,}200$$

$$\frac{n \times 100}{100} = \frac{787{,}200}{100}$$

$$n = 7872$$

No matter which method is used, we find that 123 percent of 6400 is 7872.

➠ Trial Run

_____ 1. Find 8.5 percent of $164.

_____ 2. A survey showed that 73 percent of those persons polled preferred Twinkle brand toothpaste. If 12,500 persons were polled, how many chose Twinkle brand?

_____ 3. At an appliance store all merchandise has been reduced by 30 percent. If a refrigerator regularly sells for $735, by how much has the price been reduced? What is the reduced price?

_____ 4. There were 7000 applicants for the first Summer Space Camp at Huntsville. If only 20 percent of the applicants could be accepted, how many campers were accepted? How many were rejected?

_____ 5. If Kentucky has a 5 percent sales tax, how much tax will Natalie pay on a stereo that costs $589.95? What will she pay altogether?

_____ 6. The total revenue from the lottery in Michigan last year was $527,400,000. If the state receives 41.5 percent of the revenue, what was the state's share?

Answers are on page 383.

▶ Examples You Completed

Example 11. In a freshman class of 4100 students, 2337 are receiving financial aid. What percent of the freshman class is receiving aid?

Solution. The ratio we must use is

$$\frac{\text{freshmen receiving aid}}{\text{all freshmen}}$$

and our percent proportion becomes

$$\frac{2337}{4100} = \frac{p}{100}$$

$$2337 \times 100 = 4100 \times p$$

$$233{,}700 = 4100 \times p$$

$$\frac{233{,}700}{4100} = \frac{4100 \times p}{4100}$$

$$57 = p$$

Therefore, 57 percent of the freshman class is receiving aid.

Answers to Trial Runs

page 373   1. 20%   2. 32%   3. 60%   4. 55%   5. 46%   6. 22.2%

page 376   1. 80   2. $135   3. 37   4. $993.33   5. $175.65   6. $9

page 378   1. $3547.50   2. 285   3. $243.54   4. 7,120,800   5. 90   6. $6.12

page 380   1. $66.\overline{6}\%$ 2. 425   3. 1204   4. $16.\overline{6}\%$ 5. $88,000; $312,000   6. 650

page 382   1. $13.94   2. 9125   3. $220.50; $514.50   4. 1400; 5600   5. $29.50; $619.45
             6. $218,871,000

EXERCISE SET 9.2

Solve. In your answers, round percents to the tenths place.

_____ 1. A local movie theater has 400 seats. For last Sunday's matinee only 122 persons bought tickets. What percent of the seats were filled for the matinee?

_____ 2. A baseball player had 127 hits in 351 times at bat. What percent of the times at bat did he get a hit?

_____ 3. There are 48 students enrolled in Andrew's history class. On the day before spring break 12 students were absent. What percent of the class missed the day before spring break?

_____ 4. Jason saves $200 each month from his monthly salary of $1250. What percent of his salary does Jason save?

_____ 5. Of $239,000 worth of property reported to the police as stolen, about $41,000 worth was recovered. What percent of the stolen property was recovered?

_____ 6. For one game last season, Memorial Stadium was only filled to 85 percent capacity. If attendance at that game was 17,085, what is the total seating capacity of Memorial Stadium?

_____ 7. Fifty-five percent of the math books in stock at the bookstore are used books. If there are 341 used math books at the bookstore, how many math books are in stock?

_____ 8. The cost of a seat for the concert series this year has risen by 12 percent. If the increase is $5.82, what was the cost of a seat last year?

_____ 9. Staton contributes 8 percent of his annual salary to charity. Last year his contributions to charity totaled $1807.20. What was his salary last year?

_____ 10. The state of Florida has 18,282,000 acres of forest land, which is 50 percent of the total area. What is the total area of Florida in acres?

_____ 11. Mike buys a car for $9835 in a state that has a 5.5 percent sales tax. How much sales tax (to the nearest cent) will he pay? How much will Mike pay altogether?

_____ 12. A stadium seats 57,200 persons. If it was filled to 80 percent capacity for a recent game, how many seats were filled?

_____ 13. The average annual precipitation for a certain city is 38.7 inches. Last year it received 115 percent of its average precipitation. How many inches of precipitation did it receive last year? (Round to tenths place.)

_____ 14. Sixty-three percent of the employees at the refrigerator factory voted to accept the union contract. If there are 4500 employees, how many voted to accept the contract?

_____ 15. Ella weighs 184 pounds, but her doctor says she weighs about 115 percent of what she should weigh. How much should Ella weigh? How many pounds should Ella lose?

_____ 16. When 2000 students at an elementary school were examined for vision defects, 12.4 percent needed glasses. How many children needed glasses?

_____ 17. The Division of Weights and Measures made 28,130 inspections of scales used to weigh foods. They found that 26,165 were accurate. What percent of the scales were found to be accurate? (Round to whole percent.)

_____ 18. During the month of March a factory manufactured an average of 480 small radio parts per day. The production level in April was 130 percent of the March production. How many small radio parts were produced per day in April?

_____ 19. A survey of university students showed that 82 percent use at least one of the services provided by the student center. If 12,300 use the facilities, how many students are enrolled at the university?

_____ 20. A dining room suite regularly selling for $1028 is advertised at 25 percent off. How much will be saved by purchasing the dining room suite on sale? What will be the sale price of the suite?

_____ 21. The National Safety Council reported that 52,000 out of 106,000 accidental deaths were related to motor vehicle accidents. Find what percentage of the deaths were related to motor vehicles. (Round to whole percent.)

_____ 22. The population of a city is estimated to be 98,000. In 10 years it is estimated that the population will increase by 13,200. Find the percentage increase.

_____ 23. In 1989, *Sports Illustrated* had a circulation of 2,284,800. If their circulation increased by 12 percent during the next year, how many *more* copies were sold? How many copies were sold in 1990?

_____ 24. In a certain precinct in the governor's race, Collins received 4645 votes. If 8764 persons voted, find what percent of the vote Collins received. (Round to whole percent.)

_____ 25. A clothing store advertises a student discount of 15 percent off the selling price. What will be the discount on a sweater that regularly sells for $35.50? What will be the discounted price?

☆ Stretching the Topics _____

_____ 1. A survey shows that $24\frac{3}{4}$ percent of the residents in a certain viewing area watch the local news show. If the viewing area has approximately 600,000 residents, how many of the residents do not watch the local news?

_____ 2. The drop count in an intravenous solution bottle was 24 drops per minute before the doctor ordered an increase in the drop count of $33\frac{1}{3}$ percent. What was the new drop count?

_____ 3. In a survey of 1500 teachers and 2200 parents, the participants were asked to rank the possible causes of student difficulties in school. Sixty-two percent of the teachers and 59 percent of the parents agreed that parents leave their children alone too much after school. How many participants in the survey thought children were left alone too much?

Check your answers in the back of your book.

If you can complete **Checkup 9.2,** then you are ready to do the **Review Exercises** for Chapter 9.

 **CHECKUP 9.2**

Solve.

_____ 1. When Greg bought a new car for $9600 he received a $672 rebate. What percent of the price of the car was the rebate?

_____ 2. During the 1981-1982 season the Philadelphia 76-ers attempted 2457 field goals and made 1817. Find the percentage of field goals the team made. (Round to whole percent.)

_____ 3. George receives a 15 percent commission on each house he sells. If his commission on his last sale was $7275, what was the selling price of the house?

_____ 4. When Maurice and his date went out to dinner, the restaurant bill was $26.45. If they left the usual 15 percent tip, how much tip, to the nearest cent, did they leave? How much did they spend altogether?

_____ 5. A survey of 10,500 persons revealed the fact that 62 percent attended at least one movie last year. How many of those surveyed attended at least one movie last year?

Check your answers in the back of your book.

If You Missed Problems:	You Should Review Examples:
1, 2	1–3
3	4–6
4, 5	7–9

Summary

In this chapter we became familiar with the idea of **percents.** Understanding that the word *percent* means "per hundred" helped us learn how to change among percents, fractions, and decimal numbers. Sometimes it was easier to make such changes using the **percent proportion:**

$$\frac{\text{numerator}}{\text{denominator}} = \frac{p}{100}$$

To Change	To	We Can	Examples
a percent	a fraction	Use a denominator of 100.	$17\% = \dfrac{17}{100}$
a percent	a decimal number	Move the decimal point 2 places to the left and drop the percent (%) symbol.	$6\% = 0.06$ $129\% = 1.29$ $0.5\% = 0.005$
a decimal number	a percent	Move the decimal point 2 places to the right and attach a percent (%) symbol.	$0.19 = 19\%$ $0.3 = 30\%$ $0.091 = 9.1\%$
a fraction	a percent	Use the percent proportion, letting p stand for the unknown percent. Solve for p and attach a percent (%) symbol.	$\dfrac{3}{4} = \dfrac{p}{100}$ $3 \times 100 = 4 \times p$ $300 = 4 \times p$ $75 = p$ So $\dfrac{3}{4} = 75\%$

Having learned to put the numbers in their proper places in the percent proportion, we were able to use that proportion to solve *any* percent problem. We practiced switching from word statements about percents to percent problems that could be solved using the percent proportion. As before, we stated our answer to each word problem in sentence form.

❏ Speaking the Language of Mathematics ———————

Complete each statement with the appropriate word or phrase.

1. The word percent means _____ _____ .

2. To change p percent to a fraction, we rewrite it as _____ .

3. To change a percent to a decimal number, we move the decimal point _____ places to the _____ and _____ the percent (%) symbol.

4. To change a decimal number to a percent, we move the decimal point _____ places to the _____ and _____ the percent (%) symbol.

5. The statement $\dfrac{\text{numerator}}{\text{denominator}} = \dfrac{p}{100}$ is called the _____ _____ .

△ Writing About Mathematics _____

Write your response to each question in complete sentences.

1. Show and explain the steps for using the percent proportion to change $\frac{3}{8}$ to a percent.

2. Outline the procedure needed to find a reduced price if the reduction is given as a percent of the original price.

REVIEW EXERCISES for Chapter 9

Change the percents to fractions.

_____ 1. 38%   _____ 2. 125%   _____ 3. $45\frac{1}{2}$%

_____ 4. $23\frac{1}{3}$%   _____ 5. 28.7%   _____ 6. 0.55%

_____ 7. 15.25%   _____ 8. 0.125%

Change the percents to decimal numbers.

_____ 9. 82%   _____ 10. 35.7%   _____ 11. 17%

_____ 12. 0.05%   _____ 13. 253%   _____ 14. 45.75%

Change the decimal numbers to percents.

_____ 15. 0.08   _____ 16. 0.039   _____ 17. 0.375

_____ 18. 3.5   _____ 19. 1.68   _____ 20. 4.325

Change the fractions to percents. (Round answers to the tenths place.)

_____ 21. $\frac{84}{100}$ _____ 22. $\frac{5}{100}$ _____ 23. $\frac{235}{100}$

_____ 24. $\frac{5}{8}$ _____ 25. $\frac{13}{25}$ _____ 26. $\frac{28}{7}$

_____ 27. Of the 425 runners who started the marathon, only 306 finished. What percent of those starting finished the marathon?

_____ 28. The average annual precipitation for New Orleans is 56.8 inches. Last year the total precipitation in New Orleans was 65.32 inches. What percent of the average precipitation was last year's precipitation?

_____ 29. Next year Mr. Tucker will receive a 6 percent cost-of-living raise. If his raise will be $3120, what is his salary this year? What will be his salary next year?

_____ 30. The population of a town has increased by 5445 in the past 10 years. If this represents a 15 percent increase, what was the population 10 years ago? What is the population now?

_____ 31. The Department of Health inspected 1480 restaurants and refused operating permits to 15 percent. How many restaurants did not get permits?

_____ 32. Dennis uses 25 percent of his take-home pay for rent. If his take-home pay is $953, how much rent does he pay?

Check your answers in the back of your book.

If You Missed Exercises:	You Should Review Examples:	
1, 2	Section 9.1	1–6
3, 4, 5		7–9
9–14		10–15
15–20		16, 17
21–26		18–23
27, 28	Section 9.2	1–3
29, 30		4–6
31, 32		7–9

If you have completed the **Review Exercises** and corrected your errors, you are ready to take the **Practice Test** for Chapter 9.

PRACTICE TEST for Chapter 9

		Section	Examples

Change the percents to fractions.

		Section	Examples
_____ 1.	9%	9.1	2, 3
_____ 2.	175%	9.1	5, 6
_____ 3.	$13\frac{3}{4}\%$	9.1	7, 8
_____ 4.	2.25%	9.1	9

Change the percents to decimal numbers.

_____ 5.	23%	9.1	10
_____ 6.	75.2%	9.1	11
_____ 7.	0.05%	9.1	14
_____ 8.	345%	9.1	15

Change the decimal numbers to percents.

_____ 9.	0.5	9.1	16
_____ 10.	0.048	9.1	16
_____ 11.	0.39	9.1	16
_____ 12.	1.25	9.1	17

Change the fractions to percents. (Round answers to the tenths place.)

_____ 13.	$\frac{83}{100}$	9.1	18
_____ 14.	$\frac{3}{5}$	9.1	20
_____ 15.	$\frac{8}{15}$	9.1	21
_____ 16.	$\frac{276}{828}$	9.1	23
_____ 17.	A survey showed that 3800 of 5000 parents preferred to send their children to public rather than private schools. What percent of those surveyed preferred public schools?	9.2	2

Change the percents to fractions.   SECTION   EXAMPLES

_____ **18.** In Gum Grove 80 percent of the offenses reported to the police   9.2   4
department were cleared up by the arrest of the offender. If 7252
arrests were made, how many offenses were reported to the
department? How many offenses resulted in no arrest of the offender?

_____ **19.** Out of the 3,966,716 vehicles that passed over a toll bridge, 75   9.2   7
percent were passenger cars. How many passenger cars passed over
the toll bridge?

_____ **20.** In April of last year, Evansville had 126 percent of its normal April   9.2   6
precipitation. If the precipitation last April was 15.8 inches, what is
the normal April precipitation? (Round answer to the hundredths
place.)

Check your answers in the back of your book.

SHARPENING YOUR SKILLS after Chapters 1–9

		SECTION
_____	**1.** Write the word form for 28,370.	1.1
_____	**2.** Change $15\frac{7}{8}$ to an improper fraction.	3.1
_____	**3.** Reduce $\frac{90}{117}$ to lowest terms.	3.2
_____	**4.** Change $\frac{7}{9}$ to an equivalent fraction with a denominator of 135.	5.2
_____	**5.** Use $<$, $>$, or $=$ to compare $3\frac{2}{3}$ and $\frac{57}{15}$.	5.4
_____	**6.** Write 0.379 as a fraction.	6.1
_____	**7.** Change $\frac{26}{3}$ to a decimal number and round to the tenths place.	7.3
_____	**8.** Use cross products to decide whether the ratios $\frac{3}{4}$ and $\frac{9}{16}$ are equal.	8.2
_____	**9.** Change $12\frac{1}{4}$ percent to a fraction.	9.1
_____	**10.** Change $\frac{6}{5}$ to a percent.	9.1

Perform the indicated operations.

		SECTION
_____	**11.** $\dfrac{7}{12} \times 56$	4.1
_____	**12.** $9\dfrac{1}{3} \times 1\dfrac{7}{8}$	4.1
_____	**13.** $6\dfrac{2}{5} \div 5\dfrac{1}{3}$	4.2
_____	**14.** $\dfrac{9}{7} + \dfrac{2}{3} + \dfrac{7}{15}$	5.3
_____	**15.** $9\dfrac{3}{20} - 4\dfrac{2}{5}$	5.3
_____	**16.** 18.7×13.2	7.1
_____	**17.** $9 - 0.324 - 3.782$	6.2
_____	**18.** Find $38.7 \div 0.23$ and round to the hundredths place.	7.2
_____	**19.** Find the missing part in the proportion $\dfrac{75}{n} = \dfrac{5}{3}$.	8.2
_____	**20.** In January, Julie charged purchases of $28, $115, and $73. If she paid $125 on her account on February 1, how much did she still owe for January purchases?	1.4

_____ 21. A rectangular garden 52 feet by 39 feet is to be shared equally by 3 families for
vegetable gardens. How many square feet of garden space will each family have?

_____ 22. For a play, 531 tickets were sold in advance and the rest were sold at the door. If 720
tickets were sold altogether, what fractional part of the tickets were sold in advance?

_____ 23. If the area of a floor is $82\frac{1}{3}$ square feet and its length is $12\frac{2}{3}$ feet, what is its width?

_____ 24. This year Scarlet planted $\frac{1}{4}$ of the cropland on her farm in wheat, $\frac{3}{5}$ in soybeans, and $\frac{1}{8}$
in corn. What fractional part of her cropland did she leave unplanted?

_____ 25. The yearly premium per $1000 of life insurance for a 45-year-old female is $23.18. If
Tillaya is 45 years old, find her premium for a $60,000 life insurance policy.

_____ 26. Nezam's spending and saving are in the ratio of 7 : 3. If he saves $175 this month,
how much will he spend? (Round answer to nearest cent.)

Check your answers in the back of your book.

10

Using Percents
to Solve Problems

When Stacey bought a $12 souvenir outside her home state, she was charged $0.78 sales tax. Help Stacey find the sales tax rate in this state.

We stated many percent problems in words throughout Chapter 9. The key to solving such problems was to decide where the numbers in the problem belonged in the **percent proportion.**

Now we must turn our attention to some word problems in which we may have to use more than one step to answer the question being asked. Although we will continue to use the percent proportion, we may be required to perform some other calculations as well.

In this chapter we learn how to

1. Work with markup and markdown.
2. Use percents to describe changes.
3. Compute taxes using percents.
4. Compute simple interest and compound interest.

The methods for using a calculator to solve problems of the types encountered in Chapters 9 and 10 are discussed in Section 16.3 of Chapter 16.

10.1 Working with Markup and Markdown

As consumers, we should become familiar with the calculations used by retail stores to determine the prices we pay for the goods we purchase.

Working with Markup

When a retail store buys goods from a wholesaler, the store does not sell those goods to us (the consumers) at the same price it paid to the wholesaler. Instead, the retailer **marks up** the goods to pay for the expenses of running the store and to make some profit. The actual selling price of any item is therefore equal to the cost paid to the wholesaler *plus* the markup.

> Selling price = Cost + Markup

Let's be certain that we understand the three quantities in this important equation.

Cost is the amount paid by the retailer to the wholesaler.

Markup is the amount added on by the retailer.

Selling price is the amount paid by the consumer to the retailer.

In general, you should agree that the amount of markup is the *difference* between the selling price and the cost.

> Markup = Selling price − Cost

Suppose the Video Shop buys a TV from Zenith for $250 and sells it to Jim for $375. What is the amount of markup on this TV? In this situation,

Wholesaler	Retailer	Consumer
Zenith	Video Shop	Jim

The *cost* of the TV is the amount that the Video Shop pays to Zenith, the *markup* is the amount added on by the Video Shop, and the *selling price* is the amount Jim pays to the Video Shop for the TV.

$$\text{Selling price} = \$375$$

$$\text{Cost} = \$250$$

$$\text{Markup} = \text{Selling price} - \text{Cost}$$
$$= \$375 \quad - \$250$$
$$= \$125$$

The Video Shop has marked up the TV by $125.

Most of the time retail stores decide to mark up their items by a certain **percent of the cost.** Can you figure out what percent of the cost was used by the Video Shop in marking up Jim's TV?

To find the markup as a percent of the cost, we must consider the *ratio* of markup to cost and use the percent proportion.

$$\left.\frac{\text{Markup}}{\text{Cost}}\right\} \quad \frac{125}{250} = \frac{p}{100}$$

$$\frac{1}{2} = \frac{p}{100} \qquad \text{Reduce the fraction on the left.}$$

$$1 \times 100 = 2 \times p \qquad \text{Write the cross products.}$$

$$100 = 2 \times p \qquad \text{Find the product on the left.}$$

$$\frac{100}{2} = \frac{2 \times p}{2} \qquad \text{Divide both sides by 2.}$$

$$50 = p \qquad \text{Find the quotients.}$$

We conclude that the Video Shop used a markup that was 50 percent of the cost.

You complete Example 1.

Example 1. The Jeans Barn buys Levis for \$12 a pair and sells them for \$16 a pair. Find the amount of markup. Then find the markup as a percent of the cost.

Solution. Here we know the selling price (\$16) and the cost (\$12), so we can find the amount of markup.

$$\text{Markup} = \text{Selling price} - \text{Cost}$$

$$= \$\underline{\hspace{1cm}} - \$\underline{\hspace{1cm}}$$

$$\text{Markup} = \$\underline{\hspace{1cm}}$$

Now to compute the markup as a percent of the cost, we use the percent proportion.

$$\left.\frac{\text{Markup}}{\text{Cost}}\right\} \quad \frac{4}{12} = \frac{p}{100}$$

$$\frac{1}{3} = \frac{p}{100}$$

$$1 \times 100 = 3 \times p$$

$$\underline{\hspace{1cm}} = p$$

The Jeans Barn uses a markup that is _____ percent of the cost.

Check your work on page 404. ▶

Suppose you work in a retail record store that uses a standard markup that is 39 percent of the cost. How can you find the dollar amount of markup on a record that costs $5 from the wholesaler? Once again, you can use the percent proportion. We know here that

$$\left.\begin{array}{c}\text{Markup} \\ \text{Cost}\end{array}\right\} \quad \frac{M}{5} = \frac{39}{100}$$

$$M \times 100 = 5 \times 39$$

$$M \times 100 = 195$$

$$\frac{M \times 100}{100} = \frac{195}{100}$$

$$M = 1.95$$

This record will be marked up $1.95 before it is sold to the customer. Can you figure out the *selling price* for this record?

$$\text{Selling price} = \text{Cost} + \text{Markup}$$

$$\text{Selling price} = \$5 \quad + \$1.95$$

$$= \$6.95$$

So this record will sell for $6.95.

Discount stores and large grocery stores usually use a much lower percentage markup than department stores or specialty shops because they expect to sell more merchandise.

Example 2. The Foodway Super Market buys tomato juice from Libby's at 92¢ per can. If Foodway uses a markup that is 6 percent of the cost, find the selling price of a can of Libby's tomato juice.

Solution. First we find the amount of markup.

$$\left.\begin{array}{c}\text{Markup} \\ \text{Cost}\end{array}\right\} \quad \frac{M}{92} = \frac{6}{100}$$

$$M \times 100 = 92 \times 6$$

$$M \times 100 = 552$$

$$\frac{M \times 100}{100} = \frac{552}{100}$$

$$M = 5.52¢$$

A retail store will probably round this amount *up* to the nearest cent (6¢).

$$\text{Selling price} = \text{Cost} + \text{Markup}$$

$$= 92¢ + 6¢$$

$$\text{Selling price} = 98¢$$

Libby's tomato juice will sell for 98 cents per can.

⫸ Trial Run

1. The Carter Lumber Company buys decorative vinyl paneling for $10.36 per panel and sells it for $12.95. Find the amount of markup. Then find the markup as a percent of the cost.

_____ 2. Thorton's Used Auto Sales advertised a 1978 Plymouth station wagon for $1525. If Mr. Thorton paid the original owner $925 for the wagon, find the amount of markup. Then find the markup as a percent of the cost. (Round answer to the tenths place.)

_____ 3. Floyd's Market buys chickens for 33¢ per pound and marks them up by 30 percent. Find the selling price per pound, rounding _up_ to the nearest cent.

_____ 4. Reburn's Jewelry Store has a ruby ring in the window. If Mr. Reburn bought the ring for $365.30 and marked it up 150 percent of the cost, find the selling price of the ring.

Answers are on page 405.

Working with Markdown

If a sign in a store reads

> **SALE**
> All coats marked down
> $15

you would expect all the coats in the store to be selling for a new price that is $15 _less_ than the original price. You expect

$$\text{New price} = \text{Old price} - \$15$$

We call this $15 reduction in the price the **markdown,** and we agree that

> New price = Old price − Markdown

If you know the new and old prices, you can figure the amount of markdown because

> Markdown = Old price − New price

Example 3. If a chair that regularly sells for $315 is ''on sale'' for $210, find the amount of markdown.

Solution. Here we know the old price ($315) and the new price ($210), so we find

$$\text{Markdown} = \text{Old price} - \text{New price}$$
$$= \$315 - \$210$$
$$\text{Markdown} = \$105$$

The chair has been marked down $105.

In Example 3, suppose we wish to find the **percent of markdown** represented by the dollar markdown of $105. To compute the percent of markdown, we must compare the dollar markdown to the *old* price and use the percent proportion.

$$\left.\begin{array}{c}\text{Markdown} \\ \hline \text{Old price}\end{array}\right\} \quad \frac{105}{315} = \frac{p}{100}$$

$$\frac{1}{3} = \frac{p}{100} \qquad \text{Reduce the fraction on the left.}$$

$$1 \times 100 = 3 \times p \qquad \text{Write the cross products.}$$

$$100 = 3 \times p \qquad \text{Find the product on the left.}$$

$$\frac{100}{3} = \frac{3 \times p}{3} \qquad \text{Divide both sides by 3.}$$

$$33.\overline{3} = p \qquad \text{Find the quotients.}$$

We see that the chair has been marked down $33.\overline{3}$ percent.

Now complete Example 4.

Example 4. If a jacket that regularly sells for $33 is "on sale" for $24.75, find the percent of markdown.

Solution. First we find the dollar amount of markdown.

$$\text{Markdown} = \text{Old price} - \text{New price}$$

$$= \$\text{\_\_\_\_} \quad - \$\text{\_\_\_\_}$$

$$= \$\text{\_\_\_\_}$$

Now we use the percent proportion.

$$\left.\begin{array}{c}\text{Markdown} \\ \hline \text{Old price}\end{array}\right\} \quad \frac{8.25}{33} = \frac{p}{100}$$

$$\text{\_\_\_\_} \times 100 = \text{\_\_\_\_} \times p$$

$$825 = 33 \times p$$

$$\frac{825}{33} = \frac{33 \times p}{33}$$

$$\text{\_\_\_\_} = p$$

The jacket has been marked down ____ percent.

Check your work on page 405. ▶

Suppose a newspaper ad states

SALE
All stereos marked down
30%

What would you expect to be true? You would expect all the stereos to be selling for a new price that is 30 percent *less* than the old price.

Example 5. If a stereo had been selling for $420 before this "sale" what would you expect the new price to be?

Solution. First we compute the *dollar amount* of markdown.

$$\left.\frac{\text{Markdown}}{\text{Old price}}\right\} \quad \frac{M}{420} = \frac{30}{100}$$

$$M \times 100 = 420 \times 30$$

$$M \times 100 = 12{,}600$$

$$\frac{M \times 100}{100} = \frac{12{,}600}{100}$$

$$M = 126$$

Then we find the new price.

$$\text{New price} = \text{Old price} - \text{Markdown}$$

$$= \$420 \quad - \$126$$

$$= \$294$$

The stereo should be "on sale" for $294.

Example 6. Both the Shoe Shack and the Footery sell a certain pair of shoes at a regular price of $23. This week the Shoe Shack advertises a $5 markdown on all shoes and the Footery advertises a 25 percent markdown. Where will the $23 pair of shoes be selling for less? What will be the cheaper selling price?

Solution. First we decide which store offers the large dollar markdown.

$$\textit{Shoe Shack:} \qquad \text{Markdown} \quad = \$5$$

$$\textit{Footery:} \qquad \left.\frac{\text{Markdown}}{\text{Old price}}\right\} \quad \frac{M}{23} = \frac{25}{100}$$

$$M \times 100 = 23 \times 25$$

$$M \times 100 = 575$$

$$\frac{M \times 100}{100} = \frac{575}{100}$$

$$M = 5.75$$

Since the markdown at the Footery ($5.75) is larger than the markdown at the Shoe Shack ($5), the $23 pair of shoes will sell for less at the Footery. To find the new price at the Footery we must use

$$\text{New price} = \text{Old price} - \text{Markdown}$$

$$= \$23.00 \quad - \$5.75$$

$$\text{New price} = \$17.25$$

ⅢⅢ➡ Trial Run

_____ **1.** A garden tiller that is regularly priced $995.87 is on sale for $875.29. Find the amount of the markdown.

_____ 2. If a radial tire that regularly sells for $80.20 is on sale for $44.11, find the amount of the markdown. Then find the percent of markdown.

_____ 3. For their Labor Day weekend sale, French's Market marked down hot dog buns from 93 cents to 62 cents per package. Find the percent of markdown.

_____ 4. At its grand opening sale, Waterbed Land has all merchandise marked down 35 percent for the first 3 days. If a waterbed regularly sells for $485, find the amount of the markdown. Then find the sale price of the waterbed.

_____ 5. Glazer's is having a Father's Day sale on recliners. A recliner that regularly sells for $370 can be purchased at "18 percent off." Find the sale price of the recliner.

Answers are on page 405.

▶ **Examples You Completed** _____

Example 1. The Jeans Barn buys Levis for $12 a pair and sells them for $16 a pair. Find the amount of markup. Then find the markup as a percent of cost.

Solution. Here we know the selling price ($16) and the cost ($12), so we can find the amount of markup.

$$\text{Markup} = \text{Selling price} - \text{Cost}$$
$$= \$16 \qquad - \$12$$
$$\text{Markup} = \$4$$

Now to compute the markup as a percent of cost, we use the percent proportion.

$$\left.\frac{\text{Markup}}{\text{Cost}}\right\} \quad \frac{4}{12} = \frac{p}{100}$$

$$\frac{1}{3} = \frac{p}{100}$$

$$1 \times 100 = 3 \times p$$

$$100 = 3 \times p$$

$$\frac{100}{3} = \frac{3 \times p}{3}$$

$$33.\overline{3} = p$$

The Jeans Barn uses a markup that is $33.\overline{3}$ percent of cost.

Example 4. If a jacket that regularly sells for $33 is "on sale" for $24.75, find the percent of markdown.

Solution. First we find the dollar amount of markdown.

$$\text{Markdown} = \text{Old price} - \text{New price}$$
$$= \$33.00 \quad - \$24.75$$
$$= \$8.25$$

Now we use the percent proportion.

$$\left.\frac{\text{Markdown}}{\text{Old price}}\right\} \quad \frac{8.25}{33} = \frac{p}{100}$$

$$8.25 \times 100 = 33 \times p$$

$$825 = 33 \times p$$

$$\frac{825}{33} = \frac{33 \times p}{33}$$

$$25 = p$$

The jacket has been marked down 25 percent.

Answers to Trial Runs

page 400   **1.** $2.59; 25%   **2.** $600; 64.9%   **3.** 43¢   **4.** $913.25

page 403   **1.** $120.58   **2.** $36.09; 45%   **3.** 33.$\overline{3}$%   **4.** $169.75; $315.25   **5.** $303.40

EXERCISE SET 10.1

_____ 1. The Lollipop Tree buys children's dresses for $16 each and sells them for
_____ $24 each. Find the amount of markup. Then find the markup as a percent of
               the cost.

_____ 2. The Bootery buys leather boots for $36 a pair and sells them for $63 a pair.
_____   Find the amount of markup. Then find the markup as a percent of the cost.

_____ 3. Bingham's Furniture Store uses a 40 percent markup on all appliances. Find
_____   the dollar amount of markup on an electric range costing $325 from the
               wholesaler. Then find the selling price of the range.

_____ 4. If a paint store uses a 60 percent markup, find the dollar amount of markup
_____   on a gallon of latex paint costing $6 from the wholesaler. Then find the
               selling price per gallon of the paint.

_____ 5. The I.G.A. Market buys homegrown tomatoes from the Amish farmers at 50
               cents per pound. If the market uses a markup that is 30 percent of cost, find
               the selling price of a pound of homegrown tomatoes.

_____ 6. Key Market buys spareribs from the meat packing company at $1.20 per
               pound. If the market uses a markup that is 40 percent of cost, find the
               selling price per pound of spareribs.

_____ 7. The Patio Shop advertises a set of redwood patio furniture for $200. If the
               Patio Shop paid the wholesaler $125 for the set, find the markup as a percent
               of the cost.

_____ 8. The Sport's Studio advertises an exercise bicycle for $435. If the Sport's
               Studio paid the wholesaler $200 for the bicycle, find the markup as a percent
               of the cost.

_____ 9. The Curio Cubbyhole buys carved cuckoo clocks for $398 from the
               wholesaler. If the markup at the Curio Cubbyhole is 85 percent of the cost,
               find the selling price of a cuckoo clock.

_____ 10. The Gawking Gallery bought an oil painting from a struggling artist for
               $190. If the markup at the gallery is 125 percent, find the selling price of the
               painting.

_____ 11. A pair of roller skates that regularly sells for $42.90 is on sale for $38.61.
_____   Find the dollar amount of the markdown. Then find the percent markdown.

_____ 12. A 32-piece set of stainless steel tableware is on sale for $115.50. If the
_____   tableware regularly sells for $150, find the dollar markdown. Then find the
               percent markdown to the nearest percent.

_____ 13. At the pre-inventory sale at a sporting goods store, a fishing rod regularly
_____   selling for $58.95 is marked down 20 percent. Find the amount of the
               markdown. Then find the sale price of the fishing rod.

_____ 14. In August, Sal buys an air conditioner regularly selling for $385 that is
_____   marked down 25 percent. Find the amount of the markdown. Then find the
               sale price of the air conditioner.

_____ 15. At Holt's semi-annual clearance sale all merchandise is "30 percent off."
               Find the sale price of a set of queen size sheets that regularly sells for
               $23.90.

_____ 16. At its January clearance sale, the Specialty Shop marked down all winter
               dresses 50 percent. Find the sale price of a dress regularly selling for
               $80.98.

_____ 17. A TV scratched in shipping is marked down 15 percent. If the TV regularly
               sells for $788, find its sale price.

_____ 18. At the bakery, all day-old items are marked down 33 percent. If 3 loaves of
               bread regularly sell for $2, find the price when they are a day old.

_____ 19. Aggie bought an oak washstand for $80 at an auction. The next day in her
               antique shop, the washstand was selling for $240. Find the percent of
               markup on the washstand.

_____ 20. A year's membership at the Ship-Shape Spa regularly costs $280, but during
               their special introductory offer a year's membership can be purchased for
               $210. Find the percent of markdown on a year's membership purchased
               during the introductory offer.

☆ Stretching the Topics _____

_____ 1. An appliance store uses a markup of 30 percent on all merchandise. What is
_____   the selling price of a refrigerator that costs the retailer $600? If the store has a
              sale advertising all merchandise at 10 percent off the selling price, what is the
              sale price of the refrigerator?

_____ 2. A clothing store advertises a 30 percent markdown on a sweater that regularly
_____   sells for $69.95. What is the sale price of the sweater? After a week, the
_____   sweater has not been sold. The owner then advertises 20 percent off the sale
_____   price. After the second markdown, what is the new price of the sweater?
              What is the total dollar amount of the two markdowns? If the store owner had
              marked the sweater down by that amount originally, what would have been
              the percent markdown? (Round to the nearest percent.)

Check your answers in the back of your book.

If you can complete **Checkup 10.1,** you are ready to go on to Section 10.2.

 **CHECKUP 10.1**

_____ **1.** Fan World buys ceiling fans for $90 and sells them for $157.50. Find the
_____   amount of markup. Then find the markup as a percent of the cost.

_____ **2.** Sparkle Jewelers uses a 125 percent markup on all rings. Find the dollar
_____   amount of markup on a ring costing $195 from the wholesaler. Then find the
             selling price of the ring.

_____ **3.** The Window Place is having an end-of-summer sale on all draperies.
_____   Draperies regularly selling for $267 are on sale for $178.89. Find the dollar
             markdown. Then find the percent markdown.

_____ **4.** For a limited time, Stewart's Vision Center offered all prescription eyeware at
_____   20 percent off the regular price. Find the dollar markdown on glasses that
             regularly sell for $128. Then find the new selling price.

_____ **5.** A telephone regularly selling for $58 is marked down 35 percent. Find the
             new selling price.

Check your answers in the back of your book.

If You Missed Problems:	You Should Review Examples:
1, 2	1, 2
3	3, 4
4, 5	5, 6

10.2 Using Percents to Describe Changes

If Antonia's monthly salary changes from $900 to $963, we know that the amount of change will be the *difference* between the new salary and the old salary.

$$\boxed{\text{Change} = \text{New amount} - \text{Old amount}}$$

For Antonia we find

$$\text{Change} = \text{New salary} - \text{Old salary}$$
$$\text{Change} = \$963 \qquad - \$900$$
$$= \$63$$

Antonia's salary has *increased* by $63.

Working with Changes as Percents

If Antonia wishes to calculate her **percent change** in salary, she must use a ratio that compares the amount of change to her *old* salary. Then she can use a percent proportion to find her percent increase.

$$\left.\begin{array}{c}\text{Change} \\ \hline \text{Old salary}\end{array}\right\} \quad \frac{63}{900} = \frac{p}{100}$$

$$\frac{7}{100} = \frac{p}{100} \qquad \text{Reduce the fraction on the left.}$$

$$7 = p \qquad \begin{array}{l}\text{Since the two fractions have the same} \\ \text{denominator, their numerators must be} \\ \text{the same.}\end{array}$$

Antonia has received an increase of 7 percent in her monthly salary.

The important fact to remember in finding a percent change is that we must compare the amount of the change to the *old* amount.

Finding a Percent Change

1. Find the amount of change.
2. Use the percent proportion

$$\frac{\text{Change}}{\text{Old amount}} = \frac{p}{100}$$

and find p.
3. The percent change is p.

Example 1. Last year's enrollment at City Junior College was 5920. This year's enrollment is 6660. Find the percent change in enrollment.

Solution. First we find the amount of change.

$$\text{Change} = \text{New enrollment} - \text{Old enrollment}$$

$$= 6660 \qquad\qquad - 5920$$

$$\text{Change} = 740$$

Now we use a percent proportion.

$$\left.\frac{\text{Change}}{\text{Old enrollment}}\right\} \quad \frac{740}{5920} = \frac{p}{100}$$

$$\frac{74}{592} = \frac{p}{100} \qquad\qquad \text{Reduce the fraction on the left.}$$

$$\frac{37}{296} = \frac{p}{100} \qquad\qquad \text{Reduce again.}$$

$$37 \times 100 = 296 \times p \qquad \text{Write the cross products.}$$

$$3700 = 296 \times p$$

$$\frac{3700}{296} = \frac{296 \times p}{296} \qquad \text{Divide both sides by 296.}$$

$$12.5 = p$$

Enrollment has *increased* by 12.5 percent.

Example 2. Last year, a VCR sold for $900. This year, the same VCR sells for $750. Find the percent change in price.

Solution. Notice here that the price has *decreased*. We must subtract the new (lower) price from the old (higher) price.

$$\text{Change} = \text{Old price} - \text{New price}$$

$$= \$900 \qquad - \$750$$

$$\text{Change} = \$150$$

To find the percent change we still use the same percent proportion.

$$\left.\frac{\text{Change}}{\text{Old price}}\right\} \quad \frac{150}{900} = \frac{p}{100}$$

$$\frac{15}{90} = \frac{p}{100}$$

$$\frac{3}{18} = \frac{p}{100}$$

$$3 \times 100 = 18 \times p$$

$$300 = 18 \times p$$

$$\frac{300}{18} = \frac{18 \times p}{18}$$

$$16.\overline{6} = p$$

The price of this VCR has *decreased* by $16.\overline{6}$ percent.

You complete Example 3.

Example 3. Before going on his diet Barney weighed 280 pounds. He now weighs 240 pounds. By what percent has Barney's weight changed? (Round to the tenths place.)

Solution. Barney's weight has decreased, so we find the amount of change by

$$\text{Change} = \text{Old weight} - \text{New weight}$$

$$= \underline{\hspace{1cm}} \quad - \underline{\hspace{1cm}}$$

$$= \underline{\hspace{1cm}} \text{ lb}$$

Now we use the percent proportion.

$$\left.\frac{\text{Change}}{\text{Old weight}}\right\} \quad \frac{40}{280} = \frac{p}{100}$$

$$\frac{1}{7} = \frac{p}{100}$$

$$1 \times 100 = 7 \times p$$

$$\underline{\hspace{1cm}} \doteq p$$

Barney's weight has _____ by about _____ percent.

Check your work on page 415. ▶

⫸ Trial Run

_____ 1. The attendance last year at the first basketball game of the season was 8460. This year the attendance was 9729. Find the percent change in attendance.

_____ 2. On July 1, the amount of tax withheld from Doug's paycheck changed from $34.50 to $31.74 per week. Find the percent change in withholding tax.

_____ 3. Mr. Meacham bought a truckload of feeder calves whose average weight was 638 pounds. After feeding them for awhile, he sold them. The average weight at the time of sale was 797.5 pounds. Find the percent change in average weight.

_____ 4. Three years ago the Ramers planned to build a home that would have cost $53,500. When they decided to build this year, they found the cost of building the same home was estimated to be $81,855. Find the percent change in the estimated cost of building a home.

Answers are on page 416.

If your current hourly wage is $3.50 and your employer *increases* your hourly wage by $0.10, then you can find your new hourly wage by *adding* ($3.50 + $0.10 = $3.60).

New amount = Old amount + Increase

On the other hand, if your employer decides to *decrease* your hourly wage by $0.05, then you can find your new hourly wage by *subtracting* ($3.50 − $0.05 = $3.45).

$$\boxed{\text{New amount} = \text{Old amount} - \text{Decrease}}$$

Suppose instead of telling you the amount of change, your employer says that he plans to increase or decrease your present hourly wage by some percent. How can you figure your new hourly wage?

Problems of this type require two steps. First you must find the amount of change using the percent change given to you. Then you must add (or subtract) the change to (or from) the old amount to find the new amount.

Example 4. If your hourly wage of $4.50 is increased by 6 percent, what will be your new wage?

Solution. First you must find the *amount* of increase (or change). Let's use two methods to find the change.

Proportion Method

$$\left.\begin{array}{c}\text{Change}\\\hline\text{Old wage}\end{array}\right\} \quad \frac{c}{4.50} = \frac{6}{100}$$

$$c \times 100 = 4.50 \times 6$$

$$c \times 100 = 27$$

$$\frac{c \times 100}{100} = \frac{27}{100}$$

$$c = \$0.27$$

Alternate Method

Increase = 6% of Old wage

= 6% of $4.50

= 0.06 × $4.50

Increase = $0.27

Now you can *add* this amount to your old wage.

$$\text{New wage} = \text{Old wage} + \text{Increase}$$
$$= \$4.50 \quad + \$0.27$$
$$\text{New wage} = \$4.77$$

You complete Example 5.

Example 5. The population of New York City *decreased* by 5 percent between 1970 and 1975. If the 1970 population was 7,900,000, what was the population in 1975?

Solution. First we find the amount of change by the alternate method.

Change = 5% of Old population

= 5% of 7,900,000

= _____ × 7,900,000

Change = _____

Since this change represents a *decrease* in population, we must _____ it from the 1970 population.

$$1975 \text{ population} = 1970 \text{ population} - \text{Change}$$

$$= \underline{\qquad} \quad - \underline{\qquad}$$

$$= \underline{\qquad}$$

The population in 1975 was _____ .

Check your work on page 415. ▶

Working with the Rate of Inflation

The **rate of inflation** that you read about in the newspapers deals with the *percent change* in the Consumer Price Index (CPI) and gives a good estimate of the change in prices of things that the average consumer purchases. When we say that a price has "kept up with the rate of inflation", we mean that the price has increased by the *same* percent as the percent that describes the rate of inflation.

Example 6. If prices at the Snack Shop have kept up with the 9 percent rate of inflation for this year, what will be the new prices of these items?
  (a) A 50-cent Coke.
  (b) A $1.69 hamburger.

Solution. We must find the amount of increase in each price using a percent change equal to the rate of inflation. Then we add the increase to the old price.

(a) For a 50-cent Coke,

Increase = 9% of 50¢

= 0.09 × 50¢

= 4.5¢

Increase $\doteq$ 5¢

New price = Old price + Increase

= 50¢ + 5¢

New price = 55¢

(b) For a $1.69 hamburger,

Increase = 9% of $1.69

= 0.09 × $1.69

= $0.1521

Increase $\doteq$ $0.16

New price = Old price + Increase

= $1.69 + 0.16

New price = $1.85

Notice that a retailer will usually round *up* to the next cent in its increases.

Example 7. Patrick's weekly salary was increased from $320 to $350 at the end of this year. If the rate of inflation this year was 10.3 percent, is his salary keeping up with the rate of inflation?

Solution. Before deciding whether his salary is keeping up with the rate of inflation, let's find the amount of increase received by Patrick.

$$\text{Increase} = \text{New salary} - \text{Old salary}$$

$$= \$350 \quad - \$320$$

$$\text{Increase} = \$30$$

Now we can compute Patrick's percent increase.

$$\left.\begin{array}{r}\text{Increase}\\\text{Old salary}\end{array}\right\}\quad \frac{30}{320}=\frac{p}{100}$$

$$\frac{3}{32}=\frac{p}{100}\qquad\text{Reduce the fraction on the left.}$$

$$3\times 100 = 32\times p\qquad\text{Write cross products.}$$

$$300 = 32\times p\qquad\text{Find the product on the left.}$$

$$\frac{300}{32}=\frac{32\times p}{32}\qquad\text{Divide both sides by 32.}$$

$$9.4\doteq p\qquad\text{Find the quotients.}$$

We find that Patrick's salary has increased by about 9.4 percent, so his salary is *not* keeping up with the 10.3 percent rate of inflation.

You complete Example 8.

Example 8. In Example 7, what should be Patrick's new weekly salary in order for him to keep up with the 10.3 percent rate of inflation?

Solution. We must find the amount of increase, using 10.3 percent as the percent increase.

$$\text{Increase} = 10.3\% \text{ of } \$320$$

$$= \underline{\hspace{1cm}} \times \$320$$

$$\text{Increase} = \underline{\hspace{1cm}}$$

Now we find the new salary

$$\text{New salary} = \text{Old salary} + \text{Increase}$$

$$= \$\underline{\hspace{1cm}} + \$\underline{\hspace{1cm}}$$

$$\text{New salary} = \$\underline{\hspace{1cm}}$$

Check your work on page 416. ▶

Trial Run

_____ 1. The enrollment at Hillcrest High School increased by 12.5 percent between 1988 and 1989. If the 1988 enrollment was 2648 students, what was the enrollment in 1989?

_____ 2. In 1989 the corn acreage in Webster County was 573,200. If in 1990 the corn acreage was reduced by 32 percent, how many acres were planted in corn that year?

_____ 3. At Nathan's Shoe Hospital, the cost of replacing the heels on a pair of shoes has risen from $3.50 to $3.75 this past year. If the rate of inflation for the year was 6 percent, are Nathan's prices keeping up with the rate of inflation?

_____ 4. If Jameria's weekly salary is now $316, what should be her weekly salary next year to keep up with the 5.5 percent rate of inflation for this year?

Answers are on page 416.

▶ **Examples You Completed** _____

Example 3. Before going on his diet Barney weighed 280 pounds. He now weighs 240 pounds. By what percent has Barney's weight changed? (Round to the tenths place.)

Solution. Barney's weight has decreased, so we find the amount of change by

$$\text{Change} = \text{Old weight} - \text{New weight}$$
$$= 280 \quad\quad\;\; - 240$$
$$= 40 \text{ lb}$$

Now we use the percent proportion.

$$\left.\frac{\text{Change}}{\text{Old weight}}\right\} \quad \frac{40}{280} = \frac{p}{100}$$

$$\frac{1}{7} = \frac{p}{100}$$

$$1 \times 100 = 7 \times p$$

$$100 = 7 \times p$$

$$\frac{100}{7} = \frac{7 \times p}{7}$$

$$14.3 \doteq p$$

Barney's weight has *decreased* by about 14.3 percent.

Example 5. The population of New York City *decreased* by 5 percent between 1970 and 1975. If the 1970 population was 7,900,000, what was the population in 1975?

Solution. First we find the amount of change by the alternate method.

$$\text{Change} = 5\% \text{ of Old population}$$
$$= 5\% \text{ of } 7,900,000$$
$$= 0.05 \times 7,900,000$$
$$\text{Change} = 395,000$$

Since this change represents a *decrease* in population, we must *subtract* it from the 1970 population.

$$1975 \text{ population} = 1970 \text{ population} - \text{Change}$$
$$= 7,900,000 \quad\quad\;\; - 395,000$$
$$= 7,505,000$$

The population in 1975 was 7,505,000.

Example 8. In Example 7, what should be Patrick's new weekly salary in order for him to keep up with the 10.3 percent rate of inflation?

Solution. We must find the amount of increase using 10.3 percent as the percent increase.

$$\text{Increase} = 10.3\% \text{ of } \$320$$
$$= 0.103 \times \$320$$
$$\text{Increase} = \$32.96$$

Now we find the new salary.

$$\text{New salary} = \text{Old salary} + \text{Increase}$$
$$= \$320 \quad\quad + \$32.96$$
$$\text{New salary} = \$352.96$$

Answers to Trial Runs

page 411   1. 15% increase   2. 8% decrease   3. 25% increase   4. 53% increase

page 414   1. 2979   2. 389,776   3. yes   4. $333.38

EXERCISE SET 10.2

_____ 1. The average nightly attendance at the Victory Theater for the months of June, July, and August last year was 235. For those same months this year the average attendance was 329. Find the percent change in average nightly attendance.

_____ 2. Before she began her diet, Callie's hips measured 48 inches. After 5 weeks of dieting her hip measurement was 42 inches. Find the percent change in her hip measurement.

_____ 3. Beginning the first of the year, the amount withheld from Art's paycheck for hospitalization insurance changed from $26.75 to $28.89. Find the percent change in the amount withheld for hospitalization insurance.

_____ 4. Last year there were 536 armed robberies reported to the police department. This year there were only 488 reported. Find, to the nearest percent, the percent change in the number of reported armed robberies.

_____ 5. From a study done by the traffic department it was learned that 2368 motor vehicles passed through the intersection of Powell and Main streets each week before a new factory opened. After the opening of the factory, 5920 vehicles used the intersection each week. Find the percent change in traffic at that intersection.

_____ 6. The height of the Ohio River at Shawneetown was 16.9 feet. After a week of no rain, the river height was 15.8 feet. Find the percent change in the river height. (Round answer to nearest tenth of a percent.)

_____ 7. The yearly subscription rate for the Daily News was $31 last year. This year the subscription rate is $38.75. Find the percent change in the subscription rate.

_____ 8. In 1980 the United States produced 8,597,000 barrels of crude oil. In 1981, the production of oil had decreased to 8,572,000 barrels. Find the percent change in oil production to the nearest tenth of a percent.

_____ 9. Last year a home computer sold for $1295. This year the same home computer sells for $1165.50. Find the percent change in the price.

_____ 10. Before switching to the sandwich plan, the school cafeteria was serving school lunches to an average of 648 students per day. After the switch, an average of 810 students ate lunch daily in the cafeteria. Find the percent change.

_____ 11. The enrollment of Craig Elementary School decreased by 15 percent between 1980 and 1985. If the 1980 enrollment was 1340, find the enrollment in 1985.

_____ 12. For the first football game of the season the 52,000-seat stadium was sold out. After the team lost its first game, the attendance decreased by 18 percent. Find the attendance for the second game.

_____ 13. If prices at the Dairy Dip have kept up with the 6 percent rate of inflation for this year, what will be the new price of a $2.15 burger basket?

_____ 14. Vinnie's weekly salary of $320 as a night watchman was just increased by 8.5 percent. What is Vinnie's weekly salary now?

_____ 15. If Mr. Cable presently earns $18,350 a year as a high school teacher, how much raise in salary should he receive to keep up with the 6.3 percent rate of inflation?

_____ 16. At Green's Grocery the price of chickens has risen from 43 cents to 59 cents per pound during the past year. If the rate of inflation was 7 percent, is the price of chickens keeping up with the rate of inflation?

_____ 17. In 1970, the population of Dayton was 853,000. If by 1980 the population had decreased by about 3 percent, find the population of Dayton in 1980.

_____ 18. In 1960 the average life expectancy at birth for a male in the United States was 65.6 years. By 1980 the life expectancy had increased by 6.7 percent. Find the life expectancy for a male in the United States in 1980. (Round to the tenths place.)

_____ 19. Tuition per semester at Union University last year was $3280. If there will be a 5.5 percent increase in tuition for next semester, how much should a student expect to pay?

_____ 20. The mayor has announced that the Parks and Recreation Department budget of $239,000 will be cut by 12 percent in 1993. How much money should the department plan to spend that year?

☆ Stretching the Topics _____

_____ 1. Marty's weekly salary will increase from $890 to $940 at the end of this year. Suzie expects to get the same percent increase in salary as Marty. If her present weekly salary is $1052, what will be Suzie's weekly salary next year? (Round answer to nearest dollar.)

_____ 2. The value of a new car decreased by 26 percent during the first year. If the amount of first-year decrease was $3770, what was the original value of the car?

Check your answers in the back of your book.

If you can complete **Checkup 10.2,** then you are ready to go on to Section 10.3.

✓ CHECKUP 10.2

_____ **1.** At Movac Manufacturing Company the production level changed from 1200 items to 1596 items per day when music was played over the intercom. Find the percent change in production.

_____ **2.** Daaron's employer has decided to increase his hourly wage of $6.20 per hour by 15 percent. Find Daaron's new hourly wage.

_____ **3.** The attendance at PTA meetings decreased by 40 percent when the meeting time was changed from nights to afternoons. If the average attendance at night meetings was 86 persons, find the average attendance after the time change.

_____ **4.** This year a lawn tractor costs $1095. If the rate of inflation is 8 percent, find (to the nearest dollar) how much you would expect to pay for the same model tractor next year.

_____ **5.** If Max's weekly salary of $385 is increased to $410 at the end of the year, is his salary keeping up with the 6.3 percent inflation rate?

Check your answers in the back of your book.

If You Missed Problems:	You Should Review Examples:
1	1–3
2	4
3	5
4, 5	6–8

419

10.3 Computing Taxes Using Percents

Sometimes it seems as though every time we turn around, we are required to pay some sort of tax. Most states now charge a sales tax on purchases, and the federal government (along with most states and some cities) taxes our income. In addition, we must pay Social Security tax, taxes on our utility and telephone bills, property taxes, and license taxes. Obviously it is important for us to be able to compute taxes so that we are not taken by surprise when our bills (or paychecks) are more (or less) than we expected.

Computing Taxes from Percent Rates

Sales taxes, Social Security (F.I.C.A.) taxes, utility taxes, and occupational taxes are usually stated using percent rates.

Calculating the Amount of Tax from a Percent Rate

1. Change the percent rate to a decimal number.
2. Multiply that decimal number times the quantity being taxed.
3. Round the amount of tax to the nearest cent.

Example 1. Find the total amount to be paid for a $12.95 meal in Connecticut where the sales tax rate is 8 percent.

Solution

$$\text{Amount of tax} = 8\% \text{ of } \$12.95$$
$$= 0.08 \times 12.95$$
$$= \$1.04$$

$$\text{Total amount} = \text{Selling price} + \text{Sales tax}$$
$$= \$12.95 \quad + \$1.04$$
$$= \$13.99$$

Now we can solve the problem stated at the beginning of this chapter.

Example 2. When Stacey bought a $12 souvenir outside her home state, she was charged $0.78 sales tax. Help Stacey find the sales tax rate in this state.

Solution. We must compare the amount of tax to the amount of purchase.

$$\left.\frac{\text{Amount of tax}}{\text{Amount of purchase}}\right\} \quad \frac{0.78}{12} = \frac{p}{100}$$

$$0.78 \times 100 = 12 \times p$$
$$78 = 12 \times p$$
$$\frac{78}{12} = \frac{12 \times p}{12}$$
$$6.5 = p$$

The sales tax rate is 6.5 percent.

Social Security taxes are subtracted from a worker's paycheck before the worker receives the check. In 1991, the Social Security tax rate is 7.65 percent of the original (**gross**) amount of the paycheck.

Example 3. Find the dollar amount of Social Security taxes subtracted in 1991 from a $1540 monthly paycheck.

Solution

$$\text{S.S. tax amount} = 7.65\% \text{ of } \$1540$$
$$= 0.0765 \times \$1540$$
$$= \$117.81$$

From a $1540 monthly paycheck, $117.81 will be subtracted for Social Security tax.

Other kinds of taxes are also subtracted from the gross amount of a paycheck.

Example 4. When Doug moved to a new city, he was surprised to find that $24 had been taken out of his gross paycheck of $1600 for a city occupational tax. Compute the city occupational tax rate as a percent.

Solution. We must consider the proportion

$$\left. \frac{\text{City tax}}{\text{Gross pay}} \right\} \quad \frac{24}{1600} = \frac{p}{100}$$

$$\frac{3}{200} = \frac{p}{100}$$

$$3 \times 100 = 200 \times p$$

$$300 = 200 \times p$$

$$\frac{300}{200} = \frac{200 \times p}{200}$$

$$1.5 = p$$

The city occupational tax rate is 1.5 percent.

⫸ Trial Run

_____ 1. Find the total amount to be paid for a $49.99 raincoat in North Dakota where the sales tax rate is 5.5 percent.

_____ 2. Mr. Shouse bought his wife a $3500 diamond ring while on a business trip and he was charged $192.50 in sales tax. Find the sales tax rate for the state in which he purchased the ring.

_____ 3. Find the dollar amount of Social Security tax subtracted in 1991 from a weekly paycheck of $523. (Round to nearest cent.)

_____ 4. In some states there is an entertainment tax. An $8 ticket to a concert in one state costs $9.20. Compute the state's entertainment tax rate as a percent.

Answers are on page 426.

Using Percents on Federal Income Tax Forms

There are several parts of the Internal Revenue Service 1040 tax return that require you to work with percents. If you itemize your deductions, you may wish to fill in the portion on Medical and Dental Expenses shown here.

Medical and Dental Expenses

(Do not include expenses reimbursed or paid by others.)

(See Instructions on page 19.)

1a	Prescription medicines and drugs, insulin, doctors, dentists, nurses, hospitals, insurance premiums you paid for medical and dental care, etc.	1a
b	Transportation and lodging	1b
c	Other (list—include hearing aids, dentures, eyeglasses, etc.) ▶ ---------------------------------	1c
2	Add lines 1a through 1c, and enter total here	2
3	Multiply the amount on Form 1040, line 31, by 7.5% (.075) . .	3
4	Subtract line 3 from line 2. If zero or less, enter -0-. Total medical and dental . . ▶	4

As you can see, you must compute a percent in line 3, but first you must locate an amount on line 31 of Form 1040. Line 31 contains the adjusted gross income for the person filing the tax return.

Example 5. Jane Barton's adjusted gross income is $17,200. She spent $233 for prescription drugs and medicines, $720 for insurance premiums, and $1845 for doctors' and dentists' bills. For contact lenses and dental bridgework, she spent $600. She calculated her expenses for transportation to doctors' and dentists' offices to be $17.80. Find the amount that Jane can deduct for medical and dental expenses.

Solution. Let's list Jane's expenses as they should appear in her schedule of deductions.

Line 1a: $233 + $720 + $1845 = $2798

Line 1b: $17.80

Line 1c: $600

To find line 2, we sum lines 1a, 1b, and 1c.

$$\begin{aligned} \$2798.00 \\ 17.80 \\ + \ 600.00 \\ \hline \$3415.80 \end{aligned}$$

To find line 3, we calculate

7.5% of $17,200

$= 0.075 \times \$17,200$

$= \$1290$

Now let's put our information into Jane's form and see how much she can deduct.

Medical and Dental Expenses

(Do not include expenses reimbursed or paid by others.)

(See Instructions on page 19.)

1a	Prescription medicines and drugs, insulin, doctors, dentists, nurses, hospitals, insurance premiums you paid for medical and dental care, etc.	1a	2798 00
b	Transportation and lodging	1b	17 80
c	Other (list—include hearing aids, dentures, eyeglasses, etc.) ▶ *Contact lenses, bridgework* ---------------	1c	600 00
2	Add lines 1a through 1c, and enter total here	2	3415 80
3	Multiply the amount on Form 1040, line 31, by 7.5% (.075) . .	3	1290 00
4	Subtract line 3 from line 2. If zero or less, enter -0-. Total medical and dental . . ▶	4	2125.80

From line 4, Jane concludes that she can deduct $2125.80 for medical and dental expenses.

Example 6. If Carlos has an adjusted gross income of $39,350, calculate 7.5 percent of his adjusted gross income.

Solution

$$7.5\% \text{ of } \$39,350$$
$$= 0.075 \times \$39,350$$
$$= \$2951.25$$

After a taxpayer has figured all the possible deductions and adjustments to be made, he or she finally arrives at the **taxable income.** The taxpayer's federal taxes are based upon this taxable income. One of the tables that can be used by married couples to find the amount of tax is reproduced here.

Taxable Income	Amount of Tax
Between $0 and $29,750	15% of taxable income
Over $29,750	$4462.50 plus 28% of the taxable income over $29,750

Suppose Mr. and Mrs. Vail have a combined taxable income of $37,300. We first locate their income in the proper row. The Vails' income is over $29,750, so we must look at row 2 in the table. That row tells us that the Vails must pay taxes equal to $4462.50 plus 28 percent of their income *over* $29,750.

We first figure how much *more* than $29,750 the Vails earned.	Now we find 28 percent of that extra income
$37,300 - $29,750 = $7550	28% of $7550 = 0.28 × $7550 = $2114

The tax that the Vails must pay is $4462.50 + $2114 = $6576.50.

You complete Example 7.

Example 7. Compute the federal income tax to be paid by Mr. and Mrs. Cutler whose combined taxable income is $63,250.

Solution. We locate the row that contains the Cutlers' income (row _____).

Then we find how much they earned *over* $29,750.	Now we find _____% of that amount.
$63,250 - $_____ = _____	28% of $33,500 = _____ × $33,500 = _____

The total amount of federal tax to be paid by the Cutlers is

$$\$\underline{\quad} + \$\underline{\quad} = \$\underline{\quad}$$

Check your work on page 425. ▶

Example 8. Graduate students Henry and Claire Booker have a taxable income of $12,300. Compute their federal taxes.

Solution. We locate the Bookers' income in row 1. Their tax is 15 percent of $12,300.

$$0.15 \times \$12,300$$
$$= \$1845$$

Since there is no amount to be added to this figure, the Bookers' total tax bill is $1845.

⟶ Trial Run

_____ 1. If Wallace has an adjusted gross income of $15,600, calculate 7.5 percent of his adjusted gross income.

2. Complete the following form for James Scott, using this information.

Adjusted gross income	$34,000
Prescription drugs and medicine	295
Insurance premiums	1140
Doctor and dentist fees	750
Hospital bills	530
Hearing aid and glasses	490

Medical and Dental Expenses (Do not include expenses reimbursed or paid by others.) (See Instructions on page 19.)				
1a	Prescription medicines and drugs, insulin, doctors, dentists, nurses, hospitals, insurance premiums you paid for medical and dental care, etc. . . .	1a		
b	Transportation and lodging	1b		
c	Other (list—include hearing aids, dentures, eyeglasses, etc.) ▶ _____ _____	1c		
2	Add lines 1a through 1c, and enter total here	2		
3	Multiply the amount on Form 1040, line 31, by 7.5% (.075) . .	3		
4	Subtract line 3 from line 2. If zero or less, enter -0-. Total medical and dental . . ▶	4		

_____ 3. Compute the tax to be paid by Jane and Eddie Donosky, whose taxable income is $49,600.

_____ 4. Compute the tax to be paid by the Heavrins, whose taxable income is $15,300.

Answers are on page 426.

▶ Example You Completed

Example 7. Compute the federal income tax to be paid by Mr. and Mrs. Cutler, whose combined taxable income is $63,250.

Solution. We locate the row that contains the Cutlers' income (row 2).

Then we find how much they earned *over* $29,750.

$$\$63,250 - \$29,750$$
$$= \$33,500$$

Now we find 28% of that amount.

28% of $33,500
= 0.28 × $33,500
= $9380

The total amount of federal tax to be paid by the Cutlers is

$$\$4462.50 + \$9380 = \$13,842.50$$

Answers to Trial Runs

page 422   **1.** $52.74   **2.** 5.5%   **3.** $40.01   **4.** 15%

page 425   **1.** $1170
2.

Medical and Dental Expenses					
Medical and Dental Expenses (Do not include expenses reimbursed or paid by others.)	**1a** Prescription medicines and drugs, insulin, doctors, dentists, nurses, hospitals, insurance premiums you paid for medical and dental care, etc.	**1a**	2715	00	
	b Transportation and lodging	**1b**	—		
	c Other (list—include hearing aids, dentures, eyeglasses, etc.) ► *hearing aid, glasses*	**1c**	490	00	
(See Instructions on page 19.)	**2** Add lines 1a through 1c, and enter total here	**2**	3205	00	
	3 Multiply the amount on Form 1040, line 31, by 7.5% (.075) . .	**3**	2550		
	4 Subtract line 3 from line 2. If zero or less, enter -0-. Total medical and dental . . ►	**4**		655	00

3. $10,020.50   **4.** $2295

EXERCISE SET 10.3

_____ 1. Find the amount paid for a $62.95 bathrobe purchased in Virginia, where the sales tax rate is 3.5 percent.

_____ 2. If the sales tax rate is 7 percent, find the total amount paid for a coat purchased for $189.98.

_____ 3. The sales tax on a hat purchased in Indianapolis was $4.75. If the price of the hat was $95, find the sales tax rate in Indiana.

_____ 4. Find the sales tax rate if the tax on a $59 pair of shoes is $4.72.

_____ 5. Find the amount (to the nearest cent) of Social Security taxes subtracted from a $1850 monthly salary if the Social Security tax rate is 7.65 percent of the gross amount.

_____ 6. If Clara's weekly salary is $382, find the amount (to the nearest cent) that is withheld from her check for Social Security taxes. (Use a 7.65 percent rate.)

_____ 7. If the state where Tom lives has an income tax rate of 6 percent of gross income, find the state income tax withheld from a monthly income of $1325.

_____ 8. If there is a 3 percent federal tax added to your telephone bill, find the tax (to the nearest cent) on monthly phone charges of $19.93.

_____ 9. The school tax on personal property in Clinton County is 1.25 percent of the assessed value. Find the school tax (to the nearest cent) on property valued at $175,000.

_____ 10. The city occupational tax rate is 1.5 percent of the gross amount of the salary for anyone working within the city limits. Find the city occupational tax on a weekly salary of $520.

_____ 11. If Proctor has an adjusted gross income of $38,900, calculate 7.5 percent of his adjusted gross income.

_____ 12. If Louise has an adjusted gross income of $8750, calculate 7.5% of her adjusted gross income.

13. Complete the next tax form for Dempsey Orr, using this information.

Adjusted gross income	$43,000
Medical and dental expenses	3,235
Transportation and lodging	108
Glasses	215

Medical and Dental Expenses (Do not include expenses reimbursed or paid by others.) (See Instructions on page 19.)	1a	Prescription medicines and drugs, insulin, doctors, dentists, nurses, hospitals, insurance premiums you paid for medical and dental care, etc.	1a			
	b	Transportation and lodging	1b			
	c	Other (list—include hearing aids, dentures, eyeglasses, etc.) ▶ -- --	1c			
	2	Add lines 1a through 1c, and enter total here	2			
	3	Multiply the amount on Form 1040, line 31, by 7.5% (.075) . .	3			
	4	Subtract line 3 from line 2. If zero or less, enter -0-. Total medical and dental . . ▶	4			

14. Complete the next tax form for Isaac Omer, using this information.

Adjusted gross income	$28,500
Prescription medicine	486
Insurance premiums	985
Doctor fees	563
Dentist fees	85
Hearing aid	379

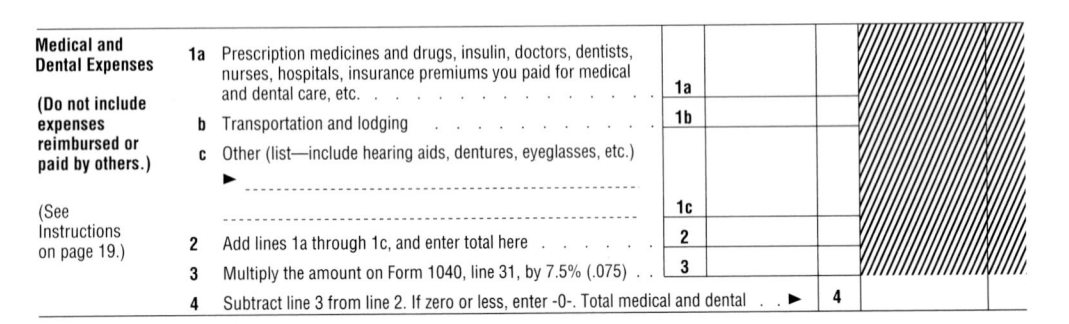

15. Use the tax table to compute the tax to be paid by Senator and Mrs. Clements, whose taxable income is $55,400.

16. Use the tax table to compute the tax to be paid by Reverend and Mrs. Babb, whose taxable income is $21,400.

17. Use the tax table to compute the tax to be paid by social workers Bob and Helen Allen, whose combined taxable income is $10,390.

18. Use the tax table to compute the tax to be paid by Mr. Collins and his wife Dr. Anderson, whose combined taxable income is $42,500.

☆ Stretching the Topics

1. Clarissa can purchase a certain model TV for $790 in her hometown, where the sales tax rate is 6 percent. She can purchase the same model for $800 in a nearby state where the sales tax rate is 4.5 percent. Clarissa calculates her expenses for driving to the nearby town at $0.12 per mile for the 35-mile round-trip. Where should Clarissa buy her new TV?

2. Jim earns $22,500 per year and his wife Gayle earns $26,000 per year. If Jim receives a $2000 raise this year and the rate of inflation is 8.2 percent, is he keeping up with the rate of inflation? If Gayle receives a $2000 raise, is her salary keeping up with the rate of inflation? Is their combined family income keeping up with the rate of inflation?

Check your answers in the back of your book.

If you can work the problems in **Checkup 10.3,** then you are ready to go on to Section 10.4.

✓ CHECKUP 10.3

_____ 1. Find the total amount paid for a $798 refrigerator purchased in Kentucky where the sales tax is 5 percent.

_____ 2. Find the sales tax rate in Mississippi if the sales tax on a $1600 gold necklace is $96.

_____ 3. Find the amount of Social Security tax subtracted from a $1530 monthly paycheck if the Social Security tax rate is 7.65 percent of the gross amount.

_____ 4. If Otis has an adjusted gross income of $9380, calculate 7.5 percent of his adjusted gross income.

_____ 5. Use the tax table to compute the tax to be paid by the Wathens, whose taxable income is $36,500.

Check your answers in the back of your book.

If You Missed Problems:	You Should Review Exercises:
1	1
2	2
3	3
4	6
5	7, 8

10.4 Computing Interest

Whenever someone is allowed to use someone else's money for a period of time, the one using the money is often required to pay a charge for that privilege. Such a charge is called **interest.**

When Jessica borrows money from the credit union, she must pay *interest* to the credit union for the privilege of using money belonging to the credit union.

When Carlton deposits money in a savings account at First National Bank, the bank pays Carlton *interest* for the privilege of using his money.

Stated simply,

> **Interest** is a payment made for the privilege of using someone else's money for a certain period of time.

Computing Simple Interest

The dollar amount of interest paid or received depends on the **rate of interest** and that rate is usually stated as *some percent per year*.

If your savings bank offers an interest rate of 6 percent per year, the bank is promising that at the end of 1 year you will receive an amount of interest equal to 6 percent of the amount you deposited in the bank. Suppose you deposit $200 in this bank. Then, after 1 year, you will receive interest of

$$6\% \text{ of } \$200$$
$$= 0.06 \times \$200$$
$$= \$12$$

How much interest would you receive after just *6 months?* after *3 years?* Before we can find the amounts of interest in these situations, we must identify the important quantities that will be used in our computations.

> **Principal (*P*)** = amount deposited
> or
> amount borrowed
> **Rate (*r*)** = percent of interest per year
> **Time (*t*)** = number of years

	Principal (*P*)	Rate (*r*)	Time (*t*)
For $200 deposited at 6 percent for one year	$200	6% per year	1 year
For $200 deposited at 6 percent for 6 months	$200	6% per year	$\frac{1}{2}$ year
For $200 deposited at 6 percent for 3 years	$200	6% per year	3 years

In each of these situations, we can find the amount of interest (*i*) by *multiplying* principal × rate × time. Interest calculated in this way is called **simple interest.**

> **Simple Interest**
> Amount of interest = Principal × Rate × Time
> $$i = P \times r \times t$$

Let's use this formula to calculate the amount of interest for the 3 situations we have described. Remember that since the interest *rate* is stated as percent per *year*, the *time* must always be stated in *years*.

	P	r	t	$i = P \times r \times t$
For 1 year	$200	6% = 0.06	1 year	$i = \$200 \times 0.06 \times 1$ $= \$12$
For 6 months	$200	6% = 0.06	$\frac{1}{2}$ year	$i = \$200 \times 0.06 \times \frac{1}{2}$ $= \$6$
For 3 years	$200	6% = 0.06	3 years	$i = \$200 \times 0.06 \times 3$ $= \$36$

Example 1. Find the simple interest that will be charged if Ted borrows $1250 at 11 percent per year for 4 years. How much must Ted repay at the end of 4 years?

Solution. First we find the amount of simple interest Ted must pay.

$$i = P \times r \times t$$
$$i = \$1250 \times 0.11 \times 4$$
$$= \$137.50 \times 4$$
$$= \$550$$

Ted must repay the amount he borrowed ($1250) *plus* the interest he was charged ($550).

$$\text{Total} = \$1250 + \$550$$
$$= \$1800$$

Ted must repay a total of $1800.

Example 2. How much interest will Emile receive after 9 months if he deposits $6040 in his savings account paying $6\frac{1}{2}$ percent per year? How much will be in his account after 9 months?

Solution. Here we know that

$$P = \$6040$$

$$r = 6\frac{1}{2}\% = 6.5\% = 0.065 \qquad \text{Change percent to decimal number.}$$

$$t = \frac{9}{12} \text{ yr} = \frac{3}{4} \text{ yr} \qquad \text{Change months to fractional years.}$$

$$i = P \times r \times t$$

$$= \$6040 \times 0.065 \times \frac{3}{4}$$

$$= \$392.60 \times \frac{3}{4}$$

$$= \frac{\overset{98.15}{\cancel{\$392.60}}}{1} \times \frac{3}{\underset{1}{\cancel{4}}}$$

$$= \$294.45$$

Emile will have his principal ($6040) plus his interest ($294.45) in his account.

$$\text{Total} = \$6040 + \$294.45$$

$$= \$6334.45$$

▥➡ Trial Run

———— 1. Find the simple interest that will be charged if Maurice borrows $1750 at 13 percent per year for 3 years.

———— 2. Find the simple interest that will be charged if $800 is borrowed at 11.5 percent per year for 3 months.

———— 3. If Daphne lends her brother $2000 at 9 percent per year for $1\frac{1}{2}$ years, how much interest will she receive? What is the total amount her brother must repay at the end of $1\frac{1}{2}$ years?

———— 4. Roderick invested $25,000 in savings certificates at 9.5 percent per year for 6 months. How much interest will Roderick's investment earn?

Answers are on page 435.

Computing Compound Interest

After working the examples in the last section, you see that simple interest is paid only on the original amount deposited. Although it is important for you to understand simple interest, you must also be aware of the fact that banks do *not* pay their depositors simple interest. Instead they pay what is called **compound interest.** When a bank pays compound interest to a depositor, the bank is paying interest on the original deposit, but it is also paying *interest on any interest* received along the way.

The time period after which the bank records the interest on a savings account is called the **compounding period.** We will only consider situations in which the bank's compounding period is 1 year. When a bank records the interest once each year, we say the interest is **compounded annually.**

Let's consider a situation in which Juan deposits $1000 in a savings account paying 6 percent compounded annually. Suppose Juan leaves his money in the bank for 3 years.

	Principal	Interest	Total Amount
Year 1	$1000	$i = \$1000 \times 0.06 \times 1$ $= \$60$	$1000 + $60 = $1060
Year 2	$1060	$i = \$1060 \times 0.06 \times 1$ $= \$63.60$	$1060 + $63.60 = $1123.60
Year 3	$1123.60	$i = \$1123.60 \times 0.06 \times 1$ $= \$67.416$ $\doteq \$67.42$	$1123.60 + $67.42 = $1191.02

At the end of the third year, Juan has $1191.02 in his account. He has received $191.02 ($1191.02 − $1000) in interest. Let's compare this amount of interest to the amount Juan would have received if the bank had paid *simple interest* for 3 years instead of compound interest.

$$i = \$1000 \times 0.06 \times 3$$
$$= \$60 \times 3$$
$$i = \$180$$

As you can see, a compound interest account pays a higher amount of interest. Why? Because it pays interest on the original principal *and* on any interest received in earlier compounding periods.

You complete Example 3.

Example 3. Find the amount after 2 years if $3500 is deposited in a bank paying 8 percent compounded annually. Then find the total amount of interest.

Solution

	Principal	Interest	Total Amount
Year 1	$3500	$i = \$3500 \times$ _____ $\times 1$ $= \$$_____	$3500 + $_____ = $_____
Year 2	$3780	$i = \$3780 \times$ _____ $\times 1$ $= \$$_____	$3780 + $_____ = $_____

The amount after 2 years is $_____ . The total amount of interest is $4082.40 − $3500 = $_____ .

Check your work on page 435. ▶

⫸ Trial Run

_____ **1.** Find the amount after 3 years if $5000 is deposited in a bank paying 9 percent compounded annually.

_____ **2.** Susannah deposits $3500 in a savings and loan association. How much will she _____ have on deposit at the end of 2 years if the rate of interest is 8.5 percent compounded annually? How much of the total amount is interest?

Answers are below.

▶ Example You Completed

Example 3 (_Solution_)

	Principal	Interest	Total Amount
Year 1	$3500	$i = \$3500 \times 0.08 \times 1$ $= \$280$	$\$3500 + \$280$ $= \$3780$
Year 2	$3780	$i = \$3780 \times 0.08 \times 1$ $= \$302.40$	$\$3780 + \$302.40$ $= \$4082.40$

The amount after 2 years is $4082.40. The total amount of interest is $4082.40 − $3500 = $582.40.

Answers to Trial Runs

page 433   **1.** $682.50   **2.** $23   **3.** $270; $2270   **4.** $1187.50

page 435   **1.** $6475.15   **2.** $4120.29; $620.29

EXERCISE SET 10.4

_____ 1. Find the simple interest that will be charged on a loan of $1500 borrowed at 12 percent per year for 2 years.

_____ 2. Find the simple interest that will be charged if $850 is borrowed at 10.5 percent per year for 3 years.

_____ 3. Find the simple interest that will be charged if $3000 is borrowed at 12.5 percent interest for 4 months.

_____ 4. Find the simple interest that will be charged on $2800 borrowed at 9.5 percent interest for 3 months.

_____
_____ 5. If Sara lends her friend Joan $3000 at 9 percent simple interest per year for 15 months, how much interest will Sara receive? What is the total amount Joan must repay at the end of 15 months?

_____
_____ 6. The Quick Credit Loan Company will lend $250 at 20 percent simple interest per year for 18 months. How much interest will Quick Credit collect? What is the total amount that must be repaid at the end of 18 months?

_____ 7. Cassandra invested $16,000 in savings certificates at 10.25 percent simple interest per year. How much interest will Cassandra's investment earn in 3 months?

_____ 8. Find the amount after 3 years if $5000 is deposited in a savings account paying 8 percent interest compounded annually.

_____ 9. Find the amount after 4 years if $10,000 is invested in a savings association at an interest rate of $4\frac{1}{2}$ percent compounded annually.

_____
_____ 10. Ronald deposits $5000 in his credit union. How much will he have on deposit at the end of 3 years if the interest rate is 5 percent compounded annually? How much of the total amount is interest?

☆ Stretching the Topics _____

_____
_____
_____ 1. Denise has $3000 that she wishes to invest for 5 years. Federal Savings will pay 9 percent simple interest per year. Mercantile Bank will pay 8 percent interest compounded annually. What would be the total amount of interest at the end of 5 years at Federal Savings; at Mercantile Bank? Which is the better investment?

Check your answers in the back of your book.

If you can complete **Checkup 10.4,** then you are ready to do the **Review Exercises** for Chapter 10.

✓ CHECKUP 10.4

_____ **1.** Find the simple interest that will be charged on a loan of $3000 borrowed at 12.5 percent for 4 years.

_____ **2.** Find the simple interest earned on an investment of $8000 at 13 percent for 3 months.

_____ **3.** Find the amount after 2 years if $6000 is deposited in a savings account
_____   paying 8 percent compounded annually. How much of the total amount is interest?

Check your answers in the back of your book.

If You Missed Problems:	You Should Review Examples:
1	1
2	2
3	3

Summary

In this chapter we discovered that we may be called upon to compute with percents in many areas of our lives.

In the area of retail sales, we discussed **markup** (the amount added by the retailer to the cost to arrive at the selling price) and **markdown** (the amount subtracted by the retailer from an old price to arrive at a new price).

To Find	We Use the Equation
Selling price	Selling price = Cost + Markup
Markup as a percentage (p) of cost	$\dfrac{\text{Markup}}{\text{Cost}} = \dfrac{p}{100}$
New price	New price = Old price − Markdown
Markdown as a percentage (p) of old price	$\dfrac{\text{Markdown}}{\text{Old price}} = \dfrac{p}{100}$

We learned to use two steps to express any change as a percent. First we found the amount of change and then we used the percent proportion to find the percent (p).

$$\text{Change} = \text{New quantity} - \text{Old quantity}$$

$$\frac{\text{Change}}{\text{Old quantity}} = \frac{p}{100}$$

When tax rates were expressed as percents, we learned to compute those taxes by multiplying the percent rate (expressed as a decimal number) times the dollar amount being taxed. Then we studied some parts of federal income tax forms and tables in which percents were used in computations.

Finally we discussed **interest,** the payment made for the privilege of using someone else's money for a certain period of time. We learned to compute **simple interest** using the formula

$$\text{Interest} = \text{Principal} \times \text{Rate} \times \text{Time}$$

Then we learned to calculate **compound interest,** which is interest paid on the principal *and* on any interest received in previous compounding periods.

❏ Speaking the Language of Mathematics ⎯⎯⎯⎯⎯⎯⎯⎯⎯⎯

Complete each sentence with the appropriate word or phrase.

1. The amount added by a retailer to the cost of an item to arrive at its selling price is called the ⎯⎯⎯⎯⎯ .

2. The amount subtracted by a retailer from the old price of an item to arrive at its new price is called the ⎯⎯⎯⎯⎯ .

3. To express a markup as a percentage of cost or a markdown as a percentage of the old price, we can use the ⎯⎯⎯⎯⎯ ⎯⎯⎯⎯⎯ .

4. In using the percent proportion to find a percentage change, the numerator of the first fraction contains the _____ and the denominator of the first fraction contains the _____ quantity.

5. In using the formula $i = P \times r \times t$ to calculate simple interest, if the rate (r) is expressed as some percent per year, then the time (t) must be expressed in _____ .

6. Interest received on the principal *and* on any interest received in previous compounding periods is called _____ interest.

△ Writing About Mathematics

Write your response to each question in complete sentences.

1. Discuss the relationship among the wholesaler, retailer, and consumer in the pricing of a product. Be sure to mention cost, markup, and selling price.

2. If you know your last year's salary and this year's salary, tell how you would find your percentage raise.

3. Explain the difference between simple interest and interest compounded annually, over a period of 2 years.

REVIEW EXERCISES for Chapter 10

_____ 1. Lowe's Hardware Store buys gas-powered chain saws for $90 and sells them
_____   for $126. Find the amount of markup. Then find the markup as a percent of
            the cost.

_____ 2. Fan World advertises a ceiling fan for $78. If Fan World paid the wholesaler
            $60 for the fan, find the markup as a percent of the cost.

_____ 3. Ford's Furniture uses a 40 percent markup on all merchandise. Find the
_____   dollar amount of markup on a swivel rocker costing $185 from the
            wholesaler. Then find the selling price of the rocker.

_____ 4. Key Market buys peanut butter from the wholesaler at $1.10 per 18-ounce
            jar. If the market uses a markup that is 35 percent of the cost, find the
            selling price of an 18-ounce jar of peanut butter.

_____ 5. If running shoes that regularly sell for $45 are on sale for $27 find the
_____   amount of markdown. Then find the percent of markdown.

_____ 6. If a refrigerator that regularly sells for $690 is on sale for $460, find the
            percent of markdown.

_____ 7. If a compact disc player selling for $280.60 is marked down 25 percent,
            what will be the sale price?

_____ 8. After the first of the year, a car dealer lowered the price of a new car from
            $14,600 to $12,848. Find the percent change in the price.

_____ 9. Due to an increase in demand for its product, Movac Manufacturing
            Company increased its work force from 3540 employees to 3693 employees.
            Find the percent change in the number of employees. (Round to the tenths
            place.)

_____ 10. Because of an oversupply, the price of a can of green beans dropped from 59
             cents to 39 cents per can. By what percent did the price of green beans
             change? (Round to the tenths place.)

_____ 11. The employees of the Mane Tamers beauty shop negotiated a 12 percent
             salary increase. If their former hourly wage was $6.85, find their new hourly
             wage.

_____ 12. The enrollment of Southwest Community College decreased by 5 percent
             between 1985 and 1990. If the 1985 enrollment was 23,400, what was the
             enrollment in 1990?

_____ 13. If prices at the Burger Barn have kept up with the 5.2 percent rate of
             inflation for this year, what will be the new price of a $1.85 superburger?

_____ 14. The price of a radio increased from $34.50 to $39 this year. If the rate of
             inflation this year was 6.3 percent, did the price keep up with the rate of
             inflation?

_____ 15. If Ahman's weekly salary is now $986, what should be her weekly salary
             next year to keep up with a 5.6 percent rate of inflation?

_____ 16. Find the total amount to be paid for a $99 water heater if the sales tax rate is 5.5 percent.

_____ 17. If the Social Security tax rate is 7.65 percent, find the dollar amount subtracted from a $1485 monthly paycheck.

_____ 18. If $11.40 was taken out of Salvadore's $900 weekly salary for a city occupational tax, find the tax rate as a percent. (Round to the tenths place.)

19. Complete the form for Kenny Chandler, using this information.

Adjusted gross income	$65,450
Prescription drugs and medicine	984
Insurance premiums	1956
Doctor and dentist bills	2655
Contact lenses	389

Medical and Dental Expenses (Do not include expenses reimbursed or paid by others.) (See Instructions on page 19.)				
	1a	Prescription medicines and drugs, insulin, doctors, dentists, nurses, hospitals, insurance premiums you paid for medical and dental care, etc.	1a	
	b	Transportation and lodging	1b	
	c	Other (list—include hearing aids, dentures, eyeglasses, etc.) ▶ -- --	1c	
	2	Add lines 1a through 1c, and enter total here	2	
	3	Multiply the amount on Form 1040, line 31, by 7.5% (.075) . .	3	
	4	Subtract line 3 from line 2. If zero or less, enter -0-. Total medical and dental . . ▶	4	

_____ 20. Compute the federal income tax to be paid by Mr. and Mrs. Gonzales-Martin, whose taxable income is $83,520.

_____ 21. Compute the federal income tax to be paid by Sherry Ladd, whose taxable income is $16,385.

_____
_____ 22. Find the interest that will be charged if Brett borrows $1500 at $12\frac{1}{2}$ percent simple interest for 5 years. How much will Brett repay at the end of 5 years?

_____
_____ 23. How much interest will Donetta receive after 6 months if she deposits $5250 into her savings account paying 7 percent simple interest per year? How much will be in her account after 6 months?

_____
_____ 24. Find the amount after 2 years if $6500 is deposited in a bank paying 8 percent interest compounded annually. Then find the total amount of interest.

_____
_____ 25. Martina deposits $2500 in a savings and loan institution that pays 5 percent interest compounded annually. How much will she have in her account after 3 years? How much of the total amount is interest?

Check your answers in the back of your book.

If You Missed Exercises:	You Should Review Problems:	
1, 2	Section 10.1	1
3, 4		2
5, 6		3, 4
7		5
8–10	Section 10.2	1–3
11, 12		4, 5
13–15		7, 8
16	Section 10.3	1, 2
17		3
18		4
19		5
20, 21		6–8
22	Section 10.4	1
23		2
24, 25		3

If you have completed the **Review Exercises** and corrected your errors, you are ready to take the **Practice Test** for Chapter 10.

PRACTICE TEST for Chapter 10

	SECTION	EXAMPLE

_____ 1. Bingham's Appliance Store buys dishwashers from a wholesaler for
_____ $196 and sells them for $260.68. Find the amount of markup. Then
             find the markup as a percent of the cost.   **10.1   1**

_____ 2. A gold chain that regularly sells for $376 is on sale for $225.60.
_____   Find the amount of markdown. Then find the percent of markdown.   **10.1   3**

_____ 3. Ace Hardware is having a sale with all merchandise marked down
             35 percent. Find the sale price of a 5-gallon can of driveway sealer
             that regularly sells for $5.99.   **10.1   5**

_____ 4. Last year 5685 students were enrolled in math courses at Seabrook
             University. This year's enrollment is 7959. Find the percent change
             in enrollment in math courses.   **10.2   1**

_____ 5. This year Maxine's salary is $23,485. Next year her salary will be
             $25,363.80. Find the percent change in salary.   **10.2   2**

_____ 6. The value of an antique ring has increased by 15 percent since
             Amanda purchased it last year. If Amanda paid $250, what is the
             value of the ring now?   **10.2   4**

_____ 7. The number of violent crimes in Henshaw decreased by 12 percent
             between 1980 and 1985. If there were 625 violent crimes in 1980,
             how many violent crimes were committed in 1985?   **10.2   5**

_____ 8. If the prices at Pizza Shack have kept up with the 6 percent rate of
             inflation for this year, what will be the new price of a $9.85 10-inch
             pizza?   **10.2   6**

_____ 9. Carol's yearly salary will increase from $35,650 to $36,858.50 at the
             end of this year. If the rate of inflation this year was 6.3 percent, is
             her salary keeping up with the rate of inflation?   **10.2   7**

_____ 10. Find the total amount to be paid for a leaf blower that sells for $70,
             if the sales tax rate is 7.25 percent.   **10.3   1**

_____ 11. Find the dollar amount of Social Security taxes subtracted in 1991
             from a $988 monthly paycheck if the Social Security tax rate is 7.65
             percent.   **10.3   3**

	SECTION	EXAMPLE

12. Complete the form for Tony White, using this information.   10.3   5

Adjusted gross income	$23,000
Prescription drugs and medicines	780
Doctor and dentist bills	586
Insurance premiums	960
Glasses and dentures	835
Travel to doctor's office	53

Medical and Dental Expenses		
Medical and Dental Expenses (Do not include expenses reimbursed or paid by others.) (See Instructions on page 19.)	**1a** Prescription medicines and drugs, insulin, doctors, dentists, nurses, hospitals, insurance premiums you paid for medical and dental care, etc. **1a**	
	b Transportation and lodging **1b**	
	c Other (list—include hearing aids, dentures, eyeglasses, etc.) ▶ -------------------------------- -------------------------------- **1c**	
	2 Add lines 1a through 1c, and enter total here **2**	
	3 Multiply the amount on Form 1040, line 31, by 7.5% (.075) . . **3**	
	4 Subtract line 3 from line 2. If zero or less, enter -0-. Total medical and dental . . ▶	**4**

_____ **13.** Compute the federal income tax to be paid by Andrea Kus, whose   10.3   8
taxable income is $23,250.

_____ **14.** Compute the federal income tax to be paid by Mr. and Mrs. Kyle   10.3   7
Kline, whose combined taxable income is $105,300.

_____ **15.** Find the simple interest that will be charged if Jon borrows $850 at
_____ 12 percent interest for 3 years. How much will Jon repay at the end of   10.4   1
3 years?

_____ **16.** Find the amount after 2 years if $2500 is deposited in a bank paying
_____ 9 percent interest compounded annually. Then find the total amount   10.4   3
of interest.

Check your answers in the back of your book.

SHARPENING YOUR SKILLS after Chapters 1–10

SECTION

_____ 1. In the whole number 387,965, name the thousands digit. 1.1

_____ 2. Reduce $\frac{545}{630}$ to lowest terms. 3.2

_____ 3. Use $<$, $>$, or $=$ to compare $17\frac{2}{3}$ and $\frac{89}{5}$. 5.4

_____ 4. Write $37.85 in words as it would appear on a check. 6.3

_____ 5. Compare 15.703 and 15.073 using $<$ or $>$. 6.1

_____ 6. Round 0.57045 to the thousandths place. 6.1

_____ 7. Change $\frac{87}{15}$ to a decimal number. 7.3

_____ 8. Use a shortcut to find $8.7 \div 10,000$. 7.3

_____ 9. Find the missing part of the proportion $\dfrac{x}{65} = \dfrac{5}{4\frac{1}{3}}$ 8.2

_____ 10. Change 275 percent to a fraction. 9.1

_____ 11. Change 0.04 percent to a decimal number. 9.1

_____ 12. Change 0.052 to a percent. 9.1

_____ 13. Change $\frac{4}{5}$ to a percent. 9.1

Perform the indicated operations.

_____ 14. $25,000 - 13,976$ 1.3

_____ 15. $17,032 \times 405$ 2.1

_____ 16. $134,590 \div 789$ 2.2

_____ 17. $3\frac{3}{8} \times 12\frac{2}{3}$ 4.1

_____ 18. $15\frac{3}{4} \div 21$ 4.2

_____ 19. $\dfrac{8}{7} + \dfrac{3}{4} + \dfrac{7}{15}$ 5.3

_____ 20. $18.312 + 9.7 - 4.896$ 6.2

_____ 21. 34.26×3.9 7.1

_____ 22. $9.376 \div 0.231$ (Round to hundredths place.)   7.2

_____ 23. There were 540 applicants for the Governor's Scholars Program. If only 20 percent of the applicants were accepted, how many were rejected?   9.2

_____ 24. The ratio of a certain baseball player's hits to times at bat is 3 to 10. If he was at bat 370 times, how many hits did he get?   8.3

_____ 25. The Dayberrys bought 2 plots of land. One plot was $7\frac{3}{8}$ acres and the other was $5\frac{1}{6}$ acres. How many acres of land did they buy?   5.5

_____ 26. Find the total amount paid for a $980 gas range purchased in a state where the sales tax is 6.5 percent.   10.3

_____ 27. If a $16\frac{3}{4}$-ounce can of green beans costs 63 cents, how much do green beans cost per ounce? (Round to hundredths of a cent.)   4.3

Check your answers in the back of your book.

11

Working with Statistics

Margarita's boss told her that 648 of her co-workers at the factory earn less money than she does. The boss also mentioned that her salary ranks at the 72nd percentile. How many workers are employed at Margarita's factory?

In this age of computers and mass media, it is nearly impossible to get through a day without reading or hearing some **statistics** about everything from batting averages to unemployment to cola tests. As consumers and citizens it is important that we are able to understand statistics so that we can interpret what they mean to us.

In this chapter we learn how to

1. Find the mean, the median, and the mode.
2. Find the range and the midrange.
3. Work with percentiles.

11.1 Working with Measures of the Average

We often read statements such as

The average life expectancy for females born in 1965 is 73.7 years.

The median height for an 18-year-old male in the United States is 70.2 inches.

The most common legal driving age among the states in the United States is 16.

Each of these statements describes the *average* or most likely or "most normal" number to expect when looking at a group of numbers. Some of the numbers in the group will be higher than the average number, and some of the numbers will be lower than the average. The average just tells us about the middle or center of the group of numbers. Such averages are also called **measures of central tendency.** There are several different measures of central tendency. We will consider each one.

Finding the Mean

The first measure that you ever learned for finding the middle of a group of numbers was the **mean.** In fact, when most people talk about the "average" of a group of numbers, they are thinking of the mean.

To find the mean of a group of numbers, you probably remember that you must add up all the numbers in the group and then divide your sum by however many numbers there are in the group.

For instance, suppose we find the mean for the following group of weekly salaries: $500, $400, $400, $370, $300, $250, $200, $180, $140, and $130. We note that there are 10 salaries in the group.

First we add the salaries.   Then we divide this sum by 10.

$$
\begin{array}{r}
\$\ 500 \\
400 \\
400 \\
370 \\
300 \\
250 \\
200 \\
180 \\
140 \\
+\quad 130 \\
\hline
\$2870
\end{array}
$$

$$\text{Mean} = \$2870 \div 10$$

$$= \frac{2870}{10}$$

$$\text{Mean} = \$287$$

The mean salary for the group is $287.

Notice that no person in the group actually earns $287. It is important to realize that the mean of a group of numbers may not be one of the numbers actually in the group.

Let's write the steps we use to find the mean of a group of numbers.

Finding the Mean

1. Find the sum of the quantities in the group.
2. Divide that sum by the number of quantities in the group.
3. The quotient is the mean.

Example 1. Find the mean of the following group of children's weights (in pounds): 83, 80, 82, 87, 75, 79, 79, 90, and 76.

Solution. We note that there are 9 weights in the group and find the sum of the weights.

$$
\begin{array}{r}
83 \\
80 \\
82 \\
87 \\
75 \\
79 \\
79 \\
90 \\
+76 \\
\hline
731
\end{array}
$$

Mean $= 731 \div 9$

$$
\begin{array}{r}
81 \\
9\overline{)731} \\
72 \\
\hline
11 \\
9 \\
\hline
2
\end{array}
$$

Mean $= 81\frac{2}{9}$

The mean weight is $81\frac{2}{9}$ pounds.

You complete Example 2.

Example 2. Find the mean for the following group of annual precipitation measurements (in inches) for 11 U.S. cities. (Round your answer to the hundredths place.)

Phoenix	7.05	Miami	59.80
Los Angeles	14.05	Atlanta	48.34
Denver	15.51	Chicago	34.44
San Francisco	19.53	New Orleans	56.77
Washington, D.C.	38.89	Boston	42.52
Detroit	30.96		

Solution. We note that there are 11 measurements in the group and find their sum.

$$
\begin{array}{r}
7.05 \\
14.05 \\
15.51 \\
19.53 \\
38.89 \\
30.96 \\
59.80 \\
48.34 \\
34.44 \\
56.77 \\
+42.52 \\
\hline
367.86
\end{array}
$$

Mean $= \underline{\hspace{1cm}} \div 11$

$$11\overline{)\phantom{\underline{\hspace{1cm}}}}$$

Mean $\doteq \underline{\hspace{1cm}}$

The mean precipitation measurement is about _____ inches.

Check your work on page 462. ▶

Example 3. The turkeys for sale at Burns' Butcher Shop weigh the following amounts (in pounds): $16\frac{1}{2}$, $13\frac{3}{4}$, $15\frac{1}{2}$, $14\frac{1}{8}$, and $17\frac{1}{4}$. What is the mean weight for the group?

Solution. We note that there are 5 weights in the group and find the sum of the weights. Remember that we add mixed numbers by adding the whole number parts and the fractional parts.

$$16\frac{1}{2} = 16 + \frac{1}{2} = 16 + \frac{4}{8}$$

$$13\frac{3}{4} = 13 + \frac{3}{4} = 13 + \frac{6}{8}$$

$$15\frac{1}{2} = 15 + \frac{1}{2} = 15 + \frac{4}{8}$$

$$14\frac{1}{8} = 14 + \frac{1}{8} = 14 + \frac{1}{8}$$

$$17\frac{1}{4} = 17 + \frac{1}{4} = 17 + \frac{2}{8}$$

$$75 + \frac{17}{8}$$

$$75 + \frac{17}{8} = 75 + 2 + \frac{1}{8}$$

$$= 77\frac{1}{8}$$

$$\text{Mean} = 77\frac{1}{8} \div 5$$

$$= \frac{617}{8} \times \frac{1}{5}$$

$$= \frac{617}{40}$$

$$\begin{array}{r} 15 \\ 40\overline{)617} \\ \underline{40} \\ 217 \\ \underline{200} \\ 17 \end{array}$$

$$\text{Mean} = 15\frac{17}{40}$$

The mean weight for the turkeys is $15\frac{17}{40}$ pounds.

Sometimes one or more numbers may appear more than once in a group. To obtain an accurate mean for such a group, each of the repeated numbers must be counted as many times as it occurs. The number of times that a certain number appears is called its **frequency** (or its **weight**). Before finding the sum of all the numbers in such a group, we multiply each number times its frequency. Then we add all those products and divide by the sum of *all* the frequencies to find the mean of the group. The mean of such a group is sometimes called a **weighted mean.**

Finding a Weighted Mean

1. Find the product of each number in the group and its corresponding frequency.
2. Find the sum of those products.
3. Find the sum of all the frequencies in the group.
4. Divide the sum of the products by the sum of the frequencies.
5. The quotient is the weighted mean of the group.

Example 4. Find the mean age for the students in an English class containing two 18-year-olds, seven 19-year-olds, five 20-year-olds, three 22-year-olds, one 26-year-old, one 29-year-old, and one 42-year-old.

Solution

Number	Frequency	Product
18	2	$18 \times 2 = 36$
19	7	$19 \times 7 = 133$
20	5	$20 \times 5 = 100$
22	3	$22 \times 3 = 66$
26	1	$26 \times 1 = 26$
29	1	$29 \times 1 = 29$
42	1	$42 \times 1 = 42$
	Sum: 20	Sum: 432

The sum of the products of the numbers and their corresponding frequencies is 432. The sum of the frequencies is 20. The mean age for the 20 students in the class is

$$\frac{432}{20} = 21.6 \text{ years}$$

A student's grade point average (**GPA**) is a good example of a weighted mean. Each letter grade is worth a specified number of points (A, 4 points; B, 3 points; C, 2 points; D, 1 point; F, 0 points), and each number of points is weighted by the number of course hours in which the corresponding grade was earned. For instance, a grade of B in a four-hour course would be worth

$$3 \times 4 = 12 \text{ points}$$

The GPA for a term (semester, quarter, trimester, and so on) is the weighted mean of the points earned.

Example 5. Last semester Tim earned a grade of B in his 3-hour English course, a C in his 5-hour mathematics course, an A in his 3-hour psychology course, a C in his 3-hour chemistry course, and a B in his 2-hour laboratory. What was Tim's GPA last semester? (Round your answer to the hundredths place.)

Solution

Course and Grade		Points	Frequency (Hours)	Total Points
English	B	3	3	$3 \times 3 = 9$
Mathematics	C	2	5	$2 \times 5 = 10$
Psychology	A	4	3	$4 \times 3 = 12$
Chemistry	C	2	3	$2 \times 3 = 6$
Laboratory	B	3	2	$3 \times 2 = 6$
			Sum: 16	Sum: 43

The GPA is the weighted mean, found by dividing the sum of the total points by the sum of the frequencies (hours).

$$\frac{43}{16} = 2.6875$$

Tim's GPA last semester was about 2.69.

⫸ Trial Run

_____ 1. Find the mean for the following group of monthly salaries.

$2000, $1480, $1000, $1650, $980, $1548

_____ 2. Find the mean for the following scores on a math test.

98, 79, 83, 86, 93, 55, 75, 86, 93, 75

_____ 3. During July the following daily high temperatures were recorded for the given number of days in Evansville. Find the mean daily high temperature for that month. (Round answer to the nearest hundredth degree.)

93 (7 days), 98 (3 days), 97 (5 days), 92 (10 days)
95 (4 days), 99 (1 day), 101 (1 day)

_____ 4. At the weekly meeting of Weight Worriers, the following losses (in pounds) were reported. Find the mean weight loss for the group.

$1\frac{1}{2}$, $\frac{3}{4}$, $2\frac{3}{4}$, $3\frac{1}{8}$, $4\frac{1}{2}$,   3, $2\frac{1}{8}$,   2

_____ 5. At a track meet, the following times (in seconds) were recorded for the participants in the men's 50-yard running event. Find the mean time for the runners in this event.

6.10, 6.06, 5.24, 5.36, 5.95, 5.55

Answers are on page 463.

Sometimes when we are asked to find the mean for a group of numbers, we already know the sum of the numbers in the group.

For instance, suppose a bag containing 12 apples weighs 16 ounces and we wish to find the mean weight of 1 apple. In this case we already know the _sum_ of the apple weights is 16. To find the mean weight we simply divide.

$$\text{Mean} = 16 \div 12$$

$$= \frac{16}{12}$$

$$= \frac{4}{3}$$

$$\text{Mean} = 1\frac{1}{3}$$

We expect the weight of each apple to be "about" $1\frac{1}{3}$ ounces.

You complete Example 6.

Example 6. If Matt worked a total of 150 hours in 20 working days last month, find the mean number of hours he worked each day.

Solution. There are 20 numbers and we are told that their sum is 150.

$$\text{Mean} = \frac{150}{20}$$ Divide the sum of the hours by the number of days.

$$= \underline{\hspace{2cm}}$$ Reduce the fraction.

$$\text{Mean} = \underline{\hspace{2cm}} \text{ hours}$$ Divide.

So, "on the average," Matt worked _____ hours per day.

Check your work on page 463. ▶

When Matt worked 150 hours in 20 days, we found his mean number of daily hours to be $7\frac{1}{2}$. What should Matt's total hours have been for 20 days if he wanted a mean of *8 hours* per day? Remember that

$$\text{Mean} = \frac{\text{total number of hours}}{\text{number of days}}$$

Here, we know the mean is to be 8 hours and the number of days is 20, so letting t stand for the total number of hours, we write

$$8 = \frac{t}{20}$$

Now we must find t. What number divided by 20 equals 8? Remembering (from Chapter 2) the fact that every division statement corresponds to a multiplication statement, we see that

$$8 = \frac{t}{20} \quad \text{means} \quad 20 \times 8 = t$$

$$160 = t$$

So Matt should have worked a total of 160 hours if he wanted his mean number of daily hours to be 8.

Example 7. What must be the total number of students in 15 classes of Math 101 if the mean class size is to be 33?

Solution. We know the mean is 33 and we know the number of classes is 15. We must find the total number of students, t.

$$\text{Mean} = \frac{\text{total number of students}}{\text{number of classes}}$$

$$33 = \frac{t}{15}$$

$$15 \times 33 = t$$

$$495 = t$$

There must be 495 students enrolled in Math 101.

You complete Example 8.

Example 8. What must be the sum of your points on 12 quizzes if you wish the mean of your points to be 85?

Solution. We know the mean is 85 and the number of quizzes is 12. We must find the total number of points, t.

$$\text{Mean} = \frac{\text{total number of points}}{\text{number of quizzes}}$$

$$85 = \frac{t}{12}$$

$$\underline{\quad\quad} \times \underline{\quad\quad} = t$$

$$\underline{\quad\quad} = t$$

The sum of all your quiz points must be _____ .

Check your work on page 463. ▶

Suppose you have already taken 11 of your 12 quizzes and you have 957 points so far. How many points must you receive on your last quiz if you wish to have a mean of 85 points for all the quizzes?

From Example 6, we already know that you must have a *total* of 1020 points on all the quizzes to have a mean of 85 points. Since you have already reached 957 points on the first 11 quizzes, the number of points you need on the last quiz is just the *difference* between the total needed and the amount you already have.

$$\begin{array}{r} 1020 \\ - 957 \\ \hline 63 \end{array}$$

You must receive 63 points on the last quiz. Let's check this answer to be sure we are correct.

$$\text{Mean} = \frac{\text{total number of points}}{\text{number of quizzes}}$$

$$= \frac{\overbrace{957}^{\text{11 quizzes}} + \overbrace{63}^{\text{1 quiz}}}{12}$$

$$= \frac{1020}{12}$$

$$\text{Mean} = 85 \quad \checkmark$$

Example 9. Stephanie will receive a bonus from her company at the end of the year if her mean monthly sales are at least $2500. During the first 10 months of the year, she has total sales of $21,500. How much must she sell during the last 2 months if she wishes to receive a bonus?

Solution. We must find what Stephanie's total sales for the year must be.

$$\text{Mean} = \frac{\text{total sales}}{\text{number of months}}$$

$$2500 = \frac{t}{12}$$

$$12 \times 2500 = t$$

$$30,000 = t$$

Her total sales for the year must be $30,000. Since she has already reached $21,500, her sales for the last 2 months must be $30,000 − $21,500 = $8500.

⇒ Trial Run

_____ 1. A crate containing 16 heads of cabbage weighs 40 pounds. Find the mean weight of each head of cabbage.

_____ 2. If Charlotte watched a total of 28 hours of television last week, find the mean number of hours watched each day.

_____ 3. What must be the sum of your points on 5 tests if you wish the mean of your points to be 93?

_____ 4. How many pounds must you lose over an 8-week period if you want your weekly mean weight loss to be $2\frac{1}{2}$ pounds?

_____ 5. The Fountain Square Players are performing "Look Homeward Angel" for 5 nights. During the first 4 performances, a total of 976 persons attended the play. How many persons must attend the fifth night so that the mean attendance will be 250 persons?

Answers are on page 463.

Finding the Median

The **median** is another popular measure of the middle of a group.

> The **median** of a group of numbers is the number that appears in the *middle* when the numbers have been put in order.

In an earlier example, we found the *mean* for the following group of children's weights (in pounds) 83, 80, 82, 87, 75, 79, 79, 90, and 76. The mean was $81\frac{2}{9}$ pounds. Before looking for the *median,* we must *put the numbers in order*. It does not matter whether we list them in increasing order or in decreasing order, but they *must* be in order. We'll list them with the heaviest weight at the top of the list and the lightest weight at the bottom.

There are 9 numbers in this group of weights. The weight in the *middle* should have as many weights *above* it as it has weights *below* it. Can you see that 80 pounds is the median weight?

$$
\left.\begin{array}{c} 90 \\ 87 \\ 83 \\ 82 \end{array}\right\} \text{4 above}
$$

80 ← **Median**

$$
\left.\begin{array}{c} 79 \\ 79 \\ 76 \\ 75 \end{array}\right\} \text{4 below}
$$

Notice that the mean and the median are not necessarily the same number.

The nice thing about working with the median is that it requires no arithmetic except counting.

Finding the Median

1. Put the numbers in order.
2. Find the number in the middle of the group.
3. That middle number is the median.

Example 10. Find the median salary for the 7 people employed at the Dental Care Clinic.

Dentist Smith	$58,000
Dentist Jones	73,000
Dental assistant Whitby	16,000
Dental assistant Malone	18,000
Secretary Alberts	14,000
Clerk Manning	9,000
Maintenance person Kraft	3,000

Solution. We must put the salaries in order before finding the median.

$$
\left.\begin{array}{c} 73,000 \\ 58,000 \\ 18,000 \end{array}\right\} \text{3 above}
$$

16,000 ← **Median**

$$
\left.\begin{array}{c} 14,000 \\ 9,000 \\ 3,000 \end{array}\right\} \text{3 below}
$$

The median salary is $16,000.

Suppose we compute the *mean* salary for this group of employees.

73,000
58,000
18,000
16,000
14,000
9,000
3,000
191,000

Mean = 191,000 ÷ 7

$$
\begin{array}{r}
27,285 \\
7)\overline{191,000} \\
\underline{14} \\
51 \\
\underline{49} \\
2\,0 \\
\underline{1\,4} \\
60 \\
\underline{56} \\
40 \\
\underline{35} \\
5
\end{array}
$$

Mean ≐ $27,286

Notice that this mean of $27,286 is very different from the median of $16,000.

It is often true that the mean and the median will be far apart, especially when the group contains a few very large numbers or a few very small numbers.

In the case of the Dental Care Clinic, the median probably describes the "middle" salary of the group much better than the mean does. The mean salary is far above the salaries of 5 out of the 7 employees. It does not provide a very real picture of salaries at the clinic because of the 2 very high salaries that affect the mean.

Example 11. Find the median salary for the following group.

$500, $400, $400, $370, $300, $250, $200, $180, $140, $130

Solution. Since the salaries are already in order, we look for the median.

$500, $400, $400, $370, $300, $250, $200, $180, $140, $130
↑
Median

Because there is an *even* number of salaries in the group, there is no middle salary in the list. The median falls halfway between 250 and 300. To find the median we use the *mean* of these 2 salaries.

$$\frac{250 + 300}{2} = \frac{550}{2}$$
$$= 275$$

The median is $275. Notice that there are 5 salaries above the median and 5 salaries below the median.

In an earlier section, we found that the *mean* for this same set of salaries was $287. Notice that in this case the mean ($287) and the median ($275) are not very far apart. Why? Because there are no salaries that are much higher or much lower than all the other salaries in the group.

You complete Example 12.

Example 12. Find the median age for the 20 students in an English class if there are two 18-year-olds, seven 19-year-olds, five 20-year-olds, three 22-year-olds, one 26-year-old, one 29-year-old, and one 42-year-old.

Solution. We find the median age by listing all the students' ages in order. Remember, we must list each age *as many times as it occurs* in the group.

18, 18, 19, 19, 19, 19, 19, 19, 19, 20, 20, 20, 20, 20, 22, 22, 22, 26, 29, 42

The median age falls between the tenth and eleventh ages. Since both those ages are _____ , we know that the median age is _____ years.

Check your work on page 463. ▶

▥➡ **Trial Run**

_____ 1. Find the median height for Western's basketball team. The heights (in inches) are 80, 82, 74, 79, 83, 80, 82, 81, 78, 76, 83, and 79.

_____ 2. Find the median weekly salary for the following group.

$418, $520, $342, $275, $195, $478, $394, $216, $500

_____ 3. A farmer's records show that the numbers of bushels per acre produced in a certain field over the past 6 years were 198, 185, 163, 175, 172, and 180. Find the median number of bushels.

_____ 4. The numbers of speeding tickets written by the city police department during the first 15 days of January were 22, 24, 18, 22, 16, 24, 20, 20, 23, 17, 29, 27, 19, 21, and 22. Find the median number of tickets written.

Answers are on page 463.

In working with the median for a large set of numbers, it is helpful to remember that roughly half the numbers will fall above the median and roughly half will fall below the median. For instance, if we read that the median age for a person in the United States is 29 years, we conclude that roughly half the people in the United States are younger than 29 and half are older than 29.

Finding the Mode

There are times when neither the mean nor the median is very useful in making a prediction about a group of numbers.

Suppose a department head wants to decide how many sections of a certain course to offer this fall by looking back at fall enrollments in that course for the past few years.

Suppose a cola manufacturer wants to decide what size bottle is the most popular by looking at the number of bottles of each size sold during the past few years.

Perhaps you can see that the mean and median for these groups of numbers will not provide the answers to the questions being asked. Instead, the department head and the cola manufacturer should probably pay more attention to the number that appears *most often* in their groups of numbers. That number is called the **mode** for the group.

> The **mode** for a group of numbers is the number that appears *most often* in the group.

Example 13. The numbers of sections of Math 101 filled during the past 6 fall semesters were 13, 14, 15, 15, 16, and 15. Find the mode for this group.

Solution. The number of sections filled most often was 15. The *mode* is 15 sections.

You complete Example 14.

Example 14. The owner of the Pop Shop has decided to stock just one size of bottled soft drinks. He has kept a record for the past month of the sizes sold.

10-ounce	120 bottles
12-ounce	250 bottles
16-ounce	790 bottles

Use the mode to help the Pop Shop owner decide what size bottle to stock.

Solution. Since the size sold most often (the mode) is the _____-ounce size, we advise him to stock only _____-ounce bottles.

Check your work on page 463. ▶

If each number in a group of numbers appears just once, we say that group has *no mode*. Sometimes a group of numbers has *more* than one mode.

Example 15. Find the mode for the following group of salaries.

$500, $400, $400, $370, $300, $250, $250

Solution. Both $400 and $250 are modes for this group. Each appears twice.

Example 16. Find the mode for the following group of test scores.

83, 92, 67, 74, 71, 60, 95, 98, 80, 76

Solution. Since no test score appears more than once, there is *no* mode for this group.

▥▶ Trial Run

_____ 1. The numbers of bolts found defective in each of the first 6 batches produced were 20, 22, 19, 22, 20, and 21. Find the mode for this group.

_____ 2. A grocer has decided to stock just one brand of cat food. For one month he kept a record of the brands sold.

Cutey Cat	38 bags
Mighty Meow	51 bags
Chewy Chomp	46 bags

Use the mode to help the grocer decide which cat food to stock.

_____ 3. Find the mode of the following group of test scores.

100, 93, 85, 76, 97, 65, 85, 72, 84, 78, 72

_____ 4. Find the mode for the following weights (in pounds) of Little League baseball players.

53, 55, 49, 61, 48, 43, 50, 41, 32

Answers are on page 463.

▶ Examples You Completed

Example 2. Find the mean for the following group of normal annual precipitation measurements for 11 U.S. cities. (Round your answer to the hundredths place.)

Solution. We note that there are 11 measurements in the group and find their sum.

7.05	Mean = 367.86 ÷ 11
14.05	
15.51	33.441
19.53	11)367.860
38.89	33
59.80	37
48.34	33
34.44	4 8
56.77	4 4
42.52	46
30.96	44
367.86	20
	11
	9

Mean $\doteq$ 33.44

The mean precipitation measurement is about 33.44 inches.

Example 6. If Matt worked a total of 150 hours in 20 working days last month, find the mean number of hours he worked each day.

Solution. There are 20 numbers and we are told that their sum is 150.

$$\text{Mean} = \frac{150}{20}$$

$$= \frac{15}{2}$$

$$\text{Mean} = 7\frac{1}{2} \text{ hours}$$

So, "on the average," Matt worked $7\frac{1}{2}$ hours per day.

Example 8. What must be the sum of your points on 12 quizzes if you wish the mean of your points to be 85?

Solution. We know the mean is 85 and the number of quizzes is 12. We must find the total number of points, t.

$$\text{Mean} = \frac{\text{total number of points}}{\text{number of quizzes}}$$

$$85 = \frac{t}{12}$$

$$12 \times 85 = t$$

$$1020 = t$$

The sum of all your quiz points must be 1020.

Example 12 (*Solution*). We find the median age by listing all the students' ages in order. Remember, we must list each age as many times as it occurs in the group.

$$18, 18, 19, 19, 19, 19, 19, 19, 19, 20, \;\; 20, 20, 20, 20, 22, 22, 22, 26, 29, 42$$
$$\uparrow$$

The median age falls between the tenth and eleventh ages. Since both those ages are 20, we know that the median age is 20 years.

Example 14 (*Solution*). Since the size sold most often (the mode) is the 16-ounce size, we advise him to stock only 16-ounce bottles.

Answers to Trial Runs

page 454   **1.** $1443   **2.** 82.3   **3.** 94.52°   **4.** $2\frac{31}{64}$ pounds   **5.** 5.71 seconds

page 457   **1.** $2\frac{1}{2}$ pounds   **2.** 4   **3.** 465   **4.** 20 pounds   **5.** 274

page 460   **1.** 80 inches   **2.** $394   **3.** $177\frac{1}{2}$ **4.** 22

page 462   **1.** 22   **2.** Mighty Meow   **3.** 85 and 72   **4.** no mode

EXERCISE SET 11.1

_____ 1. The following amounts were awarded to the plaintiffs in a group action personal injury lawsuit. Find the mean amount awarded.

$25,000, $18,500, $21,000, $50,000, $32,000, $24,500, $20,000, $19,000, $23,700, $42,000

_____ 2. At a radio factory, a quality control inspector checks each radio for defects. Her records show that the following numbers of radios were rejected each day last week. Find the mean number rejected.

23, 14, 15, 17, 11

_____ 3. The following weights (in pounds) were recorded for the babies born at the Medical Center over the weekend. Find the mean weight.

$10\frac{3}{4}$, $6\frac{1}{2}$, 8, $7\frac{1}{4}$, $5\frac{1}{8}$, $4\frac{1}{2}$, $8\frac{1}{4}$, $7\frac{5}{8}$

_____ 4. A sample of 20 transcripts was selected from the freshman class to study grade point averages. Find the mean grade point average to the nearest hundredth.

1.32, 4.00, 2.35, 3.64, 2.75, 1.95, 0.75, 3.27, 3.83, 1.72, 2.03, 2.00, 2.84, 3.12, 2.38, 3.15, 3.42, 1.52, 2.25, 1.86

_____ 5. Midterm examination scores for 15 students in an English class are as follows. Find the mean score to the nearest whole number.

50, 75, 80, 35, 96, 75, 84, 76, 82, 92, 43, 71, 67, 63, 75

_____ 6. During a 3-day antique show and sale, the daily attendance figures were 836, 920, and 746. Find the mean daily attendance.

_____ 7. On a recent multiple-choice quiz, 5 students missed 0 questions, 4 missed 1, 6 missed 2, 3 missed 4, 7 missed 5, and 3 missed 7. Find the mean number of questions missed.

_____ 8. In a secretarial pool, the monthly salary figures are as follows: 8 earn $1500, 6 earn $1300, and 5 earn $1150. Find the mean monthly salary to the nearest dollar.

_____ 9. Last summer, Juanita earned an A in her 3-hour history course, a B in her 4-hour physics course, and a C in her 2-hour physical education course. Find her GPA for the summer session (to the hundredths place).

_____ 10. For the babies born at City Hospital last week, the following lengths were recorded: 12 were $20\frac{1}{2}$ inches long, 8 were 19 inches long, 10 were $19\frac{1}{2}$ inches long, 7 were 20 inches long, and 3 were 21 inches long. Find the mean length for the babies.

_____ 11. The police department responded to 371 calls for help last week. Find the mean number of calls answered each day.

_____ 12. Last year, a used car dealer sold a total of 780 cars. Find the mean number of cars sold each week.

_____ 13. What must be the total attendance figure for 6 home football games if the coach wishes to have mean home attendance of 22,250?

_____ 14. What must be the total weight of a box containing 24 chickens if the mean weight of the chickens is to be $1\frac{3}{4}$ pounds?

_____ 15. Six communities will receive federal funding for local projects. If the total amount already granted to the first 5 applicants is $4,260,000, how much will the sixth applicant receive if the mean amount granted is $800,000?

_____ 16. During the first 5 weeks of his 6-week diet plan, Chad has lost 13 pounds. How many pounds must he lose during the sixth week if he wishes to reach his goal of a weekly mean weight loss of $2\frac{1}{2}$ pounds?

_____ 17. The mathematics department's records show that the numbers of students enrolled in 7 sections of calculus last semester were 36, 28, 32, 43, 35, 39, and 30. Find the median number of students in a calculus section.

_____ 18. During the first 10 days of July, the daily high temperatures recorded in Fahrenheit degrees were 99, 98, 102, 105, 98, 93, 89, 95, 100, and 101. Find the median temperature.

_____ 19. Find the median price per pound for ground beef if a survey shows that the following prices per pound are charged at the 6 grocery stores closest to campus.

$1.83, $1.79, $1.94, $1.87, $1.75, $1.88

_____ 20. The following test scores were recorded for the students in a Math 109 class. Find the median test score.

79, 83, 45, 97, 65, 69, 75, 83, 89, 91, 67, 83, 52, 88, 78, 70, 82, 93, 78, 80

_____ 21. A publishing company found the following numbers of defective books during the first 10 days of production of a new mystery. Find the mode.

34, 26, 31, 24, 31, 25, 26, 22, 26, 30

_____ 22. A group of college freshmen reported the following numbers of hours spent studying each night. Find the mode of the numbers reported.

1, 3, $2\frac{1}{2}$, 5, 4, 3, $1\frac{1}{2}$, 2, 4, 3, 1, 0, $3\frac{1}{2}$, 4, 2, 3, 1, 2, 5, 3

_____ 23. A realtor plans to build a new apartment complex. He kept a record for the past month of requests for one-, two-, and three-bedroom apartments.

One-bedroom	35 requests
Two-bedroom	52 requests
Three-bedroom	24 requests

Use the mode to help the realtor decide what size apartments he should build.

_____ 24. Find the mode for the following group of shoe sizes worn by the girls' basketball team: $7\frac{1}{2}$, 10, $9\frac{1}{2}$, $8\frac{1}{2}$, 8, 7, 8, 11, 10, and 8.

☆ Stretching the Topics _____

_____ 1. On her first 6 tests Carlotta earned scores of 83, 75, 87, 68, 91, and 85. If a mean score of 80 is needed for a grade of B in the course, what is the lowest score Carlotta can earn on the last test and still earn a B in the course? A mean grade of 90 is required for an A in the course. If the highest possible score on any test is 100 points, is it possible for Carlotta to earn an A?

_____ 2. A set of data is separated into 2 groups. The first group has 15 scores with a mean of 52.3 and the second group has 20 scores with a mean of 64.5. What is the mean if the 2 groups of data are combined? (Round answer to the tenths place.)

Check your answers in the back of your book.

If you can do the problems in **Checkup 11.1,** then you are ready to go on to Section 11.2.

✓ CHECKUP 11.1

_____ 1. The following deductions for taxes were made from the checks of the employees at Genault: $158, $176, $205, $163, $192, $132, $168, and $179. Find the mean amount deducted (to the nearest dollar).

_____ 2. The students on the varsity baseball team were enrolled for the following numbers of credit hours: 8 for 12 hours, 5 for 13 hours, 4 for 14 hours, 3 for 15 hours, and 2 for 17 hours. Find the mean number of credit hours (to the nearest tenth) for a baseball player.

_____ 3. Find the total number of students enrolled in Introductory Astronomy if there are 12 classes and the mean class size is 32 students.

_____ 4. Salespersons at the Swank Shop are expected to sell a daily mean of $750 worth of merchandise. How much must Polly sell on Friday if her sales during the first 4 days of the week totaled $2872? (Polly works 5 days each week.)

_____ 5. The students in Ramon's exercise class were able to do the following numbers of pushups at the first session. Find the median number of push-ups.

12, 23, 19, 15, 12, 25, 18, 13, 17, 10

_____ 6. Find the mode for the following numbers of times the students in a class of English 100 had to take the "pass-fail" exam before scoring 70 percent or above.

2, 3, 1, 2, 3, 2, 2, 3, 3, 1, 5, 1, 3, 2, 4, 1, 3, 3, 1, 3, 4

Check your answers in the back of your book.

If You Missed Problems:	You Should Review Examples:
1	1–3
2	4
3	7, 8
4	9
5	10, 11
6	13–16

11.2 Finding the Midrange and Range

So far we have discussed 3 measures of central tendency for a group of numbers.

Mean	The sum of all quantities in the group divided by the number of quantities in the group.
Median	The middle number in the group.
Mode	The number appearing most often in the group.

There is one more measure of central tendency that is especially useful with large groups of numbers.

Finding the Midrange

When a group contains many numbers, it can take a great deal of time to calculate the mean or to find the median. If you wish to find the middle of a large group of numbers quickly, you can use the **midrange.**

> The **midrange** for a group of numbers is the *mean* of the largest and smallest numbers in the group.

The midrange will not always be the same as the mean or the median, but it will give you a good estimate of the middle of the group without difficult calculation.

For instance, suppose 1979 average incomes per person were listed for all the 50 states in the United States. If Alaska had the highest average income of $11,219 and Mississippi had the lowest average income of $6178, we can use the midrange to estimate the middle of the group of 50 incomes.

To find the midrange we add the highest number ($11,219) and the lowest number ($6178) and divide by 2.

$$\begin{array}{r} 11{,}219 \\ +\ 6{,}178 \\ \hline 17{,}397 \end{array} \qquad \begin{array}{l} \text{Midrange} = 17{,}397 \div 2 \\ \qquad\qquad = \$8698.50 \end{array}$$

By the way, the actual mean for the group of 50 incomes is $8773, so you can see that the midrange gives a pretty good "guess" without long calculations.

Example 1. Of the first 40 U.S. presidents, Theodore Roosevelt took office at the youngest age, 42, and Ronald Reagan took office at the oldest age, 69. Find the midrange for the group of presidents' ages.

Solution. We find the midrange by calculating the mean of the youngest and oldest ages.

$$\text{Midrange} = \frac{69 + 42}{2}$$

$$= \frac{111}{2}$$

$$\text{Midrange} = 55.5 \text{ years}$$

By the way, the mean age for *all 40* presidents is 54.8 years. Once again, the midrange is a good estimate.

You complete Example 2.

Example 2. Among the National League batting champions between 1930 and 1980, the highest batting average was held by Ted Williams: 0.406 in 1941. The lowest batting average was held by Carl Yastrzemski: 0.301 in 1968. Find the midrange for this group of 51 batting champions. (Round your answer to the thousandths place.)

Solution. To find the midrange, we calculate the mean of the highest and lowest batting averages in the group.

$$\text{Midrange} = \frac{ + }{2}$$

$$= \frac{}{2}$$

$$\text{Midrange} \doteq \underline{}$$

Check your work on page 474. ▶

The actual mean batting average for the 51 batting champions is 0.348, so once again our estimate was fairly close.

Keep in mind that the midrange is a useful measure of central tendency when we are working with a large group of numbers and we want a quick estimate of the "average" for the group. For example, weatherpersons find the average temperature for a day by finding the midrange of the group of temperatures recorded throughout that day.

⭢ **Trial Run**

_____ 1. In 1981, a listing of payments by the federal government to the states for conservation purposes shows the lowest was $107,000 to Rhode Island and the highest was $15,445,000 to Texas. Find the midrange for these payments.

_____ 2. Of the 58 counties in the state of California, San Francisco is the smallest with a land area of 46 square miles and San Bernadino is the largest with a land area of 20,064 square miles. Find the midrange for the areas of the California counties.

_____ 3. A listing of the 20 NBA scoring leaders during the period from 1962 through 1981 shows that Wilt Chamberlain scored the most with 4029 points and Dave Bing had the least with 2142 points. Find the midrange for the number of points scored by the NBA leaders.

_____ 4. Mr. Torance sold 50 chickens, the largest weighing $3\frac{3}{4}$ pounds and the smallest weighing $2\frac{1}{2}$ pounds. Find the midrange for the weights of the chickens.

_____ 5. When the contents of 100 twelve-ounce boxes of cereal were weighed, the actual weights were between 12.35 ounces and 11.72 ounces. Find the midrange of the weights of the 100 boxes of cereal.

Answers are on page 474.

Finding the Range

Let's find the _mean_ for each of the two groups of numbers listed here.

Group I	_Group II_
1000	700
700	600
500	550
400	450
300	400
+ 100	+ 300
3000	3000

$$\text{Mean} = \frac{3000}{6} \qquad \text{Mean} = \frac{3000}{6}$$

$$= 500 \qquad\qquad = 500$$

Each group has a mean of 500 but the groups don't look much alike. In Group I, the numbers are much more ''spread out'' than the numbers in Group II. To give a better picture of these two groups to someone who has not seen them, we can state what the ''spread'' is. The word used in statistics to describe the ''spread'' for a set of numbers is the **range.**

> The **range** for a group of numbers is the _difference_ between the largest and the smallest numbers in the group.

In Group I, we find | In Group II, we find

Range = 1000 − 100 | Range = 700 − 300

       = 900 | = 400

By looking at the range for a group we can see how ''spread out'' the numbers are in that group. A small range tells us the numbers are close together. A large range tells us the numbers are more spread out.

Example 3. If the oldest U.S. president was 69 when he took office and the youngest was 42, what is the range for the ages of U.S. presidents?

Solution

$$\text{Range} = \text{largest number} - \text{smallest number}$$

$$= 69 - 42$$

$$\text{Range} = 27 \text{ years}$$

You complete Example 4.

Example 4. Find the range for the following group of salaries at the Dental Care Clinic: $58,000, $73,000, $16,000, $18,000, $14,000, $9,000, and $3,000.

Solution

$$\text{Range} = \text{largest number} - \text{smallest number}$$

$$= \$\_\_\_\_ - \$\_\_\_\_$$

$$= \$\_\_\_\_$$

Check your work on page 474. ▶

Let's consider a few examples that contain all the ideas we have discussed in this chapter.

Example 5. The following numbers of points were scored by Hometown U. in its 12 home basketball games of the season.

83, 92, 77, 89, 71, 82, 78, 74, 68, 78, 90, 78

For this set of points, find (a) the range, (b) the mean, (c) the median, (d) the mode, and (e) the midrange.

Solution. Let's put the points in order before we begin.

92
90
89
83
82
78
78 ←
78
77
74
71
+68
960

(a) Range $= 92 - 68$
$= 24$ points

(b) Mean $= 960 \div 12$
$= 80$ points

(c) Median $= 78$ points

(d) Mode $= 78$ points

(e) Midrange $= \dfrac{92 + 68}{2}$

$= \dfrac{160}{2}$

$= 80$ points

Because this set of points does not contain any unusually high or unusually low numbers, we found that the 4 measures of central tendency were close together.

Example 6. The deposits in the 10 leading U.S. commercial banks (as of 1980) are listed here, rounded to the nearest million dollars.

Bank of America	$85,070,000,000
Citibank	72,430,000,000
Chase Manhattan Bank	56,173,000,000
Manufacturers Hanover Trust	39,001,000,000
Morgan Guaranty Trust	35,486,000,000

Chemical Bank	30,522,000,000
Continental Illinois National Bank	25,667,000,000
Bankers Trust Company	22,122,000,000
First National Bank of Chicago	20,785,000,000
Security Pacific National Bank	19,591,000,000

For this group of deposits, find (a) the range, (b) the mean, (c) the median, (d) the mode, and (e) the midrange.

Solution

$85,070,000,000
72,430,000,000
56,173,000,000
39,001,000,000
→ 35,486,000,000
30,522,000,000
25,667,000,000
22,122,000,000
20,785,000,000
+ 19,591,000,000
$406,847,000,000

(a) Range $= \$85,070,000,000 - \$19,591,000,000$

$= \$65,479,000,000$

(b) Mean $= \dfrac{\$406,847,000,000}{10}$

$= \$40,684,700,000$

(c) Median $= \dfrac{\$35,486,000,000 + \$30,522,000,000}{2}$

$= \dfrac{\$66,008,000,000}{2}$

$= \$33,004,000,000$

(d) Mode: none

(e) Midrange $= \dfrac{\$85,070,000,000 + \$19,591,000,000}{2}$

$= \dfrac{\$104,661,000,000}{2}$

$= \$52,330,500,000$

Notice here that the deposits in the Bank of America and Citibank are much larger than the deposits in the other banks in the group. Therefore the measures of central tendency are very different.

▮▶ Trial Run

_____ 1. When shopping for a refrigerator, Mrs. Bailey visited several appliance stores to look at the prices of comparable refrigerators. The highest price she found was $799.95 and the lowest was $639.89. What was the range of the prices?

_____ 2. Find the range for the following group of salaries of administrators in a county school system.

$45,000, $38,000, $29,000, $25,000, $23,000

_____ 3. Weight charts show that a 5-foot 2-inch female with a medium frame should
_____ weigh between 122 pounds and 130 pounds. Find the range and midrange.

_____ 4. The following scores were made on the first test in Math 102.
_____ 82, 71, 89, 75, 94, 83, 78, 74, 78, 68, 90, 78
_____ For this set of scores, find (a) the range, (b) the mean, (c) the median, (d) the
_____ mode, and (e) the midrange.

Answers are on page 474.

► **Examples You Completed** ————————————————

Example 2. Among the National League batting champions between 1930 and 1980, the highest batting average was held by Ted Williams: 0.406 in 1941. The lowest batting average was held by Carl Yastrzemski: 0.301 in 1968. Find the midrange for this group of 51 batting champions.

Solution. To find the midrange, we calculate the mean of the highest and lowest batting averages in the group.

$$\text{Midrange} = \frac{0.406 + 0.301}{2}$$

$$= \frac{0.707}{2}$$

$$\text{Midrange} \doteq 0.354$$

Example 4. Find the range for the following group of salaries at the Dental Care Clinic: $58,000, $73,000, $16,000, $18,000, $14,000, $9000, and $3000.

Solution

$$\text{Range} = \text{largest number} - \text{smallest number}$$

$$= \$73,000 - \$3000$$

$$= \$70,000$$

Answers to Trial Runs ————————————————

page 470   **1.** $7,776,000   **2.** 10,055 square miles   **3.** 3085.5   **4.** $3\frac{1}{8}$ pounds   **5.** 12.035 ounces

page 473   **1.** $160.06   **2.** $22,000   **3.** 8 pounds; 126 pounds
4. (a) 26   (b) 80   (c) 78   (d) 78   (e) 81

EXERCISE SET 11.2

_____ 1. On the Professional Bowlers Association winter tour in 1981, the most
_____   successful winner earned $40,000 and the least successful earned $13,000.
             Find the range and midrange for PBA bowlers' earnings.

_____ 2. The height of the tallest player on Central's basketball team is 86 inches, and
_____   the shortest player is 71 inches tall. Find the range and midrange for the
             heights of Central's players.

_____ 3. For the years 1948 through 1981, the fastest speed recorded by a winner of
_____   the Indianapolis 500 was 162.026 miles per hour and the slowest winning
             speed was 119.814 miles per hour. Find the range and midrange for the
             winning speeds.

_____ 4. Find the range and midrange for the following times (in seconds) recorded
_____   by the participants in a men's 50-yard running event.

                5.75, 6.03, 6.73, 5.83, 5.52, 6.13, 6.02, 5.78, 5.67

_____ 5. Find the range and midrange for the following low temperatures recorded in
_____   Spring Grove during the first 10 days of 1989.

                38, 32, 30, 27, 19, 9, 15, 20, 28, 25

_____ 6. On the first day of school the following enrollments were recorded for the 6
_____   elementary schools in Johnson County: 836, 582, 647, 912, 596, and 728.
             Find the range and midrange for the enrollments.

_____ 7. For the following weekly earnings of salespersons working in a clothing
_____   store, find the mean, median, and mode.
_____
_____ $235, $252, $250, $195, $205, $240, $235, $200, $248, $195

_____ 8. The largest amount of rainfall ever recorded in a certain city for the month
_____   of July was 5.36 inches and the least amount was 0.73 inches. Find the
             range and midrange for July rainfall.

_____ 9. Find the mean (to the nearest whole number), median, mode, range, and
_____   midrange for the following scores on a test.
_____
_____   75, 90, 80, 85, 95, 75, 80, 60, 80, 90, 95, 70, 80, 75, 80, 80, 90, 80,
_____   70, 65, 80, 85, 95, 90, 95, 90
_____

_____ 10. At a lake trout fishing tournament, the following weights in pounds were
_____   reported. Find the mean, median, mode, range, and midrange.
_____
_____   45, 47, 50, 52, 46, 62, 45, 36, 59, 40, 57
_____

☆ Stretching the Topics

_____ 1. The average speeds (in miles per hour) of the cars that won the Daytona 500
_____   from 1972 through 1982 were 161.550, 157.205, 140.894, 153.649, 152.810,
_____   153.218, 159.730, 143.977, 177.602, 169.651, and 153.991. Find the mean,
_____   mode, median, range, and midrange for this set of speeds.
_____

Check your answers in the back of your book.

If you can do the problems in **Checkup 11.2,** then you are ready to go on to Section 11.3.

✓ **CHECKUP 11.2**

_____ **1.** During the 1981–1982 NBA season, Denver scored the highest average number of points per game with 126.5. Atlanta had the lowest average of 101 points. Find the midrange of average points scored by NBA teams that season.

_____ **2.** Calvin and Gloria found the prices of a certain brand of jeans at several department stores varied from a low of $19.98 to a high of $42.95. Find the range of the prices.

_____ **3.** The following numbers represent the scores made by students on a diagnostic
_____ math test. Find the mean, median, mode, midrange, and range.
_____
_____   18, 23, 15, 17, 10, 8, 12, 15, 17, 22
_____

Check your answers in the back of your book.

If You Missed Problems:	You Should Review Examples:
1	1, 2
2	3, 4
3	5, 6

11.3 Working with Percentiles

So far we have discussed statistical measures that help us describe the "middle" of a group of numbers (mean, median, mode, and midrange) and a measure that describes the "spread" of a group of numbers (range). Now we would like to look at a measure that helps us describe the **position** of a particular number in a group of numbers. The measure that we shall use to describe position is the **percentile** rank.

You have taken many standardized tests in your life, and when you received your scores on these tests, they were probably presented as *percentiles*. What is meant by these statements?

Gretchen's score on the PSAT was at the 96th percentile.

Any child who scores at or above the 85th percentile on the Youth Fitness Test will receive the Presidential Physical Fitness Award.

First of all, you must understand that a percentile does *not* tell what percent of correct answers a person had on a test. Instead, a percentile rank for a test score tells what percent of all the people taking the test made scores that were *below* that score.

For instance, if Gretchen's PSAT score was at the 96th percentile, we know that 96 percent of *all* the students taking the PSAT had scores that were *below* Gretchen's score. If Jamie scored at the 85th percentile on the Youth Fitness Test, then we know that 85 percent of *all* the children taking the test had scores *below* Jamie's score.

In general, we say that

> If some number is at the *p*-th **percentile** in a set of numbers, then we know that *p* percent of the numbers in the set are *below* that number.

Example 1. A score of 15 on the ACT Exam is at the 50th percentile. What does that mean?

Solution. If a score of 15 on the ACT Exam is at the 50th percentile, that means that 50 percent of *all* the students taking the ACT Exam have scores *below* 15.

Example 2. At his high school, Carlo's grade point average of 3.2 is at the 75th percentile. What does that mean?

Solution. We conclude that 75 percent of *all* the students at Carlo's high school have grade point averages *below* 3.2.

In Example 2, suppose we know that there are 800 students in Carlo's high school. How can we figure out how many students have a grade point average below 3.2? We know that 75 percent of all the students have a grade point average below 3.2. Therefore, we can use the percent proportion to find the unknown number, *n*.

$$\left. \begin{array}{c} \text{number below} \\ \hline \text{total number} \end{array} \right\} \quad \frac{n}{800} = \frac{75}{100} \qquad \text{Write the percent proportion.}$$

$$n \times 100 = 800 \times 75 \qquad \text{Write the cross products.}$$

$$n \times 100 = 60{,}000 \qquad \text{Multiply on the right side.}$$

$$\frac{n \times 100}{100} = \frac{60{,}000}{100} \qquad \text{Divide both sides by 100.}$$

$$n = 600$$

Thus, we see that there are 600 students in Carlo's high school with a grade point average below 3.2.

As you will recall from Chapter 9, the percent proportion is very useful because it allows you to find any missing part in *any* percent statement. In working with a percentile (*p*), we will always be looking at the following percent proportion (which we might call the **percentile proportion**).

Percentile Proportion

$$\frac{\text{number below}}{\text{total number}} = \frac{p}{100}$$

Example 3. Of the 120,000 children taking the Youth Fitness Test, Juan scored at the 87th percentile. How many children scored below Juan?

Solution. Let's use the percentile proportion.

$$\frac{\text{number below}}{\text{total number}} = \frac{p}{100}$$

Here we do not know the "number below" Juan, so our proportion becomes

$$\frac{n}{120{,}000} = \frac{87}{100}$$

$$n \times 100 = 120{,}000 \times 87$$

$$n \times 100 = 10{,}440{,}000$$

$$\frac{n \times 100}{100} = \frac{10{,}440{,}000}{100}$$

$$n = 104{,}400$$

We see that 104,400 children scored below Juan.

Example 4. If 69 of the 150 soldiers in Edna's training company received a rating below hers, what was Edna's percentile rank in the company?

Solution. We use the percentile proportion to find the unknown percentile, *p*.

$$\left.\begin{array}{l}\text{number below}\\ \text{total number}\end{array}\right\} \quad \frac{69}{150} = \frac{p}{100}$$

$$69 \times 100 = 150 \times p$$

$$6900 = 150 \times p$$

$$\frac{6900}{150} = \frac{150 \times p}{150}$$

$$46 = p$$

We find that 46 percent of the company is below Edna, so she ranks at the 46th percentile.

Let's look at the problem stated at the beginning of this chapter.

Example 5. Margarita's boss told her that 648 of her co-workers at the factory earn less money than she does. The boss also mentioned that her salary ranks at the 72nd percentile. How many workers are employed at Margarita's factory?

Solution. We can use the percentile proportion to solve for the total number t.

$$\left.\frac{\text{number below}}{\text{total number}}\right\} \quad \frac{648}{t} = \frac{72}{100}$$

$$648 \times 100 = t \times 72$$

$$64800 = t \times 72$$

$$\frac{64800}{72} = \frac{t \times 72}{72}$$

$$900 = t$$

There are 900 workers at Margarita's factory.

If we wish to find a percentile rank for a particular number in a group, we may do so by looking at the original list of numbers.

Consider this list of normal monthly temperatures for the city of Chicago.

Before we can discuss percentiles, we must put the temperatures *in order*, from highest to lowest.

Jan.	24° F		75° F
Feb.	27		74
Mar.	37		71
Apr.	50		66
May	60		60
June	71		55
July	75		50
Aug.	74		40
Sept.	66		37
Oct.	55		29
Nov.	40		27
Dec.	29		24

Let's find the percentile rank for a temperature of 37 degrees. We know we would like to use the percentile proportion to find the unknown percentile, p.

$$\frac{\text{number below}}{\text{total number}} = \frac{p}{100}$$

We know the total number of temperatures is 12. But how can we find the *number* of temperatures *below* our temperature of 37 degrees? By counting, of course! In our list, we see **3** temperatures *below* 37 degrees. Our proportion becomes

$$\left.\frac{\text{number below}}{\text{total number}}\right\} \quad \frac{3}{12} = \frac{p}{100}$$

$$3 \times 100 = 12 \times p$$

$$300 = 12 \times p$$

$$\frac{300}{12} = \frac{12 \times p}{12}$$

$$25 = p$$

We find that 25 percent of the monthly temperatures are below 37 degrees, so that temperature is at the 25th percentile.

You complete Example 6.

Example 6. For Chicago's monthly temperatures, find the percentile rank for a temperature of 55 degrees.

Solution. We count the number of temperatures *below* our temperature of 55 degrees and find that there are _____ temperatures below 55 degrees. Our percentile proportion becomes

$$\left.\begin{array}{c} \text{number below} \\ \hline \text{total number} \end{array}\right\} \quad \frac{6}{12} = \frac{p}{100}$$

$$\underline{\qquad} \times 100 = \underline{\qquad} \times p$$

$$600 = 12 \times p$$

$$\underline{\qquad} = p$$

Since _____ percent of the temperatures are below 55 degrees, we conclude that 55 degrees is at the _____th percentile.

Check your work on page 484. ▶

Example 7. Find the Chicago monthly temperature that ranks at the 75th percentile.

Solution. This problem will take 2 steps. In the percentile proportion

$$\frac{\text{number below}}{\text{total number}} = \frac{p}{100}$$

we know the percentile (p) is 75, and we know the total number of temperatures in the list is 12. We do *not* know the *number of temperatures below* the temperature we're looking for. We must find n in the percentile proportion.

$$\frac{n}{12} = \frac{75}{100}$$

$$\frac{n}{12} = \frac{3}{4}$$

$$n \times 4 = 12 \times 3$$

$$n \times 4 = 36$$

$$\frac{n \times 4}{4} = \frac{36}{4}$$

$$n = 9$$

75° F	
74	
71	
66	
60	
55	
50	
40	9 below
37	
29	
27	
24	

Now we know there are 9 temperatures *below* the temperature we are looking for. How do we find that temperature? *By counting up 9 from the bottom.*

There are 9 temperatures *below* the temperature of 71 degrees. Therefore that temperature is at the 75th percentile. What this means to an observer is that 75 percent of the time the normal monthly temperature will be *below* 71 degrees. A person thinking about moving to Chicago might be interested in this fact.

⫸ Trial Run

_____ 1. Of the 5250 students in the freshman class, Maria ranked at the 74th percentile at the end of the year. How many freshmen ranked below Maria?

_____ 2. If 136 of the 200 students enrolled in an exercise class were rated below him, what was Claude's percentile rank in the class?

_____ 3. Mr. Chinn's supervisor has told him that 408 of his co-workers at the factory produce at a lower level than he does. If Mr. Chinn's production level ranks at the 96th percentile, how many persons are employed at this factory?

Consider the following list of weight losses (in pounds) for the members of Weight Worriers during a 6-month period.

Ann	38	Dede	50	Ginger	59	Juan	34
Ben	46	Eula	48	Henri	26	Keith	23
Cassie	14	Fred	37	Ioona	41	Luana	30

_____ 4. Find Luana's percentile rank in this group of dieters.

_____ 5. Find the weight loss that ranks at the 50th percentile.

Answers are on page 484.

Remember that a percentile rank for a certain number in a group tells us only what percent of all the numbers in the group are _below_ that certain number. The percentile rank tells _nothing_ about the percent of all the numbers that are _above_ that certain number. Look at the following group of weights (in pounds) for 10 people.

Suppose Becky weighs 120 pounds. How many weights are _below_ hers? There are 7 weights below Becky's weight. To find her percentile rank we use

120
120
120
118
117
115
114
112
111
110

$$\frac{7}{10} = \frac{p}{100}$$

$$7 \times 100 = 10 \times p$$

$$700 = 10 \times p$$

$$\frac{700}{10} = \frac{10 \times p}{10}$$

$$70 = p$$

Since 70 percent of the weights are below hers, we say that Becky's weight is at the 70th percentile in this group. May we conclude that 30 percent of the weights are _above_ Becky's weight? Certainly _not_. No weight is above Becky's weight. Each person who weighs 120 pounds is at the 70th percentile. No one has a higher percentile rank.

People often draw incorrect conclusions from percentile ranks. If you learn what a percentile rank means (and what it does _not_ mean), you will avoid errors in interpreting percentiles.

▶ Example You Completed

Example 6. For Chicago's monthly temperatures, find the percentile rank for a temperature of 55 degrees.

Solution. We count the number of temperatures *below* our temperature of 55 degrees and find there are 6 temperatures below 55 degrees. Our percentile proportion becomes

$$\left.\frac{\text{number below}}{\text{total number}}\right\} \frac{6}{12} = \frac{p}{100}$$

$$6 \times 100 = 12 \times p$$

$$600 = 12 \times p$$

$$\frac{600}{12} = \frac{12 \times p}{12}$$

$$50 = p$$

Since 50 percent of the temperatures are below 55 degrees, we conclude that 55 degrees is at the 50th percentile.

Answers to Trial Run

page 483   **1.** 3885   **2.** 68th percentile   **3.** 425   **4.** 25th percentile   **5.** 38 pounds

EXERCISE SET 11.3

_____ 1. If 625 students took the pretest in English and Irene ranked at the 84th percentile, how many students taking the test scored below Irene?

_____ 2. Of the 150,000 teenagers taking the Teen Physical Fitness Test, Gordon scored at the 35th percentile. How many teenagers scored below Gordon on the test?

_____ 3. If 272 students in Aletha's graduating class of 425 have grade point averages below hers, what is Aletha's percentile rank in the class?

_____ 4. In a survey of the reading habits of college students, responses were received from 5000 persons. If 4100 students had read fewer books than Gracie, what was her percentile rank?

_____ 5. Mr. Crowe's superintendent told him that 504 teachers in the school system earn less money than he does. Mr. Crowe's salary ranks at the 56th percentile. How many teachers are employed in the school system?

_____ 6. At the insurance company where Karla works, 36 of the secretaries type fewer words per minute than Karla does. Karla also knows that her typing ranks at the 75th percentile. How many secretaries work for the insurance company?

In answering questions 7 and 8, consider the following list of average numbers of bushels of corn per acre produced by farmers in Union County on experimental plots.

Anderson	185	Foster	209
Boswell	196	Greenwell	174
Casey	205	Henshaw	183
Dempsey	195	Ingle	207
Ewing	215	Jackson	191

_____ 7. Find Boswell's percentile rank in this group of farmers.

_____ 8. Find the farmer whose production ranks at the 80th percentile.

In answering questions 9 and 10, consider this list of the numbers of defective spark plugs turned out one day by the different production teams at the factory where Sam works.

Aries	41	Leo	41	Sagittarius	40
Taurus	36	Virgo	23	Capricorn	37
Gemini	45	Libra	43	Aquarius	32
Cancer	48	Scorpio	30	Pisces	30

_____ 9. If Sam is a member of the Libra production team, what is his team's percentile rank?

_____ 10. Which team ranks at the 25th percentile?

☆ Stretching the Topics _____

———— 1. Roberto is ranked 75th in his senior class of 425. Mesia is in the same senior class and ranks at the 80th percentile. Which student has the higher standing in the class?

———— 2. If Jamal ranks first in a class of 150, what is his percentile rank? (Round to the tenths place.)

———— 3. In the Chrystal's 10-K Classic, Tony finished 15th in a field of 240. Tom finished 35th in a field of 500 in the Rally's 10-K Run. Who had the better relative (percentile) position in his own race?

Check your answers in the back of your book.

If you can complete **Checkup 11.3,** you are ready to do the **Review Exercises** for Chapter 11.

✔ CHECKUP 11.3

_____ 1. If 325 persons work with Paul and he ranks at the 52nd percentile in seniority, how many persons rank below Paul?

_____ 2. Of the 85 salespersons at Tucker Auto Sales, 68 of them have sold fewer cars this year than Joe. What is Joe's percentile rank among the salespersons?

_____ 3. Linda's principal told her that 462 seniors in the class have a lower grade point average than she has. Linda ranks at the 88th percentile in her class. How many students are in Linda's class?

In answering questions 4 and 5, consider this list of the number of accidental deaths in 1989 by months.

January	8100	May	8500	September	7800
February	7000	June	8850	October	8350
March	7500	July	9500	November	7850
April	7700	August	9300	December	8550

_____ 4. What is the percentile rank for October accidents?

_____ 5. Which month ranks at the 90th percentile?

Check your answers in the back of your book.

If You Missed Problems:	You Should Review Examples:
1	3
2	4
3	5
4, 5	6

Summary

In this chapter, we discussed several basic statistical measures that can be used to describe a group of numbers. We learned that there are 4 measures of central tendency (**mean, median, mode,** and **midrange**) that give us information about the "middle" of a group of numbers. We also learned to use the **range** to describe the "spread" of such a group.

We summarize these 5 statistical measures in the next table, using the following group of numbers: 12, 11, 10, 10, 9, 8, 5, 4, and 3.

To Obtain the	We Must Find	Example
Mean	The sum of all the quantities, divided by the number of quantities in the group.	$\frac{12 + 11 + 10 + 10 + 9 + 8 + 5 + 4 + 3}{9} = \frac{72}{9} = 8$
Median	The number in the middle of the group, when the numbers are listed *in order*.	The fifth number: 9
Mode	The number that appears most often in the group.	The only number appearing twice: 10
Midrange	The sum of the group's largest and smallest numbers, divided by 2.	$\frac{12 + 3}{2} = \frac{15}{2} = 7.5$
Range	The difference between the largest and smallest numbers in the group.	$12 - 3 = 9$

To describe the "position" of a particular number in a group, we introduced the idea of **percentile rank.** If some number is ranked at the p-th percentile in a group of numbers, we know that p percent of the numbers in the group are *below* that number. In working with percentiles, we learned to use the **percentile proportion**

$$\frac{\text{number below}}{\text{total number}} = \frac{p}{100}$$

to find any unknown piece of information.

❑ Speaking the Language of Mathematics

Complete each sentence with the appropriate word or phrase.

1. The mean, median, mode, and midrange are all measures of _____ _____ .

2. To find the mean for a group of 15 numbers, we must find the _____ of all the numbers and then divide by _____ .

3. To find the median for a group of numbers, we first write the numbers in _____ .

4. If a test score ranks at the 73rd percentile, we know that _____ percent of all the scores are _____ that score.

△ Writing About Mathematics _____

Write your response to each question in complete sentences.

1. Explain how to calculate the mean, the median, the mode, and the midrange for a group of 25 ages.

2. If you have decided to stock just 1 size bag of potato chips instead of 3 different sizes at your snack bar, which measure of central tendency would you use to make your decision? Why?

3. Your friend Melinda has scored at the 63rd percentile on a standardized test of mechanical ability. Explain what her score means.

REVIEW EXERCISES for Chapter 11

_____ 1. The highest temperatures ever recorded in 6 southeastern cities were 106° F, 102° F, 98° F, 102° F, 112° F, and 110° F. Find the mean for this set of temperatures.

_____ 2. The daily sales at the Petal Pushers Flower Shop for a 6-day week were $384.70, $236.55, $576.89, $222.82, $472.54, and $351.61. Find the mean daily sales for the week.

_____ 3. In Ellen's math class students made the following test scores with the indicated frequencies: 90(3), 85(6), 80(7), 75(10), 70(6), 65(2), and 60(3). Find the mean score to the tenths place.

_____ 4. The total number of speeding tickets issued by a city police department for the month of September was 840. Find the mean number of tickets per day.

_____ 5. Quarterback Jeff Cesarone completed 289 passes in 11 games. What was the mean number of completions per game? (Round to the tenths place.)

_____ 6. What must be the total attendance figure for a play that runs 5 nights if the producer wishes to have a mean nightly attendance of 385?

_____ 7. What will be the total weight of a carton of 36 T-bone steaks if the mean weight of the steaks is $12\frac{2}{5}$ ounces?

_____ 8. During the first 11 weeks of his 12-week summer vacation, Jerome earned $4048 mowing lawns. How much must he earn during the 12th week if he wishes to reach his goal of earning a weekly mean of $375?

_____
_____ 9. The number of cars sold by a used car dealer over a 10-day period was 3, 5, 6, 2, 9, 8, 12, 15, 6, and 4. What is the median number of cars sold per day? What is the mean number of cars sold?

_____ 10. The amounts spent annually on advertising by the 8 leading motels in a certain city are $45,000, $87,000, $75,000, $89,000, $62,000, $58,000, $49,000, and $46,500. For this set of expenditures, find the median.

_____ 11. The heights (in inches) for 12 basketball players at Western University are 77, 73, 76, 77, 76, 78, 79, 77, 80, 75, 81, and 78. Find the mode of the players' heights.

_____ 12. The Medical Center reported the following number of patients treated in the emergency room during the first 15 days of January.

   28, 31, 22, 15, 17, 23, 22, 19, 21, 23, 17, 19, 22, 20, 18

Find the mode number of patients.

_____
_____ 13. At the Indian Hills Golf Club the length of the longest hole is 490 yards and the shortest is 150 yards. Find the range and midrange.

_____
_____ 14. Find the range and midrange for the following numbers of strikes recorded by 12 bowlers in a Sunnyvale Lanes bowling tournament.

   2, 5, 4, 2, 3, 6, 7, 5, 5, 6, 3, 4

_____ 15. The Jabar's electric bills for the last 6 months were $43.59, $53.72, $37.18,
_____ $32.29, $28.72, and $48.23. Find (to the nearest cent) the mean, median,
_____   mode, range, and midrange for these bills.
_____
_____

_____ 16. Claude ranked at the 85th percentile on an exam given to 200 prospective
             employees. How many persons taking the test ranked below Claude?

_____ 17. In a survey of 460 employee salaries at a large department store, Katrina
             learned that 276 employees earn less than she earns. What is Katrina's
             percentile rank?

_____ 18. Sam's psychology teacher told him that 58 students in the class made a
             lower score on the first test than he did. If Sam ranks at the 29th percentile,
             how many students are in Sam's psychology class?

*In answering questions 19 and 20, consider this list of times (in minutes) recorded by the top
10 finishers in the one-mile run.*

Charles	4.6	Roger	4.1
Jules	5.3	Jim	4.7
Sydney	4.8	Filbert	5.2
Gunder	4.4	Herb	4.3
Arne	5.7	Derek	4.8

_____ 19. Find Jim's percentile rank in this group of runners.

_____ 20. Who finished at the 90th percentile?

Check your answers in the back of your book.

If You Missed Exercises:	You Should Review Examples:	
1, 2	Section 11.1	1–3
3		4, 5
4, 5		6
6, 7		7, 8
8		9
9, 10		10–12
11, 12		13–16
13, 14	Section 11.2	1–4
15		5, 6
16	Section 11.3	3
17		4
18		5
19, 20		6, 7

If you have completed the **Review Exercises** and corrected your errors, you are ready to take
the **Practice Test** for Chapter 11.

PRACTICE TEST for Chapter 11

	SECTION	EXAMPLES

_____ 1. Find the mean for the following group of weights (in pounds) recorded at the Thursday night Weight Worriers meeting.
        11.1   1

          138, 179, 154, 149, 163, 142, 135, 191, 187, 201

_____ 2. In 15 days Betty collected 386 aluminum cans for recycling. Find the mean number of cans collected each day. (Round to nearest whole number.)
        11.1   4

_____ 3. What must be the total number of points that Central College's opponents have scored in 6 football games if the mean number of points per game given up by the Central defense is 8?
        11.1   5

_____ 4. Nicole will receive a bonus from her employer if the mean of her sales for 6 months is at least $1500. What must be her sales for the sixth month if the total of her sales for the first 5 months is $6750?
        11.1   7

_____ 5. The circulation manager of a small weekly newspaper has recorded the following numbers of copies sold during the past 8 weeks.
        11.1   9

         5729, 4623, 3987, 5022, 4973, 3875, 5326, 4678

Find the median number of papers sold.

_____ 6. A survey of 10 department heads at a university requested information concerning the ideal number of students in a class. Four said 15, 4 said 20, 5 said 25, 3 said 35, and 4 said 40. Find the mean and mode for these responses.
        11.1   11

_____ 7. When the heights of all 10-year-old girls in Warren Elementary School were recorded, the shortest was 127.5 centimeters and the tallest was 149.5 centimeters. Find the midrange for this group.
        11.2   1, 2

_____ 8. At Union High School, the highest teacher's salary is $45,382 and the lowest is $23,755. Find the range of salaries at U.H.S.
        11.2   4

_____ 9. The following numbers of cars passed through the Audubon Parkway tollbooth in 1989.
        11.2   6

January	19,300	July	30,150
February	16,250	August	24,230
March	17,420	September	23,460
April	18,740	October	20,700
May	23,400	November	19,210
June	28,510	December	28,370

Find the range, mean, median, mode, and midrange.

_____ 10. In a class of 44 students, Carmen's score ranks at the 75th percentile. How many people made grades lower than Carmen's?
        11.3   3, 4

	SECTION	EXAMPLE

_____ 11. If 34 of the 85 persons employed at the sewing factory produce shirt collars at a rate below hers, what is Eunice's percentile rank for production?   11.3   4

_____ 12. Janie has been told by her physical education teacher that 781 students can do fewer push-ups than she can. He also told her that she ranks at the 65th percentile. How many students are taking physical education?   11.3   5

In answering questions 13–16, consider the following group of IQ scores for a class of 20 ninth graders.

76, 138, 78, 130, 79, 125, 80, 82, 120, 117,
84, 90, 95, 100, 116, 115, 110, 108, 97, 103

_____ 13. Find the percentile rank for a score of 115.   11.3   6

_____ 14. Find the percentile rank for a score of 79.   11.3   6

_____ 15. What score ranks at the 80th percentile?   11.3   7

_____ 16. What score ranks at the 25th percentile?   11.3   7

Check your answers in the back of your book.

SHARPENING YOUR SKILLS after Chapters 1–11

SECTION

_____ 1. Classify 173 as prime or composite. If the number is composite, write it as a product of prime numbers.

2.3

_____ 2. Change $14\frac{7}{8}$ to an improper fraction.

3.1

_____ 3. Reduce $\frac{1350}{9270}$ to lowest terms.

3.2

_____ 4. Write 34.072 in words.

6.1

_____ 5. Change $\frac{35}{49}$ to a decimal number. (Round to the thousandths place.)

7.3

_____ 6. Find the missing part in the proportion $\dfrac{3.6}{c} = \dfrac{0.9}{12}$.

8.2

_____ 7. Change $\frac{5}{8}$ to a percent.

9.1

_____ 8. Change $15\frac{3}{4}\%$ to a fraction.

9.1

_____ 9. Find 125% of 385.

9.2

_____ 10. Find the mean for the temperatures 93°, 98°, 97°, 92°, 95°, 99°, and 101°. (Round to the hundredths place.)

11.1

Perform the indicated operations.

_____ 11. $8\dfrac{2}{3} \times 2\dfrac{7}{13} \times \dfrac{1}{2}$

4.1

_____ 12. $28\dfrac{1}{3} \div 5\dfrac{5}{6}$

4.2

_____ 13. $\dfrac{3\frac{2}{3}}{1\frac{5}{6}}$

4.2

_____ 14. $\dfrac{11}{12} - \dfrac{1}{8} - \dfrac{2}{3}$

5.3

_____ 15. $0.7238 - 0.6564$

6.2

_____ 16. $6.83 \times 1000 \times 0.503$

7.1

_____ 17. $36.42 \div 2.7$

7.2

_____ 18. Find the area of a triangle with a base of $12\frac{3}{5}$ centimeters and a height of $7\frac{1}{3}$ centimeters.

4.3

_____ 19. Mr. Estevez has a rectangular piece of plywood that is $3\frac{3}{4}$ feet wide and $6\frac{2}{3}$ feet long. How many feet longer is the piece of plywood than it is wide?

5.5

_____ 20. At the cafeteria, Sarna had an appetizer for $2.25, a salad for $1.75, an entree for $6.50, a drink for $1.15, and a dessert for $1.35. The sales tax was $0.86. How much change did she receive from a twenty-dollar bill?

6.3

_____ 21. The tax on property is computed by multiplying its assessed value by the tax rate. If the tax rate for Franklin County is 0.0125, find the tax bill on property evaluated at $75,500.

7.4

_____ 22. The ratio of students to faculty at Benton High School is 28 : 1. If there are 73 faculty members, how many students are enrolled at Benton High?

8.3

_____ 23. Of the 45 students who began a certain math class, only 27 completed the course. What percent of those starting finished the course?

9.2

_____ 24. If a refrigerator that regularly sells for $720 is "on sale" for $504, find the amount of markdown. Find the percent of markdown.

10.1

_____ 25. Last year there were 439 armed robberies reported to the police department. This year there were only 368 reported. Find the percentage change in the number of armed robberies. (Round to the nearest percent.)

10.2

_____ 26. Find the simple interest that will be charged if Ruby borrows $1250 at 10.5 percent interest for 3 years. How much will Ruby repay at the end of 3 years?

10.4

_____ 27. In a class of 60 students, Tonya's score ranks at the 75th percentile. How many people made scores lower than Tonya's?

11.3

_____ 28. In a class of 16 students, the scores on a recent test were 76, 88, 78, 50, 79, 95, 80, 82, 68, 97, 84, 90, 95, 54, 76, and 87. Find the percentile rank for a score of 82.

11.3

Check your answers in the back of your book.

12

Working
with Measurement

A public health clinic staff expects to give a 0.5 milliliter allergy shot to each one of 2100 patients this week. How many liters of allergy medicine must they have ready for the week?

In earlier chapters we have discussed measurements of many kinds. We have solved problems involving length, distance, weight, temperature, and time. Now we must take a closer look at the topic of measurement and become more familiar with metric, as well as American, units of measure.

In this chapter we learn how to

1. Measure length and distance, using American and metric units.
2. Measure weight, using American and metric units.
3. Measure capacity, using American and metric units.

The methods for using a calculator to solve problems of the types encountered in Chapters 12 and 13 are discussed in Section 16.4 of Chapter 16.

12.1 Measuring Length and Distance

When you find your height, the distance from your house to the movie theater, or the width of a piece of furniture, you are dealing with **length,** or **linear measure.**

Linear measure is used to find length or distance.

Using American Units to Measure Length

You would probably measure your height in feet and inches, the distance to the movie theater in miles, and the width of a piece of furniture in inches. Feet, inches, and miles are all **units** of linear measure used in the United States. They are referred to as **American units** of measure.

You should be familiar with the most basic American units of linear measure and their relationships to each other.

Units	Abbreviations
1 foot = 12 inches	1 ft = 12 in.
1 yard = 3 feet	1 yd = 3 ft
1 mile = 5280 feet	1 mi = 5280 ft

When we measure length we must be sure to include the proper *units* as a part of our answer.

Example 1. Mark must pick up Suzanne on his way to the movies. The distance from Mark's house to Suzanne's house is 3.2 miles, and the distance from Suzanne's house to the movies is 6.5 miles. How far must Mark drive from his house to the movies?

Solution. We must add the distances

$$
\begin{array}{r}
3.2 \text{ mi} \\
+\,6.5 \text{ mi} \\
\hline
9.7 \text{ mi}
\end{array}
$$

Mark must drive 9.7 miles.

In Example 1, we notice that the unit used to measure both distances was the mile. We added the distances and attached the unit "mile" to the answer, because *the units matched.*

To add linear measures, the units must be the same.

What do we do if the units to be added are *not* the same? Somehow we must make them match. Look at this problem.

Carpet for stairs is sold by the "running foot". In other words, it is sold according to the length measured in feet. If Hector needs 13 yards of carpet for his stairway, how can we change this length to feet? We know that 1 yard = 3 feet, so 13 yards should be *13 times* as long as 1 yard.

$$13 \text{ yd} = 13 \times 3 \text{ ft}$$
$$= 39 \text{ ft}$$

Hector needs 39 feet of carpet for his stairs.

What we have done here is to **convert** our units from yards to feet. There is a handy way to convert units with just 1 step of arithmetic. Before we learn that handy method, we must spend a minute laying some groundwork. When we say that 1 yard = 3 feet, we are saying that 1 yard and 3 feet are exactly the same length. If that is so, then

$$\frac{1 \text{ yd}}{3 \text{ ft}} = 1 \qquad \text{and} \qquad \frac{3 \text{ ft}}{1 \text{ yd}} = 1$$

Why? Because we know that if the numerator and denominator of a fraction are the same, then the fraction is equal to 1.

Now let's return to Hector's problem of converting 13 yards to feet. See if you can follow this calculation.

$$13 \text{ yd} = 13 \times 1 \text{ yd}$$

$$= \frac{13}{1} \times \frac{1 \text{ yd}}{1} \times \frac{3 \text{ ft}}{1 \text{ yd}} \qquad \text{Multiply on the right side by 1.}$$

$$= \frac{13}{1} \times \frac{\overset{1}{\cancel{1 \text{ yd}}}}{1} \times \frac{3 \text{ ft}}{\underset{1}{\cancel{1 \text{ yd}}}} \qquad \text{Reduce the fractions.}$$

$$= 13 \times 3 \text{ ft}$$

$$13 \text{ yd} = 39 \text{ ft} \qquad \text{Find the product.}$$

There are 2 fractions that involve feet and yards. The trick in choosing the correct fraction for multiplying is to see which fraction will cause the unwanted unit to *divide out,* leaving only the unit you want. For instance, let's convert 24 feet to yards.

$$24 \text{ ft} = 24 \times 1 \text{ ft}$$

$$= \frac{24}{1} \times \frac{1 \text{ ft}}{1} \times \frac{1 \text{ yd}}{3 \text{ ft}}$$ Multiply on the right side by 1. (We don't want feet; we do want yards.)

$$= \frac{24}{1} \times \frac{\overset{1}{\cancel{1 \text{ ft}}}}{1} \times \frac{1 \text{ yd}}{\underset{3}{\cancel{3 \text{ ft}}}}$$ Reduce the fractions.

$$= \frac{\overset{8}{\cancel{24}} \times 1 \text{ yd}}{\underset{1}{\cancel{3}}}$$ Reduce again.

$$= 8 \times 1 \text{ yd}$$

$$24 \text{ ft} = 8 \text{ yd}$$ Find the product.

The fractions that we use to make such conversions are called **conversion factors.** The most common conversion factors for American units of length are

$\dfrac{12 \text{ in.}}{1 \text{ ft}}$	$\dfrac{1 \text{ ft}}{12 \text{ in.}}$	$\dfrac{3 \text{ ft}}{1 \text{ yd}}$	$\dfrac{1 \text{ yd}}{3 \text{ ft}}$	$\dfrac{5280 \text{ ft}}{1 \text{ mi}}$	$\dfrac{1 \text{ mi}}{5280 \text{ ft}}$

Remember that *each of these fractions is equivalent to the number 1.*

Example 2. Use a conversion factor to change 4 miles to feet.

Solution. We want the *feet* unit in our answer.

$$4 \text{ mi} = 4 \times 1 \text{ mi} \times \frac{5280 \text{ ft}}{1 \text{ mi}}$$ Choose the conversion factor. (We don't want miles; we do want feet.)

$$= \frac{4}{1} \times \frac{\overset{1}{\cancel{1 \text{ mi}}}}{1} \times \frac{5280 \text{ ft}}{\underset{1}{\cancel{1 \text{ mi}}}}$$ Reduce the fractions.

$$= 4 \times 5280 \text{ ft}$$

$$4 \text{ mi} = 21,120 \text{ ft}$$ Find the product.

You complete Example 3.

Example 3. Use a conversion factor to change 1760 feet to miles.

Solution. We want the *mile* unit in our answer.

$$1760 \text{ ft} = 1760 \times 1 \text{ ft} \times \frac{1 \text{ mi}}{5280 \text{ ft}}$$

Choose the conversion factor.
(We don't want _____; we do want _____.)

$$= \frac{1760}{1} \times \frac{\overset{1}{\cancel{1 \text{ ft}}}}{1} \times \frac{1 \text{ mi}}{\underset{5280}{\cancel{5280 \text{ ft}}}}$$

Reduce the fractions.

$$= \frac{1760 \times 1 \text{ mi}}{5280}$$

Reduce again.

$$= \text{\_\_\_\_} \times 1 \text{ mi}$$

$$1760 \text{ ft} = \text{\_\_\_\_} \text{ mi}$$

Find the product.

Check your work on page 508. ▶

The nice thing about conversion factors is that you can use more than one such factor in the same problem if necessary. For instance, to convert 7 yards to inches, we may write

$$7 \text{ yd} = 7 \times 1 \text{ yd} \times \frac{3 \text{ ft}}{1 \text{ yd}}$$

This will "get rid of" the yard unit, but we'll be left with the *feet* unit. Since we want the *inch* unit in our answer, we must use another conversion factor.

$$7 \text{ yd} = 7 \times 1 \text{ yd} \times \frac{3 \text{ ft}}{1 \text{ yd}} \times \frac{12 \text{ in.}}{1 \text{ ft}}$$

$$= \frac{7}{1} \times \frac{\overset{1}{\cancel{1 \text{ yd}}}}{1} \times \frac{\overset{3}{\cancel{3 \text{ ft}}}}{\underset{1}{\cancel{1 \text{ yd}}}} \times \frac{12 \text{ in.}}{\underset{1}{\cancel{1 \text{ ft}}}}$$

$$= 7 \times 3 \times 12 \text{ in.}$$

$$7 \text{ yd} = 252 \text{ in.}$$

If you ever plan to take a science course, you will find this work with conversion factors to be very helpful.

Example 4. Use conversion factors to change $1\frac{1}{4}$ miles to yards.

Solution. Let's change the mixed number to an improper fraction before we convert.

$$1\frac{1}{4} \text{ mi} = \frac{5}{4} \text{ mi}$$

Change mixed number to improper fraction.

$$= \frac{5}{4} \times 1 \text{ mi} \times \frac{5280 \text{ ft}}{1 \text{ mi}} \times \frac{1 \text{ yd}}{3 \text{ ft}}$$

Choose the conversion factors. (We don't want miles; we do want yards.)

$$= \frac{5}{4} \times \frac{\overset{1}{\cancel{1 \text{ mi}}}}{1} \times \frac{\overset{1760}{\cancel{5280 \text{ ft}}}}{\underset{1}{\cancel{1 \text{ mi}}}} \times \frac{1 \text{ yd}}{\underset{1}{\cancel{3 \text{ ft}}}}$$

Reduce the fractions.

$$= \frac{5 \times \overset{440}{\cancel{1760}} \times 1 \text{ yd}}{\underset{1}{\cancel{4}}}$$

Reduce again.

$$1\frac{1}{4} \text{ mi} = 2200 \text{ yd}$$

Find the product.

Example 5. Which is longer, a surfboard that is 100 inches long or a surfboard that is $8\frac{1}{2}$ feet long?

Solution. To compare the lengths, we must make our units match. Let's write both measurements in inches.

First board   100 in. = 100 in.

Second board $8\frac{1}{2}$ ft $= \frac{17}{2}$ ft

$$= \frac{17}{\underset{1}{\cancel{2}}} \times \frac{\overset{1}{\cancel{1 \text{ ft}}}}{1} \times \frac{\overset{6 \text{ in.}}{\cancel{12 \text{ in.}}}}{\underset{1}{\cancel{1 \text{ ft}}}}$$

$$= 17 \times 6 \text{ in.}$$

$$8\frac{1}{2} \text{ ft} = 102 \text{ in.}$$

The second surfboard is longer (by 2 inches).

⫸ **Trial Run**

_____ **1.** Use a conversion factor to change 3 miles to feet.

_____ **2.** Use a conversion factor to change 4 feet to inches.

_____ **3.** Use a conversion factor to change 108 feet to yards.

_____ **4.** Use a conversion factor to change 880 yards to miles.

_____ **5.** Use conversion factors to change $9\frac{1}{2}$ yards to inches.

_____ **6.** Which is longer, a ski that is 81 inches long or a ski that is $6\frac{1}{2}$ feet long?

Answers are on page 508.

Using Metric Units to Measure Length

In most countries, the **metric** system of measurement is used instead of the American system. The nice thing about the metric system is that it is much easier to convert among different units because the system is based on the number 10.

Using the metric system, you would probably find your height in meters or centimeters, the distance to the movie theater in kilometers, and the width of a piece of furniture in centimeters. The basic unit for linear measure in the metric system is the **meter.** If you have ever seen a meter stick, you have probably noticed that it is just a little longer than a yardstick.

> One meter is slightly longer than 1 yard.

All the other metric units of length are whole number or fractional number *multiples* of the meter. The basic metric units of linear measure and their relationships to each other are listed here for your reference.

Units	Abbreviations
1 kilometer = 1000 meters	1 km = 1000 m
1 hectometer = 100 meters	1 hm = 100 m
1 dekameter = 10 meters	1 dkm = 10 m
1 decimeter = $\frac{1}{10}$ meter	1 dm = $\frac{1}{10}$ m
(10 decimeters = 1 meter)	(10 dm = 1 m)
1 centimeter = $\frac{1}{100}$ meter	1 cm = $\frac{1}{100}$ m
(100 centimeters = 1 meter)	(100 cm = 1 m)
1 millimeter = $\frac{1}{1000}$ meter	1 mm = $\frac{1}{1000}$ m
(1000 millimeters = 1 meter)	(1000 mm = 1 m)

The linear units that you should become *most* familiar with are those that are used most often: kilometer, meter, centimeter, and millimeter. You should memorize the relationships for these important units.

> 1 km = 1000 m   1000 mm = 1 m
> 100 cm = 1 m   10 mm = 1 cm

Let's try to become more familiar with these metric units by relating them to some familiar objects. Look at the following centimeter scale.

|++++++++|———+———+———+———+———+———+———+———|
0   1   2   3   4   5   6   7   8   9   10   **centimeters**

Now look at your hand and find some part of it that is close to 1 centimeter long or wide. Perhaps it will be the width of your little fingernail or some part of a ring that you wear. Whatever you measure to be 1 centimeter, that is what you should think of when you try to picture 1 centimeter.

1 centimeter is about as wide as my _____ .

The 10 marks between the centimeter markings on the scale are the millimeter marks. A millimeter is approximately the width of a paper clip wire or a ball point pen point or a sharpened pencil point. Choose something familiar that is about 1 millimeter wide and remember that

1 millimeter is about as wide as _____ .

Now you have some handy ways to remember lengths of 1 meter, 1 centimeter, and 1 millimeter. What about 1 kilometer? We know that 1 km = 1000 m, but it is hard to picture 1000 meter sticks stretched out on the ground. The speedometers in many newer automobiles are marked off in miles per hour *and* in kilometers per mile.

The outer scale measures miles per hour and the inner scale measures kilometers per hour. From this speedometer, we notice that

$$40 \text{ km} \doteq 25 \text{ mi}$$

$$80 \text{ km} \doteq 50 \text{ mi}$$

and we conclude that

$$8 \text{ km} \doteq 5 \text{ mi}$$

Or you might prefer to remember that

1 kilometer is a little more than half a mile.

Now that you have decided upon your personal scales for estimating lengths with metric units, let's practice measuring a few familiar objects.

Example 6. Use your personal scale to estimate the indicated distance in metric units.

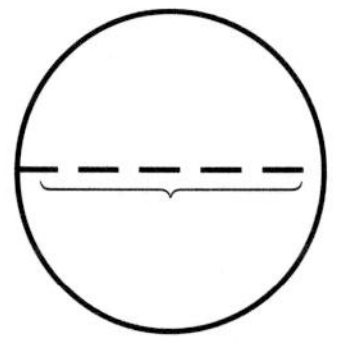

(a) the width of a nickel   **(b)** the thickness of a nickel   **(c)** the length of a large
                                                                              paper clip

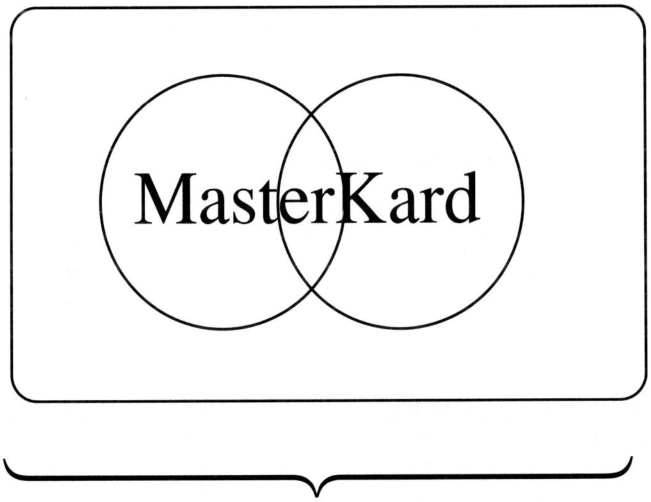

MasterKard

(d) the width of a credit card   **(e)** the width of a matchbook

Solution. **(a)** The width of a nickel is about 2.5 centimeters.
  (b) The thickness of a nickel is about 2 millimeters.
  (c) The length of the paper clip is about 5 centimeters.
  (d) The width of the credit card is about 9 centimeters.
  (e) The width of the matchbook is about 4 centimeters.

Example 7. Estimate the following distances using metric units.
(a) The length of a tablecloth that is a little over 2 yards long.
(b) A distance of approximately 10 miles.

Solution. **(a)** Since 1 meter is a little longer than 1 yard, the tablecloth is about 2 meters
              long.
  (b) Since 5 miles $\doteq$ 8 kilometers, we estimate that 10 miles $\doteq$ 16 kilometers.

⫸ Trial Run

Use your personal scale to estimate the indicated distances in metric units.

———————— 1. The length of a nail file.

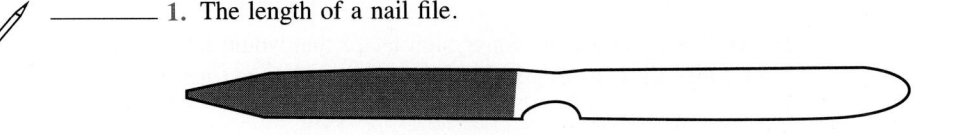

_____ **2.** The width of a postage stamp.

_____ **3.** The thickness of a breath mint.

Estimate the following distances using metric units.

_____ **4.** The width of a sheet of typing paper.

_____ **5.** The height of a door that is about 7 feet tall.

_____ **6.** A distance that is about 25 miles.

Answers are on page 508.

Now that we have a "feel" for the size of the basic metric units of length, we can practice converting from one metric unit to another. We will use **conversion factors** here, just as we did in converting American units of length. From our table of basic metric facts, we know the following conversion factors are all equal to the number 1.

$\dfrac{1000 \text{ m}}{1 \text{ km}}$	$\dfrac{100 \text{ cm}}{1 \text{ m}}$	$\dfrac{1000 \text{ mm}}{1 \text{ m}}$	$\dfrac{10 \text{ mm}}{1 \text{ cm}}$
$\dfrac{1 \text{ km}}{1000 \text{ m}}$	$\dfrac{1 \text{ m}}{100 \text{ cm}}$	$\dfrac{1 \text{ m}}{1000 \text{ mm}}$	$\dfrac{1 \text{ cm}}{10 \text{ mm}}$

Let's convert 2.3 meters to *centimeters*.

$$2.3 \text{ m} = 2.3 \times 1 \text{ m} \times \frac{100 \text{ cm}}{1 \text{ m}}$$

Choose the conversion factor.
(We don't want meters; we do want centimeters.)

$$= \frac{2.3}{1} \times \frac{\overset{1}{\cancel{1 \text{ m}}}}{1} \times \frac{100 \text{ cm}}{\underset{1}{\cancel{1 \text{ m}}}}$$

Reduce the fractions.

$$= 2.3 \times 100 \text{ cm}$$

$$2.3 \text{ m} = 230 \text{ cm}$$

Find the product.

Notice that we chose the conversion factor that would relate centimeters to meters and, at the same time, allow us to get rid of the unwanted units (meters).

You complete Example 8.

Example 8. Change 516 millimeters to meters.

Solution

$$516 \text{ mm} = 516 \times 1 \text{ mm} \times \frac{1 \text{ m}}{1000 \text{ mm}}$$

Choose the conversion factor. (We don't want _____ ; we do want _____.)

$$= \frac{516}{1} \times \frac{\overset{1}{\cancel{1 \text{ mm}}}}{1} \times \frac{1 \text{ m}}{\underset{1000}{\cancel{1000 \text{ mm}}}}$$

Reduce the fractions.

$$= \frac{516 \times 1 \text{ m}}{1000}$$

Multiply numerators and denominators.

$$= \underline{} \times 1 \text{ m}$$

$$516 \text{ mm} = \underline{} \text{ m}$$

Use the shortcut for dividing by 1000. (Move the decimal point 3 places to the left.)

Check your work on page 508. ▶

Example 9. Change 0.009 kilometers to centimeters.

Solution. We do not have *one* conversion factor that relates kilometers to centimeters. As with American units, we may use *two* conversion factors.

$$0.009 \text{ km} = 0.009 \times 1 \text{ km} \times \frac{1000 \text{ m}}{1 \text{ km}} \times \frac{100 \text{ cm}}{1 \text{ m}}$$

$$= \frac{0.009}{1} \times \frac{\overset{1}{\cancel{1 \text{ km}}}}{1} \times \frac{\overset{1000}{\cancel{1000 \text{ m}}}}{\underset{1}{\cancel{1 \text{ km}}}} \times \frac{100 \text{ cm}}{\underset{1}{\cancel{1 \text{ m}}}}$$

$$= 0.009 \times 1000 \times 100 \text{ cm}$$

$$= 9 \times 100 \text{ cm}$$

$$0.009 \text{ km} = 900 \text{ cm}$$

Example 10. If Art ran 1.53 km and Susan ran 1500 m, who ran farther? How much farther?

Solution. Before comparing distances, we must make our units match. Let's change Art's distance to meters.

Art's distance
$$1.53 \text{ km} = \frac{1.53}{1} \times \frac{\overset{1}{\cancel{1 \text{ km}}}}{1} \times \frac{1000 \text{ m}}{\underset{1}{\cancel{1 \text{ km}}}}$$

$$= 1.53 \times 1000 \text{ m}$$

$$1.53 \text{ km} = 1530 \text{ m}$$

Susan's distance   1500 m

Since 1530 > 1500, we conclude that Art ran farther (by 30 meters).

Trial Run

_______ 1. Change 3.8 meters to centimeters.

_______ 2. Change 0.015 kilometers to meters.

_______ 3. Change 785 millimeters to centimeters.

_______ 4. Change 116 centimeters to kilometers.

_______ 5. Change 5 meters to millimeters.

_______ 6. If an Olympic size pool is 100 meters long and a football field is 0.0914 kilometers long, which is longer? How much longer?

Answers are given below.

We have learned to measure length and distance using American and metric units of linear measure. In the next chapter we will see how to apply linear measurement to geometric figures.

Examples You Completed

Example 3. Use a conversion factor to change 1760 feet to miles.

Solution. We want the *mile* unit in our answer.

$$1760 \text{ ft} = 1760 \times 1 \text{ ft} \times \frac{1 \text{ mi}}{5280 \text{ ft}}$$

$$= \frac{1760}{1} \times \frac{\overset{1}{\cancel{1 \text{ ft}}}}{1} \times \frac{1 \text{ mi}}{\underset{5280}{\cancel{5280 \text{ ft}}}}$$

$$= \frac{\overset{1}{\cancel{1760}} \times 1 \text{ mi}}{\underset{3}{\cancel{5280}}}$$

$$= \frac{1}{3} \times 1 \text{ mi}$$

$$1760 \text{ ft} = \frac{1}{3} \text{ mi}$$

Example 8. Change 516 millimeters to meters.

Solution

$$516 \text{ mm} = 516 \times 1 \text{ mm} \times \frac{1 \text{ m}}{1000 \text{ mm}}$$

$$= \frac{516}{1} \times \frac{\overset{1}{\cancel{1 \text{ mm}}}}{1} \times \frac{1 \text{ m}}{\underset{1000}{\cancel{1000 \text{ m}}}}$$

$$= \frac{516 \times 1 \text{ m}}{1000}$$

$$= 0.516 \times 1 \text{ m}$$

$$516 \text{ mm} = 0.516 \text{ m}$$

Answers to Trial Runs

page 502 1. 15,840 ft 2. 48 in. 3. 36 yd 4. $\frac{1}{2}$ mi 5. 342 in. 6. The first ski (by 3 inches)

page 505 1. About 10 cm 2. About 2 cm 3. About 4 mm 4. About 22 cm 5. About 2 m 6. About 40 km

page 508 1. 380 cm 2. 15 m 3. 78.5 cm 4. 0.00116 km 5. 5000 mm 6. Olympic pool; 8.6 m

EXERCISE SET 12.1

_____ 1. Change 7 feet to inches.

_____ 2. Change 9 feet to inches.

_____ 3. Change 10 feet to yards.

_____ 4. Change 15 feet to yards.

_____ 5. Change 156 inches to feet.

_____ 6. Change 108 inches to feet.

_____ 7. Change 72 inches to yards.

_____ 8. Change 180 inches to yards.

_____ 9. Change 8 yards to inches.

_____ 10. Change 5 yards to inches.

_____ 11. Change 5 miles to feet.

_____ 12. Change 7 miles to feet.

_____ 13. Change 1320 feet to miles.

_____ 14. Change 3520 feet to miles.

_____ 15. Change 12 yards to feet.

_____ 16. Change 9 yards to feet.

_____ 17. Change 17 meters to centimeters.

_____ 18. Change 23 meters to centimeters.

_____ 19. Change 3850 meters to kilometers.

_____ 20. Change 4520 meters to kilometers.

_____ 21. Change 535 centimeters to meters.

_____ 22. Change 756 centimeters to meters.

_____ 23. Change 815 millimeters to meters.

_____ 24. Change 725 millimeters to meters.

_____ 25. Change 9.3 kilometers to meters.

_____ 26. Change 12.4 kilometers to meters.

_____ 27. Change 0.05 kilometer to centimeters.

_____ 28. Change 0.035 kilometer to centimeters.

_____ 29. Change 3.4 meters to millimeters.

_____ 30. Change 8.2 meters to millimeters.

_____ 31. David cut a piece of lumber 26 feet long into 4 equal pieces. What is the length of each piece in feet? in inches?

_____ 32. Mariah bought 58 yards of ribbon and cut it into 3 equal pieces. What is the length of each piece in yards? in inches?

_____ 33. To the nearest tenth, how many miles high is an airplane if its altitude is 48,065 feet?

_____ 34. If Delinda jogged 4400 yards this morning, how many miles (to the tenths place) did she jog?

_____ 35. The band director bought 16 yards of material to make flags. If he uses 2 feet of material for each flag, how many flags can he make?

_____ 36. Alf cut a piece of wire 7.8 meters in length into 6 equal pieces. What is the length of each piece in meters? in centimeters?

_____ 37. Wilma ran 9 times around a track that is 500 meters long. Find the distance
_____   she ran in meters; in kilometers.

_____ 38. A cattle rancher needs 10 kilometers of barbed wire for fencing. How much
                will it cost him if the wire sells for $0.50 per meter?

_____ 39. Karyn needs 604 centimeters of lace border for a tablecloth. How much will
                it cost her if the border sells for $0.75 per meter?

_____ 40. In meters, how high is a stack of 25 books if each book is 3 centimeters
                thick?

☆ Stretching the Topics _____

_____ 1. On a baseball diamond, the distance between any 2 bases is about 27 meters.
_____   How many meters does a batter run if she hits a home run? how many
_____   kilometers? about how many miles?

_____ 2. If picture framing is sold in centimeters, how many centimeters of framing
               should be purchased to frame the 2 pictures with the given dimensions?

0.55 m

0.35 m

450 mm

33.5 cm

_____ 3. If a roll of masking tape contains 600 inches, what part of a roll will be
               needed to protect the edges of a picture window that is 3 yards long and
               2 yards wide?

Check your answers in the back of your book.

If you can complete **Checkup 12.1,** then you are ready to go on to Section 12.2.

Name _____   **Date** _____

✔ CHECKUP 12.1

_____ 1. Change 7 feet to inches.

_____ 2. Change 9 inches to yards.

_____ 3. Change 1056 feet to miles.

_____ 4. Change $6\frac{1}{2}$ yards to inches.

_____ 5. Change 7 centimeters to meters.

_____ 6. Change 9 kilometers to centimeters.

_____ 7. Change 1.8 centimeters to kilometers.

_____ 8. Change 3.6 meters to millimeters.

_____ 9. Doug's triangular jogging route takes him 1.2 kilometers north, 1.6 kilometers east, and 2 kilometers back to his starting place. How many meters does he jog on this route?

_____ 10. If Gary ran $2\frac{1}{2}$ miles and Eli ran 4107 yards, who ran farther?

Check your answers in the back of your book.

If You Missed Problems:	You Should Review Examples:
1–4	2–4
5–8	8, 9
9	7
10	5

12.2 Measuring Weight

To measure weights, we again find that there are two systems that we may use, the American system and the metric system.

Using American Units to Measure Weight

To measure your weight, you would probably use pounds. To weigh a letter for mailing, you would use ounces. To measure the weight of an ocean liner, you would use tons. Each of these is an American unit of weight.

You should become familiar with the most basic American units of weight and their relationships to each other.

Units	Abbreviations
1 pound = 16 ounces	1 lb = 16 oz
1 ton = 2000 pounds	1 ton = 2000 lb

Example 1. Marie's winter coat weighs $6\frac{1}{2}$ pounds and her boots weigh $3\frac{1}{4}$ pounds. Wearing her coat and boots, Marie weighs 138 pounds. What is her weight without the coat and boots?

Solution. First we find the total weight of Marie's coat and boots by *adding*

$$6\frac{1}{2} = 6 + \frac{1}{2} = 6 + \frac{2}{4}$$

$$\underline{+3\frac{1}{4} = \underline{3 + \frac{1}{4}} = \underline{3 + \frac{1}{4}}}$$

$$9 + \frac{3}{4}$$

Her coat and boots weigh $9\frac{3}{4}$ pounds. To find Marie's weight, we must *subtract* the weight of the coat and boots from 138 pounds.

$$138 \quad = 138 \quad = 137 + 1 = 137 + \frac{4}{4}$$

$$\underline{-\ 9\frac{3}{4} = \underline{9 + \frac{3}{4}} = \underline{9 + \frac{3}{4}} = \underline{9 + \frac{3}{4}}}$$

$$128 + \frac{1}{4}$$

Marie weighs $128\frac{1}{4}$ pounds.

To change weights from one kind of American unit to another, we can use the conversion factors suggested by the table of units.

$$\frac{1\ \text{lb}}{16\ \text{oz}} \qquad \frac{16\ \text{oz}}{1\ \text{lb}} \qquad \frac{1\ \text{ton}}{2000\ \text{lb}} \qquad \frac{2000\ \text{lb}}{1\ \text{ton}}$$

Example 2. In 1976, 49,900 tons of pecans were grown in the United States. How many pounds of pecans were grown in the United States in 1976?

Solution. We must change tons to pounds.

$$49{,}900 \text{ tons} = 49{,}900 \times 1 \text{ ton} \times \frac{2000 \text{ lb}}{1 \text{ ton}}$$

Choose the conversion factor. (We don't want tons; we do want pounds.)

$$= \frac{49{,}900}{1} \times \frac{\overset{1}{\cancel{1 \text{ ton}}}}{1} \times \frac{2000 \text{ lb}}{\underset{1}{\cancel{1 \text{ ton}}}}$$

Reduce the fractions.

$$= 49{,}900 \times 2000 \text{ lb}$$

$$49{,}900 \text{ tons} = 99{,}800{,}000 \text{ lb}$$

Find the product.

In 1976, 99,800,000 pounds of pecans were grown in the United States.

Example 3. If $10\frac{1}{2}$ pounds of cheese is divided equally among 42 people, how many ounces will each person receive?

Solution. We must change pounds to ounces before finding each person's share.

$$10\frac{1}{2} \text{ lb} = \frac{21}{2} \text{ lb}$$

$$= \frac{21}{2} \times 1 \text{ lb} \times \frac{16 \text{ oz}}{1 \text{ lb}}$$

Choose the conversion factor. (We don't want pounds; we do want ounces.)

$$= \frac{21}{\underset{1}{\cancel{2}}} \times \frac{\overset{1}{\cancel{1 \text{ lb}}}}{1} \times \frac{\overset{8 \text{ oz}}{\cancel{16 \text{ oz}}}}{\underset{1}{\cancel{1 \text{ lb}}}}$$

Reduce the fractions.

$$10\frac{1}{2} \text{ lb} = 168 \text{ oz}$$

Find the product.

Now we must divide the total weight (168 ounces) by the number of people (42).

$$\frac{168 \text{ oz}}{42} = 4 \text{ oz}$$

Each person will receive 4 ounces of cheese.

Example 4. The shipping charges for a catalog order are based on the total weight of the items ordered. Find the total weight of Francine's order, which contains the following items:

dress	1 lb	3 oz
sheets	2 lb	8 oz
towels	4 lb	8 oz
shirt		12 oz
boots	3 lb	4 oz

Solution

Let's add the pounds and ounces.

1 lb	3 oz
2 lb	8 oz
4 lb	8 oz
	12 oz
+3 lb	+ 4 oz
10 lb	35 oz

The total weight is 10 lb 35 oz.

Since 35 ounces is more than one pound, we change it to pounds and ounces by dividing by 16 ounces.

$$\begin{array}{r} 2 \\ 16\overline{)35} \\ \underline{32} \\ 3 \end{array}$$

35 oz = 2 lb 3 oz

The total weight of 10 lb 35 oz becomes

$$\begin{array}{r} 10 \text{ lb} \\ + \ 2 \text{ lb } 3 \text{ oz} \\ \hline 12 \text{ lb } 3 \text{ oz} \end{array}$$

Francine's order weighs 12 lb 3 oz.

▶ Trial Run

_____ 1. Jane bought 3 packages of ground beef with the weights 3 lb 9 oz, 2 lb 10 oz, and 4 lb 13 oz. Find the total weight of the ground beef.

_____ 2. While Harold was on a diet his weight changed from 183 lb 15 oz to 164 lb 12 oz. How much weight did he lose?

_____ 3. If a bag of oranges weighs 7 lb 9 oz, how much will 6 bags weigh?

_____ 4. If $16\frac{1}{2}$ pounds of butter is divided equally among 12 people, how many ounces will each person receive?

_____ 5. In 1981, 13,704 tons of sugar beets were produced in the United States. How many pounds of sugar beets were produced?

Answers are on page 518.

Using Metric Units to Measure Weight

The metric system uses an entirely different list of units to measure weight. As with the metric system of linear measure, however, we find that it is easier to convert among metric units of weight because they are based on the number 10.

The basic measure of weight in the metric system is the **gram.** An object weighing 1 gram is not very heavy. In fact,

---

One gram is about the weight of a paper clip.

All the other metric units of weight are whole number or fractional number *multiples* of the gram. A list of those units is provided here for your reference.

Units	Abbreviations
1 kilogram = 1000 grams	1 kg = 1000 g
1 hectogram = 100 grams	1 hg = 100 g
1 dekagram = 10 grams	1 dkg = 10 g
1 decigram = $\frac{1}{10}$ gram	1 dg = $\frac{1}{10}$ g
(10 decigrams = 1 gram)	(10 dg = 1 g)
1 centigram = $\frac{1}{100}$ gram	1 cg = $\frac{1}{100}$ g
(100 centigrams = 1 gram)	(100 cg = 1 g)
1 milligram = $\frac{1}{1000}$ gram	1 mg = $\frac{1}{1000}$ g
(1000 milligrams = 1 gram)	(1000 mg = 1 g)

The units of weight that you should become *most* familiar with are the kilogram, gram, centigram, and milligram. You should learn the relationships for these important units.

---

1 kg = 1000 g

100 cg = 1 g

1000 mg = 1 g

From these relationships we know that each of the following conversion factors is equal to the number 1.

$$\frac{1 \text{ kg}}{1000 \text{ g}} \qquad \frac{1000 \text{ g}}{1 \text{ kg}} \qquad \frac{1 \text{ g}}{100 \text{ cg}} \qquad \frac{100 \text{ cg}}{1 \text{ g}} \qquad \frac{1 \text{ g}}{1000 \text{ mg}} \qquad \frac{1000 \text{ mg}}{1 \text{ g}}$$

Because the gram is a very small unit of weight, it is only used to weigh very small objects. You might use grams to describe the weight of a letter, but you would use kilograms to measure your own weight.

One kilogram is about 2.2 pounds.

To have a mental picture of something that weighs about 1 kilogram, you might think of 2 pounds of margarine plus most of another whole stick of margarine.

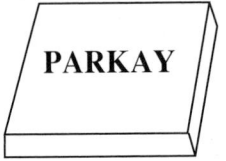  

1 kilogram

Centigrams and milligrams are used to measure *very* small weights, especially in science and medicine. You may recall seeing food labels that mention the milligrams of sodium and cholesterol contained in the product.

Example 5. If a bowl of Golden Light Flakes cereal contains 7 milligrams of cholesterol, how many grams of cholesterol does the bowl of cereal contain?

Solution. We must change milligrams to grams using the proper conversion factor.

$$7 \text{ mg} = 7 \times 1 \text{ mg} \times \frac{1 \text{ g}}{1000 \text{ mg}}$$

Choose the conversion factor.
(We don't want milligrams; we do want grams.)

$$= \frac{7}{1} \times \frac{\overset{1}{\cancel{1 \text{ mg}}}}{1} \times \frac{1 \text{ g}}{\underset{1000}{\cancel{1000 \text{ mg}}}}$$

Reduce the fractions.

$$= \frac{7}{1000} \text{ g}$$

Multiply the fractions.

$$7 \text{ mg} = 0.007 \text{ g}$$

Write the fraction as a decimal number.

Recalling that 1 gram is the weight of a paper clip, we see that 0.007 gram is a very small weight.

Example 6. One aspirin tablet contains 300 milligrams of pure aspirin. Find the weight, in grams, of the pure aspirin in a bottle containing 250 tablets.

Solution. If 1 tablet contains 300 milligrams, then 250 tablets contain

$$250 \times 300 \text{ mg} = 75{,}000 \text{ mg}$$

Now we must change this weight to grams.

$$75{,}000 \text{ mg} = 75{,}000 \times 1 \text{ mg} \times \frac{1 \text{ g}}{1000 \text{ mg}}$$

$$= \frac{75{,}000}{1} \times \frac{\overset{1}{\cancel{1 \text{ mg}}}}{1} \times \frac{1 \text{ g}}{\underset{1000}{\cancel{1000 \text{ mg}}}}$$

$$= \frac{75{,}000}{1000} \text{ g}$$

$$75{,}000 \text{ mg} = 75 \text{ g}$$

⫸ Trial Run

_____ 1. Change 1800 grams to kilograms.

_____ 2. Change 2.3 kilograms to milligrams.

_____ 3. A bottle contains 300 tablets of Vitamin C. Find the weight (in grams) of the contents if each tablet weighs 500 milligrams.

Answers are given below.

Answers to Trial Runs

page 515   **1.** 11 lb   **2.** 19 lb 3 oz   **3.** 45 lb 6 oz   **4.** 22 oz   **5.** 27,408,000 lb

page 518   **1.** 1.8 kg   **2.** 2,300,000 mg   **3.** 150 g

EXERCISE SET 12.2

_____ **1.** Change 104 ounces to pounds.

_____ **2.** Change 52 ounces to pounds.

_____ **3.** Change 9000 pounds to tons.

_____ **4.** Change 12,500 pounds to tons.

_____ **5.** Change $5\frac{1}{4}$ pounds to ounces.

_____ **6.** Change $2\frac{3}{4}$ pounds to ounces.

_____ **7.** Change $2\frac{1}{5}$ tons to pounds.

_____ **8.** Change $6\frac{1}{2}$ tons to pounds.

_____ **9.** Change 1500 grams to kilograms.

_____ **10.** Change 3250 grams to kilograms.

_____ **11.** Change 12.5 grams to centigrams.

_____ **12.** Change 15.25 grams to centigrams.

_____ **13.** Change 0.005 gram to milligrams.

_____ **14.** Change 0.0035 gram to milligrams.

_____ **15.** Change 5.25 kilograms to grams.

_____ **16.** Change 8.5 kilograms to grams.

_____ **17.** Change 1500 milligrams to grams.

_____ **18.** Change 2600 milligrams to grams.

_____ **19.** Change 18,500 milligrams to kilograms.

_____ **20.** Change 25,000 milligrams to kilograms.

_____ **21.** A mail-order store is shipping a package that contains 3 sweaters weighing 1 lb 4 oz, 1 lb 9 oz, and 1 lb 12 oz. What is the total weight of the package?

_____ **22.** A package of ground beef weighing 5 pounds 12 ounces is to be divided into 4 equal parts to be used at different times. How much will each part weigh?

_____ **23.** If a box of detergent weighs 45 ounces, what will be the weight, in pounds, of a case of 12 boxes?

_____ **24.** A garden supply store ordered 21 tons of peat moss. If peat moss comes in 70-pound bags, how many bags will be delivered?

_____ **25.** If a physician orders 0.25 gram of matromycin for a patient, how many milligrams should be administered to the patient?

_____ **26.** One zinc tablet weighs 50 milligrams. Find the weight, in grams, of 250 tablets.

_____ **27.** If a hostess plans each serving of roast beef to be about 500 grams, how large a roast (in kilograms) will she need to serve 9 people?

_____ **28.** One bar of soap weighs 135 grams. If a case of soap bars weighs 3.24 kilograms, how many bars does it contain?

_____ **29.** The required daily dosage of a certain medication is 130 milligrams. How many grams of the medication are needed for 20 days?

_____ **30.** The recommended daily allowance of protein for an adult male is 6500 centigrams. If 1 cup of cottage cheese contains approximately 32.5 grams of protein, how many cups of cottage cheese must be consumed to satisfy this allowance?

519

☆ Stretching the Topics _____

_____ 1. One grain of aspirin contains 60 milligrams of pure aspirin. If the doctor prescribes two 5-grain tablets every 4 hours for 24 hours, how many grams of aspirin would the patient consume?

_____ 2. A required dosage of morphine sulphate is 10 milligrams. If the clinic only has 1.5 kilograms of morphine sulphate, how many doses are available?

_____ 3. A dieter wants to reduce his weight from 196 pounds to 163 pounds over a 10-week period. What should be his average weekly weight loss in kilograms?

Check your answers in the back of your book.

If you can complete **Checkup 12.2,** then you are ready to go on to Section 12.3.

✓ CHECKUP 12.2

_____ 1. Change 25,125 tons to pounds.　　　_____ 2. Change $8\frac{1}{4}$ pounds to ounces.

_____ 3. Change 45 ounces to pounds.　　　_____ 4. Change 0.85 milligram to grams.

_____ 5. Change 8.5 kilograms to grams.

_____ 6. Change 25,000 milligrams to kilograms.

_____ 7. A package of sausage that weighs $3\frac{1}{2}$ pounds is to be divided into 6 equal servings. How much (in ounces) will each serving weigh?

_____ 8. If 1 tablet weighs 675 milligrams, what is the weight (in grams) of 500 tablets?

Check your answers in the back of your book.

If You Missed Problems:	You Should Review Examples:
1–3	1–4
4–6	5, 6
7	3
8	6

12.3 Measuring Capacity

When we measure the **capacity** of a can or jar or swimming pool, we are interested in finding out how much will fit inside it. As usual, capacity can be measured using American or metric units.

Using American Units to Measure Capacity

To measure the capacity of a can or jar, you would probably use cups or pints or quarts as the units of measure. To measure the capacity of a swimming pool, however, you would probably use gallons. Even smaller units of capacity are the teaspoon and tablespoon.

The units in this table of American units of capacity should be familiar to you.

Units	Abbreviations
1 tablespoon = 3 teaspoons	1 T = 3 t
1 pint = 2 cups	1 pt = 2 c
1 quart = 2 pints	1 qt = 2 pt
1 gallon = 4 quarts	1 gal = 4 qt

From this table, you should agree that we can use the following conversion factors to change among American units of capacity.

$$\frac{1\text{ T}}{3\text{t}} \qquad \frac{1\text{ pt}}{2\text{ c}} \qquad \frac{1\text{ qt}}{2\text{ pt}} \qquad \frac{1\text{ gal}}{4\text{ qt}}$$

$$\frac{3\text{ t}}{1\text{ T}} \qquad \frac{2\text{ c}}{1\text{ pt}} \qquad \frac{2\text{ pt}}{1\text{ qt}} \qquad \frac{4\text{ qt}}{1\text{ gal}}$$

Example 1. A recipe for punch calls for $1\frac{1}{2}$ cups of orange juice. If Amelia wishes to make 4 times as much punch as one recipe makes, how many pints of orange juice must she use?

Solution

First we must *multiply* $1\frac{1}{2}$ cups by 4 to find the number of cups needed when she makes 4 *times* the original recipe.

$$4 \times 1\frac{1}{2}\text{ c} = 4 \times \frac{3}{2}\text{ c}$$

$$= \frac{\overset{2}{\cancel{4}}}{1} \times \frac{3}{\underset{1}{\cancel{2}}}\text{ c}$$

$$= 6\text{ c}$$

Now we must change cups to pints.

$$6\text{ c} = 6 \times 1\text{ c} \times \frac{1\text{ pt}}{2\text{ c}}$$

$$= \frac{6}{1} \times \frac{\overset{1}{\cancel{1\text{ c}}}}{1} \times \frac{1\text{ pt}}{\underset{2}{\cancel{2\text{ c}}}}$$

$$= \frac{6}{2}\text{ pt}$$

$$6\text{ c} = 3\text{ pt}$$

Amelia must use 3 pints of orange juice.

Example 2. How many $\frac{2}{3}$-cup servings can be poured from 1 gallon of milk?

Solution

First we make our units match by changing gallons to cups.

$$1 \text{ gal} = 1 \text{ gal} \times \frac{4 \text{ qt}}{1 \text{ gal}} \times \frac{4 \text{ c}}{1 \text{ qt}}$$

$$= \frac{\overset{1}{\cancel{1 \text{ gal}}}}{1} \times \frac{\overset{4}{\cancel{4 \text{ qt}}}}{\underset{1}{\cancel{1 \text{ gal}}}} \times \frac{4 \text{ c}}{\underset{1}{\cancel{1 \text{ qt}}}}$$

$$= 4 \times 4 \text{ c}$$

$$1 \text{ gal} = 16 \text{ c}$$

To find how many $\frac{2}{3}$ cup servings can be poured from 16 cups of milk, we must *divide*.

$$16 \div \frac{2}{3} = \frac{16}{1} \times \frac{3}{2}$$

$$= \frac{\overset{8}{\cancel{16}}}{1} \times \frac{3}{\underset{1}{\cancel{2}}}$$

$$= 24$$

We can pour 24 servings of $\frac{2}{3}$ cup each.

Example 3. Instructions for mixing plant food say to mix 2 tablespoons of plant food with 3 gallons of water. If Gretchen wishes to use only 1 gallon of water, how many *teaspoons* of plant food are needed?

Solution. This looks like a ratio and proportion problem. We can let f stand for the amount of plant food needed.

$$\left.\begin{array}{l} \text{amount of plant food (T)} \\ \hline \text{amount of water (gal)} \end{array}\right\} \quad \frac{2}{3} = \frac{f}{1}$$

$$2 \times 1 = 3 \times f \qquad \text{Write cross products.}$$

$$2 = 3 \times f \qquad \text{Find the product on the left.}$$

$$\frac{2}{3} = \frac{\overset{1}{\cancel{3}} \times f}{\underset{1}{\cancel{3}}} \qquad \text{Divide both sides by 3.}$$

$$\frac{2}{3} = f \qquad \text{Reduce on the right.}$$

Gretchen must use $\frac{2}{3}$ tablespoon of plant food. But we want to measure this in *teaspoons*.

$$\frac{2}{3}\text{T} = \frac{2}{3} \times 1 \text{ T} \times \frac{3 \text{ t}}{1 \text{ T}} \qquad \begin{array}{l} \text{Choose the conversion factor.} \\ \text{(We don't want T; we do want t.)} \end{array}$$

$$= \frac{2}{\underset{1}{\cancel{3}}} \times \frac{\overset{1}{\cancel{1 \text{ T}}}}{1} \times \frac{\overset{1 \text{ t}}{\cancel{3 \text{ t}}}}{\underset{1}{\cancel{1 \text{ T}}}} \qquad \text{Reduce fractions.}$$

$$= 2 \text{ t} \qquad \text{Find the product.}$$

She should use 2 teaspoons of plant food with 1 gallon of water.

Ⅲ➡ **Trial Run**

_____ **1.** A recipe for fudge calls for $1\frac{1}{4}$ cup of milk. If Eunice wishes to triple her recipe, will 1 quart of milk be enough?

_____ **2.** Mrs. Steward has a 5-gallon container of tomato juice. How many jars will she need if she wishes to put the tomato juice in pint jars?

_____ **3.** Mr. Moman needs $2\frac{1}{2}$ gallons of paint, but the paint store has only quart cans of paint in the color he needs. How many cans will he have to buy?

_____ **4.** Cassie needs 8 cups of vinegar to make her favorite pickle recipe. How many pints of vinegar should she buy?

_____ **5.** Jim's biscuit recipe calls for 3 tablespoons of baking powder. If he wants to cut the recipe in half, how many teaspoons of baking powder should he use?

Answers are on page 527.

Using Metric Units to Measure Capacity

Ever since soft drink manufacturers began selling their products in liter bottles, Americans have become more familiar with that metric unit for measuring capacity. Most of us know that

> One liter is a little more than 1 quart.

To be more exact,

> 1 liter $\doteq$ 1.1 quart

The other metric units of capacity are all whole number or fractional number *multiples* of the liter. They are listed in the following table.

Units	Abbreviations
1 kiloliter = 1000 liters	1 kl = 1000 ℓ
1 hectoliter = 100 liters	1 hl = 100 ℓ
1 decaliter = 10 liters	1 dkl = 10 ℓ
1 deciliter = $\frac{1}{10}$ liter	1 dl = $\frac{1}{10}$ ℓ
(10 deciliters = 1 liter)	(10 dl = 1 ℓ)
1 centiliter = $\frac{1}{100}$ liter	1 cl = $\frac{1}{100}$ ℓ
(100 centiliters = 1 liter)	(100 cl = 1 ℓ)
1 milliliter = $\frac{1}{1000}$ liter	1 ml = $\frac{1}{1000}$ ℓ
(1000 milliliters = 1 liter)	(1000 ml = 1 ℓ)

The most common metric units of capacity are the kiloliter, liter, and milliliter. You should learn the relationships among them.

$$1 \text{ kl} = 1000 \; \ell$$
$$1000 \text{ ml} = 1 \; \ell$$

These relationships allow us to use conversion factors to change among metric units.

$$\frac{1 \text{ kl}}{1000 \; \ell} \qquad \frac{1000 \; \ell}{1 \text{ kl}} \qquad \frac{1 \; \ell}{1000 \text{ ml}} \qquad \frac{1000 \text{ ml}}{1 \; \ell}$$

Example 4. Use a conversion factor to change 0.55 liters to milliliters.

Solution

$$0.55 \; \ell = 0.55 \times 1 \; \ell \times \frac{1000 \text{ ml}}{1 \; \ell}$$

Choose the conversion factor. (We don't want liters; we do want milliliters.)

$$= \frac{0.55}{1} \times \frac{\overset{1}{\cancel{1 \ell}}}{1} \times \frac{1000 \text{ ml}}{\underset{1}{\cancel{1 \ell}}}$$

Reduce the fractions.

$$= 0.55 \times 1000 \text{ ml}$$

Multiply numerators and denominators.

$$0.55 \; \ell = 550 \text{ ml}$$

Find the product.

Let's solve the problem stated at the beginning of the chapter.

Example 5. A public health clinic staff expects to give a 0.5 milliliter allergy shot to each one of 2100 patients this week. How many liters of allergy medicine must they have ready for the week?

Solution. To find the total number of milliliters needed, we first *multiply*

$$2100 \times 0.5 \text{ ml} = 1050 \text{ ml}$$

They will need 1050 milliliters of medicine. We now change this amount to liters.

$$1050 \text{ ml} = 1050 \times 1 \text{ ml} \times \frac{1 \ell}{1000 \text{ ml}}$$

$$= \frac{1050}{1} \times \frac{\overset{1}{\cancel{1 \text{ ml}}}}{1} \times \frac{1 \ell}{\underset{1000}{\cancel{1000 \text{ ml}}}}$$

$$= \frac{1050 \ell}{1000}$$

$$= 1.05 \ell$$

They must have 1.05 liters of medicine for the week.

⫸ Trial Run

_____ 1. Change 5.4 kiloliters to liters.

_____ 2. Change 7500 milliliters to liters.

_____ 3. André's perfume factory bottles perfume in 7.4-milliliter containers. If André has an order for 4500 bottles this month, how many liters of perfume should he have ready to be bottled?

Answers are given below.

Answers to Trial Runs

page 524   **1.** yes   **2.** 40   **3.** 10   **4.** 4   **5.** $4\frac{1}{2}$

page 527   **1.** 5400 ℓ **2.** 7.5 ℓ **3.** 33.3

EXERCISE SET 12.3

_____ 1. Change 3.5 gallons to quarts.

_____ 2. Change $2\frac{1}{4}$ gallons to quarts.

_____ 3. Change $5\frac{1}{2}$ quarts to pints.

_____ 4. Change 4.5 quarts to pints.

_____ 5. Change 11 cups to pints.

_____ 6. Change 15 cups to pints.

_____ 7. Change $7\frac{1}{2}$ gallons to pints.

_____ 8. Change $4\frac{1}{4}$ gallons to pints.

_____ 9. Change 0.75 liter to milliliters.

_____ 10. Change 0.125 liter to milliliters.

_____ 11. Change 8.5 kiloliters to liters.

_____ 12. Change 3.25 kiloliters to liters.

_____ 13. Change 25,000 milliliters to liters.

_____ 14. Change 85,000 milliliters to liters.

_____ 15. Change 0.005 kiloliter to milliliters.

_____ 16. Change 0.009 kiloliter to milliliters.

_____ 17. If a coffee pot has a 1-quart capacity, how many cups of coffee will it hold?

_____ 18. Instructions for making a herbicide solution say to mix 2 teaspoons of herbicide with each quart of water. How many tablespoons of herbicide are used in filling a 5-gallon sprayer with the solution?

_____ 19. If Helen's family drinks 10 quarts of milk in 1 week, how many gallons should she buy for a week?

_____ 20. Dan needs $\frac{1}{2}$ pint of wood stain for each chair that he refinishes. If he has 24 chairs to refinish, how many gallons of wood stain does he need?

_____ 21. A bottle contains $2\frac{1}{2}$ liters of solution. If the solution is divided equally among 4 beakers, how many milliliters of the solution will each beaker contain?

_____ 22. A nurse plans to administer shots to 1500 patients. If each shot contains 0.75 milliliter, how many liters of medication does he need?

_____ 23. How many 500-milliliter vials can be filled with 3 kiloliters of liquid?

_____ 24. If a bottle of contact lens solution contains 60 milliliters, how many liters would be needed to fill 500 bottles?

_____ 25. A 2-liter bottle of soda is divided equally among 8 people. How many milliliters will each person receive?

☆ Stretching the Topics

_____ 1. A bottle of soda contains 360 milliliters. If an 8-bottle carton of soda costs $3.10, what is the soda's cost per liter (to the nearest cent)?

_____ 2. Nezam's compact car has a gasoline tank with a 15-gallon 1-quart capacity. If gasoline costs $0.29 per liter, how much will a full tank of gas cost?

Check your answers in the back of your book.

If you can complete **Checkup 12.3,** you are ready to do the **Review Exercises** for Chapter 12.

✓ **CHECKUP 12.3**

_____ 1. Change $7\frac{1}{2}$ gallons to quarts.

_____ 2. Change 20 pints to quarts.

_____ 3. Change 3 gallons to cups.

_____ 4. Change 0.43 liter to milliliters.

_____ 5. Change 8000 liters to kiloliters.

_____ 6. Change 8500 milliliters to liters.

_____ 7. Rachel's recipe for candy uses $1\frac{1}{2}$ cups milk. For Christmas, she wishes to make 8 times as much candy as 1 recipe makes. How many gallons of milk will she need?

_____ 8. How many liters of typewriter correction fluid will be needed to fill 2000 bottles if each bottle contains 18 milliliters of fluid?

Check your answers in the back of your book.

If You Missed Problems:	You Should Review Examples:
1–3	1–3
4–6	4, 5
7	1
8	6

Summary

In this chapter we learned to measure length and distance, weight, and capacity using American and metric units of measurement. The most important of these units are summarized in this table.

American Units		
Length	Weight	Capacity
1 ft = 12 in. 1 yd = 3 ft 1 mi = 5280 ft	1 lb = 16 oz 1 ton = 2000 lb	1 T = 3 t 1 pt = 2 c 1 qt = 2 pt 1 gal = 4 qt

Metric Units		
Length	Weight	Capacity
1 cm = 10 mm 1 m = 100 cm 1 m = 1000 mm 1 km = 1000 m	1 g = 1000 mg 1 g = 100 cg 1 kg = 1000	1 ℓ = 1000 ml 1 kl = 1000 ℓ

From this table, we learned to use **conversion factors** to change among American units and metric units. In choosing the proper conversion factor, we looked to see which factor would cause the unwanted unit to divide out, leaving the wanted unit in the answer.

To Convert	We Use the Conversion Factor(s)	Because	Examples
13 ft to yd	$\dfrac{1 \text{ yd}}{3 \text{ ft}}$	We don't want ft We do want yd	$13 \times \cancel{1\text{ ft}}^{1} \times \dfrac{1 \text{ yd}}{\cancel{3\text{ ft}}_{3}}$ $= \dfrac{13}{3} \text{ yd}$ $= 4\dfrac{1}{3} \text{ yd}$
7 yd to ft	$\dfrac{3 \text{ ft}}{1 \text{ yd}}$	We don't want yd We do want ft	$7 \times \cancel{1\text{ yd}}^{1} \times \dfrac{3 \text{ ft}}{\cancel{1\text{ yd}}_{1}}$ $= 21 \text{ ft}$
359 mg to kg	$\dfrac{1 \text{ g}}{1000 \text{ mg}} \times \dfrac{1 \text{ kg}}{1000 \text{ g}}$	We don't want mg We don't want g We do want kg	$359 \times \cancel{1\text{ mg}}^{1} \times \dfrac{\cancel{1\text{ g}}^{1}}{\cancel{1000\text{ mg}}_{1000}} \times \dfrac{1 \text{ kg}}{\cancel{1000\text{ g}}_{1000}}$ $= \dfrac{359}{1{,}000{,}000} \text{ kg}$ $= 0.000359 \text{ kg}$

☐ Speaking the Language of Mathematics ——————————

Complete each sentence with the appropriate word or phrase.

1. Feet, yards, and miles are units of length in the ——————— system of measurement.

2. Grams, milligrams, and kilograms are units of weight in the ——————— system of measurement.

3. To change from one unit of measure to another unit, we may use a ——————— ———————.

4. Pints and quarts are American units for measuring ——————— .

5. Meters and centimeters are metric units for measuring ——————— .

△ Writing About Mathematics ——————————

Write your response to each question in complete sentences.

1. Explain why many people believe that the metric system of measurement is better than the American system of measurement. What do *you* think?

2. Explain why you would use the conversion factor $\dfrac{12 \text{ in.}}{1 \text{ ft}}$ to change 5 feet to inches, but you would use the conversion factor $\dfrac{1 \text{ ft}}{12 \text{ in.}}$ to change 5 inches to feet.

3. If one measurement is made using meters and another measurement is made using centimeters, discuss how to compare the two measurements.

REVIEW EXERCISES for Chapter 12

_____ 1. Change 6 yards to inches.

_____ 2. Change 15,840 feet to miles.

_____ 3. Change 114 inches to feet.

_____ 4. Change 30 feet to yards.

_____ 5. Barry's triangular jogging route takes him 2.2 miles north, 1.8 miles east, and 3 miles back to his starting place. How far does he jog on this route?

_____ 6. Which is longer, a driveway that is 1290 feet long or a driveway that is $\frac{1}{4}$ mile long?

_____ 7. Change 423 millimeters to meters.

_____ 8. Change 800 centimeters to kilometers.

_____ 9. Change 0.165 meter to millimeters.

_____ 10. Change 9 kilometers to centimeters.

_____
_____ 11. The average monthly rainfall in a certain city is 72 centimeters. How much rain (in centimeters) would be expected during a year? How much in meters?

_____ 12. A carpenter needs 3 pieces of wood to finish a job. The lengths needed are 3 m 25 cm, 1 m 85 cm, and 2 m 40 cm. Find the total length needed (in meters).

_____ 13. Change 192 ounces to pounds.

_____ 14. Change $3\frac{1}{4}$ tons to pounds.

_____ 15. Change $15\frac{1}{2}$ pounds to ounces.

_____ 16. Change $2\frac{1}{4}$ pounds to ounces.

_____ 17. A package of $6\frac{1}{2}$ pounds of candy is divided equally among 8 children. How many ounces will each child receive?

_____ 18. A trailer is transporting 8 new hatchback cars, each weighing 1875 pounds. What is the weight (in tons) of the trailer's load?

_____ 19. Change 7.5 grams to milligrams.

_____ 20. Change 1200 grams to kilograms.

_____ 21. Change 3500 milligrams to grams.

_____ 22. Change 25,000 milligrams to kilograms.

_____ 23. The required daily dosage of a certain medication is 185 milligrams. How many grams of the medication will be needed for 5 patients for 1 week?

_____ 24. A bag of apples weighs 2.4 kilograms. If the average weight of each apple is 150 grams, how many apples are in the bag?

_____ 25. Change $5\frac{1}{2}$ gallons to quarts.

_____ 26. Change 16 cups to pints.

_____ 27. Change 15 quarts to pints.

_____ 28. Change 29 pints to gallons.

_____ 29. How many quarts of water will be needed to fill an 8-gallon wet/dry vacuum cleaner?

_____ 30. If Debbie plans to serve 1 cup of orange juice to each of her 16 brunch guests, how many quarts of orange juice should she buy?

_____ 31. Change 0.375 liter to milliliters.

_____ **32.** Change 4.73 kiloliters to liters.   _____ **33.** Change 0.032 kiloliter to milliliters.

_____ **34.** Change 85,000 milliliters to liters.

_____ **35.** How many 400-milliliter vials can be filled with 5 kiloliters of liquid?

_____ **36.** If each bottle of Eleganté perfume contains 15 milliliters, how many kiloliters of perfume would be needed to fill an order for 100,000 bottles of Eleganté?

Check your answers in the back of your book.

If You Missed Exercises:	You Should Review Examples:	
1–4	Section 12.1	2–4
5, 6		1, 5
7–10		8, 9
11, 12		10
13–16	Section 12.2	1–4
17, 18		2, 3
19–24		5, 6
25–28	Section 12.3	1–3
29, 30		2
31–36		4, 5

If you have completed the **Review Exercises** and corrected your errors, you are ready to take the **Practice Test** for Chapter 12.

PRACTICE TEST for Chapter 12

		SECTION	EXAMPLE
_____	1. Change $5\frac{1}{2}$ miles to feet.	12.1	2
_____	2. Change $2\frac{3}{4}$ feet to inches.	12.1	5
_____	3. Which is longer, a flower bed that is $10\frac{1}{2}$ feet long or one that is 120 inches long?	12.1	5
_____	4. Change 825 millimeters to meters.	12.1	8
_____	5. Change 0.0012 kilometer to centimeters.	12.1	9
_____	6. If the average rainfall in Gum Grove in the month of July is 24 centimeters and the average rainfall in Springfield for the same month is 0.254 meter, which city has more average rainfall in July?	12.1	10
_____	7. Change 725 tons to pounds.	12.2	2
_____	8. Change $15\frac{1}{4}$ pounds to ounces.	12.2	3
_____	9. Tonya bought 3 packages of sausage with the following weights: 2 lb 8 oz, 3 lb 11 oz, and 4 lb 13 oz. Find the total weight of the sausage.	12.2	4
_____	10. Change 153 milligrams to grams.	12.2	5
_____	11. Change 12,450 grams to kilograms.	12.2	6
_____	12. One gold necklace weighs 155 grams. If a jeweler orders a box of 25 necklaces, find the weight in kilograms of the box of necklaces.	12.2	6
_____	13. Change $1\frac{1}{2}$ gallons to cups.	12.3	2
_____	14. Change 15 pints to quarts.	12.3	1
_____	15. Bruce has 2 gallons 1 quart of milk. If he uses 30 cups of milk, how many quarts does he have left?	12.3	2
_____	16. Change 0.45 liter to milliliters.	12.3	4
_____	17. Change 8400 milliliters to liters.	12.3	4
_____	18. A certain solution is sold in bottles which contain 1 liter 550 milliliters. If the solution is divided equally and poured into 5 beakers, how many milliliters will be in each beaker?	12.3	5

Check your answers in the back of your book.

SHARPENING YOUR SKILLS after Chapters 1–12

		SECTION

_____ **1.** Write the word form for 290,078.   1.1

_____ **2.** Classify 936 as prime or composite. If the number is composite, write it as a product of prime numbers.   2.3

_____ **3.** Change $\frac{89}{5}$ to a mixed number.   3.1

_____ **4.** Reduce $\frac{180}{612}$ to lowest terms.   3.2

_____ **5.** Compare $\frac{55}{3}$ and $\frac{2}{5}$ using $<$ or $>$.   5.4

_____ **6.** Round 8.345 to the tenths place.   6.1

_____ **7.** Change $\frac{36}{48}$ to a decimal number.   7.3

_____ **8.** Use cross products to decide whether the ratios are equal: $\dfrac{3\frac{1}{4}}{\frac{1}{8}}$ and $\dfrac{52}{2}$.   8.2

_____ **9.** Change $16\frac{3}{4}$ percent to a fraction.   9.1

_____ **10.** Change $\frac{9}{25}$ to a percent.   9.1

_____ **11.** Change $8\frac{1}{3}$ miles to feet.   12.1

_____ **12.** Change 156 milligrams to grams.   12.2

_____ **13.** In one series of plays, the New Orleans Saints completed a 15-yard pass, a 7-yard pass, and a 12-yard run. Then the quarterback was sacked for a 6-yard loss and the team received two 15-yard penalties. What was the team's total gain or loss?   1.4

_____ **14.** Find the area of a triangular banner if the base is 24 feet and the height is 15 feet.   2.4

_____ **15.** Marco has 2 partially filled cans of blue paint. One can contains $\frac{7}{10}$ gallon and the other contains $\frac{1}{4}$ gallon. After he paints a room requiring $\frac{1}{2}$ gallon, how much paint is left?   5.5

_____ **16.** If each shelf will display 16 paperback books, how many shelves will be needed to display 147 books? Write your answer as an improper fraction and then as a mixed number.   3.1

_____ **17.** If 1008 of 3150 persons surveyed do not plan to vote in the presidential election, what fractional part of those surveyed do plan to vote?   3.3

_____ **18.** During 1 week, Rhonda drives a total of $18\frac{3}{4}$ miles going to *and* from her job. If she makes the round-trip 5 days a week, how far does Rhonda live from her workplace?   4.3

_____ **19.** Gary is purchasing framing for a rectangular window that is 1.6 meters long and 1.3 meters wide. How much framing should he buy?   6.3

_____ **20.** Miguel earns $9.83 per hour. If he worked 38.75 hours last week, find how much he   7.4
earned.

_____ **21.** If $\frac{1}{3}$ inch on a map represents 25 miles, how many inches are required to represent 375   8.3
miles?

_____ **22.** In July of last year, Uniontown had only 32 percent of its normal precipitation for   9.2
July. If the precipitation last July was 1.72 inches, what is the normal July
precipitation? (Round to the hundredths place.)

_____ **23.** Detraz Hardware is having a sale with all merchandise marked down 25 percent. Find   10.1
the sale price of a lawn mower that regularly sells for $295.98.

_____ **24.** If the prices at Chicken Queen have kept up with the 4.5 percent rate of inflation for   10.2
this year, what will be the new price of a $3.65 chicken basket?

_____ **25.** Find the total amount to be paid for an exercise bicycle that sells for $198, if the sales   10.3
tax rate is 6.5 percent.

_____ **26.** In 14 days, Amanda sold 126 boxes of Christmas cards. Find the mean number of   11.1
boxes sold per day.

_____ **27.** A survey of 9 retail merchants at the mall requested the number of full-time   11.1
employees. The responses were 40, 35, 25, 20, 25, 15, 75, 20, and 10. Find the mode
of these responses.

_____ **28.** For a barbecue Penny bought 3 packages of ribs with the weights 3 lb 7 oz, 5 lb 4 oz,   12.2
and 5 lb 12 oz. Find the total weight of the ribs.

Check your answers in the back of your book.

Working with Geometric Figures

> James has a cylindrical duffel bag with a diameter of 32 centimeters and a height of 90 centimeters. Tara has a tote bag that is 90 centimeters long, 30 centimeters wide, and 27 centimeters deep. Which bag will hold more?

Now that we have become more familiar with American and metric units of measure, we are prepared to study the branch of mathematics known as **geometry.** In particular we shall look at flat geometric figures and solid geometric figures.

In this chapter we learn how to

1. Measure perimeter and circumference.
2. Measure area.
3. Measure volume.
4. Find missing parts of a right triangle.

The methods for using a calculator to solve problems of the types encountered in Chapters 12 and 13 are discussed in Section 16.4 of Chapter 16.

13.1 Measuring Perimeter and Circumference

Suppose you must decide how many feet of fencing to buy to enclose a rectangular garden that is 10 feet long and 37 feet wide.

Suppose you wish to know whether a strip of ribbon 19 inches long will be long enough to make a hatband for a size $6\frac{1}{8}$ hat.

In each of these situations, we must be able to compute the distance around the outside of a geometric figure. In the first situation the figure is a rectangle; in the second situation the figure is a circle.

Finding Perimeter

Rectangles, squares, and triangles are familiar examples of flat, closed geometric figures with straight sides. Such figures are called **polygons.** The distance around the outside of a polygon is called its **perimeter.** As we learned in earlier chapters, we can find the perimeter of a polygon by adding the lengths of all its sides.

> The **perimeter** of a polygon is the sum of the lengths of its sides.

Let's take a look at our fencing problem in which we wish to enclose a rectangular garden that is 10 feet long and 37 feet wide. An illustration might help.

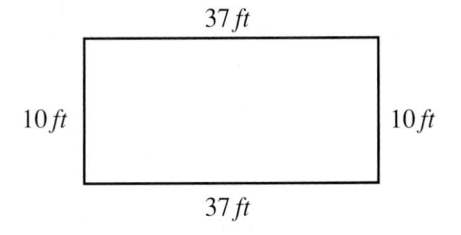

$$\text{Perimeter} = 10 \text{ ft} + 37 \text{ ft} + 10 \text{ ft} + 37 \text{ ft}$$
$$= 94 \text{ ft}$$

We must buy 94 feet of fencing.

A rectangle is just one kind of polygon. There are several kinds of polygons that we should learn to recognize. Each kind has special characteristics, but the perimeter of each is found by adding the lengths of the sides. Let's describe some basic polygons.

Name	Illustration	Characteristics
triangle		3 sides
square		4 equal sides Sides meet in right angles.
rectangle		4 sides Opposite sides are equal. Sides meet in right angles.
parallelogram		4 sides Opposite sides are equal. Opposite sides are parallel.
trapezoid		4 sides 2 opposite sides are parallel.

If a polygon has 4 sides, it can be called a **quadrilateral.** Squares, rectangles, parallelograms, and trapezoids are all examples of quadrilaterals. Some 4-sided polygons do not have the characteristics of any of these special quadrilaterals. Look at these 4-sided figures.

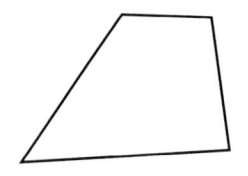 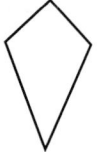

Such figures have no special names. They are simply called quadrilaterals.

Polygons with more than 4 sides are named using the Greek prefix for the number of sides. For instance

If a Polygon Has	It Is Called a
5 sides	pentagon
6 sides	hexagon
7 sides	heptagon
8 sides	octagon
9 sides	nonagon
10 sides	decagon

These polygons are mentioned here only for your general information. Now you know that the U.S. Pentagon is so named because it is a 5-sided building. Octagon soap was so named because the surface of the bar had 8 sides.

Let us return to our main purpose here, which is to learn to find perimeters of polygons. Remember that no matter what kind of polygon we are discussing, *its perimeter is the sum of the lengths of its sides*.

Example 1. Name the figure and find its perimeter.

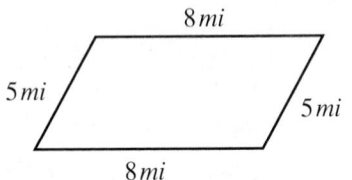

Solution. The figure is a parallelogram.

$$\text{Perimeter} = 5 \text{ mi} + 8 \text{ mi} + 5 \text{ mi} + 8 \text{ mi}$$

$$= 26 \text{ mi}$$

Example 2. Name the figure and find its perimeter.

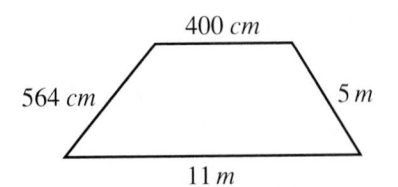

Solution. The figure is a trapezoid.

$$\text{Perimeter} = 5 \text{ m} + 400 \text{ cm} + 564 \text{ cm} + 11 \text{ m}$$

But our units do not match. Let's change all the units to centimeters.

$$5 \text{ m} = \frac{5}{1} \times \frac{1 \cancel{\text{m}}}{1} \times \frac{100 \text{ cm}}{1 \cancel{\text{m}}} \qquad\qquad 11 \text{ m} = \frac{11}{1} \times \frac{1 \cancel{\text{m}}}{1} \times \frac{100 \text{ cm}}{1 \cancel{\text{m}}}$$

$$= 500 \text{ cm} \qquad\qquad\qquad\qquad = 1100 \text{ cm}$$

Now the perimeter becomes

$$\text{Perimeter} = 500 \text{ cm} + 400 \text{ cm} + 564 \text{ cm} + 1100 \text{ cm}$$

$$= 2564 \text{ cm}$$

You try Example 3.

Example 3. Name the figure and find its perimeter in *feet*.

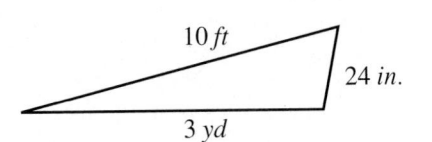

Solution. The figure is a _____ . We must change all the units to *feet* before finding the perimeter.

✎

$$10 \text{ ft} = 10 \text{ ft}$$

$$3 \text{ yd} = \frac{3}{1} \times \frac{\cancel{1 \text{ yd}}^{1}}{1} \times \frac{3 \text{ ft}}{\cancel{1 \text{ yd}}_{1}}$$

$$= \underline{\quad} \text{ ft}$$

$$24 \text{ in.} = \frac{24}{1} \times \frac{\cancel{1 \text{ in.}}^{1}}{1} \times \frac{1 \text{ ft}}{\cancel{12 \text{ in.}}_{12}}$$

$$= \frac{\cancel{24}^{2} \times 1 \text{ ft}}{\cancel{12}_{1}}$$

$$= \underline{\quad} \text{ ft}$$

$$\text{Perimeter} = 10 \text{ ft} + \underline{\quad} \text{ ft} + \underline{\quad} \text{ ft}$$

$$= \underline{\quad} \text{ ft}$$

Check your work on page 548. ▶

⫸ Trial Run

Name the figures and find the perimeters.

✎

_____ **1.**

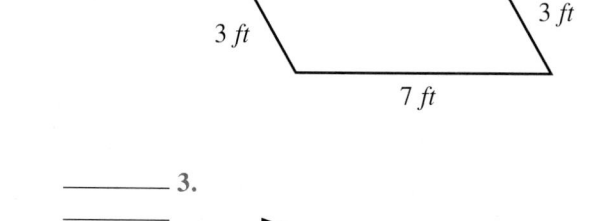

_____ **2.**

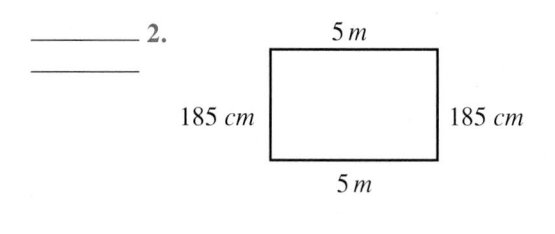

_____ **3.**

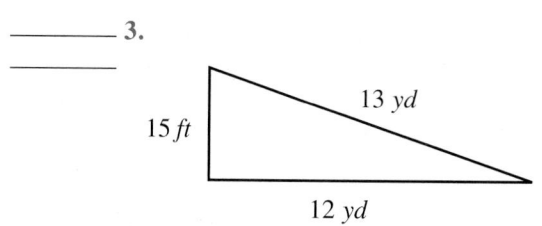

_____ **4.**

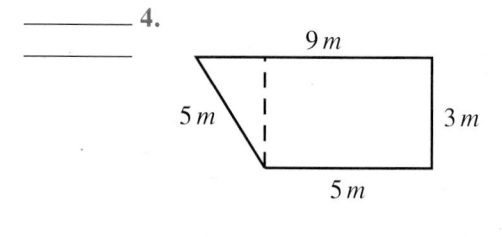

Answers are on page 548.

Before we finish our discussion of perimeter, let's take a closer look at the square and rectangle. Perhaps there is a quicker way to find their perimeters. In a *square*, we know that all 4 sides are of equal length. For instance, consider the square whose sides are each 5 inches long.

We may find the perimeter, P, by addition.

$$P = 5 \text{ in.} + 5 \text{ in.} + 5 \text{ in.} + 5 \text{ in.}$$

$$= 20 \text{ in.}$$

But we could also find this perimeter by multiplying the length of one side by 4.

$$P = 4 \times 5 \text{ in.}$$

$$= 20 \text{ in.}$$

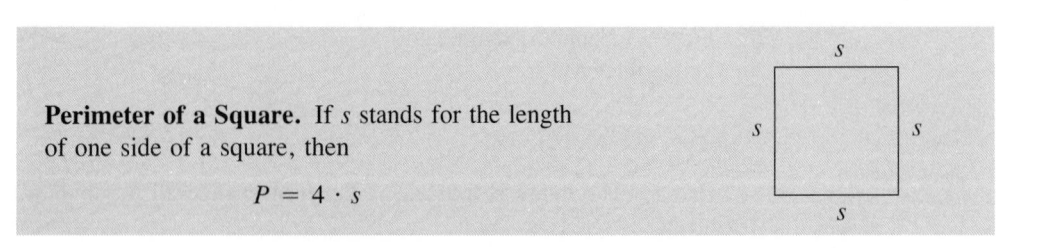

Perimeter of a Square. If s stands for the length of one side of a square, then

$$P = 4 \cdot s$$

Notice that we have used the centered dot ($\cdot$) to indicate multiplication.

You try Example 4.

Example 4. Find the perimeter of a square with each side 1.7 millimeters long.

Solution

$$\text{Perimeter} = 4 \cdot s$$

$$= 4 \, (\underline{\hspace{1cm}} \text{ mm})$$

$$= \underline{\hspace{1cm}} \text{ mm}$$

Check your work on page 548. ▶

In a *rectangle,* we know that the opposite sides are of equal length. For instance, consider the rectangle in which 2 of the sides are 5 meters long and 2 of the sides are 3 meters long.

To find the perimeter, we may add.

$$P = 3 \text{ m} + 5 \text{ m} + 3 \text{ m} + 5 \text{ m}$$

$$= 16 \text{ m}$$

But we could also find this perimeter by adding *twice* the length and *twice* the width.

$$P = (2 \cdot 3 \text{ m}) + (2 \cdot 5 \text{ m})$$

$$= 6 \text{ m} \qquad + 10 \text{ m}$$

$$= 16 \text{ m}$$

Perimeter of a Rectangle. If l stands for the length of a rectangle and w stands for the width, then

$$P = 2 \cdot l + 2 \cdot w$$

Example 5. Find the perimeter of a rectangle that is $3\frac{1}{2}$ feet long and $1\frac{1}{4}$ feet wide.

Solution. First we change our mixed numbers to improper fractions.

$$l = 3\frac{1}{2} \text{ ft} = \frac{7}{2} \text{ ft}$$

$$w = 1\frac{1}{4} \text{ ft} = \frac{5}{4} \text{ ft}$$

Now we find the perimeter.

$$P = 2 \cdot l + 2 \cdot w$$

$$= 2 \cdot \frac{7}{2} \text{ ft} + 2 \cdot \frac{5}{4} \text{ ft}$$

$$= \left(\frac{\overset{1}{\cancel{2}}}{1} \cdot \frac{7}{\underset{1}{\cancel{2}}} \text{ ft} \right) + \left(\frac{\overset{1}{\cancel{2}}}{1} \cdot \frac{5}{\underset{2}{\cancel{4}}} \text{ ft} \right) \qquad \text{Reduce fractions.}$$

$$= 7 \text{ ft} + \frac{5}{2} \text{ ft} \qquad \text{Find the products.}$$

$$= 7 \text{ ft} + 2\frac{1}{2} \text{ ft} \qquad \begin{array}{l}\text{Change improper fraction}\\\text{to a mixed number.}\end{array}$$

$$P = 9\frac{1}{2} \text{ ft} \qquad \text{Find the sum.}$$

⫸ Trial Run

_____ 1. Find the perimeter of a square with each side 2.3 centimeters long.

_____ 2. Find the perimeter of a rectangle that is $7\frac{5}{6}$ feet long and $3\frac{2}{3}$ feet wide.

_____ 3. Find the perimeter of a square with each side $5\frac{1}{3}$ feet long.

_____ 4. Find the perimeter of a rectangle that is 7.6 meters long and 253 centimeters wide.

Answers are on page 548.

Finding Circumference

Another important geometric figure is the **circle.** A circle is made up of all the points that are a certain distance from one particular point, called the **center.** The distance from any point on a circle to its center is called the **radius** (r) of the circle. The distance across the circle, through its center, is called the **diameter** (d) of a circle.

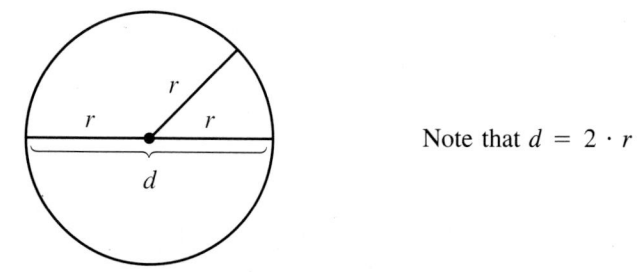

Note that $d = 2 \cdot r$

The distance around a circle is not called its perimeter. Instead it is called the **circumference.** When people first studied the circle, they wanted to relate the radius or the diameter of a circle to its circumference. After measuring the diameters and circumferences of many circles, they discovered that the circumference was always about 3.14 times as long as the diameter. Later mathematicians used more complicated methods to arrive at a more exact number than 3.14.

Unfortunately, the more exact number did not turn out to be a decimal number that terminates or repeats. Mathematicians named the exact number π and the exact measure of the circumference of a circle became

$$\text{Circumference} = \pi \cdot \text{diameter}$$

When we measure circumferences we will use 3.14 or $\frac{22}{7}$ for π. But you should understand that these numbers are just *approximations* for the exact number, π.

> **Circumference of a Circle.** If r is the radius and d is the diameter of a circle, then the circumference, C, is found by
>
> $$C = \pi \cdot d \quad \text{or} \quad C = \pi \cdot 2 \cdot r$$
> $$= 2 \cdot \pi \cdot r$$

Use the circumference formula to complete Example 6.

Example 6. Find the circumference of a circle with a diameter of 10 feet. Use $\pi \doteq 3.14$.

Solution. Here, we are given the diameter of the circle so we use

$$C = \pi \cdot d$$
$$\doteq 3.14 \,(\underline{\hspace{1cm}}\text{ ft})$$
$$\doteq \underline{\hspace{1cm}}\text{ ft}$$

Check your work on page 548. ▶

Example 7. Find the circumference of a circle with a radius of 1.3 centimeters. Use $\pi \doteq 3.14$.

Solution. Here, we are given the radius of the circle so we use

$$C = 2 \cdot \pi \cdot r$$

To avoid confusing decimal points and multiplication dots, we may put *parentheses* around the numbers being multiplied and omit the multiplication dot.

$$C \doteq 2(3.14)(1.3)$$
$$\doteq 6.28(1.3)$$
$$\doteq 8.164 \text{ cm}$$

Let's return to the hatband problem stated earlier in this chapter. Hat sizes tell the diameter (in inches) of the circular part of the hat that touches the wearer's head. If the hat size is $6\frac{1}{8}$, then we know $d = 6\frac{1}{8}$ inches. To see if 19 inches of ribbon will make a hatband for this hat, we must first find the circumference of the circle with diameter $6\frac{1}{8}$ inches. Since our diameter is a fractional number, let's use $\pi \doteq \frac{22}{7}$.

$$C = \pi \cdot d$$

$$\doteq \frac{22}{7} \cdot 6\frac{1}{8} \text{ in.}$$

$$\doteq \frac{\overset{11}{\cancel{22}}}{\underset{1}{\cancel{7}}} \cdot \frac{\overset{7}{\cancel{49}}}{\underset{4}{\cancel{8}}} \text{ in.}$$

$$\doteq \frac{77}{4} \text{ in.}$$

$$C \doteq 19\frac{1}{4} \text{ in.}$$

Our 19-inch ribbon will *not* go around the circular part of the hat. It is $\frac{1}{4}$ inch too short.

Example 8. Find the distance around the racetrack illustrated. Each end of the track is *half* a circle (called a *semicircle*). Use $\pi \doteq 3.14$.

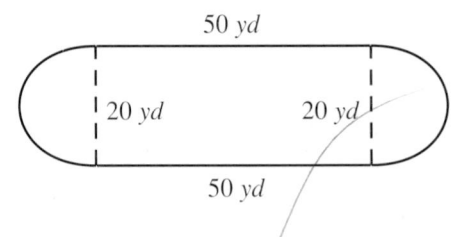

50 yd

20 yd   20 yd

50 yd

Solution. The 2 semicircles make up 1 whole circle with a diameter of 20 yards. The total distance is found by adding the lengths of the straight parts of the track to the circumference of the circle part of the track.

$$\text{Distance} = 50 \text{ yd} + 50 \text{ yd} + (\pi \cdot 20 \text{ yd})$$

lengths of   circumference
straight parts   of circle part

$$\doteq 50 \text{ yd} + 50 \text{ yd} + 3.14 \,(20 \text{ yd})$$

$$\doteq 100 \text{ yd} + 62.8 \text{ yd}$$

$$\text{Distance} \doteq 162.8 \text{ yd}$$

The distance around the racetrack is about 162.8 yards.

Example 9. Which tablecloth will require more fringe around its edge: a square measuring 52 inches on each side or a circle with a diameter of 60 inches?

Solution. We must find the distance around each tablecloth.

Square	*Circle*
Perimeter $= 4 \cdot 52$ in.	Circumference $\doteq 3.14\,(60$ in.$)$
$= 208$ in.	$\doteq 188.4$ in.

The *square* tablecloth will require more fringe.

⇒ Trial Run

_____ **1.** Find the circumference of a circle with a diameter of 12 feet. Use $\pi \doteq 3.14$.

_____ **2.** Find the circumference of a circle with a radius of 3.2 meters. Use $\pi \doteq 3.14$.

_____ **3.** Find the circumference of a circle with a radius of $4\frac{3}{8}$ inches. Use $\pi \doteq \frac{22}{7}$.

_____ **4.** Find the distance around the figure illustrated. Use $\pi \doteq 3.14$.

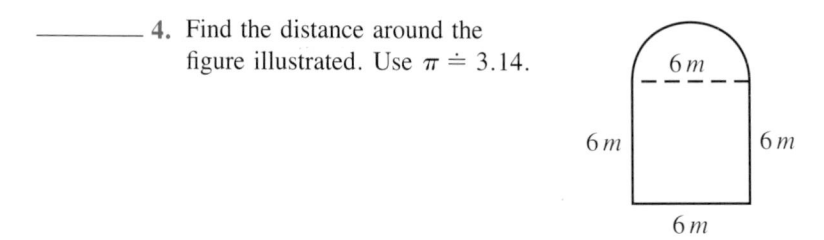

6 m

6 m   6 m

6 m

Answers are on page 548.

▶ Examples You Completed

Example 3. Name the figure and find its perimeter in *feet*.

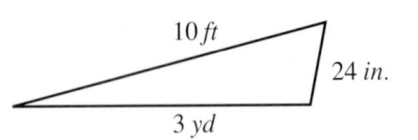

Solution. The figure is a triangle. We must change all units to *feet* before finding the perimeter.

$$10 \text{ ft} = 10 \text{ ft}$$

$$3 \text{ yd} = \frac{3}{1} \times \frac{\overset{1}{\cancel{1 \text{ yd}}}}{1} \times \frac{3 \text{ ft}}{\underset{1}{\cancel{1 \text{ yd}}}}$$

$$= 9 \text{ ft}$$

$$24 \text{ in.} = \frac{24}{1} \times \frac{\overset{1}{\cancel{1 \text{ in.}}}}{1} \times \frac{1 \text{ ft}}{\underset{12}{\cancel{12 \text{ in.}}}}$$

$$= \frac{\overset{2}{\cancel{24}} \times 1 \text{ ft}}{\underset{1}{\cancel{12}}}$$

$$= 2 \text{ ft}$$

$$\text{Perimeter} = 10 \text{ ft} + 9 \text{ ft} + 2 \text{ ft}$$
$$= 21 \text{ ft}$$

Example 4. Find the perimeter of a square with each side 1.7 millimeters long.

Solution

$$\text{Perimeter} = 4 \cdot s$$
$$= 4 \, (1.7 \text{ mm})$$
$$= 6.8 \text{ mm}$$

Example 6. Find the circumference of a circle with a diameter of 10 feet. Use $\pi \doteq 3.14$.

Solution

$$C = \pi \cdot d$$
$$\doteq 3.14 \, (10 \text{ ft})$$
$$\doteq 31.4 \text{ ft}$$

Answers to Trial Runs

page 543   **1.** Parallelogram, 20 ft   **2.** Rectangle, 1370 cm or 13.7 m   **3.** Triangle, 30 yd or 90 ft
4. Trapezoid, 22 m

page 545   **1.** 9.2 cm   **2.** 23 ft   **3.** $21\frac{1}{3}$ ft   **4.** 2026 cm

page 547   **1.** 37.68 ft   **2.** 20.096 m   **3.** $27\frac{1}{2}$ in.   **4.** 27.42

EXERCISE SET 13.1

Name each figure and find its perimeter.

_____ **1.**
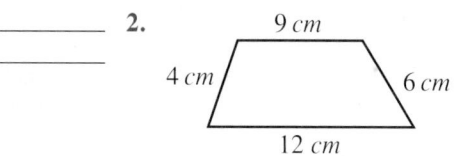

9 ft
5 ft   5 ft
9 ft

_____ **2.**
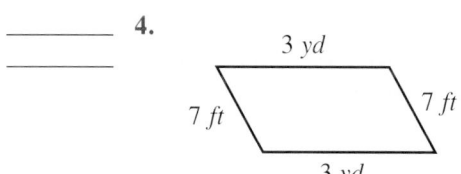

9 cm
4 cm   6 cm
12 cm

_____ **3.**
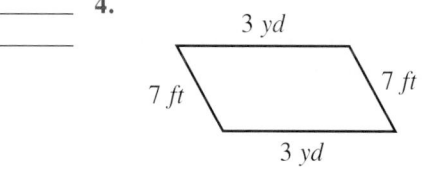

8 in.
8 in.   8 in.
8 in.

_____ **4.**
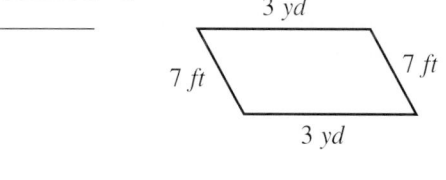

3 yd
7 ft   7 ft
3 yd

_____ **5.**
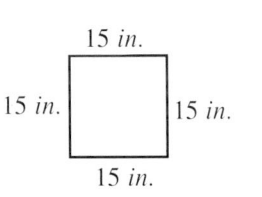

11 m
450 cm   450 cm
11 m

_____ **6.**

15 in.
15 in.   15 in.
15 in.

_____ **7.**
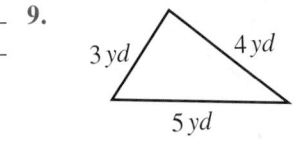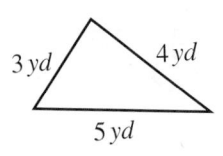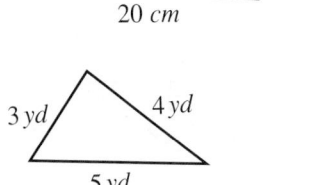

15 cm
7 cm   10 cm
20 cm

_____ **8.**
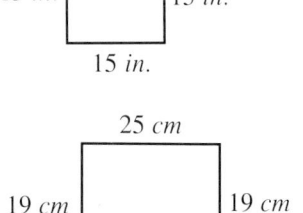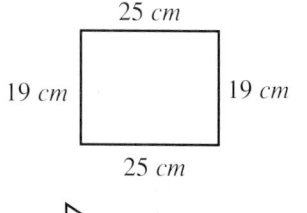

25 cm
19 cm   19 cm
25 cm

_____ **9.**
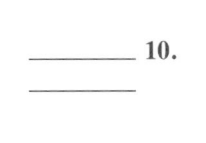

3 yd   4 yd
5 yd

_____ **10.**
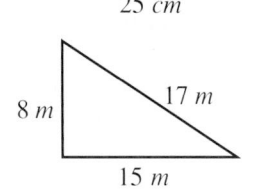

8 m   17 m
15 m

_____ **11.** Find the circumference of a circle with a diameter of 35 inches. Use
$\pi \doteq \frac{22}{7}$.

_____ **12.** Find the circumference of a circle with a diameter of 7 feet. Use $\pi \doteq \frac{22}{7}$.

_____ **13.** Find the circumference of a circle with a radius of 8.5 meters. Use
$\pi \doteq 3.14$.

_____ **14.** Find the circumference of a circle with a radius of 20.3 centimeters. Use
$\pi \doteq 3.14$. (Round answer to the tenths place.)

_____ **15.** Find the amount of fencing needed to enclose a rectangular garden that is
10 meters long and $4\frac{1}{2}$ meters wide.

_____ **16.** How much masking tape is needed to protect the edges of a picture window
that is 3 yards long and 2 yards wide?

549

_____ 17. If the radius of a nickel is 1.1 centimeters, find its circumference. Use $\pi \doteq 3.14$. (Round answer to tenths place.)

_____ 18. Find the amount of fringe needed to trim the edge of a circular tablecloth 63 inches in diameter. Use $\pi \doteq \frac{22}{7}$.

_____ 19. Find the number of inches of framing needed to frame a square picture
_____ measuring $1\frac{1}{2}$ feet on each side. How many yards will that be?

_____ 20. The bases of a baseball diamond are 90 feet apart. How many feet does a
_____ baseball player run when he hits a home run? How many yards will that be?

_____ 21. Find the perimeter of the illustrated figure. All measurements are in meters.

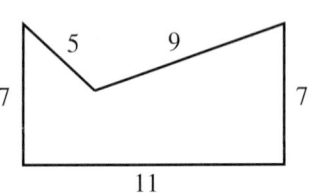

_____ 22. Find the perimeter of the illustrated figure. All measurements are in centimeters.

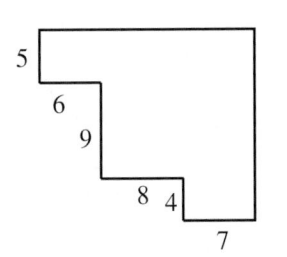

_____ 23. Doug's triangular jogging route takes him 1.2 kilometers north, 1.6 kilometers east, and 2 kilometers back to his starting place. How far does he jog on this route?

_____ 24. The banner for the Union County Band is a blue triangle with a white border. If the triangle has the dimensions illustrated, how many feet of border will be needed for the trim?

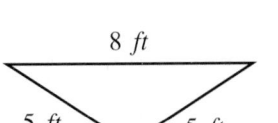

_____ 25. A pest control company sprays for termites along the floor line, charging according to the number of feet sprayed. For how many feet can the company charge when spraying a theater that is 50 feet long and 7 yards wide?

_____ 26. The front part of the Byrds' roof is a trapezoid (see illustration). If they wish to outline it with Christmas tree lights, how many feet of lighting will they need?

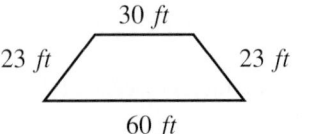

_____ 27. The Sonis owned a rectangular lot in a section of town zoned for small business. They sold triangular lots to a convenience market and a fast-food restaurant. Find the number of feet of fencing they will need to enclose the remaining property if the lots have the indicated dimensions.

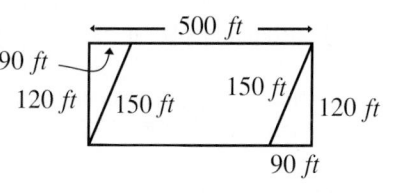

_____ 28. To install a strip of wood trim above
the paneling in a recreation room
with the illustrated floor plan, how
many feet of trim are needed?

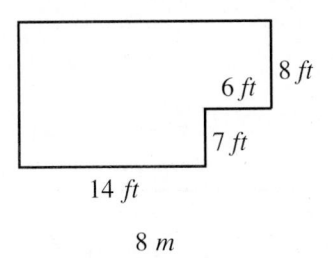

_____ 29. Find the distance around the
swimming pool in the illustration.
Use $\pi \doteq 3.14$.

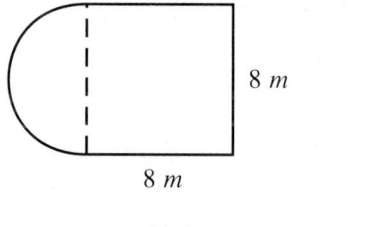

_____ 30. How long must a bicycle chain be
to fit the pair of gears in the
illustration? Use $\pi \doteq 3.14$.

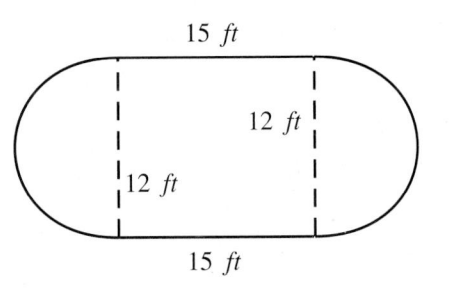

☆ Stretching the Topics _____

_____ 1. A rug has the shape and dimensions
illustrated. Find the cost of
binding the edge of the rug if the
binding costs $2.95 per yard.

_____ 2. The Edelsteins wish to build a patio
_____   at the southeast corner of their house.
If they plan to enclose the patio with
a decorative fence, which of the two
patios would require more fencing? How
many more feet of fencing does it require?

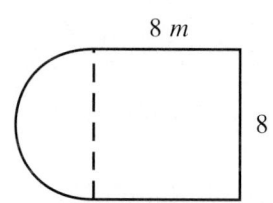

Check your answers in the back of your book.

If you can solve the problems in **Checkup 13.1,** you are ready to go on to Section 13.2.

✓ **CHECKUP 13.1**

Name the figure and find its perimeter.

_____ **1.**

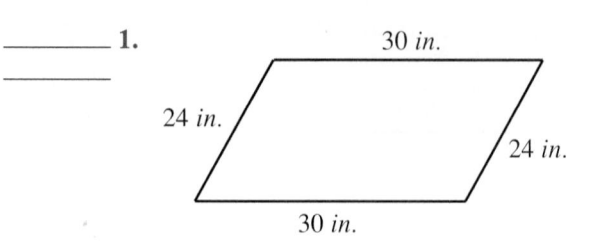

_____ **2.**

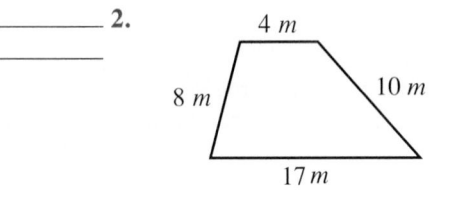

_____ **3.** Find the perimeter of a square with each side 12.3 centimeters.

_____ **4.** Find the perimeter of a rectangle that is $4\frac{3}{4}$ feet wide and $5\frac{1}{2}$ feet long.

_____ **5.** Find the circumference of a circle with a radius of 2.1 centimeters. Use $\pi \doteq 3.14$. (Round answer to the tenths place.)

_____ **6.** Which flower bed will require more fencing: a square 2 yards on each side or a circle with a diameter of 10 feet?

Check your answers in the back of your book.

If You Missed Problems:	You Should Review Examples:
1, 2	1–3
3	4
4	5
5	7
6	9

13.2 Measuring Area

When we wish to find the amount of flat surface enclosed by a geometric figure, we are looking for the **area** of that figure. The area of each of the following figures is shaded.

 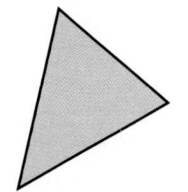 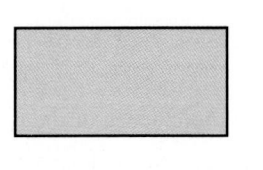 

We have worked with areas many times before in earlier chapters. We learned that area is always measured in **square units** (sq in., sq ft, sq yd, sq m, sq cm, and so on).

Finding Areas of Polygons

In Chapter 2, we discovered the formulas used to find areas of squares, rectangles, and triangles.

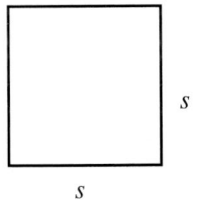 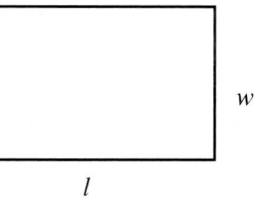 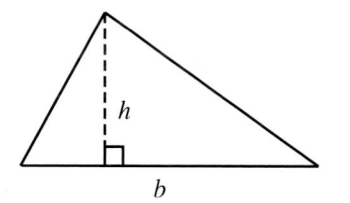

Square

Rectangle

Triangle

Area = side · side

= $s \cdot s$

$A = s^2$

Area = length · width

$A = l \cdot w$

Area = $\dfrac{\text{base} \cdot \text{height}}{2}$

$A = \dfrac{b \cdot h}{2}$

Example 1. Find the area of a square with a side $3\frac{1}{2}$ feet long.

Solution. Here, $s = 3\frac{1}{2}$ ft.

$$\text{Area} = s \cdot s$$

$$= 3\frac{1}{2} \cdot 3\frac{1}{2} \qquad \text{Substitute } 3\frac{1}{2} \text{ for } s \text{ in the formula.}$$

$$= \frac{7}{2} \cdot \frac{7}{2} \qquad \text{Change mixed numbers to improper fractions.}$$

$$= \frac{49}{4} \text{ sq ft} \qquad \text{Multiply and attach units.}$$

$$= 12\frac{1}{4} \text{ sq ft} \qquad \text{Change improper fraction to mixed number.}$$

You complete Example 2.

Example 2. Find the floor space of a rectangular room that is 3.3 meters long and 4.1 meters wide.

Solution. The floor space is the area of the floor. Here, l = 3.3 meters and w = 4.1 meters.

$$\text{Area} = l \cdot w$$

$$= (\underline{\quad})(\underline{\quad})$$

$$\text{Area} = \underline{\quad} \text{ sq m}$$

The floor space of the room is _____ square meters.

Check your work on page 560. ▶

Example 3. Find the area of a triangular sail with base 3 yards and height 11 feet.

Solution. We must make our units match, so we use a conversion factor to change yards to feet.

$$3 \text{ yd} = \frac{3}{1} \times \frac{\overset{1}{\cancel{1 \text{ yd}}}}{1} \times \frac{3 \text{ ft}}{\underset{1}{\cancel{1 \text{ yd}}}}$$

$$= 9 \text{ ft}$$

Now we know b = 9 ft and h = 11 ft.

$$\text{Area} = \frac{b \cdot h}{2}$$

$$= \frac{9 \cdot 11}{2}$$

$$= \frac{99}{2} \text{ sq ft}$$

$$= 49\frac{1}{2} \text{ sq ft}$$

The area of the sail is $49\frac{1}{2}$ square feet.

Example 4. One end of a house is to be painted. Use the picture of the end of the house to find the area to be painted.

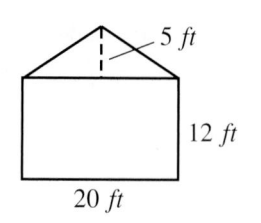

Solution. There are 2 geometric figures here for which we must find the area.

Triangle $\text{Area} = \dfrac{b \cdot h}{2}$

$$= \frac{\overset{10}{\cancel{20}} \cdot 5}{\underset{1}{\cancel{2}}}$$

$$= 50 \text{ sq ft}$$

Rectangle $\text{Area} = l \cdot w$

$$= 20 \cdot 12$$

$$= 240 \text{ sq ft}$$

Total Area = Area of triangle + Area of rectangle

$$= 50 \text{ sq ft} + 240 \text{ sq ft}$$

$$= 290 \text{ sq ft}$$

The area to be painted is 290 square feet.

⟩ Trial Run

_____ 1. Find the area of a square with each side 6.4 meters.

_____ 2. Find the area of a rectangle that is 4 yards long and $8\frac{1}{4}$ feet wide.

_____ 3. Find the area of a triangle with base 2.3 meters and height 85 centimeters.

_____ 4. Find the area of an oriental rug that is 2.7 meters long and 1.8 meters wide.

_____ 5. Find how many square inches of cardboard will be needed to make a square yard sale sign that is 30 inches on each side.

_____ 6. Find the area on a wall that will be covered by a triangular macramé wall hanging with base 18 inches and height 24 inches.

Answers are on page 560.

Finding Areas of Circles

As it was in finding the circumference, the number represented by the Greek letter π is again important in finding the *area* of a circle. If we know the length of the radius of a circle, then its area is found by

$$\text{Area} = \pi \times \text{radius} \times \text{radius}.$$

Area of a Circle. If r stands for the radius of a circle, then

$$\text{Area} = \pi \cdot r \cdot r$$
$$A = \pi \cdot r^2$$

Example 5. Find the area of a circle with a radius of 10 centimeters.

Solution. Here, $r = 10$ cm.

$$\text{Area} = \pi \cdot r \cdot r$$
$$= \pi \cdot 10 \cdot 10$$
$$= \pi \cdot 100 \text{ sq cm}$$

An approximate value for this area may be found by using 3.14 or $\frac{22}{7}$ in place of π.

$$\text{Area} \doteq 3.14 \,(100 \text{ sq cm})$$
$$\doteq 314 \text{ sq cm}$$

As before, notice that area is always measured in *square units*.

Example 6. Find the area of a circular mirror whose diameter is 7 inches.

Solution. We know the *diameter* of our circle is 7 inches, so the radius must be *half* as long.

$$r = \frac{1}{2} \text{ of diameter}$$

$$= \frac{1}{2} \cdot 7 \text{ in.}$$

$$= \frac{7}{2} \text{ in.}$$

Now we can find the area.

Let's use $\pi \doteq \dfrac{22}{7}$

$$\text{Area} = \pi \cdot r \cdot r$$

$$\doteq \frac{\overset{11}{\cancel{22}}}{\underset{1}{\cancel{7}}} \cdot \frac{\overset{1}{\cancel{7}}}{\underset{1}{\cancel{7}}} \cdot \frac{7}{2}$$

$$\doteq \frac{77}{2} \text{ sq in.}$$

$$\text{Area} \doteq 38\frac{1}{2} \text{ sq in.}$$

Example 7. Compare the area of the top of a round pizza with a *diameter* of 12 inches to the area of the top of a square pizza with $10\frac{1}{2}$-inch sides.

Solution. We must find the area of each pizza.

Circle   The radius, r, is $\dfrac{1}{2}$ of 12 in.

So $r = 6$ in. Let's use $\pi \doteq 3.14$.

$$\text{Area} = \pi \cdot r \cdot r$$

$$= \pi \cdot 6 \cdot 6$$

$$\doteq 3.14\,(36)$$

$$\doteq 113.04 \text{ sq in.}$$

Square   Here, $s = 10\dfrac{1}{2}$ in. $= \dfrac{21}{2}$ in.

$$\text{Area} = s \cdot s$$

$$= \frac{21}{2} \cdot \frac{21}{2}$$

$$= \frac{441}{4} \text{ sq in.}$$

$$= 110.25 \text{ sq in.}$$

Since $113.04 > 110.25$, the round pizza is larger. By how much? By $113.04 - 110.25 \doteq 2.79$ square inches.

⫸ Trial Run

_____ 1. Find the area of a circle with a radius of 5.2 centimeters. Use $\pi \doteq 3.14$.

_____ 2. Find the area of a circle with a diameter of $4\frac{2}{3}$ feet. Use $\pi \doteq \frac{22}{7}$.

_____ 3. Find the area of a circular flower bed with a radius of 1.75 feet. Use $\pi \doteq 3.14$. (Round answer to nearest square foot.)

_____ 4. Find the area of the top of a round table with a diameter of 52 centimeters. Use $\pi \doteq 3.14$.

_____ 5. Compare the area of a rectangular swimming pool that is 26 feet by 18 feet to the area of a circular pool with a diameter of 24 feet. Use $\pi \doteq 3.14$.

Answers are on page 560.

Converting Units of Area

The following pairs of figures might help us figure out how to change units of area from square feet to square inches or from square yards to square feet.

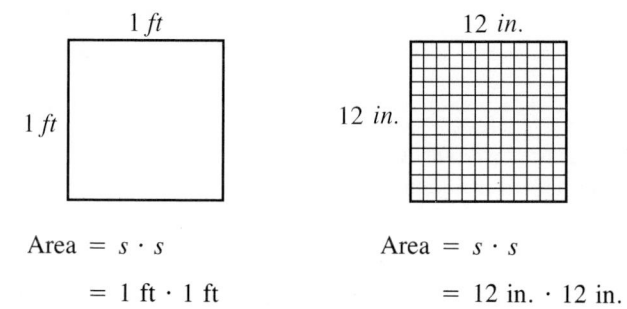

Area = $s \cdot s$
    = 1 ft $\cdot$ 1 ft
    = 1 sq ft

Area = $s \cdot s$
    = 12 in. $\cdot$ 12 in.
    = 144 sq in.

We know that these squares are identical because 1 foot = 12 inches, so their areas must be the same. We conclude that

$$1 \text{ sq ft} = 144 \text{ sq in.}$$

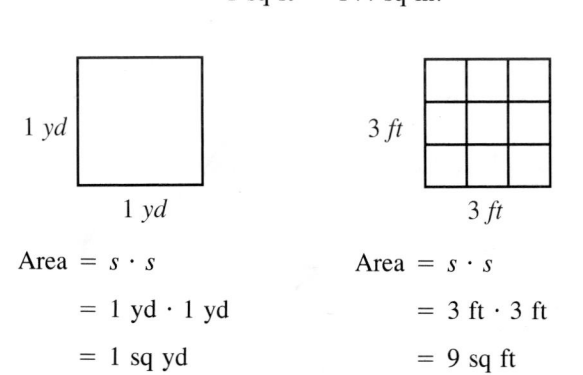

Area = $s \cdot s$
    = 1 yd $\cdot$ 1 yd
    = 1 sq yd

Area = $s \cdot s$
    = 3 ft $\cdot$ 3 ft
    = 9 sq ft

Once again, we know these squares are identical because 1 yard = 3 feet, so their areas must be the same. We conclude that

$$1 \text{ sq yd} = 9 \text{ sq ft}$$

We now have some conversion factors to use in changing among square units.

$\dfrac{1 \text{ sq ft}}{144 \text{ sq in.}}$	$\dfrac{144 \text{ sq in.}}{1 \text{ sq ft}}$	$\dfrac{1 \text{ sq yd}}{9 \text{ sq ft}}$	$\dfrac{9 \text{ sq ft}}{1 \text{ sq yd}}$

Each of these factors is equal to the number 1.

Example 8. How many square *yards* of carpet must be bought to cover the floor of a rectangular room measuring 12 feet by 18 feet?

Solution. First we find the area.

$$\text{Area} = l \cdot w$$

$$= 12 \cdot 18$$

$$\text{Area} = 216 \text{ sq ft}$$

But now we must change 216 square feet to square yards. We use the proper conversion factor to get rid of the unwanted units.

$$216 \text{ sq ft} = 216 \times 1 \text{ sq ft} \times \frac{1 \text{ sq yd}}{9 \text{ sq ft}} \qquad \begin{array}{l}\text{Choose the conversion factor.}\\ \text{(We don't want square feet; we}\\ \text{do want square yards.)}\end{array}$$

$$= \frac{216}{1} \times \frac{\overset{1}{\cancel{1 \text{ sq ft}}}}{1} \times \frac{1 \text{ sq yd}}{\underset{9}{\cancel{9 \text{ sq ft}}}} \qquad \text{Reduce the fractions.}$$

$$= \frac{216}{9} \text{ sq yd} \qquad \begin{array}{l}\text{Multiply numerators and}\\ \text{denominators.}\end{array}$$

$$= 24 \text{ sq yd} \qquad \text{Reduce the fraction.}$$

We must buy 24 square yards of carpet.

Example 9. Change 2 square yards to square inches.

Solution. Here we need 2 conversion factors.

$$2 \text{ sq yd} = 2 \times 1 \text{ sq yd} \times \frac{9 \text{ sq ft}}{1 \text{ sq yd}} \times \frac{144 \text{ sq in.}}{1 \text{ sq ft}}$$

$$= \frac{2}{1} \times \frac{\overset{1}{\cancel{1 \text{ sq yd}}}}{1} \times \frac{\overset{9}{\cancel{9 \text{ sq ft}}}}{\underset{1}{\cancel{1 \text{ sq yd}}}} \times \frac{144 \text{ sq in.}}{\underset{1}{\cancel{1 \text{ sq ft}}}}$$

$$= 2 \times 9 \times 144 \text{ sq in.}$$

$$2 \text{ sq yd} = 2592 \text{ sq in.}$$

We can use the same kind of reasoning to find conversion factors for square *metric* units. Because 1 meter = 100 centimeters, a square with $s = 1$ meter is identical to a square with $s = 100$ centimeters. Let's find the area of each.

$\text{Area} = s \cdot s$	$\text{Area} = s \cdot s$
$= 1 \text{ m} \cdot 1 \text{ m}$	$= 100 \text{ cm} \cdot 100 \text{ cm}$
$= 1 \text{ sq m}$	$= 10{,}000 \text{ sq cm}$

Since these two squares are identical, we know their areas must be the same. We conclude

$$1 \text{ sq m} = 10{,}000 \text{ sq cm}$$

Similarly, a square with $s = 1$ kilometer is identical to a square with $s = 1000$ meters. Let's compare their areas.

$\text{Area} = s \cdot s$	$\text{Area} = s \cdot s$
$= 1 \text{ km} \cdot 1 \text{ km}$	$= 1000 \text{ m} \cdot 1000 \text{ m}$
$= 1 \text{ sq km}$	$= 1{,}000{,}000 \text{ sq m}$

Once again, we know these two squares are identical (because 1 km = 1000 m), so their areas must be the same. We conclude that

$$1 \text{ sq km} = 1,000,000 \text{ sq m}$$

Now we have some conversion factors to use for square metric units.

$$\frac{1 \text{ sq m}}{10,000 \text{ sq cm}} \qquad \frac{10,000 \text{ sq cm}}{1 \text{ sq m}} \qquad \frac{1 \text{ sq km}}{1,000,000 \text{ sq m}} \qquad \frac{1,000,000 \text{ sq m}}{1 \text{ sq km}}$$

Example 10. Change an area of 9872 square centimeters to square meters.

Solution

$$9872 \text{ sq cm} = 9872 \times 1 \text{ sq cm} \times \frac{1 \text{ sq m}}{10,000 \text{ sq cm}}$$

Choose the conversion factor. (We don't want square centimeters; we do want square meters).

$$= \frac{9872}{1} \times \frac{\overset{1}{\cancel{1 \text{ sq cm}}}}{1} \times \frac{1 \text{ sq m}}{\underset{10,000}{\cancel{10,000 \text{ sq cm}}}}$$

Reduce the fractions.

$$= \frac{9872}{10,000} \text{ sq m}$$

Multiply numerators and denominators.

$$9872 \text{ sq cm} = 0.9872 \text{ sq m}$$

Perform the division.

Example 11. If a circle has a radius of 0.02 kilometers, write its area in square centimeters.

Solution. First we find the area of the circle with $r = 0.02$ km.

$$\text{Area} = \pi \cdot r \cdot r$$

$$\doteq 3.14 \, (0.02)(0.02)$$

$$\text{Area} \doteq 0.001256 \text{ sq km}$$

Now we must use conversion factors to change the area to square centimeters.

$$0.001256 \text{ sq km} = 0.001256 \times 1 \text{ sq km} \times \frac{1,000,000 \text{ sq m}}{1 \text{ sq km}} \times \frac{10,000 \text{ sq cm}}{1 \text{ sq m}}$$

$$= \frac{0.001256}{1} \times \frac{\overset{1}{\cancel{1 \text{ sq km}}}}{1} \times \frac{\overset{1,000,000}{\cancel{1,000,000 \text{ sq m}}}}{\underset{1}{\cancel{1 \text{ sq km}}}} \times \frac{10,000 \text{ sq cm}}{\underset{1}{\cancel{1 \text{ sq m}}}}$$

$$= 0.001256 \, (1,000,000 \times 10,000) \text{ sq cm}$$

$$0.001256 \text{ sq km} = 12,560,000 \text{ sq cm}$$

The area is about 12,560,000 square centimeters.

⫸ Trial Run

_____ 1. Change an area of 378 square feet to square yards.

_____ 2. Change an area of 12,300 square centimeters to square meters.

_____ 3. If a circle has a radius of 42 inches, write its area in square feet. Use $\pi \doteq \frac{22}{7}$.

_____ 4. If a mirror is a rectangle with a length of 0.8 meter and a width of 0.95 meter, write its area in square centimeters.

_____ 5. Patti is making a square tablecloth with each side 72 inches. Find how many square yards of material she needs to buy.

Answers are given below.

▶ Example You Completed

Example 2. Find the floor space of a rectangular room that is 3.3 meters long and 4.1 meters wide.

Solution. The floor space is the area of the floor. Here, $l = 3.3$ m and $w = 4.1$ m.

$$\text{Area} = l \cdot w$$
$$= (3.3)(4.1)$$
$$\text{Area} = 13.53 \text{ sq m}$$

The floor space of the room is 13.53 square meters.

Answers to Trial Runs

page 555   1. 40.96 sq m   2. 99 sq ft   3. 9775 sq cm   4. 4.86 sq m   5. 900 sq in.
6. 216 sq in.

page 556   1. 84.9056 sq cm   2. $17\frac{1}{9}$ sq ft   3. 10 sq ft   4. 2122.64 sq cm
5. The rectangular pool is larger by 15.84 square feet.

page 560   1. 42 sq yd   2. 1.23 sq m   3. 38.5 sq ft   4. 7600 sq cm   5. 4 sq yd

EXERCISE SET 13.2

_____ 1. Find the area of a square with each side 9.4 meters.

_____ 2. Find the area of a square with each side $8\frac{1}{2}$ inches.

_____ 3. Find the area of a rectangle that is $3\frac{1}{2}$ feet long and $4\frac{1}{4}$ feet wide.

_____ 4. Find the area of a rectangle that is 3.4 centimeters long and 5.8 centimeters wide.

_____ 5. Find the area of a triangle with base 15 meters and height 8 meters.

_____ 6. Find the area of a triangle with a base 8 feet and height 5 feet.

_____ 7. Find the area of a circle with a diameter of 14 centimeters. (Use $\pi \doteq \frac{22}{7}$).

_____ 8. Find the area of a circle with a diameter of 12 inches. (Use $\pi \doteq 3.14$.)

_____ 9. How many square inches of cardboard will be needed to make a square FOR RENT sign that is 18 inches on each side.

_____ 10. Find the area of a square city block if a block is 0.1 mile long.

_____ 11. How much astroturf is needed to surface a football field 100 yards long and 65 yards wide?

_____ 12. Find the amount of carpeting needed to cover the floor of a rectangular room that is 11 feet wide and 13 feet long.

_____ 13. Find the area of a yacht's triangular mainsail if the base is 80 feet and the height is 165 feet.

_____ 14. The banner for the Union County band is a blue triangle. If the triangle has base 8 feet and height 3 feet, how much fabric will be needed to make the banner?

_____ 15. Find the amount of waxed paper needed to line the bottom of a cake pan if the bottom is a circle with diameter 22 centimeters. (Use $\pi \doteq 3.14$.)

_____ 16. Find the area of the face of a quarter with diameter 2.4 centimeters. (Use $\pi \doteq 3.14$ and round answer to the tenths place.)

_____ 17. How much fabric will be needed to construct the illustrated kite?

_____ **18.** Find the area of the bottom of the illustrated swimming pool built in the shape of a square with a semicircle at one end. (Use $\pi \doteq 3.14$.)

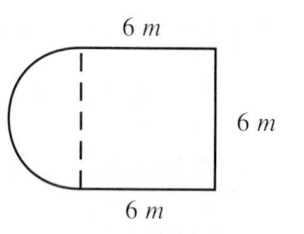

_____ **19.** Gina is sodding her front lawn. Use the drawing to determine how many square feet of sod she should order from the landscape service.

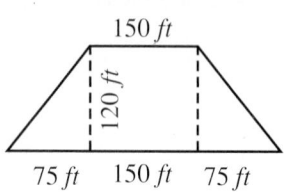

_____ **20.** The Vails wish to build a patio behind their L-shaped home in the shape of a rectangle or a quarter-circle. Which will give them more area?

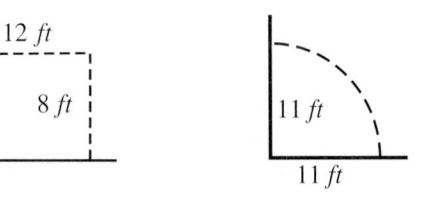

_____ **21.** Change 5 square yards to square feet.

_____ **22.** Change 8 square feet to square inches.

_____ **23.** Change 3852 square centimeters to square meters.

_____ **24.** Change 8 square kilometers to square meters.

_____ **25.** Mr. Herman has purchased a farm that is in the shape of a rectangle. The dimensions are 1.4 kilometers by 3.5 kilometers. Find the farm's area in square *meters*.

_____ **26.** The stage at Miller Theater is a rectangle with width 20 feet and length 40 feet. Find the area of the stage in square *yards*.

_____ **27.** How many square meters of a wall will be covered by a dartboard that has a diameter of 42 centimeters? (Use $\pi \doteq \frac{22}{7}$.)

_____ **28.** Which choice would give you more pizza: 2 small pizzas, each with a diameter of 10 inches, or 1 large pizza with a diameter of 15 inches? (Assume the pizzas have the same thickness.)

☆ Stretching the Topics _____

_____ **1.** Find the area of the town with the indicated boundaries.

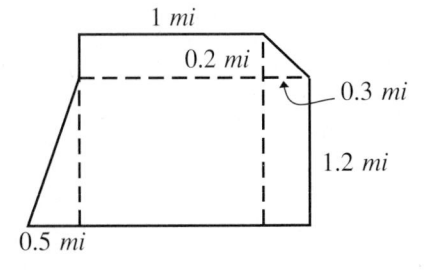

_____ 2. The Sonis owned a rectangular corner lot in a section of town zoned for small business. They sold triangular lots on each end. Find the area of the Sonis' lot after the triangular lots were sold.

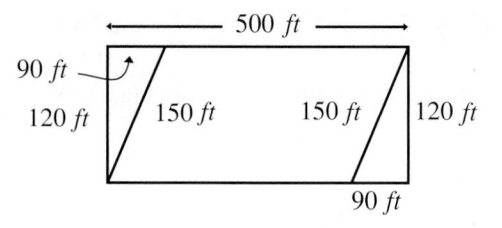

_____ 3. The Wallers have a rectangular basement that is 11.6 meters by 13.9 meters. If floor tiles are 22 centimeters by 30 centimeters, how many whole tiles should be bought to cover the basement floor?

Check your answers in the back of your book.

If you can complete **Checkup 13.2,** then you are ready to go on to Section 13.3.

Name _____   **Date** _____

✓ **CHECKUP 13.2**

_____ 1. Find the area of an index card that is 3 inches wide and 5 inches long.

_____ 2. Find the wall area that will be covered by a square picture measuring $1\frac{1}{2}$ feet on each side.

_____ 3. Find the area of a circular mirror whose diameter is 30 centimeters. (Use $\pi \doteq 3.14$.)

_____ 4. A playground has the shape of the figure at the right. Find the area in square feet.

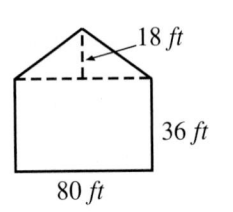

18 ft

36 ft

80 ft

_____ 5. Find the number of square feet of ceiling tiles needed for an office that is 5 yards wide and 8 yards long.

Check your answers in the back of your book.

If You Missed Problems:	You Should Review Examples:
1	2
2	1
3	6
4	4
5	8

564

13.3 Measuring Volume

So far we have studied flat geometric figures and learned how to find their perimeters and areas. Now we must look at some *solid* geometric figures. In particular we will be concerned with the *space inside* cubes, rectangular solids, cylinders, and spheres. The space enclosed by a solid figure is called its **volume.**

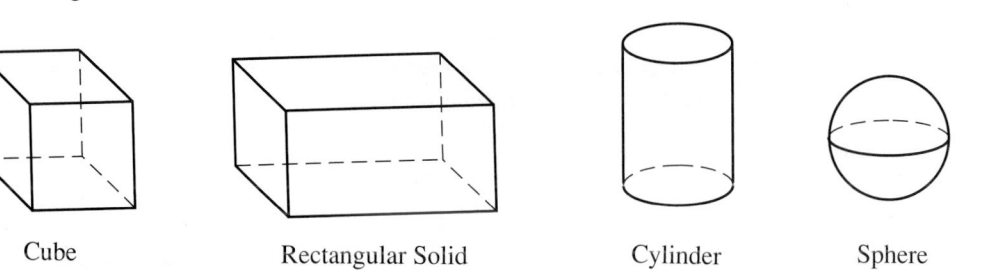

Cube   Rectangular Solid   Cylinder   Sphere

Each of these three-dimensional solids is related to the flat figures we have already studied.

Finding Volumes of Rectangular Solids

A **cube** is a box whose top, bottom, and all 4 sides are *squares*. If a cube measures 1 inch on all its edges, we say that its volume is 1 cubic inch (1 cu in.). A cube that measures 1 centimeter on all its edges has a volume of 1 cubic centimeter (1 cu cm).

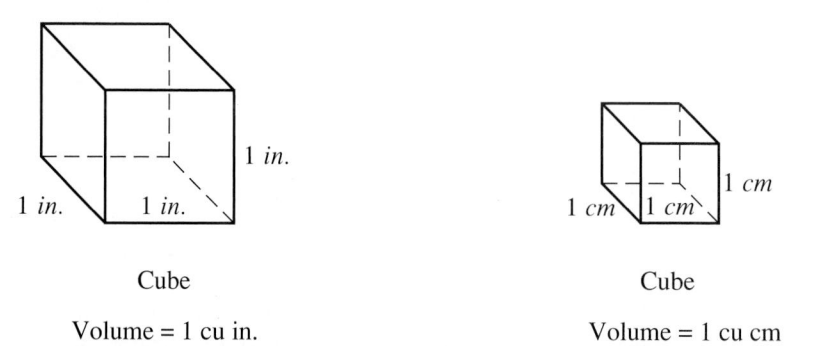

Cube   Cube

Volume = 1 cu in.   Volume = 1 cu cm

Volume is always measured in **cubic units** (cu in., cu ft, cu yd, cu cm, cu m, and so on).

Let's see if we can find the volume of the **rectangular solid** shown here. Suppose we count the cubic meters in the volume.

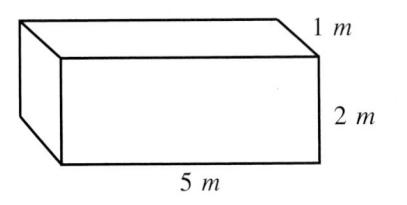
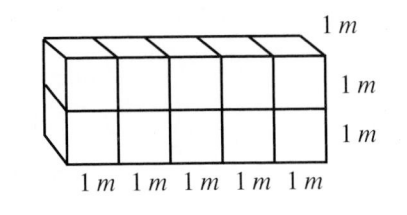

You should see that the volume is 10 cubic meters. Notice what happens if we multiply *length* times *width* times *height*.

$$\text{length} \cdot \text{width} \cdot \text{height} = 5 \cdot 1 \cdot 2$$
$$= 10 \text{ cu m}$$

Indeed, this is the way to find the volume of a rectangular solid.

> **Volume of a Rectangular Solid**
>
> Volume = length · width · height
>
> Volume = $l \cdot w \cdot h$

Since a cube is just a special case of a rectangular solid, we can use the same formula to find the volume of a cube with edge e.

> **Volume of a Cube**
>
> Volume = edge · edge · edge
>
> = $e \cdot e \cdot e$
>
> Volume = e^3

Example 1. How many cubic feet of water will be needed to completely fill a rectangular swimming pool that is 12 feet wide, 20 feet long, and 5 feet deep?

Solution. The filled pool is a rectangular solid.

$$\text{Volume} = l \cdot w \cdot h$$
$$= 20 \cdot 12 \cdot 5$$
$$\text{Volume} = 1200 \text{ cu ft}$$

We will need 1200 cubic feet of water.

You complete Example 2.

Example 2. Which box will hold more: a cube with edge 1.7 meters or a rectangular solid that is 1 meter wide, 3 meters long, and 1.6 meters high?

Solution. We must find the volume of each box.

Cube	*Rectangular Solid*
Volume = $e \cdot e \cdot e$	Volume = $l \cdot w \cdot h$
= (_____)(_____)(_____)	= (_____)(_____)(_____)
Volume = _____ cu m	Volume = _____ cu m

The box that is a _____ will hold more. How much more? _____

Check your work on page 572. ▶

Example 3. Find the volume of a rectangular piece of gutter that is 8 feet long, 6 inches wide, and $4\frac{1}{2}$ inches deep.

Solution. Before finding the volume, we must be sure all our units match. Let's change all the measurements to *inches*.

Length: $8 \text{ ft} = 8 \times \overset{1}{\cancel{1\text{ ft}}} \times \dfrac{12 \text{ in.}}{\underset{1}{\cancel{1\text{ ft}}}}$

$= 8 \times 12 \text{ in.}$

$= 96 \text{ in.}$

Width: $6 \text{ in.} = 6 \text{ in.}$

Height: $4\frac{1}{2} \text{ in.} = \frac{9}{2} \text{ in.}$

Now we can find the volume.

$\text{Volume} = l \cdot w \cdot h$

$= 96 \cdot 6 \cdot \dfrac{9}{2}$

$= \dfrac{96}{1} \cdot \dfrac{\overset{3}{\cancel{6}}}{1} \cdot \dfrac{9}{\underset{1}{\cancel{2}}}$

$\text{Volume} = 2592 \text{ cu in.}$

The volume of the gutter is 2592 cubic inches.

⇒ **Trial Run**

_____ 1. How many cubic centimeters of water will be needed to completely fill an aquarium that is 65 centimeters long, 35 centimeters wide, and 40 centimeters deep?

_____ 2. Find how many cubic feet of storage space are contained in a utility cabinet that is $2\frac{1}{2}$ feet by $1\frac{1}{3}$ feet by 6 feet.

_____ 3. A square cake pan is 12 inches by 12 inches by 2 inches. A rectangular cake pan is 11 inches by 13 inches by 2 inches. Compare the volumes of the 2 pans.

_____ 4. Find the volume of a freezer that is 1.2 meters long, 0.3 meter wide, and 0.6 meter high.

_____ 5. Find the volume of a coffee canister that is a cube with edge 16.5 centimeters.

Answers are on page 572.

Finding Volumes of Cylinders and Spheres

A **cylinder** is a solid with a top and a bottom that are circles of the same radius. Perhaps some everyday examples will help us picture the cylinder. A can of soup, a duffel bag, a hot dog, and a garden hose are all cylinders.

The volume of a cylinder depends upon its *radius* and the distance between its top and its bottom. That distance is usually called the *height* of the cylinder. The volume of a cylinder is found by the following formula.

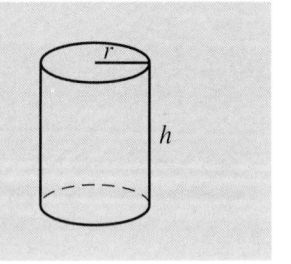

Volume of a Cylinder. If a cylinder has radius r and height h, then

$$\text{Volume} = \pi \cdot r \cdot r \cdot h$$
$$= \pi \cdot r^2 \cdot h$$

Example 4. Find the volume of a soda can that has a radius of $1\frac{1}{2}$ inches and a height of 5 inches. Use $\pi \doteq \frac{22}{7}$.

Solution. The soda can is a cylinder.

$$\text{Volume} = \pi \cdot r \cdot r \cdot h$$

$$= \pi \cdot 1\frac{1}{2} \cdot 1\frac{1}{2} \cdot 5 \qquad \text{Substitute } 1\frac{1}{2} \text{ for } r \text{ and 5 for } h.$$

$$\doteq \frac{\overset{11}{\cancel{22}}}{7} \cdot \frac{3}{\underset{1}{\cancel{2}}} \cdot \frac{3}{2} \cdot \frac{5}{1} \qquad \begin{array}{l}\text{Change mixed numbers to} \\ \text{improper fractions. Reduce.}\end{array}$$

$$\doteq \frac{495}{14} \text{ cu in.} \qquad \begin{array}{l}\text{Find the product and} \\ \text{attach units.}\end{array}$$

$$\text{Volume} \doteq 35\frac{5}{14} \text{ cu in.} \qquad \begin{array}{l}\text{Change improper fraction to} \\ \text{mixed number.}\end{array}$$

The volume of the soda can is about $35\frac{5}{14}$ cubic inches.

Example 5. If a garden hose with a diameter of 1 inch is 50 feet long, what volume of water will the hose hold? Use $\pi \doteq \frac{22}{7}$.

Solution. Before finding the volume, we need to be sure our units match. Let's change the length to *inches*.

$$50 \text{ ft} = 50 \times \overset{1}{\cancel{1\text{ ft}}} \times \frac{12 \text{ in.}}{\underset{1}{\cancel{1\text{ ft}}}}$$

$$= 600 \text{ in.}$$

We know the *diameter* of the hose is 1 inch so the *radius* must be $\frac{1}{2}$ inch. A garden hose is a cylinder, so

$$\text{Volume} = \pi \cdot r \cdot r \cdot h$$

$$\doteq \frac{\overset{11}{\cancel{22}}}{7} \cdot \frac{1}{\underset{1}{\cancel{2}}} \cdot \frac{1}{\underset{1}{\cancel{2}}} \cdot \frac{\overset{300}{\cancel{600}}}{1}$$

$$\doteq \frac{3300}{7} \text{ cu in.}$$

$$\text{Volume} \doteq 471\frac{3}{7} \text{ cu in.}$$

The garden hose will hold about $471\frac{3}{7}$ cubic inches of water.

Now let's solve the problem stated at the beginning of this chapter.

Example 6. James has a cylindrical duffel bag with a diameter of 32 centimeters and a height of 90 centimeters. Tara has a tote bag that is 90 centimeters long, 30 centimeters wide, and 27 centimeters deep. Which bag will hold *more*?

Solution. We must compare the *volume* of the duffel bag (a cylinder) to the volume of the tote bag (a rectangular solid).

For the duffel bag, the diameter is 32 cm, so the radius is 16 cm. Let's use $\pi \doteq 3.14$.

$$\text{Volume} = \pi \cdot r \cdot r \cdot h$$
$$\doteq 3.14 \cdot 16 \cdot 16 \cdot 90$$
$$\doteq 72,345.6 \text{ cu cm}$$

For the tote bag, we know the length, width, and height.

$$\text{Volume} = l \cdot w \cdot h$$
$$= 90 \cdot 30 \cdot 27$$
$$= 72,900 \text{ cu cm}$$

Since $72,900 > 72,345.6$, Tara's tote bag will hold more.

A **sphere** is a ball. The distance from any point on the outer surface of a sphere to the point in the center of the sphere is called its radius. The volume of a sphere is found by the following formula.

Volume of a Sphere. If a sphere has radius r, then

$$\text{Volume} = \frac{4}{3} \cdot \pi \cdot r \cdot r \cdot r$$

$$= \frac{4}{3} \cdot \pi \cdot r^3$$

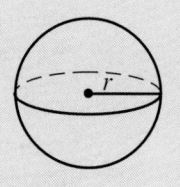

Example 7. If the radius of a golf ball is 2 centimeters, find its volume. Use $\pi \doteq \frac{22}{7}$.

Solution

$$\text{Volume} = \frac{4}{3} \cdot \pi \cdot r \cdot r \cdot r$$

$$\doteq \frac{4}{3} \cdot \frac{22}{7} \cdot 2 \cdot 2 \cdot 2$$

$$\doteq \frac{704}{21} \text{ cu cm}$$

$$\text{Volume} \doteq 33\frac{11}{21} \text{ cu cm}$$

The volume of the golf ball is about $33\frac{11}{21}$ cubic centimeters.

⇒ Trial Run

_____ **1.** Find the volume of a sphere with a radius of 9 feet. Use $\pi \doteq \frac{22}{7}$.

_____ **2.** Find the volume of a cylindrical water heater that has a diameter of 2 feet and a height of $4\frac{1}{2}$ feet. Use $\pi \doteq \frac{22}{7}$.

_____ **3.** Find the volume of a hot-air balloon that is a sphere with a diameter of 9 meters. Use $\pi \doteq 3.14$.

_____ **4.** A storage tank for gas has a radius of 180 feet and a height of 500 feet. What is its volume? Use $\pi \doteq 3.14$.

Answers are on page 572.

Relating Volume to Capacity

In Chapter 12 we discussed American and metric units of capacity (cups, quarts, gallons, milliliters, liters, and so on). The ideas of capacity and volume are very closely related. We will explore that relationship here.

Suppose we fill with water a cube measuring 1 centimeter on each edge. If we pour that water into a container marked off in milliliters, we would discover that the water would reach the 1 milliliter mark in the container. Because the volume of the cube is 1 cubic centimeter, we conclude that

$$1 \text{ cu cm} = 1 \text{ ml}$$

In the medical and health fields, these 2 measurements are used interchangeably. Doses of medicine are prescribed using either cubic centimeters or milliliters. You may know that a more common abbreviation for cubic centimeters in the medical field is **cc.** Thus

$$1 \text{ cc} = 1 \text{ ml}$$

This relationship gives us another useful conversion factor.

$$\frac{1 \text{ cc}}{1 \text{ ml}} \qquad \frac{1 \text{ ml}}{1 \text{ cc}}$$

Example 8. If a patient needs 0.1 cc of medicine every 2 hours, how many milliliters of medicine will the patient need each day?

Solution. Since the medicine is taken every 2 hours, the patient will need 12 doses in 1 day.

$$\text{Total amount} = 12 \, (0.1 \text{ cc})$$

$$= 1.2 \text{ cc}$$

Now we must change our units to milliliters. Knowing that 1 cc = 1 ml, we conclude that 1.2 cc = 1.2 ml. The patient needs 1.2 milliliters of medicine each day.

Let's consider a cube whose edges measure 10 centimeters and see if we can change the volume of this cube to milliliters.

$$\text{Volume} = e \cdot e \cdot e$$

$$= 10 \cdot 10 \cdot 10$$

$$= 1000 \text{ cu cm}$$

$$\text{Volume} = 1000 \text{ cc}$$

$$\text{Capacity} = 1000 \text{ ml} \qquad \text{Remember that 1 cc = 1 ml.}$$

$$= 1 \, \ell \qquad \text{Remember that 1000 ml = 1 } \ell.$$

So a cube with a volume of 1000 cubic centimeters has a capacity of 1 liter.

$$1000 \text{ cc} = 1 \, \ell$$

Example 9. Change 1.7 liters to cubic centimeters.

Solution. We will need to use 2 conversion factors.

$$1.7 \, \ell = 1.7 \times 1 \, \ell \times \frac{1000 \text{ ml}}{1 \, \ell} \times \frac{1 \text{ cc}}{1 \text{ ml}}$$

$$= \frac{1.7}{1} \times \frac{\cancel{1 \ell}^{\,1}}{1} \times \frac{\overset{1000}{\cancel{1000 \text{ ml}}}}{\underset{1}{\cancel{1 \ell}}} \times \frac{1 \text{ cc}}{\underset{1}{\cancel{1 \text{ ml}}}}$$

$$= 1.7 \, (1000 \text{ cc})$$

$$1.7 \, \ell = 1700 \text{ cc}$$

There are also conversion factors that can be used to change volume to capacity with American units. However, they are not used very often in everyday life. For instance, very few people know (or care) that 1 quart of water has a volume of 57.75 cubic inches. We will not concern ourselves with conversions among American units. If you are curious about them, you may look up "measure" in *Webster's Collegiate Dictionary* for a more complete list of conversions.

⫸ Trial Run

_____ 1. Change 2.3 cc to milliliters.

_____ 2. Change 3 milliliters to cubic centimeters.

_____ 3. Change 0.5 liter to cubic centimeters.

_____ 4. Change 4500 cc to liters.

_____ 5. If the volume of a perfume bottle is 29.6 cubic centimeters, how many milliliters of perfume will it hold?

_____ 6. If each bottle of contact lens solution holds 60 cc of fluid, how many liters of fluid will be needed to fill 5000 bottles?

Answers are given below.

▶ Example You Completed

Example 2. Which box will hold more: a cube with edge 1.7 meters or a rectangular solid that is 1 meter wide, 3 meters long, and 1.6 meters high?

Solution. We must find the volume of each box.

Cube	_Rectangular Solid_
Volume $= e \cdot e \cdot e$	Volume $= l \cdot w \cdot h$
$= (1.7)(1.7)(1.7)$	$= 1\,(3)(1.6)$
Volume $= 4.913$ cu m	Volume $= 4.8$ cu m

The box that is a cube will hold more. How much more?

$$4.913 - 4.800 = 0.113 \text{ cu m}$$

Answers to Trial Runs

page 567   1. 91,000 cu cm   2. 20 cu ft   3. The square pan is larger by 2 cu in.   4. 0.216 cu m
5. 4492.125 cu cm

page 570   1. $3054\frac{6}{7}$ cu ft   2. $14\frac{1}{7}$ cu ft   3. 381.51 cu m   4. 50,868 cu ft

page 572   1. 2.3 ml   2. 3 cc   3. 500 cc   4. 4.5 ℓ   5. 29.6 ml   6. 300 ℓ

EXERCISE SET 13.3

_____ 1. Find the volume of a rectangular solid with length 5 feet, width $1\frac{1}{2}$ feet, and height 3 feet.

_____ 2. Find the volume of a rectangular solid with length 8 meters, width 3.5 meters, and height 7 meters.

_____ 3. Find the volume of a cube with edge 15.3 centimeters.

_____ 4. Find the volume of a cube with edge 9 inches.

_____ 5. The Willet children have a wading pool that is 30 feet long, 15 feet wide, and 30 inches deep. Find the volume of the pool in cubic feet.

_____ 6. The Belvederes are building a rectangular patio with length 50 feet and width 20 feet. The concrete patio floor is to be 6 inches thick. Find the number of cubic feet of concrete that the Belvederes will need.

_____ 7. Karla has a cube-shaped storage box with edge 18 inches. Find the volume of the storage box.

_____ 8. Find the volume of a jewelry box that is a cube with edge 15 centimeters.

_____ 9. Florence has an antique trunk that is 3 feet long, 2 feet wide, and $2\frac{1}{2}$ feet deep. Find the volume of the trunk.

_____ 10. Carlotta has a hot tub that is 8 feet long, 6 feet wide, and $3\frac{1}{2}$ feet deep. What is the volume of the tub?

_____ 11. Find the volume of a cylinder with radius 6 centimeters and height 15 centimeters. (Use $\pi \doteq 3.14$.)

_____ 12. Find the volume of a cylinder with radius 22 inches and height 6 inches. (Use $\pi \doteq 3.14$.)

_____ 13. Find the volume of a sphere with radius 32 feet. (Use $\pi \doteq 3.14$.)

_____ 14. Find the volume of a sphere with radius 8 meters. (Use $\pi \doteq 3.14$.)

_____ 15. An aluminum can in the shape of a cylinder has diameter 4 centimeters and height 15 centimeters. Find its volume. (Use $\pi \doteq 3.14$.)

_____ 16. Mr. Greenwell has a silo on his farm in the shape of a cylinder. The diameter is 20 feet and the height is 35 feet. Find the volume of the silo. (Use $\pi \doteq \frac{22}{7}$.)

_____ 17. A spherical tank has a diameter of 8.1 meters. Find the volume of the tank. (Use $\pi \doteq 3.14$.)

_____ 18. Find the volume of a spherical balloon if it is inflated so that it has a diameter of 12 inches. (Use $\pi \doteq 3.14$.)

_____ 19. Find the volume of a jelly jar with a radius of 3.5 centimeters and a height of 10.5 centimeters. (Use $\pi \doteq 3.14$.)

_____ 20. Find the volume of a cylindrical canister with radius 6 inches and height 9.5 inches.

_____ 21. Change 4.2 cc to milliliters.

_____ 22. Change 8.3 cc to milliliters.

_____ 23. Change 0.3 liter to cubic centimeters.

_____ 24. Change 0.8 liter to cubic centimeters.

_____ 25. Change 5000 cubic centimeters to liters.

_____ 26. Change 2500 cubic centimeters to liters.

_____ 27. How many liters of water will an aquarium hold if the tank is 50 centimeters long, 42 centimeters wide, and 20 centimeters high?

_____ 28. A milk carton has a square base that is 9 centimeters on each side and a height that is 18 centimeters. What is the volume of the milk carton in liters?

_____ 29. If an eye dropper contains 0.2 cc of medication, how many milliliters of medication will be in a dosage of 10 eye droppers?

_____ 30. If each bottle of Liquid Paper contains 150 cc of correction fluid, how many liters of fluid will be needed to fill 2500 bottles?

☆ Stretching the Topics _____

_____ 1. If a ball with a diameter of 12 inches is packed in a box that is a cube with edge 12 inches, how much space in the box is empty?

_____ 2. If the volume of a sphere is 1078 cubic yards, find the volume when the radius is doubled.

_____ 3. A water bed is 3 meters long, 1.5 meters wide, and 25 centimeters deep. How many liters of water will be needed to fill the water bed?

Check your answers in the back of your book.

If you can complete **Checkup 13.3,** then you are ready to go on to Section 13.4.

✓ CHECKUP 13.3

_____ 1. Find the volume of a cubic shed measuring 8 feet on each edge.

_____ 2. Find the volume of a milk carton that is 19.5 centimeters tall with a square base having 10-centimeter sides.

_____ 3. A water tank is 21 centimeters long, 15 centimeters wide, and 6.2 centimeters deep. Find the volume of the tank.

_____ 4. A cylindrical cardboard mailing tube has diameter 3 inches and length 20 inches. Find the volume of the tube. (Use $\pi \doteq 3.14$.)

_____ 5. Find the volume of a sphere with radius 20 inches. (Use $\pi \doteq 3.14$.)

_____ 6. Change 4.5 cc to milliliters.

_____ 7. Change 2.3 liters to cubic centimeters.

_____ 8. A pharmacist had 6 liters of a certain medication at the beginning of the month. During the month he filled prescriptions totaling 3650 milliliters. How many liters did he have in stock at the end of the month?

Check your answers in the back of your book.

If You Missed Problems:	You Should Review Examples:
1	2
2, 3	1, 3
4	4, 5
5	7
6, 7	8, 9
8	8

13.4 Taking a Closer Look at Right Triangles

We have already learned quite a bit about triangles. We know how to find the perimeter and the area of any triangle. Now it is time to take a closer look at the parts of a triangle.

Learning the Parts of a Triangle

A triangle is made up of 3 straight sides that meet at 3 points. Each point is called a **vertex.** The vertices are often labeled with capital letters A, B, and C. The side opposite vertex A is labeled side a, the side opposite vertex B is labeled side b, and the side opposite vertex C is labeled side c. A triangle that is labeled in this way is named triangle ABC (or $\triangle ABC$).

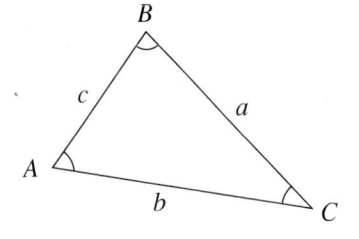

At each vertex there is an **angle** which is labeled to match the label of the vertex ($\angle A$, $\angle B$, $\angle C$).

Every angle can be measured in **degrees** using a mathematical tool called a **protractor.** To use a protractor to measure an angle, we line up its bottom edge with 1 side of the angle, being sure that the center dot is at the vertex. Then we read the number of degrees in the angle by seeing where the other side of the angle crosses the protractor. Let's measure all 3 angles in the following triangle.

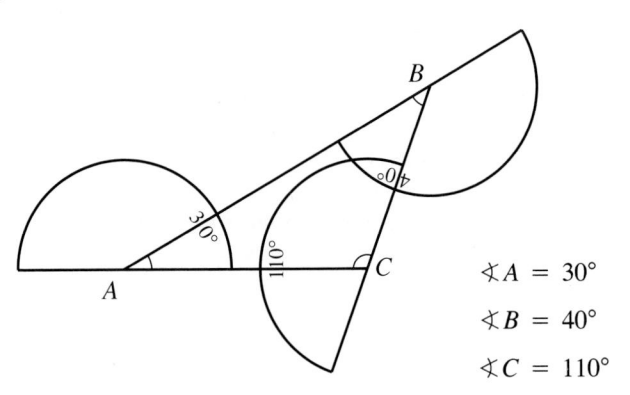

$\angle A = 30°$

$\angle B = 40°$

$\angle C = 110°$

Notice that the sum of the number of degrees in the 3 angles of this triangle is

$$30° + 40° + 110° = 180°$$

Indeed it will always be true that the sum of the number of degrees in the 3 angles in any triangle is 180°.

> For any $\triangle ABC$
>
> $\angle A + \angle B + \angle C = 180°$

Example 1. In triangle ABC, $\angle A = 55°$ and $\angle C = 47°$. Find $\angle B$.

Solution. We know that $\angle A + \angle B + \angle C = 180°$, and we are told

$$\angle A + \angle C = 55° + 47°$$
$$= 102°$$

To find the measure of the missing angle, we must find the difference between $180°$ and $102°$.

$$\angle B = 180° - 102°$$
$$= 78°$$

Let's check this result by finding the sum of the degrees in all 3 angles.

$$55° + 47° + 78° = 180° \checkmark$$

In the following triangle, let's use a protractor to measure the number of degrees in $\angle C$.

We see that $\angle C$ measures $90°$.
Such an angle is called
a **right angle.**

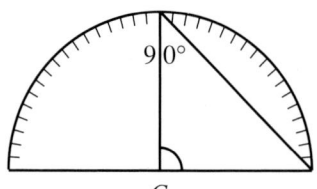

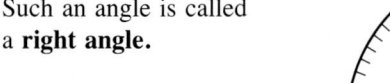

C

A **right angle** is an angle with measure $90°$. A **right triangle** is a triangle that contains a right angle.

Whenever an angle is marked in this way └┘ , you will know that it is a right angle and that its measure is $90°$.

Example 2. Find the measure of the missing angle in this right triangle.

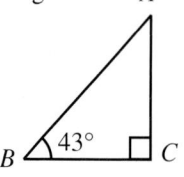

A

B $43°$ C

Solution. We know here that $\angle C = 90°$ and $\angle B = 43°$.

$$\angle C + \angle B = 90° + 43° \qquad \text{Find the sum of the given angles.}$$
$$= 133°$$

Since the sum of the degrees in all 3 angles in any triangle is $180°$, we can find angle A.

$$\angle A = 180° - 133° \qquad \text{Subtract the sum from } 180°.$$
$$= 47°$$

Example 3. In $\triangle ABC$, $\angle A = 81°$ and $\angle B = 9°$. Show that $\triangle ABC$ is a right triangle.

Solution. To show that $\triangle ABC$ is a right triangle, we must show that it contains a right angle. Since $\angle A$ and $\angle B$ are not right angles, we must see if $\angle C$ is a right angle.

$$\angle A + \angle B = 81° + 9° \qquad \text{Find the sum of the given angles.}$$
$$= 90°$$
$$\angle C = 180° - 90° \qquad \text{Subtract the sum from } 180°$$
$$= 90°$$

Since $\angle C = 90°$, we know $\angle C$ is a right angle. Therefore $\triangle ABC$ must be a right triangle.

In a *right triangle*, the parts of the triangle are given special names. Usually the right angle is labeled $\angle C$. Side c, which is opposite the right angle, is called the **hypotenuse** of the right triangle. Sides a and b are called the **legs** of the right triangle.

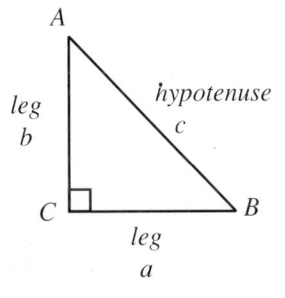

Example 4. In right triangle ABC, $\angle C = 90°$, $a = 3$ feet, $b = 4$ feet, and $c = 5$ feet. What is the length of the hypotenuse?

Solution. Side c is opposite the right angle, so side c is the hypotenuse. The length of the hypotenuse is 5 feet.

Notice that the hypotenuse is always the *longest* side in a right triangle.

⫸ Trial Run

_____ 1. In $\triangle ABC$, $\angle A = 38°$ and $\angle B = 67°$. Find the measure of $\angle C$.

_____ 2. In a right triangle ABC, $\angle C$ is the right angle and $\angle B = 54°$. Find the measure of $\angle A$.

_____ 3. In $\triangle ABC$, $\angle A = 65°$ and $\angle B = 25°$. Find the measure of $\angle C$. Is $\triangle ABC$ a
_____ right triangle?

_____ 4. In right triangle ABC, $\angle C = 90°$, $a = 5$ inches, $b = 12$ inches, and $c = 13$ inches. What is the length of the hypotenuse?

_____ 5. Is it possible to have a triangle with $\angle A = 15°$, $\angle B = 83°$, and $\angle C = 72°$? Why?

Answers are on page 582.

Finding the Sides of a Right Triangle

In this section we discuss one of the most important formulas of geometry. To introduce this formula, we take a look at the right triangle ABC in which the lengths of the 3 sides are known and $\angle C$ is the right angle. Suppose we attach a square to each side of this triangle.

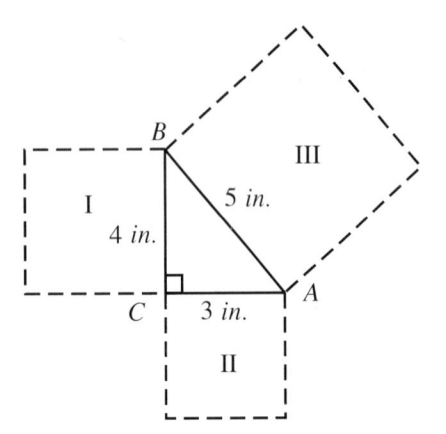

What is the area of square I?	What is the area of square II?	What is the area of square III?
Area I $= a^2$	Area II $= b^2$	Area III $= c^2$
$= 4 \times 4$	$= 3 \times 3$	$= 5 \times 5$
$= 16$ sq in.	$= 9$ sq in.	$= 25$ sq in.

Notice that

$$\text{Area I } + \text{ Area II } = \text{ Area III}$$

$$16 \text{ sq in. } + 9 \text{ sq in. } = 25 \text{ sq in.}$$

This is not just a coincidence. In fact, whenever we are dealing with any right triangle ABC, where a and b are the legs and c is the hypotenuse, it will always be true that $a^2 + b^2 = c^2$. On the other hand, if we can show that $a^2 + b^2 = c^2$, then we may conclude that triangle ABC must be a right triangle. This formula is called the **Pythagorean Formula** in honor of the Greek mathematician Pythagoras.

> **Pythagorean Formula.** In right triangle ABC, with legs a and b, and hypotenuse c,
>
> $$a^2 + b^2 = c^2$$

Example 5. Show that the triangle in which $a = 5$, $b = 12$, and $c = 13$ is a right triangle.

Solution. We must check these numbers in the Pythagorean Formula.

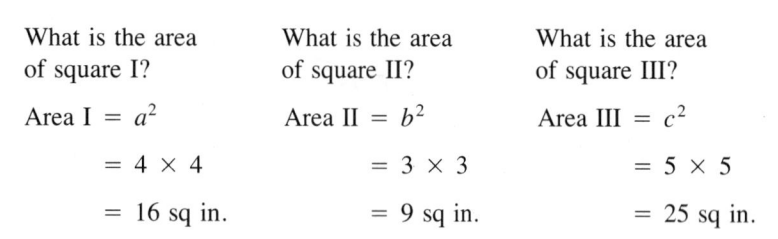

$$a^2 + b^2 \stackrel{?}{=} c^2$$

$$5^2 + 12^2 \stackrel{?}{=} 13^2$$

$$(5 \cdot 5) + (12 \cdot 12) \stackrel{?}{=} 13 \cdot 13$$

$$25 + 144 \stackrel{?}{=} 169$$

$$169 = 169$$

Since these numbers "work" in the Pythagorean Formula, we know triangle *ABC* is a right triangle.

Before we use the Pythagorean Formula again, remember that there is a list of squares for the first 26 whole numbers inside the back cover of your book. Do not try to memorize this table. But it is helpful to be able to refer to it when working problems involving squares of numbers.

We can use the Pythagorean Formula to find the length of the hypotenuse of a right triangle as long as we know the other 2 legs.

Example 6. Suppose in a right triangle leg $a = 9$ meters and leg $b = 12$ meters. Let's try to find the length of the hypotenuse, c.

Solution. We know that $a^2 + b^2 = c^2$, so

$$9^2 + 12^2 = c^2$$ Substitute the values for a and b.

$$81 + 144 = c^2$$ Find the squares in the table.

$$225 = c^2$$ Find the sum on the left.

At this point in our work we know that when the hypotenuse c is *squared,* the result is 225. Look back at the Table of Squares. What number squared is 225? The number is 15. The length of the hypotenuse, c, is 15 meters.

When we found the positive number whose square is 225, we were finding the principal **square root** of 225. The symbol used for the square root of 225 is $\sqrt{225}$. We have found that $\sqrt{225} = 15$.

Example 7. A ladder is leaning against the side of a building. The foot of the ladder is 5 feet from the building. The top of the ladder reaches a window 12 feet above the ground. How long is the ladder?

Solution. A picture can certainly help us get started on this problem.

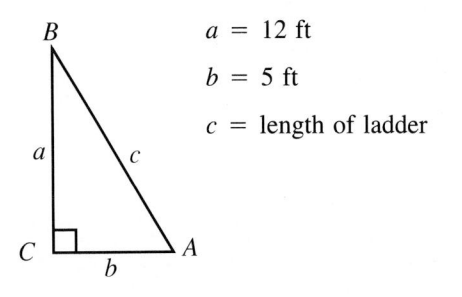

$a = 12$ ft

$b = 5$ ft

$c = $ length of ladder

Using the Pythagorean Formula, we have

$$a^2 + b^2 = c^2.$$

$$12^2 + 5^2 = c^2$$ Substitute the values for a and b.

$$144 + 25 = c^2$$ Find the squares in the table.

$$169 = c^2$$ Find the sum on the left.

$$\sqrt{169} = c$$ We must find the square root of 169.

$$13 = c$$ Use the table to find a number whose square is 169.

The ladder must be 13 feet long.

⇒ Trial Run

_____ 1. Find the value of 14^2.

_____ 2. Find the value of 25^2.

_____ 3. Using the Pythagorean Theorem, decide if the triangle with $a = 5$ centimeters, $b = 12$ centimeters, and $c = 13$ centimeters is a right triangle.

_____ 4. Find the length of the hypotenuse of a right triangle with $a = 6$ meters and $b = 8$ meters.

_____ 5. A sail on a yacht is a right triangle as shown. Find the length of the hypotenuse.

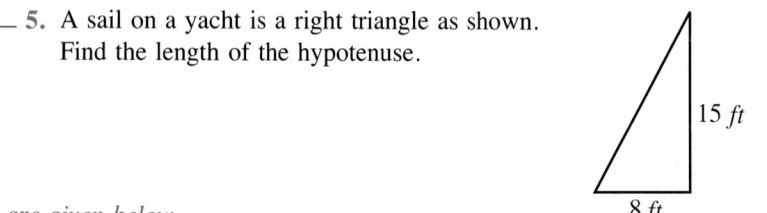

15 ft

8 ft

Answers are given below.

Answers to Trial Runs

page 579   1. 75°   2. 36°   3. 90°; yes   4. 13 in.   5. No; the sum of the angles is less than 180°.

page 582   1. 196   2. 625   3. Yes, $5^2 + 12^2 = 13^2$ 4. 10 m   5. 17 ft

EXERCISE SET 13.4

_____ 1. In △ABC, ∡A = 47° and ∡B = 54°. Find ∡C.

_____ 2. In △ABC, ∡A = 35° and ∡C = 84°. Find ∡B.

_____ 3. In right triangle ABC, ∡C is the right angle and ∡B = 38°. Find ∡A.

_____ 4. In right triangle ABC, ∡C is the right angle and ∡A = 75°. Find ∡B.

_____
_____ 5. In △ABC, ∡A = 53° and ∡B = 37°. Find ∡C. Is triangle ABC a right triangle?

_____
_____ 6. In △ABC, ∡A = 72° and ∡B = 28°. Find ∡C. Is triangle ABC a right triangle?

_____ 7. Is the triangle in which $a = 3$ feet, $b = 4$ feet, and $c = 7$ feet a right triangle?

_____ 8. Is the triangle in which $a = 12$ meters, $b = 9$ meters, and $c = 15$ meters a right triangle?

_____ 9. Find the length of the hypotenuse in a right triangle with $a = 8$ yards and $b = 15$ yards.

_____ 10. Find the length of the hypotenuse in a right triangle with $a = 12$ centimeters and $b = 16$ centimeters.

_____ 11. A ladder is leaning against the side of a building. The foot of the ladder is 7 feet from the building. The top of the ladder reaches a window 24 feet above the ground. How long is the ladder?

_____ 12. A ramp leads to a building's entrance. The foot of the ramp is 12 feet from the building. The entrance to the building is 5 feet above the ground. How long is the ramp?

_____ 13. A guy wire holds a television antenna in place. The antenna is 15 feet tall and the guy wire is anchored 8 feet from the base of the antenna. How long is the guy wire?

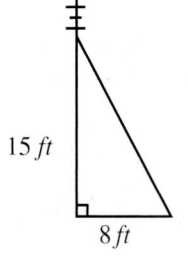

15 ft

8 ft

_____ 14. A car is driven 12 miles east and then 5 miles south. How far is the car from its starting point?

12 mi

5 mi

———— 15. The sail used on a boat is a right triangle. If two sides of the sail are 9 feet and 12 feet, find the length of the hypotenuse.

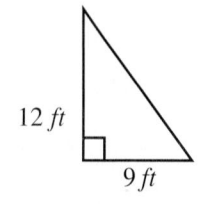

12 ft

9 ft

☆ Stretching the Topics

———— 1. A 20-foot ladder is placed against a building. The foot of the ladder is 12 feet from the base of the building. How far up on the building is the top of the ladder?

———— 2. A fence is built around a plot of land. How much fencing will be needed if the plot has the dimensions shown in the illustration?

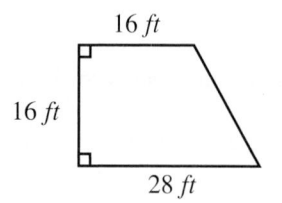

16 ft

16 ft

28 ft

———— 3. Find the perimeter and area of triangle *ABC* shown in the illustration.

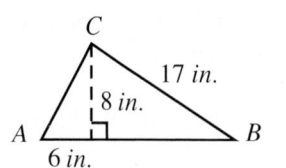

C

17 in.

8 in.

A

B

6 in.

Check your answers in the back of your book.

If you can complete **Checkup 13.4,** then you are ready to do the **Review Exercises** for Chapter 13.

✓ **CHECKUP 13.4**

_____ 1. In $\triangle ABC$, $\angle A = 93°$ and $\angle B = 37°$. Find $\angle C$.

_____ 2. In right triangle ABC, $\angle C$ is a right angle and $\angle B = 49°$. Find $\angle A$.

_____ 3. In $\triangle ABC$, $\angle A = 64°$ and $\angle B = 26°$. Find $\angle C$. Is triangle ABC a right
_____ triangle?

_____ 4. Is the triangle in which $a = 8$ meters, $b = 15$ meters, and $c = 17$ meters a right
triangle?

_____ 5. Find the length of the hypotenuse of a right triangle with $a = 5$ feet and
$b = 12$ feet.

_____ 6. If a bus travels 24 miles due east and 7 miles due south, how far is the bus from its
starting point?

Check your answers in the back of your book.

If You Missed Problems:	You Should Review Examples:
1	1
2	2
3	3
4	5
5, 6	6, 7

Summary

In this chapter we learned to find the distance around the outside of flat geometric figures (**perimeter** or **circumference**) and the **area** enclosed within them. We discovered that perimeter and circumference are always measured in *linear* units, whereas area is always measured in *square* units.

Figure	Name	Perimeter	Area
	square	$P = 4 \cdot s$	$A = s \cdot s$ $= s^2$
	rectangle	$P = 2 \cdot l + 2 \cdot w$	$A = l \cdot w$
	triangle	$P = a + b + c$	$A = \dfrac{b \cdot h}{2}$
	circle	$C = \pi \cdot d$ $= 2 \cdot \pi \cdot r$	$A = \pi \cdot r \cdot r$ $= \pi \cdot r^2$

We then looked at solid geometric figures and learned to find their **volumes,** which are always measured in *cubic* units.

Figure	Name	Volume
e ... *e* ... *e* (cube)	cube	$V = e \cdot e \cdot e$ $= e^3$
h ... *w* ... *l* (rectangular solid)	rectangular solid	$V = l \cdot w \cdot h$
r ... *h* (cylinder)	cylinder	$V = \pi \cdot r \cdot r \cdot h$ $= \pi \cdot r^2 \cdot h$
r (sphere)	sphere	$V = \dfrac{4}{3} \cdot \pi \cdot r \cdot r \cdot r$ $= \dfrac{4}{3} \cdot \pi \cdot r^3$

We discussed the close relationship between volume and capacity, especially with metric units, and we noted that

$$1 \text{ cc} = 1 \text{ ml}$$

$$1000 \text{ cc} = 1 \ \ell$$

Finally we studied the triangle in some detail and learned that in any $\triangle ABC$

$$\angle A + \angle B + \angle C = 180°$$

For a **right triangle** (a triangle containing a 90° angle) we discovered the usefulness of the Pythagorean Formula.

$$a^2 + b^2 = c^2$$

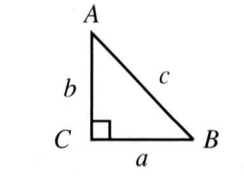

❑ Speaking the Language of Mathematics ━━━━━━━━━━

Complete each sentence with the appropriate word or phrase.

1. A flat closed figure with straight sides is called a _____ .

2. If we wish to fence in a yard, we must measure the yard's _____ . If we wish to cover the yard with sod, we must measure the yard's _____ .

3. Area is always measured in _____ units. Volume is always measured in _____ units.

4. A triangle in which the measure of one of the angles is 90° is called a _____ _____ . The side opposite the 90° angle is called the _____ .

5. The formula that states that the sum of the square of the legs of a right triangle is equal to the square of the hypotenuse is called the _____ Formula.

△ Writing About Mathematics ━━━━━━━━━━

Write your response to each question in complete sentences.

1. Tell what is *wrong* with each of these statements.
   (a) Carla needs 20 feet of carpet to cover her bedroom floor.
   (b) The area of a circle of radius r can be found by multiplying $2 \cdot \pi \cdot r$.
   (c) To find the perimeter of a rectangular field, we multiply its length times its width.
   (d) In triangle ABC, $\angle A = 17°$, $\angle B = 92°$, and $\angle C = 81°$.

2. To the question "What is the circumference of a circle with radius 7 inches?", Mary answers $C = 44$ inches and John answers $C = 43.96$ inches. Why are both of these answers incorrect?

3. State the Pythagorean Formula in your own words.

REVIEW EXERCISES for Chapter 13

Name each figure and find its perimeter.

_____ 1.

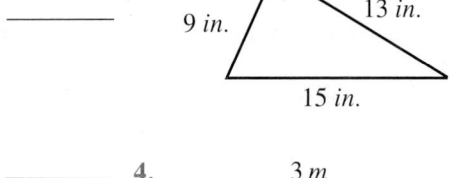

_____ 2.

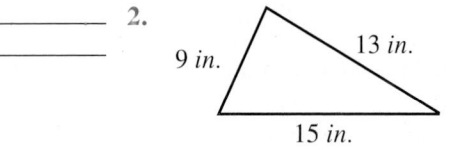

_____ 3.

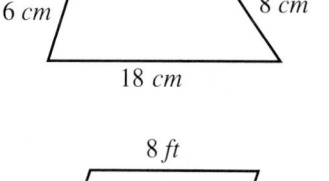

_____ 4.

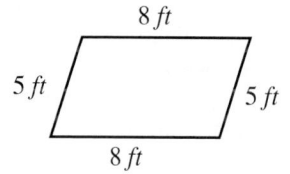

_____ 5. Find the circumference of a circle with diameter 13 feet. (Use $\pi \doteq 3.14$.)

_____ 6. Find the circumference of a circle with radius 42 centimeters. (Use $\pi \doteq \frac{22}{7}$.)

_____ 7. A chain link fence is to be put around a rectangular garden that is 36 feet wide and 25 feet long. How many feet of fencing are needed?

_____ 8. A square picture that is 12 inches on each side is to be framed. How many inches of framing will be needed?

_____ 9. Find the distance around the figure illustrated.

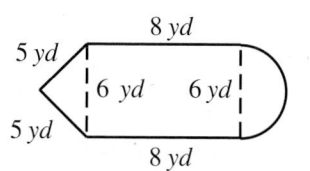

_____ 10. The base of a stereo turntable is a square 16 inches on each side. Find the area it will cover on a shelf.

_____ 11. Find the number of square feet of formica needed to cover a kitchen countertop that measures 2 feet by $10\frac{1}{2}$ feet.

_____ 12. Find the area of a triangle with base 24 meters and height 17 meters.

_____ 13. A glass top is to be cut to fit a vanity table shaped as in the figure at the right. Find the area of the top of the vanity.

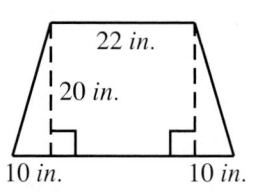

_____ 14. How many square feet of plastic are needed to cover a circular swimming pool that has a diameter of 36 feet? (Use $\pi \doteq 3.14$.)

_____ 15. Change 8 square yards to square feet.

_____ 16. Change 5285 square centimeters to square meters.

_____ 17. In problem 14, find the number of square yards of plastic that will be needed.

_____ 18. How many square meters of wall area will a square mirror cover if the mirror is 60 centimeters on each side?

_____ 19. Find the volume of a cube with edge 12 centimeters.

_____ 20. How many cubic feet of concrete are needed to pour a rectangular patio that is 15 feet long, 9 feet wide, and 4 inches thick?

_____ 21. Find the volume of a cylinder with diameter 7.8 centimeters and height 8.2 centimeters. (Use $\pi \doteq 3.14$. Round answer to the tenths place.)

_____ 22. Find the volume of a paper cup in the shape of a cylinder if the radius is 2 inches and the height is 6 inches. (Use $\pi \doteq 3.14$.)

_____ 23. The diameter of a spherical ball is 28 centimeters. Find its volume. (Use $\pi \doteq \frac{22}{7}$.)

_____ 24. Change 3.8 cc to milliliters.

_____ 25. Change 0.8 liter to cubic centimeters.

_____ 26. Change 1500 cc to liters.

_____ 27. If 1 perfume bottle contains 0.35 cc, how many liters of perfume would be needed to fill 2000 bottles?

_____ 28. In $\triangle ABC$, $\angle A = 53°$ and $\angle B = 98°$. Find $\angle C$.

_____ 29. In right triangle ABC, $\angle C$ is a right angle and $\angle B = 63°$. Find $\angle A$.

_____ 30. Find the length of the hypotenuse of a right triangle with $a = 8$ inches and $b = 6$ inches.

Check your answers in the back of your book.

If You Missed Exercises:	You Should Review Examples:	
1–4	Section 13.1	1–5
5, 6		6, 7
7–9		8
10–13	Section 13.2	1–4
14		5, 6
15–18		8–11
19, 20	Section 13.3	1–3
21, 22		4, 5
23		7
24–27		8, 9
28, 29	Section 13.4	1–3
30		6

If you have completed the **Review Exercises** and corrected your errors, you are ready to do the **Practice Test** for Chapter 13.

PRACTICE TEST for Chapter 13

		SECTION	EXAMPLES

_____ 1. Find the perimeter of a square with side 2.3 millimeters.   **13.1**   **4**

_____ 2. Find the perimeter of a rectangle that is 5 feet long and $3\frac{1}{2}$ feet wide.   **13.1**   **5**

_____ 3. Find the circumference of a circle with diameter 16 inches. (Use $\pi \doteq 3.14$.)   **13.1**   **6**

_____ 4. Find the distance around the figure illustrated.   **13.1**   **8, 9**

4 m   4 m
4 m

_____ 5. Find the area of a square with side $5\frac{1}{2}$ meters.   **13.2**   **1**

_____ 6. Find the number of square feet of carpet needed to carpet a room that is 18 feet long and 12 feet wide.   **13.2**   **2**

_____ 7. Find the area of the figure illustrated.   **13.2**   **4**

6 in.
6 in.
8 in.

_____ 8. Find the floor area covered by a circular rug with diameter 6 feet. (Use $\pi \doteq 3.14$.)   **13.2**   **5, 6**

_____ 9. Change 4 square yards to square inches.   **13.2**   **9**

_____ 10. If a circle has radius 0.06 kilometer, find its area in square meters. (Use $\pi \doteq 3.14$.)   **13.2**   **11**

_____ 11. Find the volume of a storage cube with edge 18 inches.   **13.3**   **2**

_____ 12. A truck bed is 2 yards long, $1\frac{1}{2}$ yards wide, and 2 feet deep. How many cubic yards will it hold?   **13.2**   **3**

_____ 13. Find the volume of a cylindrical bucket that has radius 7 inches and height 24 inches. (Use $\pi \doteq \frac{22}{7}$.)   **13.3**   **4**

_____ 14. Find the volume of a sphere with diameter 18 centimeters. (Use $\pi \doteq 3.14$.)   **13.3**   **7**

_____ 15. Change 1800 cc to liters.   **13.3**   **9**

		SECTION	EXAMPLE

_____ **16.** If a patient needs 0.2 cc of medicine every 3 hours, how many milliliters of medicine will the patient need during a 48-hour period?   13.3   8

_____ **17.** In $\triangle ABC$, $\angle A = 38°$ and $\angle C = 112°$. Find $\angle B$.   13.4   1

_____ **18.** In $\triangle ABC$, $\angle C$ is a right angle and $\angle B = 59°$. Find $\angle A$.   13.4   2

_____ **19.** Find the length of the hypotenuse of a right triangle with $a = 8$ meters and $b = 15$ meters.   13.4   7

_____ **20.** A loading ramp is built to the entrance of a building. Using the figure illustrated, find the length of the ramp.   13.4   8

7 ft
24 ft

Check your answers in the back of your book.

SHARPENING YOUR SKILLS after Chapters 1–13

SECTION

_____ 1. Write the word form for 100,907.

1.1

_____ 2. Classify 693 as prime or composite. If it is composite, write it as a product of prime numbers.

2.3

_____ 3. Change $15\frac{3}{5}$ to an improper fraction.

3.1

_____ 4. Reduce $\frac{42}{280}$ to lowest terms.

3.2

_____ 5. Use $<$, $>$, or $=$ to compare $\frac{7}{12}$ and $\frac{3}{5}$.

5.4

_____ 6. Write $\frac{156}{10,000}$ as a decimal number.

6.1

_____ 7. Round 0.8375 to the hundredths place.

6.1

_____ 8. Write $103.08 as it would appear on a check.

6.3

_____ 9. Change $\frac{45}{81}$ to a decimal number and round to the thousandths place.

7.3

_____ 10. Use cross products to decide if this pair of ratios is equal: $\frac{3\frac{1}{2}}{\frac{7}{8}}$ and $\frac{4}{1}$.

8.2

_____ 11. Find the missing part in the proportion $\dfrac{55}{n} = \dfrac{11}{7}$.

8.2

_____ 12. Change 3.75 percent to a fraction.

9.1

_____ 13. Change $\frac{8}{15}$ to a percent. (Round answer to the tenths place.)

9.1

_____ 14. Change $7\frac{2}{3}$ feet to inches.

12.1

_____ 15. Change 175 millimeters to meters.

12.1

_____ 16. Change 254 milligrams to grams.

12.2

_____ 17. Change 17 pints to quarts.

12.3

Perform the indicated operations.

_____ 18. $3\frac{1}{2} \times \frac{6}{7} \times 1\frac{5}{9}$

4.1

_____ 19. $28\frac{1}{3} \div 5\frac{5}{6}$

4.2

_____ 20. $\frac{7}{8} + 1\frac{5}{6} + \frac{1}{15}$

5.3

SECTION

_____ 21. $39.7 - 15.2 - 7.342$ 6.2

_____ 22. $8.732 \div 0.321$ (Round answer to the hundredths place.)   7.2

_____ 23. During 5 days Ramona walked 4, 3, 2, 6, and 5 miles. Find the average number of miles Ramona walked each day.   2.4

_____ 24. Alex had $1498.53 in his checking account before paying his monthly bills. After he wrote checks for $310, $15.85, $94.72, $28.72, and $129.07, what was his balance?   6.3

_____ 25. The tax on property is computed by multiplying its assessed value by the tax rate. If the tax rate in Jefferson County is 0.01325, find the tax bill on property that is evaluated at $85,000.   7.4

_____ 26. A nurse has been instructed to administer 40 milligrams of medication every 3 hours. How many milligrams would be administered in 24 hours?   8.3

_____ 27. There were 6500 applicants for jobs when the new factory opened in Tarrant County. If only 25 percent of the applicants could get jobs, how many applicants were hired? How many were rejected?   9.2

_____ 28. If Sam lends Carla $5000 at 8 percent simple interest for 18 months, how much interest will he receive? What is the total amount Carla must repay at the end of 18 months?   10.4

_____ 29. In a survey of the study habits of college students, responses were received from 6000 students. If 4800 students study fewer hours than Neisha, what is her percentile rank?   11.3

_____ 30. Instructions for mixing a pesticide solution say to mix 3 tablespoons of pesticide with each quart of water. How many tablespoons of pesticide should be used in filling a 4-gallon sprayer?   12.3

_____ 31. Find the volume of a cube with edge 21.5 centimeters.   13.3

_____ 32. Find the length of the hypotenuse in a right triangle with $a = 10$ centimeters and $b = 24$ centimeters.   13.4

Check your answers in the back of your book.

Working with the Numbers of Algebra

> James has been attending Weight Worriers meetings for 6 months. His monthly weight record is as follows: lost 5 pounds, lost 2 pounds, gained 3 pounds, lost 1 pound, gained 2 pounds, and lost 4 pounds. Find James's average monthly loss or gain.

To become prepared to learn some basic concepts of algebra, we must turn our attention to the arithmetic of positive and negative numbers.

In this chapter we

1. Introduce negative numbers.
2. Add and subtract positive and negative numbers.
3. Multiply and divide positive and negative numbers.
4. Review the order of operations.
5. Switch from words to numbers.

14.1   Understanding Negative Numbers

On a winter day, have you ever watched the thermometer drop to a temperature of 3° below zero? Have you ever seen your favorite football team lose 6 yards on a play? Have you ever borrowed $10?

In each of these situations the numbers we have been using do not provide us with a way to make the necessary measurement. We know that temperatures of 3° above zero and 3° below zero are very different. A gain of 6 yards is very different from a loss of 6 yards, and a credit of $10 is very different from a debt of $10. We can use a **negative sign** to help us, and

$$3° \; below \; \text{zero} \qquad \text{is written} \quad -3°$$

$$\text{a } loss \text{ of 6 yards} \qquad \text{is written} \quad -6 \text{ yards}$$

$$\text{a } debt \text{ of 10 dollars} \quad \text{is written} \quad -10 \text{ dollars}$$

Locating Negative Numbers on the Number Line

On a number line we illustrate these negative numbers by extending our line to the left, marking off each unit in a leftward direction from zero.

This new set of numbers, which includes all the whole numbers *and* their negative counterparts, is called the set of **integers.**

Integers: $\{\ldots, -6, -5, -4, -3, -2, -1, 0, 1, 2, 3, 4, \ldots\}$

Notice that this list continues indefinitely in either direction, as shown by the arrows at both ends of the number line. Units to the *right* of 0 correspond to the natural numbers (also called **positive integers**) and units to the *left* of 0 correspond to the **negative integers.**

Positive integers: $\{1, 2, 3, 4, 5, \ldots\}$
Negative integers: $\{\ldots, -4, -3, -2, -1\}$

The set of integers contains all the positive integers, all the negative integers, and the integer 0. Each integer corresponds to one and only one point on the number line.

In a similar way, we can locate positive and negative fractional numbers and decimal numbers on the number line.

The number line again gives us a handy way to compare numbers. If one number lies to the *left* of another number, we say that the first number is *less than* ($<$) the second number. If one number lies to the *right* of another number, we say that the first number is *greater than* (>) the second number.

Example 1. Compare −2 and 3.

Solution

```
←——•——+——+——+——+——•——+——→
   −2 −1  0  1  2  3  4
```

−2 lies to the left of 3, so

$$-2 < 3$$

Now try completing Example 2.

Example 2. Compare −4 and −7.

Solution

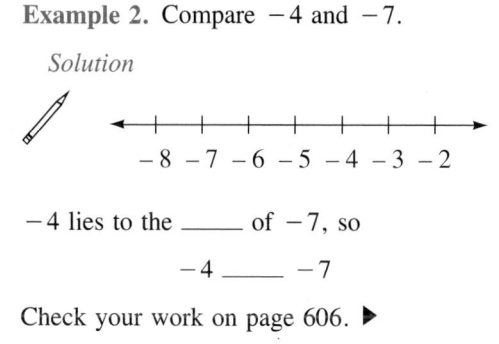

```
←——+——+——+——+——+——+——+——→
  −8 −7 −6 −5 −4 −3 −2
```

−4 lies to the _____ of −7, so

$$-4 \underline{\quad} -7$$

Check your work on page 606. ▶

Example 3. Locate −3 and locate 3 on the number line.

Solution

```
←——+——•——+——+——+——+——+——•——+——+——→
  −4 −3 −2 −1  0  1  2  3  4  5
```

If we were to take a ruler and measure the distance between −3 and 0 on our number line and then measure the distance between 0 and 3 on our number line, what would we observe? Clearly, those distances are the same. Each of the points, −3 and 3, measures 3 units from 0. Their locations are on *opposite* sides of 0, but their distances represent the same number of units. For this reason, we say

$$-3 \text{ is the } \textit{opposite} \text{ of } 3$$

$$3 \text{ is the } \textit{opposite} \text{ of } -3$$

> When two numbers measure the same distance from 0 but the two numbers are on opposite sides of 0, we say that one number is the **opposite** of the other.

Example 4. What is the opposite of 16?

Solution. −16 is the opposite of 16.

You try Example 5.

Example 5. What is the opposite of −5?

Solution. _____ is the opposite of −5.

Check your answer on page 606. ▶

Sometimes we refer to a positive number, such as 5, as +5, but the positive sign is not required. You should understand that if there is no sign in front of a number, it is considered positive. We also agree that the opposite of 0 is 0.

We can use a *negative sign* in front of a number to denote the opposite of that number.

"The opposite of 3 is −3"   can be written   −(3) = −3

"The opposite of −3 is 3"   can be written   −(−3) = 3

Notice that we can use *parentheses* around a number to avoid confusion with signs.

Writing Opposites

Opposite of a $-(a) = -a$

Opposite of $-a$ $-(-a) = a$

Opposite of 0 $-(0) = 0$

Example 6. Find $-(23)$.

Solution

$$-(23) = -23$$

Example 7. Find $-(-11)$.

Solution

$$-(-11) = 11$$

When we measure the distance between 0 and the point on the number line corresponding to a number, we are finding the **absolute value** of that number. The absolute value of a number, a, is symbolized by $|a|$.

Absolute Value. If a is a number, then $|a|$ is the distance between 0 and a on the number line.

Because absolute value represents distance, we note that

The absolute value of a number is never negative.

Example 8. Find $|-6|$.

Solution

$$|-6| = 6$$

You try Example 9.

Example 9. Find $|2.3|$.

Solution

$$|2.3| = \underline{\hspace{1cm}}$$

Check your work on page 606. ▶

Then try Example 10.

Example 10. Find $|0|$.

Solution

$$|0| = \underline{\hspace{1cm}}$$

Check your work on page 606. ▶

Example 11. Find $-|-10|$.

Solution

$$-|-10|$$
$$= -(10)$$
$$= -10$$

Adding Positive and Negative Numbers

To begin our discussion of addition, let us recall how to add numbers that are positive or 0. Remember, these are the kinds of numbers that we have discussed in earlier chapters. Let's use the number line.

Example 12. Use the number line to find $5 + 2$.

Solution. We start at 5 and move 2 units to the right.

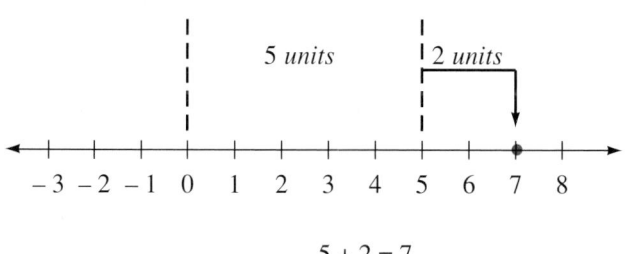

$$5 + 2 = 7$$

Notice that whenever we add two *positive* numbers, the answer will be a *positive* number.

> To add two *positive* numbers, add the units (absolute values) and give the answer a *positive* sign.

What happens when we add two *negative* numbers? The number line will help here. In locating a negative number, we must move *left* rather than right.

Example 13. Use the number line to add -3 and -2.

Solution. We start at -3 and move 2 units to the *left*.

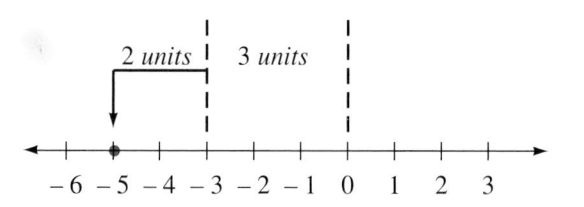

Since we ended up at -5, the sum of -3 and -2 must be -5. We write this as

$$-3 + (-2) = -5$$

Notice that in this example of adding two *negative* numbers, the number of units was the *sum* of the units in the original numbers, but the *sign* of the answer was *negative*.

> To add two *negative* numbers, add the units (absolute values) and give the answer a *negative* sign.

Example 14. Find $-13 + (-76)$.

Solution. The answer will have a *negative* sign.

$$-13 + (-76) = -89$$

You complete Example 15.

Example 15. Find $-31 + (-8)$.

Solution. The answer will have a _____ sign.

$$-31 + (-8) = \underline{}$$

Check your work on page 606. ▶

What happens if we try to add two numbers when one of the numbers is *positive* and one of the numbers is *negative*? The number line will give us a clue if we remember to move *right* for *positive* numbers and move *left* for *negative* numbers.

Example 16. Use the number line to find $-6 + 4$.

Solution. We start at -6 and move 4 units to the *right*.

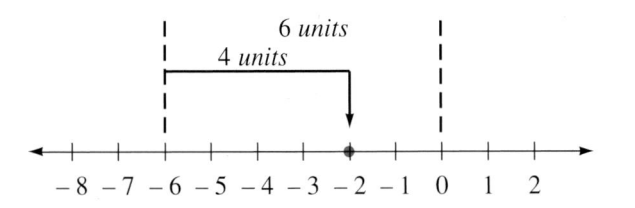

We end up at -2, so $-6 + 4 = -2$

Example 17. Use the number line to find $-3 + 8$.

Solution. We start at -3 and move 8 units to the *right*.

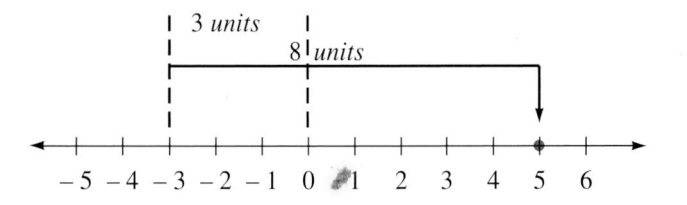

We end up at 5, so $-3 + 8 = 5$

Do you see how we could add two numbers with opposite signs without using the number line? In each example, the *units* in our answer represented the *difference* between our original units. The *sign* of the answer was the sign of the original number having *more* units. Remember, the number of units for a number is that number's distance from 0 on the number line (absolute value).

To add a *positive* number and a *negative* number, find the *difference* between their units (absolute values) and give the answer the sign of the original number having more units (larger absolute value).

Example 18. Find $-11 + 26$.

Solution. Our answer will have a positive sign. The difference between 11 units and 26 units is 15 units.

$$-11 + 26 = 15$$

You try Example 19.

Example 19. Find $-30 + 20$.

Solution. Our answer will have a _____ sign. The difference between 30 units and 20 units is _____ units.

$$-30 + 20 = \underline{\hspace{1cm}}$$

Check your work on page 606. ▶

In adding more than two numbers, **parentheses** continue to give us directions for the order in which to perform the additions. We must add the numbers within the parentheses first.

Example 20. Find $-10 + (-17 + 2)$.

Solution. We find the sum within the parentheses first.

$$-10 + (-17 + 2)$$
$$= -10 + (-15)$$
$$= -25$$

You try completing Example 21.

Example 21. Find $[5 + (-16)] + 3$.

Solution

$$[5 + (-16)] + 3$$
$$= \underline{\hspace{1cm}} + 3$$
$$= \underline{\hspace{1cm}}$$

Check your work on page 606. ▶

Example 22. Find $\dfrac{7}{8} + \left[-\dfrac{5}{8} + \dfrac{3}{8} \right]$.

Solution

$$\frac{7}{8} + \left[-\frac{5}{8} + \frac{3}{8} \right] = \frac{7}{8} + \left[\frac{-5 + 3}{8} \right]$$
$$= \frac{7}{8} + \frac{-2}{8}$$
$$= \frac{7 + (-2)}{8}$$
$$= \frac{5}{8}$$

You complete Example 23.

Example 23. Find the sum.

$$[-2.3 + (-5.1)] + 12.8$$

Solution

$$[-2.3 + (-5.1)] + 12.8$$
$$= -7.4 + 12.8$$
$$= \underline{\hspace{1cm}}$$

Check your work on page 606. ▶

⇒ Trial Run

Find the sums.

_____ 1. $8 + (-6)$

_____ 2. $-11 + 5$

_____ 3. $-3 + (-3)$

_____ 4. $[-5 + (-6)] + 10$

_____ 5. $\left(-\dfrac{12}{31} + \dfrac{2}{31}\right) + \left(-\dfrac{9}{31}\right)$ _____ 6. $-3.5 + (-7.6 + 12.8)$

Answers are on page 606.

Using Some Properties of Addition

There are several laws or properties of addition that we may observe from our work. Note these sums.

$$-15 + 8 = -7$$
$$8 + (-15) = -7$$

> It does not matter in which order we add two numbers. The sum will be the same.

This law is called the **commutative property for addition.** The commutative property gives us two ways in which to write the sum of two numbers.

Now consider these sums involving *three* numbers.

$$(-6 + 8) + (-3) = 2 + (-3) = -1$$
$$-6 + [8 + (-3)] = -6 + 5 \ \ = -1$$

In these two sums, the order in which the numbers appeared was the same, but the position of the symbols of grouping was different. In both cases the sum was the same.

> When adding three numbers, it does not matter if we add the first two and then add the third, *or* add the second two and then add the first. The sum will be the same.

This law is called the **associative property for addition.** To add several numbers when there are no brackets, the commutative and associative properties of addition allow us to arrange them in any order we choose. It is often easier to *add all the numbers with matching signs first*.

Example 24. Find the sum $-22 + (-11) + 6 + (-30) + 50$.

Solution

$-22 + (-11) + 6 + (-30) + 50$

$= [-22 + (-11) + (-30)] + [6 + 50]$ Rearrange terms (commutative and associative properties).

$= -63 + 56$ Add negative numbers; add positive numbers.

$= -7$ Find the sum.

Example 25. Find $-\dfrac{6}{11} + \dfrac{2}{11} + \left(-\dfrac{5}{11}\right)$.

Solution

$$-\dfrac{6}{11} + \dfrac{2}{11} + \left(-\dfrac{5}{11}\right) = -\dfrac{6}{11} + \left(-\dfrac{5}{11}\right) + \dfrac{2}{11}$$

$$= \dfrac{-6 + (-5) + 2}{11}$$

$$= \dfrac{-11 + 2}{11}$$

$$= -\dfrac{9}{11}$$

Consider the result when a number is added to its *opposite*.

$$9 + (-9) = 0$$

$$-62 + 62 = 0$$

These examples should help you see that

The sum of a number and its opposite is 0.

The *opposite* of a number is also called its **additive inverse.**

Example 26. Find the sum $-15 + 12 + (-11) + 14$.

Solution

$$-15 + 12 + (-11) + 14$$

$$= [-15 + (-11)] + [12 + 14] \qquad \text{Rearrange terms.}$$

$$= -26 + 26 \qquad\qquad\qquad\qquad \text{Add negative numbers; add positive numbers.}$$

$$= 0 \qquad\qquad\qquad\qquad\qquad\quad \text{Property of opposites.}$$

⇒ Trial Run

Find each sum.

_____ 1. $8 + (-6) + (-2)$ _____ 2. $-22 + (-7) + 3 + (-20) + 40$

_____ 3. $-\dfrac{3}{17} + \dfrac{5}{17} + \left(-\dfrac{8}{17}\right)$ _____ 4. $2.7 + (-3.5) + (-1.4) + 6.2$

_____ 5. $(17 + 0) + [0 + (-3)]$ _____ 6. $16 + (-23) + 4 + (-8) + (-10)$

Answers are on page 606.

▶ Examples You Completed _____

Example 2. Compare -4 and -7.

Solution

-4 lies to the right of -7, so

$$-4 > -7$$

Example 5. What is the opposite of -5?

Solution. 5 is the opposite of -5.

Example 9. Find $|2.3|$.

Solution

$$|2.3| = 2.3$$

Example 10. Find $|0|$.

Solution

$$|0| = 0$$

Example 15. Find $-31 + (-8)$.

Solution. The answer will have a negative sign.

$$-31 + (-8) = -39$$

Example 19. Find $-30 + 20$.

Solution. Our answer will have a negative sign. The difference between 30 units and 20 units is 10 units.

$$-30 + 20 = -10$$

Example 21. Find $[5 + (-16)] + 3$.

Solution

$$[5 + (-16)] + 3$$
$$= -11 + 3$$
$$= -8$$

Example 23. Find the sum.

$$[-2.3 + (-5.1)] + 12.8$$

Solution

$$[-2.3 + (-5.1)] + 12.8$$
$$= -7.4 + 12.8$$
$$= 5.4$$

Answers to Trial Runs _____

page 603   1. 2   2. -6 3. -6 4. -1 5. $-\dfrac{19}{37}$ 6. 1.7

page 605   1. 0   2. -6 3. $-\dfrac{6}{17}$ 4. 4   5. 14   6. -21

EXERCISE SET 14.1

1. Graph the points corresponding to these numbers on the number line.
   (a) -7　　　　(b) 4.5　　　　(c) 0　　　　(d) -3　　　　(e) $3\frac{1}{2}$

2. Graph the points corresponding to these numbers on the number line.
   (a) $-5\frac{1}{3}$　　　　(b) 3　　　　(c) 1　　　　(d) -8　　　　(e) 4.75

3. Identify by number the point corresponding to each letter on the number line.

4. Identify by number the point corresponding to each letter on the number line.

5. Compare each pair of numbers using $<$ or $>$.
   _____ (a) $5, 12$　　　　　　_____ (b) $-4, -10$

   _____ (c) $0.8, -1.1$　　　　_____ (d) $-\dfrac{3}{5}, \dfrac{7}{5}$

6. Compare each pair of numbers using $<$ or $>$.
   _____ (a) $3, 9$　　　　　　_____ (b) $-12, -2$

   _____ (c) $10\dfrac{1}{2}, -15\dfrac{1}{4}$　　_____ (d) $-2.3, 9.5$

Simplify by removing absolute value bars.

_____ 7. $|-5|$　　　　　　　　_____ 8. $|-7|$

_____ 9. $-|-8|$　　　　　　　_____ 10. $-|-17|$

_____ 11. $|9 + (-12)|$　　　　_____ 12. $|-10 + 8|$

_____ 13. $\left|\dfrac{9}{2}\right| + \left|-\dfrac{3}{2}\right|$　　　　_____ 14. $\left|\dfrac{11}{4}\right| + \left|-\dfrac{5}{4}\right|$

_____ 15. $|1.9| - |-0.8|$　　　_____ 16. $|1.5| - |-0.6|$

Perform the indicated operation.

_____ 17. 4 + 18

_____ 18. 12 + 5

_____ 19. −5 + (−7)

_____ 20. −9 + (−12)

_____ 21. 15 + (−3)

_____ 22. 19 + (−4)

_____ 23. 3 + (−9)

_____ 24. 2 + (−15)

_____ 25. −11.3 + 4.2

_____ 26. −12.5 + 7.3

_____ 27. $-\dfrac{2}{3} + \dfrac{5}{3}$

_____ 28. $-\dfrac{3}{5} + \dfrac{7}{5}$

_____ 29. 14 + [8 + (−2)]

_____ 30. 20 + [10 + (−3)]

_____ 31. [9 + (−7)] + 6

_____ 32. [15 + (−8)] + 7

_____ 33. [−15 + (−3)] + 1

_____ 34. [−41] + (−9)] + 1

_____ 35. −1.2 + (−7.3 + 5.4)

_____ 36. −1.3 + (−5.2 + 12.5)

_____ 37. 21 + (−12) + 9

_____ 38. 32 + (−24) + 7

_____ 39. $-\dfrac{15}{2} + \dfrac{7}{2} + \left(-\dfrac{13}{2}\right)$

_____ 40. $-\dfrac{26}{3} + \dfrac{10}{3} + \left(-\dfrac{4}{3}\right)$

_____ 41. 13 + (−8) + (−9) + 4

_____ 42. 18 + (−5) + (−17) + 4

_____ 43. −19 + (−9) + 12 + 7 + (−2)

_____ 44. −25 + (−11) + 9 + 8 + (−15)

_____ 45. −12 + 0

_____ 46. 0 + (−20)

_____ 47. −3 + 4 + (−1)

_____ 48. −10 + 15 + (−5)

_____ 49. 4.3 + (−2.6) + (−9.6) + 5.4

_____ 50. 8.6 + (−3.5) + (−9.8) + 3.7

☆ Stretching the Topics _____

Perform the indicated operations.

_____ 1. 5 × [−11 + 17 + (−3)]

_____ 2. $2 \times \left[\dfrac{18}{9} + 7 + (-1)\right] + (-4)$

_____ 3. $\dfrac{3}{5} \times \left\{-\dfrac{5}{4} + \left[\dfrac{10}{3} + \left(-\dfrac{2}{3}\right)\right]\right\}$

Check your answers in the back of your book.

If you can complete **Checkup 14.1,** you are ready to go on to Section 14.2.

✓ **CHECKUP 14.1**

_____ **1.** Compare 3 and -15 using $<$ or $>$.

_____ **2.** Simplify $-|-1.2|$ by removing absolute value bars.

Perform the indicated operation.

_____ **3.** $-9 + 4$ _____ **4.** $10 + (-2)$

_____ **5.** $-3.5 + (-8.7)$ _____ **6.** $[-3 + (-5)] + 7$

_____ **7.** $-\dfrac{3}{5} + \left[-\dfrac{4}{5} + \dfrac{2}{5}\right]$ _____ **8.** $8 + (-9) + 4 + (-3)$

_____ **9.** $\dfrac{5}{8} + \left(-\dfrac{3}{8}\right) + \dfrac{7}{8} + \left(-\dfrac{1}{8}\right)$ _____ **10.** $7.3 + (-3.6) + (-2.4) + 7.8$

Check your answers in the back of your book.

If You Missed Problems:	You Should Review Examples:
1	1, 2
2	8, 9
3–5	13–19
6, 7	21–23
8–10	24–26

14.2 Subtracting Positive and Negative Numbers

When we discussed the subtraction of whole numbers in Chapter 1, we agreed that

$$9 - 3 = 6$$

In the section on the addition of positive and negative numbers, we agreed that

$$9 + (-3) = 6$$

so it seems that $9 - 3$ and $9 + (-3)$ are two different ways of saying the same thing.

Subtraction		*Addition*
$9 - 3$	means the same as	$9 + (-3)$
9 *minus* 3	means the same as	9 *plus* the *opposite* of 3

In fact, this is the way in which *subtraction* is defined using *addition*. Since we know how to *add,* subtraction should proceed smoothly.

Example 1. Write $10 - 1$ using addition, and simplify.

Solution

$$10 - 1 = 10 + (-1)$$
$$= 9$$

Now you complete Example 2.

Example 2. Write $6 - 6$ using addition, and simplify.

Solution

$$6 - 6 = 6 + (-6)$$
$$= \underline{}$$

Check your work on page 613. ▶

Every subtraction problem can be written as an addition problem.

> To subtract two numbers, take the *opposite* of the number being subtracted, and *add.*

Example 3. Find $-7 - (-5)$.

Solution

$-7 - (-5)$	Subtraction problem.
$= -7 + 5$	Take the opposite of -5, and *add.*
$= -2$	

You try Example 4.

Example 4. Find $-11 - 19$.

Solution

$-11 - 19$	Subtraction problem.
$= -11 + (\underline{})$	Take the opposite of 19, and *add.*
$= \underline{}$	

Check your work on page 613. ▶

When expressions contain symbols of grouping, we will perform the operations within parentheses first and work from the innermost symbols of grouping outward. If you are impatient and try to hurry through these problems, you will find yourself making mistakes. It is important to get in the habit of dealing carefully with symbols of grouping.

Example 5. Find $(7 - 10) - (8 - 15)$.

Solution

$$(7 - 10) - (8 - 15)$$

$$= [7 + (-10)] - [8 + (-15)] \qquad \text{Rewrite inner subtractions as additions.}$$

$$= -3 - (-7) \qquad\qquad\qquad \text{Simplify within brackets.}$$

$$= -3 + 7 \qquad\qquad\qquad\quad \text{Rewrite subtraction as addition.}$$

$$= 4 \qquad\qquad\qquad\qquad\quad\; \text{Find the sum.}$$

You complete Example 6.

Example 6. Find $8 - (10 - 23)$.

Solution

$$8 - (10 - 23)$$

$$= 8 - [10 + (\underline{\hspace{1cm}})] \qquad \text{Rewrite inner subtraction as addition.}$$

$$= 8 - [\underline{\hspace{1cm}}] \qquad\qquad\;\; \text{Simplify within brackets.}$$

$$= 8 + \underline{\hspace{1cm}} \qquad\qquad\quad \text{Rewrite subtraction as addition.}$$

$$= \underline{\hspace{1cm}}$$

Check your work on page 613. ▶

Example 7. Find $\dfrac{5}{3} - \dfrac{11}{2}$.

Solution

$$\frac{5}{3} - \frac{11}{2} \qquad \text{LCD: } 3 \cdot 2$$

$$= \frac{5 \cdot 2}{3 \cdot 2} - \frac{11 \cdot 3}{3 \cdot 2}$$

$$= \frac{10}{6} - \frac{33}{6}$$

$$= \frac{10 + (-33)}{6}$$

$$= -\frac{23}{6}$$

Example 8. Find $2.3 - (3.51 - 7.6)$.

Solution

$$2.3 - (3.51 - 7.6)$$

$$= 2.3 - [3.51 + (-7.6)]$$

$$= 2.3 - (-4.09)$$

$$= 2.3 + 4.09$$

$$= 6.39$$

▌▐▶ **Trial Run**

Simplify.

_____ **1.** $7 - 12$ 　　　　　　　 _____ **2.** $-15 - 8$

_____ **3.** $-\dfrac{9}{10} - \left(-\dfrac{17}{10}\right)$ 　　　 _____ **4.** $9.1 - (3.7 + 1.2)$

_____ **5.** $(3 - 7) - (9 - 12)$ _____ **6.** $10 - (9 - 21)$

Answers are given below.

▶ Examples You Completed

Example 2. Write $6 - 6$ using addition, and simplify.

Solution

$$6 - 6 = 6 + (-6)$$
$$= 0$$

Example 6. Find $8 - (10 - 23)$.

Solution

$$8 - (10 - 23) = 8 - [10 + (-23)]$$
$$= 8 - [-13]$$
$$= 8 + 13$$
$$= 21$$

Example 4. Find $-11 - 19$.

Solution

$$-11 - 19 = -11 + (-19)$$
$$= -30$$

Answers to Trial Run

page 612   **1.** -5 **2.** -23 **3.** $\dfrac{4}{5}$ **4.** 4.2 **5.** -1 **6.** 22

EXERCISE SET 14.2

Write each expression using addition, and simplify.

_____ 1. $12 - 9$

_____ 2. $23 - 8$

_____ 3. $11 - 27$

_____ 4. $9 - 18$

_____ 5. $-15 - 7$

_____ 6. $-19 - 8$

_____ 7. $12 - (-3)$

_____ 8. $17 - (-5)$

_____ 9. $-9 - (-7)$

_____ 10. $-11 - (-8)$

_____ 11. $-\dfrac{2}{3} - \left(-\dfrac{1}{2}\right)$

_____ 12. $-\dfrac{4}{5} - \left(-\dfrac{3}{4}\right)$

_____ 13. $-8.6 - 3.7$

_____ 14. $-9.5 - 6.7$

Perform the indicated operations.

_____ 15. $15 + (8 - 13)$

_____ 16. $17 + (9 - 15)$

_____ 17. $21 - (3 - 7)$

_____ 18. $11 - (5 - 12)$

_____ 19. $\left(\dfrac{8}{9} - \dfrac{7}{9}\right) - \dfrac{2}{3}$

_____ 20. $\left(\dfrac{5}{8} - \dfrac{3}{8}\right) - \dfrac{3}{4}$

_____ 21. $(7 - 14) - 2$

_____ 22. $(9 - 16) - 5$

_____ 23. $(5.6 - 4.2) - (-2.6)$

_____ 24. $(10.7 - 3.5) - (-9.8)$

_____ 25. $-7 - (10 - 5)$

_____ 26. $-13 - (12 - 6)$

_____ 27. $(-5 - 2) - 3$

_____ 28. $(-8 - 4) - 9$

_____ 29. $[-5 - (-2)] - 3$

_____ 30. $[-8 - (-4)] - 9$

_____ 31. $(-3.9 - 5) - (-1.5)$

_____ 32. $(-1.2 - 7) - (-1.8)$

_____ 33. $-13 - (-3 - 11)$

_____ 34. $-12 - (-5 - 8)$

_____ 35. $3 + [-2 - (-7)]$

_____ 36. $4 + [-3 - (-8)]$

_____ 37. $-13 - 8 - 5$

_____ 38. $-17 - 9 - 5$

_____ 39. $-6 - (-6)$

_____ 40. $-11 - (-11)$

_____ 41. $12 - (-3 + 15)$

_____ 42. $26 - (-11 + 37)$

_____ 43. $(12 - 15) - (7 - 13)$

_____ 44. $(8 - 13) - (5 - 9)$

_____ 45. $\left(-\dfrac{3}{5} - \dfrac{9}{5}\right) - \left(\dfrac{7}{2} - \dfrac{3}{2}\right)$ _____ 46. $\left(-\dfrac{2}{3} - \dfrac{5}{3}\right) - \left(\dfrac{9}{4} - \dfrac{5}{4}\right)$

_____ 47. $(9 - 15) - (17 - 21) - 5$ _____ 48. $(4 - 12) - (21 - 15) - 7$

_____ 49. $[7.2 - (-3.1)] - [-1.2 - (-9.5)]$ _____ 50. $[8.3 - (-5.2)] - [-1.6 - (-7.3)]$

☆ Stretching the Topics

Perform the indicated operations.

_____ 1. $-9 - \{10 + [1 - (-2)]\}$

_____ 2. $2 \times \{10 - (-7) - [-2 + 5]\}$

_____ 3. $15 \times \left[\left(\dfrac{2}{3} \times \dfrac{3}{4}\right) - \left(-\dfrac{1}{3} - \dfrac{3}{5}\right)\right]$

Check your answers in the back of your book.

If you can complete **Checkup 14.2,** you are ready to go on to Section 14.3.

Name _____ Date _____

✔ CHECKUP 14.2

Perform the indicated operation.

_____ 1. $12 - 5$

_____ 2. $8 - (-13)$

_____ 3. $-17 - 5$

_____ 4. $-19 - (-9)$

_____ 5. $(4 - 11) - (11 - 19)$

_____ 6. $\left(-\dfrac{3}{5} - \dfrac{7}{5}\right) - \left(\dfrac{9}{5} - \dfrac{11}{5}\right)$

_____ 7. $-\dfrac{7}{3} - \left(-\dfrac{1}{4}\right)$

_____ 8. $-7.5 - (10 - 2.3)$

Check your answers in the back of your book.

If You Missed Problems:	You Should Review Examples:
1–4	3, 4
5, 6	5, 6
7, 8	7, 8

14.3 Multiplying and Dividing Positive and Negative Numbers

In Chapter 13 we observed that there are several ways to indicate the *product* of two numbers, such as 4 times 3.

$$4 \times 3 \quad 4 \cdot 3 \quad 4(3) \quad (4)(3)$$

In our work with algebra, we prefer not to use the multiplication sign **×**. Instead, we use the multiplication dot · or parentheses to indicate a product.

Multiplying Numbers

When multiplying whole numbers, we observed that multiplication was a way of performing repeated addition. Whenever we multiply two *positive* numbers, we are repeatedly adding positive numbers, so the answer must be *positive*.

$$4 \cdot 3 = 3 + 3 + 3 + 3 = 12$$

$$5 \cdot 1 = 1 + 1 + 1 + 1 + 1 = 5$$

> To find the product of *two positive* numbers, multiply the units (absolute values) and give the answer a *positive* sign.

Look at the product of a *positive* number and a *negative* number, again using repeated addition.

$$5(-3) = -3 + (-3) + (-3) + (-3) + (-3) = -15$$

$$4(-1) = -1 + (-1) + (-1) + (-1) = -4$$

In each case, we were repeatedly adding negative numbers, so the answer was *negative*.

> To find the product of a *positive* number and a *negative* number, multiply the units (absolute values) and give the answer a *negative* sign.

Example 1. Find $3(-6)$.

Solution

$$3(-6) = -18$$

You do Example 2.

Example 2. Find $-4 \cdot 11$.

Solution

$$-4 \cdot 11 = \underline{\qquad}$$

Check your work on page 625. ▶

What happens when we multiply two *negative* numbers? Look at the pattern that occurs when we consider some multiples of -3.

$$4(-3) = -12$$
$$3(-3) = -9$$
$$2(-3) = -6$$
$$1(-3) = -3$$
$$0(-3) = 0$$

As the multipliers of -3 on the left are *decreasing by 1,* notice that the products on the right are *increasing by 3.* To continue this pattern, we must let the multipliers continue to *decrease by 1.* At the same time, the products must continue to *increase by 3.*

$$-1(-3) = 3$$
$$-2(-3) = 6$$
$$-3(-3) = 9$$
$$-4(-3) = 12$$

Observe that the product of two negative numbers turns out to be a positive number.

> To find the product of *two negative* numbers, multiply the units (absolute values) and give the answer a *positive* sign.

Example 3. Find $(-1)(-1)$.

Solution

$$(-1)(-1) = 1$$

You try Example 4.

Example 4. Find $(-8)(-6)$.

Solution

$$(-8)(-6) = \underline{\quad\quad}$$

Check your work on page 625. ▶

Symbols of grouping still give us direction regarding the order of operations.

Example 5. Find $[-3(-2)](-7)$.

Solution

$$[-3(-2)](-7)$$
$$= [6](-7)$$
$$= -42$$

You complete Example 6.

Example 6. Find $-5(3) + (-2)(-1)$.

Solution

$$-5(3) + (-2)(-1)$$
$$= -15 + \underline{\quad\quad}$$
$$= \underline{\quad\quad}$$

Check your work on page 625. ▶

Example 7. Find $-\dfrac{2}{3} \cdot \dfrac{5}{6}$.

Solution

$$-\frac{2}{3} \cdot \frac{5}{6} = -\frac{\overset{1}{\cancel{2}}}{3} \cdot \frac{5}{\underset{3}{\cancel{6}}}$$

$$= -\frac{5}{3 \cdot 3}$$

$$= -\frac{5}{9}$$

Example 8. Find $-3(2.7 - 8.2)$.

Solution

$$-3(2.7 - 8.2)$$
$$= -3[2.7 + (-8.2)]$$
$$= -3(-5.5)$$
$$= 16.5$$

IIII➡ **Trial Run**

Find the products.

_____ **1.** $-7(-8)$ _____ **2.** $-6(12)$

_____ **3.** $4(-3) + 5(-2)$ _____ **4.** $-5(8.9 - 7.7)$

_____ **5.** $-\dfrac{4}{9} \cdot \dfrac{3}{8}$ _____ **6.** $[-7(-3)][-1(2)]$

Answers are on page 625.

Using Some Properties of Multiplication

There are several properties of multiplication that are similar to the properties of addition. From the products

$$-6(5) = -30$$
$$5(-6) = -30$$

you might agree that

> It does not matter in which order we multiply two numbers. The product is the same.

This property is called the **commutative property for multiplication.**

Now consider a few products involving three factors.

$$[-2 \cdot 3](-4) = -6(-4) \ = 24$$
$$-2[3(-4)] = -2(-12) = 24$$

In these products, the order in which the factors appeared was the same, but the position of the symbols of grouping was different. In both cases, the product was the same.

> When multiplying three numbers, it does not matter if we multiply the product of the first two times the third, *or* multiply the first times the product of the second two. The product will be the same.

As you may have guessed, this law is called the **associative property for multiplication.** To multiply several numbers when there are no brackets, the commutative and associative properties of multiplication allow us to arrange them in any order we choose.

Example 9. Find $3(-2)(-1)(-5)$.

Solution

$$3(-2)(-1)(-5)$$

$= -6(5)$ Multiply $3(-2)$ and multiply $(-1)(-5)$.

$= -30$ Multiply.

There are two ways to work a problem such as $8(-2 + 5)$.

First Method	Second Method
$8(-2 + 5)$	$8(-2 + 5)$
$= 8 \cdot 3$	$= 8(-2) + 8 \cdot 5$
$= 24$	$= -16 + 40$
	$= 24$

Notice that the results are the same. It seems that the first method is easier, but later we shall have to use the second method, so we note here that

To multiply a number times the sum of numbers in parentheses, we may multiply the number outside the parentheses times each of the numbers inside the parentheses and add those products together.

This new law is called the **distributive property for multiplication over addition.**

Example 10. Use the distributive property to find the product $-5(7 - 1)$.

Solution

$$-5(7 - 1)$$
$$= -5(7) - (-5)(1)$$
$$= -35 - (-5)$$
$$= -35 + 5$$
$$= -30$$

Now try Example 11.

Example 11. Use the distributive property to find the product $-10(-17 + 19)$.

Solution

$$= -10(-17 + 19)$$

Check your work on page 625. ▶

⫸ **Trial Run**

Find each product.

_____ 1. $-10(-3)(-7)$ _____ 2. $6(5)(-1)(-1)$

_____ 3. $\frac{2}{3}\left(-\frac{3}{4}\right)\left(\frac{1}{5}\right)$ _____ 4. $0.2(-1.7)(0)$

Use the distributive property to find each product.

_____ 5. $-2(-3 + 8)$ _____ 6. $-3(-7 - 6)$

Answers are on page 625.

Finding Powers

In Chapter 2 we discovered that **exponents** could be used to write products in which a factor is repeated. Recall that an exponent tells us the number of times that the base is to be used as a factor in a product.

Example 12. Find $(-2)^3$.

Solution. The base is -2.

$$(-2)^3 = (-2)(-2)(-2)$$
$$= 4(-2)$$
$$= -8$$

You try Example 13.

Example 13. Find $(-1.1)^2$.

Solution. The base is -1.1.

$$(-1.1)^2 = (-1.1)(-1.1)$$
$$= \underline{}$$

Check your work on page 625. ▶

Students often confuse the meanings of expressions such as $(-2)^2$ and -2^2.

To compute $(-2)^2$, the base is (-2).

$$(-2)^2 = (-2)(-2)$$
$$= 4$$

To compute -2^2, the negative sign tells us to find the *opposite* of 2^2.

$$-2^2 = -1 \cdot 2^2$$
$$= -1 \cdot 2 \cdot 2$$
$$= -1 \cdot 4$$
$$= -4$$

Note that an exponent applies only to the base to which it is attached, unless parentheses indicate otherwise.

Example 14. Find $-\left(\dfrac{1}{2}\right)^4$.

Solution

$$-\left(\frac{1}{2}\right)^4 = -1\left(\frac{1}{2}\right)\left(\frac{1}{2}\right)\left(\frac{1}{2}\right)\left(\frac{1}{2}\right)$$
$$= -1\left(\frac{1}{4}\right)\left(\frac{1}{4}\right)$$
$$= -\frac{1}{16}$$

You try Example 15.

Example 15. Find $(-2)^3(-5)^2$.

Solution

$$(-2)^3(-5)^2$$
$$= (-2)(-2)(-2)(-5)(-5)$$
$$= (-8)(\underline{})$$
$$= \underline{}$$

Check your work on page 625. ▶

⇒ Trial Run

Simplify.

———— 1. -7^2

———— 2. $(-5)^3$

———— 3. $\left(-\dfrac{4}{3}\right)^2$

———— 4. $(-2)^2(-6)^2$

Answers are on page 625.

Dividing Numbers

Remember that every division statement corresponds to a multiplication statement. To work a division problem, we may look at the corresponding multiplication problem.

Example 16. Find $\dfrac{123}{3}$ and write the corresponding multiplication statement.

Solution

$$\frac{123}{3} = 41$$

because $3 \cdot 41 = 123$

You try Example 17.

Example 17. Find $\dfrac{99}{9}$ and write the corresponding multiplication statement.

Solution

$$\frac{99}{9} = \text{———}$$

because $9 \cdot \text{———} = 99$

Check your work on page 625. ▶

In each of these examples we found that the quotient of *two positive* numbers was a *positive* number.

> To divide a *positive* number by a *positive* number, divide the units (absolute values) and give the answer a *positive* sign.

How do we find the quotient of two *negative* numbers? We can use multiplication to help us see that

$$\frac{-36}{-9} = 4 \quad \text{because} \quad -9(4) = -36$$

and that

$$\frac{-18}{-6} = 3 \quad \text{because} \quad -6(3) = -18$$

In each case, we found that the quotient of *two negative* numbers was a *positive* number.

> To divide a *negative* number by a *negative* number, divide the units (absolute values) and give the answer a *positive* sign.

Example 18. Simplify $\dfrac{-75}{-3}$.

Solution

$$\frac{-75}{-3} = 25$$

because $-3(25) = -75$

Complete Example 19.

Example 19. Simplify $\dfrac{-6}{-6}$.

Solution

$$\frac{-6}{-6} = \underline{\hspace{1cm}}$$

because $-6(\underline{\hspace{1cm}}) = -6$

Check your work on page 625. ▶

What happens if the dividend and divisor have *different* signs? Again we must remember our multiplication facts. You should agree that

$$\frac{-35}{7} = -5 \quad \text{because} \quad 7(-5) = -35$$

and

$$\frac{39}{-3} = -13 \quad \text{because} \quad -3(-13) = 39$$

In both cases, we found that the quotient of a *positive* number and a *negative* number is a *negative* number.

To divide a *positive* number by a *negative* number (or a *negative* number by a *positive* number), divide the units (absolute values) and give the answer a *negative* sign.

Example 20. Simplify $\dfrac{-16}{2}$.

Solution

$$\frac{-16}{2} = -8$$

Complete Example 21.

Example 21. Simplify $\dfrac{17}{-17}$.

Solution

$$\frac{17}{-17} = \underline{\hspace{1cm}}$$

Check your work on page 625. ▶

Some people prefer to learn a slightly different form of the rules for multiplying and dividing integers.

Multiplying or Dividing Two Numbers

1. If the signs are the *same*, the answer will be *positive*.
2. If the signs are *different*, the answer will be *negative*.

Example 22. Find $-4.44 \div 3.7$.

Solution. The answer will be *negative*.

$$
\begin{array}{r}
1.2 \\
3.7\,)\overline{4.4\,4} \\
3\,7 \\
\hline
7\,4 \\
7\,4 \\
\hline
0
\end{array}
$$

$$-4.44 \div 3.7 = -1.2$$

Example 23. Find

$$\left(-\frac{5}{7} \cdot \frac{3}{10}\right) \div \left(-\frac{9}{20}\right)$$

Solution

$$\left(-\frac{5}{7} \cdot \frac{3}{10}\right) \div \left(-\frac{9}{20}\right)$$

$$= \frac{-5}{7} \cdot \frac{3}{10} \cdot \frac{20}{-9} \qquad \text{Invert divisor.}$$

$$= \frac{\cancel{1} \cdot 5}{7} \cdot \frac{\cancel{3}}{2 \cdot \cancel{5}} \cdot \frac{2 \cdot \cancel{2} \cdot \cancel{5}}{\cancel{1} \cdot \cancel{3} \cdot 3} \qquad \text{Reduce fractions.}$$

$$= \frac{5 \cdot 2}{7 \cdot 3} \qquad \text{Multiply.}$$

$$= \frac{10}{21} \qquad \text{Find the product.}$$

The rules for working with *zero* in a division problem continue to apply for positive and negative numbers.

Example 24. Find $\dfrac{0}{-13}$.

Solution

$$\frac{0}{-13} = 0$$

because $-13(0) = 0$

Example 25. Find $\dfrac{-37}{0}$.

Solution. $\dfrac{-37}{0}$ is undefined.

⫸ Trial Run

Find the quotients.

_____ 1. $\dfrac{-16}{2}$

_____ 2. $\dfrac{-24}{-8}$

_____ 3. $\dfrac{-9}{0}$

_____ 4. $\dfrac{-4 + 10}{-3}$

_____ 5. $\dfrac{6.3}{-3}$

_____ 6. $-\dfrac{8}{9} \div \left(-\dfrac{2}{3}\right)$

Answers are on page 625.

▶ Examples You Completed

Example 2. Find $-4 \cdot 11$.

Solution

$$-4 \cdot 11 = -44$$

Example 6. Find $-5(3) + (-2)(-1)$.

Solution

$$-5(3) + (-2)(-1)$$
$$= -15 + 2$$
$$= -13$$

Example 13. Find $(-1.1)^2$.

Solution

$$(-1.1)^2 = (-1.1)(-1.1)$$
$$= 1.21$$

Example 17. Find $\dfrac{99}{9}$ and write the corresponding multiplication statement.

Solution

$$\frac{99}{9} = 11$$

because $9 \cdot 11 = 99$

Example 21. Simplify $\dfrac{17}{-17}$.

Solution

$$\frac{17}{-17} = -1$$

Example 4. Find $(-8)(-6)$.

Solution

$$(-8)(-6) = 48$$

Example 11. Use the distributive property to find the product $-10(-17 + 19)$.

Solution

$$-10(-17 + 19)$$
$$= -10(-17) + (-10)(19)$$
$$= 170 + (-190)$$
$$= -20$$

Example 15. Find $(-2)^3(-5)^2$.

Solution

$$(-2)^3(-5)^2$$
$$= (-2)(-2)(-2)(-5)(-5)$$
$$= (-8)(25)$$
$$= -200$$

Example 19. Simplify $\dfrac{-6}{-6}$.

Solution

$$\frac{-6}{-6} = 1$$

because $-6(1) = -6$

Answers to Trial Runs

page 619   **1.** 56   **2.** -72 **3.** -22 **4.** -6 **5.** $-\dfrac{1}{6}$ **6.** -42

page 620   **1.** -210 **2.** 30   **3.** $-\dfrac{1}{10}$ **4.** 0   **5.** -10 **6.** 39

page 622   **1.** -49 **2.** -125 **3.** $\dfrac{16}{9}$ **4.** 144

page 624   **1.** -8 **2.** 3   **3.** Undefined   **4.** -2 **5.** -2.1 **6.** $\dfrac{4}{3}$

EXERCISE SET 14.3

Perform the indicated operations.

_____ 1. $5(-6)$ _____ 2. $3(-4)$

_____ 3. $-6(-7)$ _____ 4. $-4(-8)$

_____ 5. $-3(8)$ _____ 6. $-7(9)$

_____ 7. $-0.3(-4.5)$ _____ 8. $(-3.4)(-0.5)$

_____ 9. $-\dfrac{2}{3}\cdot\dfrac{15}{16}$ _____ 10. $\dfrac{12}{13}\left(-\dfrac{26}{15}\right)$

_____ 11. $-10(3)(-5)$ _____ 12. $-5(7)(-3)$

_____ 13. $4(-3)(0)(9)$ _____ 14. $6(-5)(0)(11)$

_____ 15. $-9(3-8)$ _____ 16. $-10(5-12)$

_____ 17. $6(-10)-20$ _____ 18. $8(-9)-15$

_____ 19. $2.3-0.7(-3.2)$ _____ 20. $2.9-3.1(-0.5)$

Use the distributive property to find each product.

_____ 21. $-9(8+10)$ _____ 22. $2(-11-4)$

_____ 23. $\dfrac{5}{2}\left[\dfrac{7}{10}-\left(-\dfrac{9}{10}\right)\right]$ _____ 24. $\dfrac{3}{4}\left[\dfrac{1}{9}-\left(-\dfrac{7}{9}\right)\right]$

Perform the indicated operations.

_____ 25. $(-8)^2$ _____ 26. $(-2)^4$

_____ 27. $(-3)^3$ _____ 28. $(-4)^3$

_____ 29. $\left(-\dfrac{2}{3}\right)^2$ _____ 30. $\left(-\dfrac{3}{4}\right)^2$

_____ 31. $(-1.2)^3$ _____ 32. $(-2.1)^3$

_____ 33. $(-2)^3(-6)^2$ _____ 34. $(-5)^3(-1)^2$

_____ 35. $\dfrac{24}{-6}$ _____ 36. $\dfrac{20}{-4}$

_____ 37. $\dfrac{-16}{8}$ _____ 38. $\dfrac{-15}{5}$

_____ 39. $\dfrac{-35}{-7}$ _____ 40. $\dfrac{-48}{-6}$

_____ 41. $\dfrac{11(-4)}{22}$

_____ 42. $\dfrac{12(-3)}{4}$

_____ 43. $-\dfrac{5}{8} \div \dfrac{15}{44}$

_____ 44. $\dfrac{7}{9} \div \left(-\dfrac{14}{27}\right)$

_____ 45. $-4.32 \div 0.6$

_____ 46. $2.94 \div (-0.7)$

_____ 47. $-0.3[-6.2 - (8 - 9.3)]$

_____ 48. $-0.4[-5.3 - (7 - 8.6)]$

_____ 49. $\left(-\dfrac{3}{8} \cdot \dfrac{4}{9}\right) \div \left(-\dfrac{5}{12}\right)$

_____ 50. $\left(-\dfrac{3}{7} \cdot \dfrac{14}{15}\right) \div \left(-\dfrac{4}{15}\right)$

☆ Stretching the Topics

Perform the indicated operations.

_____ 1. $-7\left[-2(5) - \dfrac{3^2 - (-2)^2}{6 - 11}\right]$

_____ 2. $(-2)^2\{(-1)^3 - 3[8 - (-12)]\}$

_____ 3. $3^3[7 + (-2)^3] - [(-1)^5(-3) - 12] - (8 - 12)^2$

Check your answers in the back of your book.

If you can simplify the expressions in **Checkup 14.3,** you are ready to go on to Section 14.4.

✓ CHECKUP 14.3

Perform the indicated operations.

_____ 1. $6(-5)$

_____ 2. $\left(-\dfrac{3}{4}\right)\left(-\dfrac{5}{6}\right)$

_____ 3. $-3(-2) + (-6)5$

_____ 4. $-5(4.6 - 9.3)$

_____ 5. $(-3.2)^2$

_____ 6. $\left(-\dfrac{1}{2}\right)^3 \cdot (4)^2$

_____ 7. $\dfrac{-35}{7}$

_____ 8. $\dfrac{-63}{-9}$

_____ 9. $4.76 \div (-1.7)$

_____ 10. $\left(-\dfrac{2}{3} \cdot \dfrac{3}{10}\right) \div \left(-\dfrac{9}{10}\right)$

Check your answers in the back of your book.

If You Missed Problems:	You Should Review Examples:
1, 2	1–4, 7
3	6
4	8
5, 6	12–15
7, 8	18–21
9	22
10	23

14.4 Learning the Order of Operations

In earlier chapters, we learned to perform more than one operation within the same problem. Now we must see if there are some rules that tell us the *order* in which we should perform those operations.

We worked many problems in which we added several numbers together. We learned that it did not matter in which order we found such sums, because of the commutative and associative properties for addition. Likewise, the commutative and associative properties for multiplication gave us the same freedom in *multiplying* several numbers.

If we wish to *subtract* several times within the same problem, however, we must be a bit more careful. We *cannot* do our subtraction in whatever order we wish. Instead, we must work from *left to right*. For instance, consider the problem $17 - 10 - 3$.

Correct Method   *Incorrect Method*

$$17 - \underbrace{10 - 3}$$
$$= \quad 7 \quad - 3 \qquad = 17 - \quad 7$$
$$= \quad 4 \qquad\qquad = 10$$

> To find the differences of several numbers, we must do the subtractions in order *from left to right*.

Example 1. Find the difference $86 - 80 - 11$.

Solution

$$\underbrace{86 - 80} - 11$$
$$= \quad 6 \quad - 11$$
$$= \quad -5$$

If we wish to *divide* several times within the same problem, we must use the same care that we used with subtraction. We must work from *left to right*. For instance, consider the problem $48 \div 6 \div 2$.

Correct Method   *Incorrect Method*

$$\underbrace{48 \div 6} \div 2$$
$$= \quad 8 \quad \div 2 \qquad = 48 \div \quad 3$$
$$= \quad 4 \qquad\qquad = 16$$

> To find several quotients in the same problem, we must do the divisions in order *from left to right*.

Example 2. Find the quotient $75 \div 3 \div 5$.

Solution

$$\underbrace{75 \div 3} \div 5$$
$$= \underbrace{25 \quad \div 5}$$
$$= \quad 5$$

From these examples you may have decided that if you must repeat the same operation in the same problem, you should *perform the operations from left to right*.

▮▮▶ Trial Run

Perform the operations.

———— **1.** $2 + 5 + 3$ ———— **2.** $3 \cdot 8 \cdot \dfrac{1}{2}$

———— **3.** $19 - 7 - 3$ ———— **4.** $12.5 - 6.5 - 13$

———— **5.** $36 \div 6 \div 2$ ———— **6.** $50 \div 5 \div \dfrac{1}{4}$

Answers are on page 635.

What happens when you must perform several *different* operations within the same problem?

> In problems involving addition, subtraction, multiplication, and division, mathematicians have agreed to **first do the multiplications and/or divisions in order from left to right, then do the additions and/or subtractions in order from left to right.**

Example 3. Find $6 \cdot 2 + 8 \cdot 1$.

Solution

$$\underbrace{6 \cdot 2} + \underbrace{8 \cdot 1}$$
$$= \quad 12 \quad + \quad 8$$
$$= \quad 20$$

You complete Example 4.

Example 4. Find $9 + 5 \cdot 4$.

Solution

$$9 + \underbrace{5 \cdot 4}$$
$$= 9 + \underline{\quad\quad}$$
$$= \underline{\quad\quad}$$

Check your work on page 635. ▶

Example 5. Find $176 - 14 \div 2$.

Solution

$$176 - \underbrace{14 \div 2}$$
$$= 176 - \quad 7$$
$$= 169$$

You complete Example 6.

Example 6. Find $15 \div 3 + 18 \div 6$.

Solution

$$\underbrace{15 \div 3} + \underbrace{18 \div 6}$$
$$= \underline{\quad\quad} + \underline{\quad\quad}$$
$$= \underline{\quad\quad}$$

Check your work on page 635. ▶

Example 7. Find $11 + 20 \div 4 - 2 \cdot 8$.

Solution

$$11 + \underbrace{20 \div 4} - \underbrace{2 \cdot 8}$$
$$= \underbrace{11 + \quad 5} - 16$$
$$= \quad 16 \quad - \quad 16$$
$$= \quad 0$$

Example 8. Find $7 \cdot 9 \div 21 \cdot 2$.

Solution

$$\underbrace{7 \cdot 9} \div 21 \cdot 2$$
$$= \underbrace{63 \div 21} \quad \cdot 2$$
$$= \quad 3 \quad \cdot 2$$
$$= \quad 6$$

Working with several operations in the same problem is not difficult if you learn to proceed in an orderly manner. You should not try to do such arithmetic in your head.

⫸ Trial Run

Perform the indicated operations.

———— 1. $3 \cdot 5 + 7 \cdot 2$

———— 2. $13 + 9 \cdot 6$

———— 3. $45 - 3 \cdot 7$

———— 4. $126 + 48 \div 6$

———— 5. $28 \div 7 + 63 \div 7$

———— 6. $15 - 18 \div 6 + 6 \cdot 0$

Answers are on page 635.

Working with Parentheses

Mathematicians decided long ago to use symbols of grouping in a problem to tell the reader the order in which the operations are to be performed. When parentheses, ( ), and brackets, [ ], are used to group numbers in this way, they always tell us to *do the operation inside them first*.

Example 9. Find $(-3 + 8) - (2 + 7)$.

Solution

$$\underbrace{(-3 + 8)} - \underbrace{(2 + 7)}$$
$$= \quad 5 \quad - \quad 9$$
$$= \quad -4$$

Example 10. Find $(27 - 2) \div (3 - 8)$.

Solution

$$\underbrace{(27 - 2)} \div \underbrace{(3 - 8)}$$
$$= \quad 25 \quad \div \quad (-5)$$
$$= \quad -5$$

You complete Example 11.

Example 11. Find $(36 \div 3) \div (11 - 17)$.

Solution

$$(36 \div 3) \div (11 - 17)$$
$$= \quad \underline{\quad} \div \underline{\quad}$$
$$= \quad \underline{\quad}$$

Check your work on page 635. ▶

Example 12. Find $\left(\dfrac{3}{7} + \dfrac{9}{7}\right)\left(\dfrac{7}{11} + \dfrac{1}{11}\right)$.

Solution

$$\left(\frac{3}{7} + \frac{9}{7}\right)\left(\frac{7}{11} + \frac{1}{11}\right)$$
$$= \left(\frac{12}{7}\right)\left(\frac{8}{11}\right)$$
$$= \frac{96}{77}$$

How do parentheses fit into our rules for the order of operations? We have already agreed that parentheses say ''do this first.''

Order of Operations. To perform several operations within the same problem, we must

1. First do operations inside symbols of grouping.
2. Then do multiplications and divisions in order from left to right.
3. Then do additions and subtractions in order from left to right.

Example 13. Find $6 + 5(11 - 8)$.

Solution

$$6 + 5(11 - 8)$$

$$= 6 + 5(3) \qquad \text{Work inside parentheses.}$$

$$= 6 + 15 \qquad \text{Perform multiplication.}$$

$$= 21 \qquad \text{Perform addition.}$$

Example 14. Find $-3 - 2(-2 + 7)$.

Solution

$$-3 - 2(-2 + 7)$$

$$= -3 - 2(5)$$

$$= -3 - 10$$

$$= -13$$

▸ Trial Run

Perform the operations.

_____ **1.** $(9 + 5) - (6 + 2)$ _____ **2.** $(35 - 3) \div (12 - 14)$

_____ **3.** $(45 \div 3) \div (11 - 8)$ _____ **4.** $(7 - 9)(13 - 8)$

_____ **5.** $11 + 6(17 - 18)$ _____ **6.** $7 - 6(21 - 19)$

Answers are on page 635.

Fraction bars act just like parentheses in arithmetic and algebra problems. Everything above the fraction bar (the numerator) is to be treated as one quantity, and everything below the fraction bar (the denominator) is to be treated as another quantity.

Example 15. Find $\dfrac{6 + 19}{9 - 4}$.

Solution

$$\frac{6 + 19}{9 - 4} = \frac{25}{5} \qquad \text{Simplify numerator; simplify denominator.}$$

$$= 5 \qquad \text{Find the quotient.}$$

Example 16. Find $\dfrac{-7 - 3(2 + 1)}{2(11 - 9)}$.

Solution

$$\dfrac{-7 - 3(2 + 1)}{2(11 - 9)} = \dfrac{-7 - 3(3)}{2(2)} \qquad \text{Simplify within parentheses.}$$

$$= \dfrac{-7 - 9}{4} \qquad \text{Perform multiplications.}$$

$$= \dfrac{-16}{4} \qquad \text{Simplify numerator.}$$

$$= -4 \qquad \text{Find the quotient.}$$

To use an *exponent* on a quantity in parentheses, we follow our usual method and do the operation inside the parentheses first.

Example 17. Find $(6 + 9)^2$.

Solution

$$(6 + 9)^2$$
$$= 15^2$$
$$= 15 \cdot 15$$
$$= 225$$

You complete Example 18.

Example 18. Find $(5 - 7)^3$.

Solution

$$(5 - 7)^3$$
$$= (-2)^3$$
$$= (\underline{\hspace{1cm}})(\underline{\hspace{1cm}})(\underline{\hspace{1cm}})$$
$$= \underline{\hspace{1cm}}$$

Check your work on page 635. ▶

⫸ Trial Run

Perform the operations.

_____ **1.** $\dfrac{8 + 25}{7 + 4}$

_____ **2.** $\dfrac{-9 + 5(3 - 6)}{3(8 - 6)}$

_____ **3.** $(7 - 3)^3$

_____ **4.** $(5 - 8)^2$

_____ **5.** $\left(\dfrac{31}{3} - \dfrac{29}{3}\right)^4$

_____ **6.** $(9 - 11)^5$

Answers are on page 635.

Where do exponents fit into our rules for the order of operations? After doing the operations inside parentheses, exponents are the first thing to take care of.

Rules for Order of Operations

1. First do the operations inside symbols of grouping.
2. Then work with exponents.
3. Then do the multiplications and divisions in order from left to right.
4. Then do the additions and subtractions in order from left to right.

Example 19. Find $3 + 7^2$.

Solution

$$3 + 7^2 = 3 + 7 \cdot 7$$
$$= 3 + 49$$
$$= 52$$

You complete Example 20.

Example 20. Find $5(-3)^2$.

Solution

$$5(-3)^2 = 5(-3)(-3)$$
$$= 5 \cdot \underline{\quad\quad}$$
$$= \underline{\quad\quad}$$

Check your work on page 635. ▶

Example 21. Find $(32 \div 8) + 6(3 - 5)^3$.

Solution

$$(32 \div 8) + 6(3 - 5)^3$$
$$= \quad 4 \quad + 6(-2)^3$$
$$= \quad 4 \quad + 6(-2)(-2)(-2)$$
$$= \quad 4 \quad + 6(-8)$$
$$= \quad 4 \quad + (-48)$$
$$= \quad -44$$

Example 22. Find $\dfrac{5 + 8^2}{2^2 - 3^3}$.

Solution

$$\frac{5 + 8^2}{2^2 - 3^3}$$
$$= \frac{5 + 8 \cdot 8}{2 \cdot 2 - 3 \cdot 3 \cdot 3}$$
$$= \frac{5 + 64}{4 - 27}$$
$$= \frac{69}{-23}$$
$$= -3$$

⫸ **Trial Run**

Perform the operations.

_____ **1.** $5 + 8^2$

_____ **2.** $(-3)^2 - 7^2$

_____ **3.** $8 \cdot 5^2$

_____ **4.** $[35 \div (-7)] + 3(4 - 3)^5$

_____ **5.** $\dfrac{3 + 5^2}{3^2 - 2}$

_____ **6.** $3(10 \div 5)^2 - 2(10 - 8)$

Answers are on page 635.

▶ Examples You Completed

Example 4. Find $9 + 5 \cdot 4$.

Solution

$$9 + \underbrace{5 \cdot 4}$$
$$= 9 + \quad 20$$
$$= 29$$

Example 6. Find $15 \div 3 + 18 \div 6$.

Solution

$$\underbrace{15 \div 3} + \underbrace{18 \div 6}$$
$$= \quad 5 \quad + \quad 3$$
$$= \quad 8$$

Example 11. Find $(36 \div 3) \div (11 - 17)$.

Solution

$$\underbrace{(36 \div 3)} \div \underbrace{(11 - 17)}$$
$$= \quad 12 \quad \div \quad (-6)$$
$$= \quad -2$$

Example 18. Find $(5 - 7)^3$.

Solution

$$(5 - 7)^3 = (-2)^3$$
$$= (-2)(-2)(-2)$$
$$= -8$$

Example 20. Find $5(-3)^2$.

Solution

$$5(-3)^2 = 5(-3)(-3)$$
$$= 5 \cdot 9$$
$$= 45$$

Answers to Trial Runs

page 630   **1.** 10   **2.** 12   **3.** 9   **4.** -7 **5.** 3   **6.** 40

page 631   **1.** 29   **2.** 67   **3.** 24   **4.** 134   **5.** 13   **6.** 12

page 632   **1.** 6   **2.** -16 **3.** 5   **4.** -10 **5.** 5   **6.** -5

page 633   **1.** 3   **2.** -4 **3.** 64   **4.** 9   **5.** $\frac{16}{81}$ **6.** -32

page 634   **1.** 69   **2.** -40 **3.** 200   **4.** -2 **5.** 4   **6.** 8

EXERCISE SET 14.4

Perform the indicated operations.

_____ 1. $36 - 15 - 23$

_____ 2. $46 - 13 - 31$

_____ 3. $18 \cdot 3 \cdot \dfrac{1}{6}$

_____ 4. $15 \cdot 4 \cdot \dfrac{1}{3}$

_____ 5. $182 \div 7 \div 13$

_____ 6. $60 \div 4 \div 3$

_____ 7. $12 \div 4 \div \dfrac{1}{3}$

_____ 8. $54 \div 9 \div \dfrac{1}{6}$

_____ 9. $8 \cdot 3 + 7 \cdot 3$

_____ 10. $4 \cdot 3 + 9 \cdot 7$

_____ 11. $15.3 - 7.1 \div 2$

_____ 12. $18.5 - 3.21 \div 3$

_____ 13. $18 \div 3 + 27 \div (-3)$

_____ 14. $35 \div 5 + 48 \div (-6)$

_____ 15. $-13.6 + 10.5 \div 5 - 6.3$

_____ 16. $-12.2 + 21.7 \div 7 - 8.8$

_____ 17. $8 \cdot 6 \div 4 \cdot 5$

_____ 18. $9 \cdot 14 \div 21 \cdot 6$

_____ 19. $(-5 + 12) - (6 + 3)$

_____ 20. $(-8 + 15) - (7 + 3)$

_____ 21. $(38 - 2) \div (4 - 10)$

_____ 22. $(15 - 45) \div (8 - 2)$

_____ 23. $\left(\dfrac{3}{4} + \dfrac{15}{4}\right)\left(\dfrac{5}{3} - \dfrac{7}{3}\right)$

_____ 24. $\left(-\dfrac{5}{8} + \dfrac{3}{8}\right)\left(\dfrac{5}{6} - \dfrac{13}{6}\right)$

_____ 25. $15 - 3(8 - 11)$

_____ 26. $19 - 2(23 - 14)$

_____ 27. $\dfrac{2}{3} - \left(-\dfrac{1}{2} + \dfrac{5}{6}\right)$

_____ 28. $\dfrac{4}{5} - \left(-\dfrac{1}{2} + \dfrac{7}{10}\right)$

_____ 29. $-0.3[-6.2 - (8 - 9.3)]$

_____ 30. $-0.4[-5.3 - (7 - 8.6)]$

_____ 31. $\dfrac{-40 + 15}{16 - (-9)}$

_____ 32. $\dfrac{-17 + 10}{4 - (-3)}$

_____ 33. $\dfrac{2.0 + 1.5}{0.7 - 0.7}$

_____ 34. $\dfrac{-3.5 + 2}{5.2 - 5.2}$

_____ 35. $\dfrac{3(-8) + (-2)(-4)}{4(-4)}$

_____ 36. $\dfrac{5(-9) + (-3)(-2)}{3(-13)}$

_____ 37. $\dfrac{7[18 + 5(-2)]}{-8}$

_____ 38. $\dfrac{9[21 + 7(-2)]}{-21}$

_____ 39. $\dfrac{-9(7 - 1) - 8(-3)}{-5(-8 + 6)}$

_____ 40. $\dfrac{-8(10 - 3) - 9(-4)}{2(-9 + 4)}$

_____ 41. $(9 - 6)^2$

_____ 42. $(7 - 3)^2$

_____ 43. $(12 - 15)^3$

_____ 44. $(3 - 8)^3$

_____ 45. $\left(\dfrac{7}{8} - \dfrac{3}{8}\right)^5$

_____ 46. $\left(\dfrac{7}{9} - \dfrac{4}{9}\right)^4$

_____ 47. $(25 \div 5) - 3(4 - 6)^3$

_____ 48. $(48 \div 6) - 2(3 - 5)^3$

_____ 49. $\dfrac{9 + 6^2}{2^2 - 3^2}$

_____ 50. $\dfrac{17 + 8^2}{4^2 - 5^2}$

☆ Stretching the Topics

Perform the indicated operations.

_____ 1. $\left[\left(\dfrac{1}{2} - \dfrac{2}{3}\right) \div \left(\dfrac{5}{6} - \dfrac{5}{8}\right)\right]^2$

_____ 2. $(-0.1)^3[3.76(0.2)^2 + (0.3)^3(-5)]$

_____ 3. $\left(\dfrac{3}{4}\right)^2 - \left(\dfrac{7}{12} \div \dfrac{2}{3}\right) + \left(\dfrac{2}{9} \cdot \dfrac{3}{4}\right)$

Check your answers in the back of your book.

If you can do the problems in **Checkup 14.4,** you are ready to go to Section 14.5.

✓ CHECKUP 14.4

Perform the indicated operations.

_____ 1. $84 \div 6 \div 7$

_____ 2. $-13 + 21 \div 7 - 3 \cdot 5$

_____ 3. $(8.6 - 5.1) \div (2.3 - 3)$

_____ 4. $\left(\dfrac{3}{8} - \dfrac{9}{8}\right)\left(\dfrac{11}{6} - \dfrac{7}{6}\right)$

_____ 5. $-9 - 5(6 - 15)$

_____ 6. $\dfrac{12 + 15}{-17 + 8}$

_____ 7. $\dfrac{-12 + (8 - 23)}{3(-7 + 4)}$

_____ 8. $\left(-\dfrac{7}{3} + \dfrac{5}{3}\right)^{3}$

_____ 9. $-3(-5)^{2}$

_____ 10. $(75 \div 3) - (4 - 9)^{2}$

Check your answers in the back of your book.

If You Missed Problems:	You Should Review Exercises:
1	1, 2
2	7
3	10
4	12
5	13, 14
6, 7	15, 16
8	18
9, 10	20, 21

14.5 Switching from Word Expressions to Number Expressions

Now that we understand how to operate with all kinds of numbers, we must practice switching from words to expressions involving positive and negative numbers.

Example 1. The Sociology Club began the semester with $689.37 in its treasury. During the semester, dues of $5 were collected from 43 members. The club also held a banquet costing $413.79 and presented two $25 awards to outstanding students. What was the balance in the treasury at the end of the semester?

Solution. Money coming into the treasury can be represented by *positive* numbers and money spent can be represented by *negative* numbers.

Beginning balance	$689.37
Dues collected	43($5)
Banquet cost	$-$413.79
Awards cost	2($-$25)

We must find the *sum* of these amounts.

$$689.37 \, + \, 43(5) \, + \, (-413.79) \, + \, 2(-25)$$

$= 689.37 \, + \, 215 \, + \, (-413.79) \, + \, (-50)$ Perform multiplications.

$= 904.37 \qquad\qquad + \, (-463.79)$ Add positive numbers.
Add negative numbers.

$= 440.58$ Find the sum.

The balance in the club treasury is $440.58.

Let's look at the problem stated at the beginning of this chapter.

Example 2. James has been attending Weight Worriers meetings for 6 months. His monthly weight record is as follows: lost 5 pounds, lost 2 pounds, gained 3 pounds, lost 1 pound, gained 2 pounds, and lost 4 pounds. Find James's average monthly loss or gain.

Solution. Gains can be represented by positive numbers and losses by negative numbers. To find the average, we must *add* the numbers and *divide* by 6.

$$\frac{-5 + (-2) + 3 + (-1) + 2 + (-4)}{6}$$

$$= \frac{-5 + (-2) + (-1) + (-4) + 3 + 2}{6}$$ Rearrange the numbers.

$$= \frac{-12 + 5}{6}$$ Add positive numbers.
Add negative numbers.

$$= \frac{-7}{6}$$ Find the sum.

$$= -1\frac{1}{6}$$ Change improper fraction to mixed numbers.

James *lost* an average of $1\frac{1}{6}$ pounds per month.

Example 3. When Professor Temin left Wisconsin, the air temperature was $-5°$ F. When he arrived in Hawaii, the air temperature was 89°F. What was the change in air temperature experienced by Professor Temin?

Solution. To find the *change* in air temperature, we must *subtract* the original temperature from the new temperature.

$$89° - (-5°)$$

$= 89° + 5°$ Rewrite subtraction as addition.

$= 94°$ Find the sum.

The air temperature increased by 94°.

EXERCISE SET 14.5

Change each word expression to a number expression and simplify.

_____ **1.** Chelsea scored 93 on her first math test. If she scored 15 points lower on her next test, what score did she receive on her second test?

_____ **2.** At a summer sale Jason bought a stepladder that regularly sold for $41.95. If the price had been reduced by $15, how much did Jason pay for the stepladder?

_____ **3.** The Hatchers used 215 fewer kilowatt-hours of electricity in April than they used in March. If they used 1503 kilowatt-hours in April, how many kilowatt-hours did they use in March?

_____ **4.** At Floyd's Market, Shannon's total grocery bill was $89.73, but after the total value of his coupons was deducted, he paid only $81.93. How much did he save by using the coupons?

_____ **5.** A dinner for 2 at the Round Table cost $29.74. If Dan and Martina split the bill evenly, how much will each pay?

_____ **6.** If Myron's car averages 18.5 miles per gallon of gas, how many gallons of gas will he need to drive 296 miles?

_____ **7.** If Gwen drives at an average speed of 63 miles per hour, how far can she drive in 5.5 hours?

_____ **8.** Sherri bought 13.7 gallons of gas at $1.05 per gallon. Find the cost of the gasoline.

_____ **9.** Ms. Graff uses $\frac{1}{3}$ cup of fertilizer for each rose bush in her yard. If she has 28 rose bushes, how many cups of fertilizer does she need?

_____ **10.** Ken uses $2\frac{1}{3}$ cups of flour for each jam cake he bakes. If he is baking 5 jam cakes for the Jaycee Bake Sale, how many cups of flour does he need?

_____ **11.** Erin played a 25-cent slot machine 9 times, winning the 75-cent jackpot 4 times. How much did Erin win or lose at the slot machine?

_____ **12.** If Whitney works 8 hours at $10.35 per hour and has $13.46 deducted for taxes, what will be his take-home pay?

_____ **13.** If taxi fare is $1.95 for the first mile and $1.25 for each additional mile, what will be the fare for a 12-mile trip?

_____ **14.** A club sold 175 dance tickets for $5.25 each and had expenses of $315.75. What was the club's profit?

_____ **15.** Santos borrowed $7.50 from his sister twice last month. This month he paid her $5.25 and borrowed $4.25. What is the present condition of Santos' account with his sister?

_____ **16.** The estate of wealthy Uncle Linus is to be divided equally among his 5 nieces. If the estate includes a house worth $150,000 and 3 automobiles worth $11,000 each, together with debts of $9000 and $29,000, how much will each niece inherit?

_____ 17. Nezam plans to fence his rectangular yard which measures $152\frac{1}{4}$ feet by $242\frac{2}{3}$ feet, but there will be 2 gates where no fencing is needed. One gate is $3\frac{1}{3}$ feet wide and the other is $6\frac{1}{2}$ feet wide. How many feet of fencing will Nezam need?

_____ 18. Jake uses $\frac{3}{10}$ of his salary for rent, $\frac{1}{8}$ for food, and $\frac{1}{5}$ for utilities. What fractional part of his salary is left for other expenses?

_____ 19. Kalani works in the morning for 3 hours at $4.15 per hour and in the afternoon for 4 hours at $4.95 per hour. How much does she earn in 1 day?

_____ 20. Mrs. Ladusaw works 8 hours a day at $9.86 per hour and hires a baby-sitter for 9 hours a day at $2.95 per hour. After she pays the baby-sitter, how much does Mrs. Ladusaw earn each day?

_____ 21. For lunch Christy had a Big Burger for $1.79, a salad for $1.29, and a
_____   carton of milk for $0.65. If the sales tax rate is 4.5 percent, find the tax on
_____   Christy's lunch. Find the total cost of her lunch. How much change should she receive from a 10-dollar bill?

_____ 22. Emile and his roommate split equally rent and food expenses. This month they spent $310 for rent and $115.82 for food. If Emile has $315.75, how much will he have left after paying his share of the expenses?

_____ 23. On an achievement test, a student receives 2 points for every correct answer, loses 2 points for every incorrect answer, and loses 1.5 points for every unanswered question. If Miles answered 27 questions correctly, answered 13 questions incorrectly, and left 10 questions unanswered, find his score on the test.

_____ 24. At the Greenwood Theater the price of an adult's ticket is $5.50. A child's ticket sells for half as much as an adult's ticket, and a senior citizen's ticket sells for $0.75 less than an adult's ticket. For one performance, 75 adults' tickets, 23 children's tickets, and 14 senior citizens' tickets were sold. Find the amount received at the ticket window.

☆ Stretching the Topics _____

Change each word expression to a number expression and simplify.

_____ 1. Profits or losses are split equally among the 5 owners of the Pizza Shack. During one week, the shack took in $1873.95. Supplies cost $928.17, 8 workers were paid $150 each, and advertising cost $53.75. What was each owner's share of the profits or losses?

_____ 2. Emily Wright bought a home for $95,000. She sold the house a year later for $116,000. During the year she had spent $796.83 for taxes and $585 for insurance. She paid 6.5 percent of the selling price to a real estate agent. How much profit did Emily make on the transaction?

Check your answers in the back of your book.

If you can do the problems in **Checkup 14.5,** you are ready to do the **Review Exercises** for Chapter 14.

✓ **CHECKUP 14.5**

Change each word expression to a number expression and simplify.

_____ **1.** Angela has a balance of $875.82 in her checking account. If she writes a $123.73 check at the grocery store and a $31.85 check at the cleaners, what will be her balance?

_____ **2.** If a balloon rises from the ground at a rate of $3\frac{1}{4}$ feet per second for 24 seconds and then falls at a rate of $2\frac{1}{2}$ feet per second for 12 seconds, what is its final height?

_____ **3.** During one year, a farm with 3 equal-sharing owners had sales of $826,736 and expenses of $215,927. What is each owner's share of the profits?

_____ **4.** After a heavy rainfall the Wabash River reached a height of 48 feet. During the next 5 hours, readings indicated that the river rose $\frac{2}{9}$ foot, rose $\frac{1}{6}$ foot, remained steady, fell $\frac{1}{5}$ foot, and fell $\frac{3}{4}$ foot. What was the river's height after the last reading?

_____ **5.** At the local VFW bingo game, Monty lost $5 per week for 3 weeks, won $7 per week for 2 weeks, and lost $2.50 per week for 4 weeks. Find Monty's average weekly gain or loss.

Check your answers in the back of your book.

If You Missed Problems:	You Should Review Examples:
1	1
2	2
3	1
4	2
5	1–3

Summary

In this chapter we learned to work with integers, decimal numbers, and fractional numbers that are positive or negative or 0. After noting that every such number corresponds to a point on the number line, we learned to compare them and find their absolute values.

Symbol	Words	Meaning	Examples
$A < B$	A is less than B	A lies to the left of B on the number line	$0 < 3$ $-7 < -4$
$A > B$	A is greater than B	A lies to the right of B on the number line	$9 > 1$ $-3.7 > -4$
$\lvert A \rvert$	Absolute value of A	The distance between 0 and A on the number line	$\lvert -3 \rvert = 3$ $\left\lvert \dfrac{17}{5} \right\rvert = \dfrac{17}{5}$ $\lvert -6.84 \rvert = 6.84$

Then we turned our attention to performing operations with positive and negative numbers.

Operation	Examples
To *add* two positive numbers, add the units (absolute values). The sum will be positive.	$13 + 17 = 30$
To *add* two negative numbers, add the units (absolute values). The sum will be negative.	$-8 + (-11) = -19$
To *add* a positive number and a negative number, find the difference in the units (absolute values). Give the answer the sign of the original number with more units.	$-12 + 3 = -9$ $-6 + 10 = 4$
To *subtract* numbers, rewrite the difference as a sum (using the opposite) and add.	$5 - 8 = 5 + (-8)$ $\qquad = -3$ $-7 - (-12) = -7 + 12$ $\qquad = 5$
To *multiply* two numbers with the same sign, multiply the units (absolute values). The product will be positive.	$6(12) = 72$ $-3(-8) = 24$
To *multiply* two numbers with different signs, multiply the units (absolute values). The product will be negative.	$5(-6) = -30$ $-4(7) = -28$
To *divide* two numbers with the same sign, divide the units (absolute values). The quotient will be positive.	$\dfrac{15}{3} = 5$ $\dfrac{-18}{-6} = 3$
To *divide* two numbers with different signs, divide the units (absolute values). The quotient will be negative.	$\dfrac{-45}{5} = -9$ $\dfrac{44}{-4} = -11$

We found that the **commutative** and **associative** properties for addition and multiplication allow us to rearrange the terms of a sum or the factors of a product in convenient order. Then we discussed the rules for the **order of operations** illustrated in the next example.

Example	*Steps*
$13 - 2(5 - 2)^2 + \dfrac{15}{5}$	
$= 13 - 2(3)^2 + \dfrac{15}{5}$	Do operation inside parentheses.
$= 13 - 2 \cdot 9 + \dfrac{15}{5}$	Deal with the exponent.
$= 13 - 18 + 3$	Do multiplication and division from left to right.
$= -5 + 3$	Do addition and subtraction from left to right.
$= -2$	

☐ Speaking the Language of Mathematics

Complete each sentence with the appropriate word or phrase.

1. All negative numbers are located to the _____ of 0 on the number line.

2. The distance between a number and 0 on the number line is called the _____ _____ of the number.

3. The numbers 5 and -5 are called _____ of each other.

4. When we add 2 negative numbers, the sum is _____ ; when we multiply 2 negative numbers, the product is _____ .

5. In the expression 2^3, we call 2 the _____ and we call 3 the _____ .

6. When we state that $5(3 + 7) = 5 \cdot 3 + 5 \cdot 7$, we are using the _____ property for multiplication over addition.

△ Writing About Mathematics

Write your response to each question in complete sentences.

1. Explain to someone, in words, how to use the number line to find the sum $-2 + (-3)$.

2. Translate the distributive property statement $3(-2 + 7) = 3(-2) + 3 \cdot 7$ into words.

3. Summarize, in words, the rules for the sign of a product or quotient when multiplying or dividing positive and/or negative numbers.

REVIEW EXERCISES for Chapter 14

1. Compare each pair of numbers using $<$ or $>$.

_____ **(a)** $-6, -8$

_____ **(b)** $-3.5, 5.7$

_____ **(c)** $\dfrac{8}{5}, -\dfrac{3}{5}$

_____ **(d)** $4, -7$

2. Simplify by removing absolute value bars.

_____ **(a)** $|-3|$

_____ **(b)** $\left|\dfrac{3}{2} - \dfrac{7}{2}\right|$

_____ **(c)** $-|3.9|$

_____ **(d)** $|3| - |-12|$

Perform the indicated operations.

_____ **3.** $-8 + (-6)$

_____ **4.** $12 + (-15)$

_____ **5.** $-12.3 + 7.8$

_____ **6.** $[9 + (-12)] + (-7)$

_____ **7.** $-\dfrac{3}{8} + \dfrac{7}{8} + \left(-\dfrac{5}{8}\right)$

_____ **8.** $9 + (-3) + (-7) + 5$

_____ **9.** $3[12 + (-4)]$

_____ **10.** $3[(-8) + 5 \cdot 2]$

Write each expression using addition, and simplify.

_____ **11.** $6 - 15$

_____ **12.** $1.8 - (-0.7)$

_____ **13.** $-\dfrac{3}{4} - \left(-\dfrac{1}{4}\right)$

_____ **14.** $-13 - 9$

Perform the indicated operations.

_____ **15.** $20 - (4 - 8)$

_____ **16.** $-9 - (12 - 6)$

_____ **17.** $[-9 - (-4)] - 8$

_____ **18.** $5[(-3) - (-12)]$

_____ **19.** $(1.3 - 6.2) - (5 - 3.4)$

_____ **20.** $\left(-\dfrac{1}{2} - \dfrac{3}{4}\right) - \left(\dfrac{3}{2} - \dfrac{5}{4}\right)$

_____ **21.** $9(-3)$

_____ **22.** $(-2.3)(-4.5)$

_____ **23.** $-6(-4) + (-7)(5)$

_____ **24.** $-\dfrac{3}{5}\left(\dfrac{2}{9} - \dfrac{7}{9}\right)$

_____ **25.** $(-4.1)^2$

_____ **26.** $(0.2)^3(5)^2$

_____ **27.** $\dfrac{-45}{9}$

_____ **28.** $\dfrac{-54}{-9}$

_____ **29.** $(-4.16) \div (1.3)$

_____ **30.** $(-9 \cdot 8) \div (-12)$

_____ **31.** $72 \div 4 \div (-3)$

_____ **32.** $-15 + 36 \div 9 - 4 \cdot 3$

_____ 33. $(48 - 15) \div (12 - 23)$

_____ 34. $\left(\dfrac{5}{6} - \dfrac{13}{6}\right)\left(\dfrac{1}{3} - \dfrac{5}{3}\right)$

_____ 35. $-12 - 6(13 - 8)$

_____ 36. $\dfrac{17 + 32}{-18 + 11}$

_____ 37. $\dfrac{-28 + (7 - 15)}{3(-12 + 9)}$

_____ 38. $\left(-\dfrac{7}{2} + 2\right)^3$

_____ 39. $-5(-2)^3$

_____ 40. $(125 \div 5) - (7 - 12)^2$

Change each word expression to a number expression and simplify.

_____ 41. At 7 P.M. the temperature was 16°. It dropped 3° each hour for the next 7 hours. What was the temperature at 2 A.M.?

_____ 42. Marquita worked at a restaurant for 8 hours at $5.35 per hour but had $6.87 deducted for what she ate during that time. How much did she earn that day?

_____ 43. A hot-air balloon rises from the ground at $3\frac{1}{2}$ feet per second for 20 seconds, then falls at $2\frac{1}{4}$ feet per second for 12 seconds. What is its height after 32 seconds?

_____ 44. Yesterday Rudy's checking account showed a balance of $68.20. Today he deposited $125 and wrote 2 checks for $25 each and others for $17.83 and $25.72. What is his balance now?

_____ 45. Carlos opened a hot dog stand at the carnival. During the first 3 days he lost $40 per day. As business picked up, he made a profit of $110 per day for the next 4 days, but during a 2-day cold spell, he lost $30 per day. On the carnival's closing day, he made a profit of $230. What was Carlos's average daily profit or loss?

Check your answers in the back of your book.

If You Missed Exercises:	You Should Review Examples:	
1	Section 14.1	1, 2
2		8–11
3–5		17–19
6–10		20–26
11–14	Section 14.2	1–4
15–20		5–8
21, 22	Section 14.3	1–4
23, 24		5–8
25, 26		11–15
27–30		16–23
31, 32	Section 14.4	2–8
33–35		9–12
36, 37		15, 16
38–40		17, 18
41–45	Section 14.5	1–3

If you have completed the **Review Exercises** and corrected your errors, you are ready to take the **Practice Test** for Chapter 14.

PRACTICE TEST for Chapter 14

	SECTION	EXAMPLES

1. Compare each pair of numbers using $<$ or $>$.   14.1 1, 2

     (a) $-15, -1$ (b) $\dfrac{1}{2}, -3$

_____ 2. Simplify $-|-8|$.   14.1 11

Perform the indicated operations.

_____ 3. $[3 + (-11)] + (-12)$ 14.1 21

_____ 4. $-\dfrac{9}{5} + \dfrac{3}{5} + \left(-\dfrac{2}{5}\right)$ 14.1 25

_____ 5. $-19 - (-23)$ 14.2 3

_____ 6. $[-10 + 6] - [5 - (-9)]$ 14.2 5

_____ 7. $3.7 - (1.5 - 2.2)$ 14.2 8

_____ 8. $(-12)(-7)$ 14.3 4

_____ 9. $[(-7)(2)](-3)$ 14.3 5

_____ 10. $-5(8.3 - 10.2)$ 14.3 8

_____ 11. $\dfrac{-72}{-8}$ 14.3 18

_____ 12. $(-32.2) \div (0.23)$ 14.3 22

_____ 13. $\dfrac{3}{8} + \dfrac{3}{4} - \dfrac{1}{2}$ 14.4 1

_____ 14. $12 + 15 \div 3 - 7.5$ 14.4 7

_____ 15. $\left(\dfrac{5}{7} + \dfrac{3}{7}\right)\left(\dfrac{4}{9} - \dfrac{11}{9}\right)$ 14.4 12

_____ 16. $-2.3 - (7 - 6.1)$ 14.4 14

_____ 17. $\dfrac{(-6)(-3) - 18}{-9 + 4}$ 14.4 16

_____ 18. $(3 - 8)^3$ 14.4 18

_____ 19. After attending Dieter's Delight Club for 8 weeks, Clem's record 14.5 2
showed that he had lost 2.5 pounds twice, gained 1.5 pounds twice,
lost 4 pounds 3 times, and stayed the same once. What was Clem's
average weekly loss?

	SECTION	EXAMPLES

_____ **20.** At the track Bob bet $2 on each of the 10 races. Twice he won   14.5   1
$6.70 and once he won $15.80. How much did Bob win or lose that
day?

Check your answers in the back of your book.

SHARPENING YOUR SKILLS after Chapters 1–14

SECTION

_____ 1. In the whole number 856,729, name the thousands digit.   1.1

_____ 2. Reduce $\frac{234}{273}$ to lowest terms.   3.2

_____ 3. Compare 5.305 and 5.035 using $<$ or $>$.   6.1

_____ 4. A 20-ounce box of oat bran cereal sells for $3.40. Find the ratio of cents to ounces.   8.1

_____ 5. Change 1.75 to a percent.   9.1

_____ 6. Change $2\frac{1}{4}$ miles to feet.   12.1

_____ 7. Change 0.0015 kilometer to centimeters.   12.1

_____ 8. Change $8\frac{1}{2}$ pounds to ounces.   12.2

_____ 9. Change 2500 milliliters to liters.   12.3

_____ 10. Compare $-3\frac{1}{2}$ and $-\frac{2}{3}$ using $<$ or $>$.   14.1

Perform the indicated operations.

_____ 11. $10{,}904 \div 29$ 2.2

_____ 12. $4\frac{2}{3} \times 1\frac{13}{14}$ 4.1

_____ 13. $9\frac{2}{7} - 6\frac{3}{4}$ 5.3

_____ 14. $350 - 29.73 + 12.86$ 6.2

_____ 15. 0.4357×800 7.1

_____ 16. $[-12 + 8] - [9 - (-13)]$ 14.2

_____ 17. $\dfrac{(-5)(-3) - 15}{-12 + 7}$ 14.4

_____ 18. The roof of a storage shed consists of 4 triangular sections. The base of each triangle is 20 feet and the height is 12 feet. What is the area of the roof?   2.4

_____ 19. Chad's math class meets 64 hours during the semester. If Chad cut class 12 hours during the semester, what fractional part of the class time did he attend?   3.3

_____ 20. If one allows 8 ounces of roast per person, how many pounds will be needed to serve 15 people?   4.3

_____ 21. Estavez is fencing a rectangular lot which measures $145\frac{3}{4}$ feet by $230\frac{1}{3}$ feet, but there will be a gate $5\frac{1}{2}$ feet wide where no fencing will be needed. How many feet of fencing will he need?   5.5

_____ 22. If 64 ounces of laundry detergent costs $4.70, find the cost per ounce of the detergent. (Round answer to the nearest cent.)   7.4

_____ 23. A picture that is to be enlarged measures $3\frac{3}{4}$ inches wide and $4\frac{1}{2}$ inches long. If the enlargement is to have a length of 18 inches, what will be its width?   8.3

_____ 24. Next year Mr. Bush will receive a $4\frac{1}{2}$ percent cost-of-living raise. If his raise will be $1305, what is his salary this year? What will be his salary next year?   9.2

_____ 25. If Mr. McGee has an adjusted gross income of $28,375, calculate 7.5 percent of his adjusted gross income.   10.3

_____ 26. Find the mean for the following group of normal annual precipitation measurements (in inches) for 8 U.S. cities.   11.1

      7.05, 14.13, 15.4, 38.09, 12.72, 59.8, 10.72, 30.96

_____ 27. A farmer needs 8 kilometers of barbed wire for fencing. How much will the fencing cost if the wire sells for $0.55 per meter?   12.1

_____ 28. Find the distance around a circular swimming pool that has a diameter of 36 feet. (Use $\pi \doteq 3.14$.)   13.1

_____ 29. How many square yards of carpet must be bought to cover the floor of a rectangular room measuring 15 feet by 18 feet?   13.2

_____ 30. A promoter sold 3528 tickets for a concert at $18 per ticket. He paid the band $25,000 and had other expenses totaling $18,785. What was the promoter's profit?   14.4

Check your answers in the back of your book.

15

Working with Basic Algebra

Suppose Bradley has 42 yards of fencing to use around his rectangular garden. If he wants the garden to be twice as long as it is wide, find the length and width of the garden.

We have spent a great deal of time becoming familiar with the four basic operations of arithmetic (addition, subtraction, multiplication, and division). Now we are ready to put our arithmetic skills to use in studying another branch of mathematics called **algebra.**

If you have mastered the skills of arithmetic, you will find that the study of basic algebra is the next logical step in your mathematical development. Your understanding of the basic ideas of algebra depends upon your understanding of arithmetic. But the skills you develop in algebra will provide you with powerful tools for dealing with problems that would be difficult to solve using arithmetic alone.

In this chapter we learn how to

1. Work with variables.
2. Work with equations.
3. Switch from word statements to equations.

15.1 Working with Variables

In the study of algebra, we must work with known and with unknown numbers. Known numbers (such as 3, 0, $\frac{1}{2}$, 5.7, π, and so on) are called **constants.** They are the familiar numbers we have used throughout our study of arithmetic. Unknown numbers are called **variables,** and we usually let some letter of the alphabet (such as *l, w, p,* or *x*) stand for a variable.

You should recall that we let *l* and *w* stand for unknown lengths and widths in perimeter and area problems. We let *p* stand for the unknown percent in solving percent proportions. Although we did not say so at the time, we were working with algebra when we solved such problems.

Evaluating Algebraic Expressions

When we combine constants and variables using the operations of addition, subtraction, multiplication, and division, we are working with **algebraic expressions.** We continue to use familiar symbols to show what operation is to be done.

$7 + x$ means add 7 and the variable *x*.

$y - 2$ means subtract 2 from the variable *y*.

$\dfrac{a}{5}$ means divide the variable *a* by 5.

$6 \cdot x$ or $6x$ means multiply 6 times the variable *x*.

$4(x + 1)$ means multiply 4 times the sum of the variable *x* and 1.

Example 1. Write an algebraic expression that means "add the variable *y* and 14."

Solution. $y + 14$

Example 2. Write an algebraic expression that means "find the sum of *x* and -9, then divide the sum by 2."

Solution. $[x + (-9)] \div 2$ or $\dfrac{x - 9}{2}$

Notice that we used brackets in Example 2 to show that we must add *first* and then divide.

Algebraic expressions have different values depending upon what number we let the variable be. Finding the value of an algebraic expression when the variable is allowed to be some particular number is called **evaluating** the expression. We evaluate an expression by replacing the variable with the particular number. For instance, we can evaluate $5x$ when $x = 2$ by replacing *x* by 2 in the expression.

$$5x = 5(2)$$
$$= 10$$

Replacing the variable with a particular number is also called **substituting** the number for the variable.

Example 3. Evaluate $a + 3$ when $a = 18$.

Solution

$a + 3$

$= 18 + 3$ Substitute 18 for *a*.

$= 21$ Find the sum.

You try Example 4.

Example 4. Evaluate $2l + 2w$ when $l = 7$ and $w = 9$.

Solution

$2l + 2w$

$= 2(\underline{\hspace{1cm}}) + 2(\underline{\hspace{1cm}})$

$= \underline{\hspace{1cm}} + \underline{\hspace{1cm}}$

$= \underline{\hspace{1cm}}$

Check your work on page 661. ▶

Example 5. Evaluate $x + 2(5 + x)$ when $x = 3$.

Solution

$$x + 2(5 + x)$$
$$= 3 + 2(5 + 3)$$
$$= 3 + 2(8)$$
$$= 3 + 16$$
$$= 19$$

You complete Example 6.

Example 6. Evaluate $y^2 - 19$ when $y = 7$.

Solution

$$y^2 - 19$$
$$= (\underline{})^2 - 19$$
$$= \underline{} \cdot \underline{} - 19$$
$$= \underline{} - 19$$
$$= \underline{}$$

Check your work on page 661. ▶

Notice that we continue to follow our rules for the order of operations.

⟫ Trial Run

———— **1.** Write an algebraic expression that means "add the variable x and 15."

———— **2.** Write an algebraic expression that means "subtract y from 8, then multiply by 10."

———— **3.** Evaluate $a - 15$ when $a = 27$.

———— **4.** Evaluate $a^2 + b^2$ when $a = 3$ and $b = 4$.

———— **5.** Evaluate $2m + 2n$ when $m = 8$ and $n = 11$.

———— **6.** Evaluate $x + 3(x - 2)$ when $x = 9$.

Answers are on page 661.

Combining Like Terms

The parts being added or subtracted in an algebraic expression are called the **terms** of the expression. In the expression $x + 3$, x is a term and 3 is a term. In the expression $y + 5x + 6$, y is a term, $5x$ is a term, and 6 is a term.

If a term is just a constant by itself, we call it a **constant term.** If a term contains a variable, we call it a **variable term.** Every variable term has a constant part (also called the **numerical coefficient**) and a variable part. In the variable term $5x$, the 5 is the numerical coefficient and the x is the variable part.

variable   *constant*
term   *term*
↓   ↓
$5x$ $+$ 9
↗ ↖
numerical *variable*
coefficient   *part*

If two variable terms contain *exactly the same variable part*, they are called **like terms.**

$5x$ and $9x$ are like terms.

$3y$ and $10y$ are like terms.

y and $2x$ are *not* like terms.

x^2 and $6x$ are *not* like terms.

Like terms are terms whose variable parts are *exactly* the same.

Example 7. Which terms in $5x + 2y + x + 7y$ are like terms?

Solution. $5x$ and x are like terms.

$2y$ and $7y$ are like terms.

Suppose we want to add or subtract the terms in an algebraic expression.

To add (or subtract) like terms, we add (or subtract) the numerical coefficients and keep the same variable part.

For instance,

$$5x + 3x = 8x$$

Example 8. Find $9a + 10a + a$.

Solution. All the terms are like terms. But what is the numerical coefficient in the last term, a? You should agree that $a = 1a$.

$$9a + 10a + a = \underbrace{9a + 10a} + 1a$$
$$= 19a + 1a$$
$$= 20a$$

Example 9. Find $2x + 3y + 5 + x + 18y + 3$.

Solution. We must decide which terms are like terms.

$2x$ and x are like terms.

$3y$ and $18y$ are like terms.

5 and 3 are like (constant) terms.

Now we may add the like terms.

$$2x + 3y + 5 + x + 18y + 3$$
$$= 2x + x + 3y + 18y + 5 + 3 \qquad \text{Rearrange terms.}$$
$$= \underbrace{2x + 1x} + \underbrace{3y + 18y} + \underbrace{5 + 3} \qquad \text{Rewrite } x \text{ as } 1x.$$
$$= 3x + 21y + 8 \qquad \text{Add like terms.}$$

You try Example 10.

Example 10. Find $7x - 4x$.

Solution

$$7x - 4x = \underline{\hspace{2cm}}$$

Check your work on page 661. ▶

Example 11. Find $19x - 20x - x$.

Solution

$$19x - 20x - x = \underbrace{19x - 20x}_{} - 1x$$
$$= \underbrace{-1x}_{} - 1x$$
$$= -2x$$

Example 12. Find $7 + x - 11 - x$.

Solution

$$7 + x - 11 - x = \underbrace{7 - 11}_{} + \underbrace{1x - 1x}_{}$$
$$= \underbrace{-4}_{} + \underbrace{0}_{}$$
$$= -4$$

The process of adding or subtracting terms that are alike is often called **combining like terms.**

⮕ Trial Run

Combine like terms.

$\underline{\hspace{2cm}}$ **1.** $9x - 5x$

$\underline{\hspace{2cm}}$ **2.** $18a - 12a - a$

$\underline{\hspace{2cm}}$ **3.** $7x + 3y - 2x - x - y$

$\underline{\hspace{2cm}}$ **4.** $13 + x - 19 - x$

$\underline{\hspace{2cm}}$ **5.** $6a + 7b - 2a - 9b$

$\underline{\hspace{2cm}}$ **6.** $6m + 2n + 4p - 2n - 4m + 5p$

Answers are on page 661.

Multiplying and Dividing Algebraic Expressions by Constants

Now that we know how to add and subtract algebraic expressions, we must learn to multiply and divide expressions by constants. Suppose we wish to find a product such as $2 \cdot 3x$. Recalling that $3x$ means $3 \cdot x$, we can rewrite our problem as

$$2 \cdot 3x = 2 \cdot 3 \cdot x \qquad \text{Rewrite } 3x \text{ as } 3 \cdot x.$$
$$= 6 \cdot x \qquad \text{Use associative property and multiply } 2 \cdot 3.$$
$$= 6x \qquad \text{Rewrite } 6 \cdot x \text{ as } 6x.$$

> **Multiplying a Constant Times a Variable Term.** To multiply a constant times a variable term, we multiply the constant times the coefficient of the term and keep the same variable part.

You complete Example 13.

Example 13. Find $6(-5x)$.

Solution

$$6(-5x) = 6(-5)x$$

$$= \underline{\hspace{1cm}}$$

Check your work on page 661. ▶

Example 14. Find $\frac{1}{2}(12a)$.

Solution

$$\frac{1}{2}(12a) = \frac{1}{2} \cdot 12a$$

$$= \frac{1}{\cancel{2}} \cdot \frac{\overset{6}{\cancel{12}}}{1} \cdot a$$

$$= 6a$$

To *divide* a variable term by a constant we may write the division problem as a fraction. Then we can reduce the fraction by dividing numerator and denominator by common factors.

$$10x \div 5 = \frac{10 \cdot x}{5}$$

$$= \frac{\overset{2}{\cancel{10}} \cdot x}{\underset{1}{\cancel{5}}}$$

$$= 2x$$

Dividing a Variable Term by a Constant. To divide a variable term by a constant, we divide the coefficient of the term by the constant and keep the same variable part.

In algebra, division problems are usually stated as fractions.

Example 15. Find $\frac{84a}{7}$.

Solution

$$\frac{84a}{7} = \frac{\overset{12}{\cancel{84}} \cdot a}{\underset{1}{\cancel{7}}}$$

$$= 12a$$

You complete Example 16.

Example 16. Find $\frac{-3a}{3}$.

Solution

$$\frac{-3a}{3} = \frac{\overset{-1}{-\cancel{3}} \cdot a}{\underset{1}{\cancel{3}}}$$

$$= \underline{\hspace{1cm}} \cdot a$$

$$= \underline{\hspace{1cm}}$$

Check your work on page 661. ▶

Example 17. Find $\dfrac{20x}{16}$.

Solution

$$\frac{20x}{16} = \frac{\overset{5}{\cancel{20}} \cdot x}{\underset{4}{\cancel{16}}}$$

$$= \frac{5}{4}x$$

Notice that we do *not* write improper fractions as mixed numbers in algebra.

⯈ Trial Run

Find the product or quotient.

_____ 1. $3(7x)$

_____ 2. $\dfrac{1}{3}(18a)$

_____ 3. $8\left(\dfrac{1}{2}x\right)$

_____ 4. $-72m \div 9$

_____ 5. $\dfrac{5a}{5}$

_____ 6. $\dfrac{21x}{-15}$

Answers are on page 661.

In Chapter 14, we discovered that a product such as $2(3 + 5)$ could be found in 2 ways. The second way of finding that product (using the distributive property) is especially important to us now in our study of algebraic expressions. Recall how to use the distributive property.

$$2(3 + 5) = 2 \cdot 3 + 2 \cdot 5 \qquad \begin{array}{l}\text{Multiply 2 times 3.}\\ \text{Multiply 2 times 5.}\end{array}$$

$$= \underline{6 \ + \ 10} \qquad \text{Find each product.}$$

$$= 16 \qquad \text{Find the sum.}$$

Using the Distributive Property. To multiply a constant times a sum (or difference) in parentheses, we multiply the constant times each term in the sum (or difference).

In using the distributive property, sometimes it helps to use arrows to remind us what we are doing.

$$3(2x + 5) = 3(2x + 5) \qquad \text{Multiply 3 times } 2x. \text{ Multiply 3 times 5.}$$
$$= 3(2x) + 3(5) \qquad \text{Write the products.}$$
$$= 6x + 15 \qquad \text{Find the products.}$$

You complete Example 18.

Example 18. Find $4(5y - 1)$.

Solution

$$4(5y - 1) = 4(5y - 1)$$
$$= 4(\underline{\quad}) - 4(\underline{\quad})$$
$$= \underline{\quad} - \underline{\quad}$$

Check your work on page 661. ▶

Example 19. Find $-7(x + 9)$.

Solution

$$-7(x + 9) = -7(1x + 9)$$
$$= -7(1x + 9)$$
$$= -7(1x) - 7(9)$$
$$= -7x - 63$$

Example 20. Find $\dfrac{1}{2}(2a - 6p)$.

Solution

$$\frac{1}{2}(2a - 6p) = \frac{1}{2}(2a - 6p)$$
$$= \frac{1}{2}(2a) - \frac{1}{2}(6p)$$
$$= 1a - 3p$$
$$= a - 3p$$

Example 21. Find $-3(x - 2y)$.

Solution

$$-3(x - 2y) = -3(x - 2y)$$
$$= -3x - (-3)(2y)$$
$$= -3x - (-6y)$$
$$= -3x + 6y$$

⫸ Trial Run ═══════════════════════════════════════

Find the products.

_____ 1. $2(3x + 7)$

_____ 3. $-8(x + 7)$

_____ 5. $5(x - y)$

_____ 2. $5(2y - 3)$

_____ 4. $\dfrac{1}{3}(3a - 6b)$

_____ 6. $-9(2 - 3n)$

Answers are on page 661.

▶ Examples You Completed

Example 4. Evaluate $2l + 2w$ when $l = 7$ and $w = 9$.

Solution

$$2l + 2w = 2(7) + 2(9)$$
$$= 14 + 18$$
$$= 32$$

Example 6. Evaluate $y^2 - 19$ when $y = 7$.

Solution

$$y^2 - 19 = 7^2 - 19$$
$$= 7 \cdot 7 - 19$$
$$= 49 - 19$$
$$= 30$$

Example 10. Find $7x - 4x$.

Solution

$$7x - 4x = 3x$$

Example 13. Find $6(-5x)$.

Solution

$$6(-5x) = 6(-5)x$$
$$= -30x$$

Example 16. Find $\dfrac{-3a}{3}$.

Solution

$$\frac{-3a}{3} = \frac{-\overset{-1}{\cancel{3}} \cdot a}{\underset{1}{\cancel{3}}}$$
$$= -1 \cdot a$$
$$= -a$$

Example 18. Find $4(5y - 1)$.

Solution

$$4(5y - 1) = 4(5y - 1)$$
$$= 4(5y) - 4(1)$$
$$= 20y - 4$$

Answers to Trial Runs

page 655   1. $x + 15$ 2. $10(8 - y)$ 3. 12 4. 25 **5.** 38 6. 30

page 657   1. $4x$ 2. $5a$ 3. $4x + 2y$ 4. -6 5. $4a - 2b$ 6. $2m + 9p$

page 659   1. $21x$ 2. $6a$ 3. $4x$ 4. $-8m$ 5. a 6. $-\dfrac{7}{5}x$

page 660   1. $6x + 14$ 2. $10y - 15$ 3. $-8x - 56$ 4. $a - 2b$ 5. $5x - 5y$ 6. $-18 + 27n$

EXERCISE SET 15.1

Write an algebraic expression for each word statement or expression.

_____ 1. Add 12 to x.

_____ 2. Add -5 to y.

_____ 3. Multiply $\frac{1}{3}$ times n.

_____ 4. Multiply $\frac{2}{5}$ times n.

_____ 5. Subtract 6.5 from x.

_____ 6. Subtract 3.2 from y.

_____ 7. The sum of x and -3, divided by 7.

_____ 8. The sum of y and -4, divided by 9.

_____ 9. 9 more than the product of $\frac{1}{6}$ and n.

_____ 10. 6 more than the product of $\frac{2}{3}$ and m.

Evaluate each expression.

_____ 11. $y - 3$, when $y = 10$

_____ 12. $y - 5$, when $y = 12$

_____ 13. $\frac{m}{5}$, when $m = -25$

_____ 14. $\frac{m}{7}$, when $m = -21$

_____ 15. $3x + 2y$, when $x = -3$ and $y = 4$

_____ 16. $5x + 3y$, when $x = -4$ and $y = 5$

_____ 17. $x + 2(3x + 1)$, when $x = -2$

_____ 18. $y + 3(2y + 3)$ when $y = -4$

_____ 19. $x^2 - 15$, when $x = 3$

_____ 20. $x^2 - 25$, when $x = 4$

Combine like terms.

_____ 21. $3a - 5a$

_____ 22. $5a - 7a$

_____ 23. $2x - 8x + 9x$

_____ 24. $3x - 11x + 10x$

_____ 25. $a + 10b - 9a + 7$

_____ 26. $2a - 11b - 6a + 3$

_____ 27. $4x - 3y + 13 + 7x + y - 6$

_____ 28. $2x + 5y - 6 + 3x - y - 3$

_____ 29. $\frac{2}{3}a - 5b + 4c + \frac{1}{3}a - 6b$

_____ 30. $\frac{3}{4}m - 3n + p + \frac{1}{4}m - 8n$

Find each product or quotient.

_____ 31. $3(-7x)$

_____ 32. $-4(5x)$

_____ 33. $-\frac{1}{3}(-15a)$

_____ 34. $-\frac{1}{5}(-20a)$

_____ 35. $\frac{16x}{-2}$

_____ 36. $\frac{28x}{-7}$

_____ 37. $\dfrac{-24a}{-8}$ _____ 38. $\dfrac{-26a}{-13}$

_____ 39. $\dfrac{36y}{24}$ _____ 40. $\dfrac{48y}{30}$

Find each product.

_____ 41. $4(x - 2)$ _____ 42. $5(x - 3)$

_____ 43. $-2(x - 2y)$ _____ 44. $-3(x - 5y)$

_____ 45. $\dfrac{1}{4}(12x + 4y)$ _____ 46. $\dfrac{1}{3}(6x + 12y)$

_____ 47. $1.5(6x - 2y)$ _____ 48. $0.5(2x - 4y)$

_____ 49. $9(-2m - 3n)$ _____ 50. $8(-3m - 5n)$

☆ Stretching the Topics _____

_____ 1. Evaluate $\dfrac{7(3x - 5y)}{4z + y}$ when $x = -2$, $y = 1$, and $z = 2$.

Simplify each expression.

_____ 2. $\dfrac{1}{2}x - 5y + \dfrac{4}{5}x - \dfrac{2}{3}y + 2z - 2x - \dfrac{1}{4}z$

_____ 3. $2.4(2x + 3y) - 0.5(-y - 3) + 1.2(2x - 1)$

Check your answers in the back of your book.

If you can complete **Checkup 15.1,** you are ready to go on to Section 15.2.

✓ **CHECKUP 15.1**

Write an algebraic expression for each word expression.

_____ **1.** Add y to -12.

_____ **2.** The sum of x and 9, divided by 5.

Evaluate each expression.

_____ **3.** $y + 3(y - 5)$, when $y = -2$.

_____ **4.** $x^2 - 7$, when $x = 3$

Combine like terms.

_____ **5.** $7a - 10a + 2a$

_____ **6.** $5a - 3b + 6 - 2a - 7b + 1$

Find each product or quotient.

_____ **7.** $\frac{1}{2}(-8x)$

_____ **8.** $\frac{-27x}{18}$

Find each product.

_____ **9.** $-3(x - 7)$

_____ **10.** $\frac{1}{3}(6x - 3y)$

Check your answers in the back of your book.

If You Missed Problems:	You Should Review Examples:
1, 2	1, 2
3, 4	5, 6
5, 6	8, 9
7	14
8	16, 17
9, 10	19, 20

15.2 Working with Equations

An **equation** is an algebra statement that says that one quantity is equal to another quantity. Every equation can be stated in words.

Equation Form	Word Form
$x + 3 = 7$	When 3 is added to the variable x, the sum is 7.
$x - 5 = 6$	When 5 is subtracted from the variable x, the difference is 6.
$-3x = 27$	When the variable x is multiplied by -3, the product is 27.
$\dfrac{x}{5} = 4$	When the variable x is divided by 5, the quotient is 4.

Example 1. Write an equation that says "when 2 is multiplied times the sum of the variable x and 5, the result is 3."

Solution. $2(x + 5) = 3$.

An equation is true or false, depending upon the number we choose for the variable. For instance, if we let $x = 6$, then the equation $x + 3 = 7$ is *not* a true statement, because $6 + 3 \neq 7$. But if $x = 4$, then the equation $x + 3 = 7$ becomes $4 + 3 = 7$, which is a true statement. The left-hand side is equal to the right-hand side.

When we find the value of the variable that makes an equation true, we have **solved** the equation. The correct value of the variable is called a **solution** for the equation. For the equation $x + 3 = 7$, 4 is a solution but 6 is not.

Using Addition and Subtraction to Solve Equations

We need to develop some efficient methods for finding solutions for equations. In solving an equation our main goal is to get the variable by itself on one side of the equation. The most important thing to remember about equations is

> Whatever we do to one side of an equation we must also do to the other side of the equation.

Consider this equation.

$$x + 3 = 7$$

To get x by itself, we would like to get rid of the 3 that is being added to the x. We can do this by *subtracting* 3 from the left-hand side of the equation. But if we subtract 3 on the left, we must also subtract 3 on the right.

$x + 3 = 7$	
$x + 3 - 3 = 7 - 3$	Subtract 3 from both sides.
$x + 0 = 4$	Find the differences.
$x = 4$	Find the sum on the left side.

We have found the solution. We can **check** our solution by substituting it for the variable x in the original equation to see if a true statement results.

$$x + 3 = 7 \qquad \text{Write the original equation.}$$

$$4 + 3 \overset{?}{=} 7 \qquad \text{Substitute } 4 \text{ for } x.$$

$$7 = 7 \qquad \text{Find the sum on the left side.}$$

This example illustrates one of the properties that we can use in solving equations.

We may *subtract* the same number from both sides of an equation.

You complete Example 2.

Example 2. Solve $x + 7 = 19$ and check.

Solution

$$x + 7 = 19 \qquad\qquad \text{CHECK:} \quad x + 7 = 19$$
$$x + 7 - \underline{\quad} = 19 - \underline{\quad} \qquad\qquad \underline{\quad} + 7 \overset{?}{=} 19$$
$$x + \underline{\quad} = \underline{\quad} \qquad\qquad\qquad 19 = 19$$
$$x = \underline{\quad}$$

Check your work on page 671. ▶

Example 3. Solve $14 + y = 17$ and check.

Solution

$$14 + y = 17 \qquad\qquad \text{CHECK:} \quad 14 + y = 17$$
$$14 - 14 + y = 17 - 14 \qquad\qquad 14 + 3 \overset{?}{=} 17$$
$$0 + y = 3 \qquad\qquad\qquad 17 = 17$$
$$y = 3$$

Suppose we wish to get the variable by itself in the equation

$$x - 2 = 9$$

We notice that 2 is being *subtracted* from the variable x. We may get rid of the 2 being subtracted by *adding* 2 on the left-hand side. Of course if we add 2 on the left we must also add 2 on the right.

$$x - 2 = 9$$
$$x - 2 + 2 = 9 + 2 \qquad \text{Add 2 to both sides.}$$
$$x + 0 = 11 \qquad \text{Find the sums.}$$
$$x = 11 \qquad \text{Simplify the left side.}$$

Let's check our solution.

$$x - 2 = 9$$ Write the original equation.

$$11 - 2 \overset{?}{=} 9$$ Substitute 11 for x.

$$9 = 9$$ Simplify the left side.

This example illustrates another property that can be used to solve equations.

> We may *add* the same number to both sides of an equation.

You complete Example 4.

Example 4. Solve $a - 13 = 10$ and check.

Solution

$$a - 13 = 10$$ CHECK: $a - 13 = 10$

$$a - 13 + \underline{\hspace{1cm}} = 10 + \underline{\hspace{1cm}}$$ $$\underline{\hspace{1cm}} - 13 \overset{?}{=} 10$$

$$a + \underline{\hspace{1cm}} = \underline{\hspace{1cm}}$$ $$10 = 10$$

$$a = \underline{\hspace{1cm}}$$

Check your work on page 671. ▶

Example 5. Solve $x - 2.7 = -6.9$ and check.

Solution

$$x - 2.7 = -6.9$$ CHECK: $x - 2.7 = -6.9$

$$x - 2.7 + 2.7 = -6.9 + 2.7$$ $$-4.2 - 2.7 \overset{?}{=} -6.9$$

$$x + 0 = -4.2$$ $$-6.9 = -6.9$$

$$x = -4.2$$

⇒ Trial Run

Solve each equation and check.

_____ **1.** $x + 3 = 8$ 　　　　　 _____ **2.** $3 + m = 3$

_____ **3.** $y - 4 = 6$ 　　　　　 _____ **4.** $a - 7 = 15$

_____ **5.** $x - 3.25 = 8.5$ 　　　 _____ **6.** $y + 19.5 = 12$

Answers are on page 671.

Using Multiplication and Division to Solve Equations

We have seen how to use addition to undo subtraction and how to use subtraction to undo addition in solving an equation. How can we undo multiplication or division in solving equations? To solve the equation

$$3x = 15$$

we would like to get rid of the 3 that is multiplied times the x. Perhaps we should try to use *division* to undo *multiplication*. Let's divide both sides of the equation by 3.

$$3x = 15$$

$$\frac{3x}{3} = \frac{15}{3} \qquad \text{Divide both sides by } 3.$$

$$1x = 5 \qquad \text{Simplify the quotients.}$$

$$x = 5 \qquad \text{Simplify the left side.}$$

To be sure this solution is correct, we can check it in the original equation.

$$3x = 15 \qquad \text{Write the original equation.}$$

$$3(5) \overset{?}{=} 15 \qquad \text{Substitute 5 for } x.$$

$$15 = 15 \qquad \text{Find the product.}$$

Our solution checks. This example illustrates another property that we can use in solving equations.

> We may divide both sides of an equation by the same number (as long as that number is not 0).

You complete Example 6.

Example 6. Solve $9x = -36$ and check.

Solution

$$9x = -36 \qquad\qquad \text{CHECK:} \quad 9x = -36$$

$$\frac{9x}{9} = \frac{-36}{9} \qquad\qquad 9(\underline{}) \overset{?}{=} -36$$

$$\underline{} = -36$$

$$x = \underline{}$$

Check your work on page 671. ▶

Example 7. Solve $4x = 6$ and check.

Solution

$$4x = 6 \qquad\qquad \text{CHECK:} \quad 4x = 6$$

$$\frac{4x}{4} = \frac{6}{4} \qquad\qquad 4\left(\frac{3}{2}\right) \overset{?}{=} 6$$

$$x = \frac{3}{2} \qquad\qquad \frac{\overset{2}{\cancel{4}}}{1} \cdot \frac{3}{\underset{1}{\cancel{2}}} \overset{?}{=} 6$$

$$2 \cdot 3 \overset{?}{=} 6$$

$$6 = 6$$

Example 8. Solve $-0.5x = 11$ and check.

Solution

$$-0.5x = 11 \qquad\qquad \text{CHECK:} \quad -0.5x = 11$$

$$\frac{-0.5x}{-0.5} = \frac{11}{-0.5} \qquad\qquad -0.5(-22) \overset{?}{=} 11$$

$$x = -22 \qquad\qquad 11 = 11$$

We must still consider an equation such as

$$\frac{a}{4} = 3$$

in which the variable a is being *divided* by 4. Let's try to use *multiplication* to undo the *division*. To get rid of the 4 dividing the a, let's multiply both sides of the equation by 4.

$$\frac{a}{4} = 3$$

$$4 \cdot \frac{a}{4} = 4 \cdot 3 \qquad \text{Multiply both sides by 4.}$$

$$\frac{\overset{1}{\cancel{4}}}{1} \cdot \frac{a}{\underset{1}{\cancel{4}}} = 12 \qquad \text{Find the products.}$$

$$a = 12$$

We should check this solution in the original equation.

$$\frac{a}{4} = 3 \qquad \text{Write the original equation.}$$

$$\frac{12}{4} \overset{?}{=} 3 \qquad \text{Substitute 12 for } a.$$

$$3 = 3 \qquad \text{Find the quotient.}$$

Our solution checks. This example illustrates the last property that we shall use to solve equations.

> We may multiply both sides of an equation by the same number (as long as that number is not 0).

You complete Example 9.

Example 9. Solve $\dfrac{x}{5} = 8$ and check.

Solution

$$\frac{x}{5} = 8 \qquad\qquad \text{CHECK:} \quad \frac{x}{5} = 8$$

$$\underline{\quad\quad} \cdot \frac{x}{5} = \underline{\quad\quad} \cdot 8 \qquad\qquad \frac{\underline{\quad}}{5} \overset{?}{=} 8$$

$$\frac{5}{1} \cdot \frac{x}{5} = \underline{\quad\quad} \qquad\qquad \underline{\quad\quad} = 8$$

$$x = \underline{\quad\quad}$$

Check your work on page 671. ▶

Example 10. Solve $\dfrac{x}{7.1} = -2.9$ and check.

Solution

$$\frac{x}{7.1} = -2.9 \qquad\qquad \text{CHECK:} \quad \frac{x}{7.1} = -2.9$$

$$7.1\left(\frac{x}{7.1}\right) = 7.1(-2.9) \qquad\qquad \frac{-20.59}{7.1} \overset{?}{=} -2.9$$

$$\frac{\overset{1}{\cancel{7.1}}}{1} \cdot \frac{x}{\underset{1}{\cancel{7.1}}} = -20.59 \qquad\qquad -2.9 = -2.9$$

$$x = -20.59$$

⬛▶ **Trial Run**

Solve each equation and check.

_____ 1. $4x = 20$ _____ 2. $-9a = 6$

_____ 3. $0.2x = 7$ _____ 4. $\dfrac{a}{7} = -5$

_____ 5. $\dfrac{y}{3} = 2$ _____ 6. $\dfrac{x}{4.2} = 0.5$

Answers are on page 671.

In solving equations we have used four basic properties of equations that will continue to be important in our study of algebra.

Properties of Equations

1. We may add the same number to both sides of an equation.
2. We may subtract the same number from both sides of an equation.
3. We may multiply both sides of an equation by the same number (except 0).
4. We may divide both sides of an equation by the same number (except 0).

Performing any of these operations on *both sides* of an equation will yield an equation that is equivalent to the original equation.

▶ Examples You Completed

Example 2. Solve $x + 7 = 19$ and check.

Solution

$$x + 7 = 19$$
$$x + 7 - 7 = 19 - 7$$
$$x + 0 = 12$$
$$x = 12$$

CHECK: $x + 7 = 19$

$12 + 7 \overset{?}{=} 19$

$19 = 19$

Example 4. Solve $a - 13 = 10$ and check.

Solution

$$a - 13 = 10$$
$$a - 13 + 13 = 10 + 13$$
$$a + 0 = 23$$
$$a = 23$$

CHECK: $a - 13 = 10$

$23 - 13 \overset{?}{=} 10$

$10 = 10$

Example 6. Solve $9x = -36$ and check.

Solution

$$9x = -36$$
$$\frac{9x}{9} = \frac{-36}{9}$$
$$x = -4$$

CHECK: $9x = -36$

$9(-4) \overset{?}{=} -36$

$-36 = -36$

Example 9. Solve $\frac{x}{5} = 8$ and check.

Solution

$$\frac{x}{5} = 8$$

$$5 \cdot \frac{x}{5} = 5 \cdot 8$$

$$\frac{\overset{1}{\cancel{5}}}{1} \cdot \frac{x}{\cancel{5}} = 40$$

$$x = 40$$

CHECK: $\frac{x}{5} = 8$

$\frac{40}{5} \overset{?}{=} 8$

$8 = 8$

Answers to Trial Runs

page 667 **1.** $x = 5$ **2.** $m = 0$ **3.** $y = 10$ **4.** $a = 22$ **5.** $x = 11.75$ **6.** $y = -7.5$

page 670 **1.** $x = 5$ **2.** $a = -\frac{2}{3}$ **3.** $x = 35$ **4.** $a = -35$ **5.** $y = 6$ **6.** $x = 2.1$

EXERCISE SET 15.2

Write an equation for each word statement.

_____ 1. When 5 is added to x, the result is -7.

_____ 2. When -3 is added to y, the result is 9.

_____ 3. The product of -4 and x is 24.

_____ 4. The product of 5 and y is -25.

_____ 5. When 3 is multiplied times the sum of x and -2, the result is 12.

_____ 6. When -2 is multiplied times the sum of y and -1, the result is 5.

Solve each equation and check.

_____ 7. $x + 5 = 12$ _____ 8. $x + 9 = 16$ _____ 9. $x - 9 = 3$

_____ 10. $x - 8 = 7$ _____ 11. $y + 2 = -6$ _____ 12. $y + 3 = -8$

_____ 13. $1.6 + y = 4.2$ _____ 14. $1.9 + y = 5.3$ _____ 15. $m - 11 = -6$

_____ 16. $m - 9 = -8$ _____ 17. $-4 + a = 6$ _____ 18. $-6 + a = 12$

_____ 19. $12 = k + 6$ _____ 20. $13 = k + 9$ _____ 21. $8 + x = 8$

_____ 22. $-9 + x = -9$ _____ 23. $-\dfrac{1}{4} = m + \dfrac{9}{4}$ _____ 24. $-\dfrac{1}{3} = m + \dfrac{7}{3}$

_____ 25. $y + 12 = 4$ _____ 26. $y + 9 = 2$ _____ 27. $3x = 21$

_____ 28. $4x = 28$ _____ 29. $-2x = 35$ _____ 30. $-5x = 46$

_____ 31. $-3.6 = -6x$ _____ 32. $-4.8 = -6x$ _____ 33. $4x = 0$

_____ 34. $7x = 0$ _____ 35. $-x = 6.5$ _____ 36. $-x = 9.2$

_____ 37. $\dfrac{x}{2} = 6$ _____ 38. $\dfrac{x}{4} = 12$ _____ 39. $-3 = \dfrac{x}{4}$

_____ 40. $-9 = \dfrac{x}{2}$ _____ 41. $\dfrac{x}{-7} = 8$ _____ 42. $\dfrac{x}{-8} = 9$

_____ 43. $\dfrac{x}{5} = 0$ _____ 44. $\dfrac{x}{9} = 0$ _____ 45. $\dfrac{x}{5} = -8$

_____ 46. $\dfrac{x}{11} = -4$ _____ 47. $-9 = \dfrac{x}{-7}$ _____ 48. $-7 = \dfrac{x}{-6}$

_____ 49. $\dfrac{1}{2} = \dfrac{x}{12}$ _____ 50. $\dfrac{1}{3} = \dfrac{x}{15}$

☆ Stretching the Topics _____

Solve each equation.

_____ 1. $-\dfrac{2}{3} = x - \dfrac{5}{4}$

_____ 2. $0.32x = -0.48$

_____ 3. $y + c = d$, for y

_____ 4. $ax = m$, for x

Check your answers in the back of your book.

If you can solve the equations in **Checkup 15.2,** you are ready to go on to Section 15.3.

✓ **CHECKUP 15.2**

_____ **1.** Write an equation that says that when 5 is multiplied times the sum of x and -3, the result is 11.

Solve each equation.

_____ **2.** $x + 9 = 8$ _____ **3.** $x - 9 = -2$

_____ **4.** $1.2 = x - 3.6$ _____ **5.** $x + 11 = 11$

_____ **6.** $6x = 54$ _____ **7.** $4x = 0$

_____ **8.** $\dfrac{x}{8} = 3$ _____ **9.** $\dfrac{x}{2} = 4.8$

_____ **10.** $\dfrac{x}{6} = -10$

Check your answers in the back of your book.

If You Missed Problems:	You Should Review Examples:
1	1
2–5	2–5
6, 7	6–8
8–10	9, 10

15.3 Solving More Equations

Sometimes it is necessary to do several steps before we can find the solution to an equation. We may need to combine like terms or undo more than one operation.

Using More Than One Operation to Solve Equations

In the equation

$$3x + 5 = 17$$

we must get rid of the 5 that is being added to the variable term and we must also get rid of the 3 that is multiplying x. Let's solve this equation.

$$3x + 5 = 17$$

$$3x + 5 - 5 = 17 - 5 \qquad \text{Subtract 5 from both sides.}$$

$$3x = 12 \qquad \text{Find the differences.}$$

$$\frac{3x}{3} = \frac{12}{3} \qquad \text{Divide both sides by 3.}$$

$$x = 4 \qquad \text{Find the quotients.}$$

We check our solution in the original equation.

$$3x + 5 = 17 \qquad \text{Write the original equation.}$$

$$3(4) + 5 \stackrel{?}{=} 17 \qquad \text{Substitute 4 for } x.$$

$$12 + 5 \stackrel{?}{=} 17 \qquad \text{Find the product.}$$

$$17 = 17 \qquad \text{Find the sum.}$$

In solving an equation that needs more than one operation, we proceed as follows:

> First we do any necessary additions or subtractions.
> Then we do any necessary multiplications or divisions.

You complete Example 1.

Example 1. Solve $11 + 4x = 19$ and check.

Solution

$$11 + 4x = 19$$

$$11 - \underline{\quad} + 4x = 19 - \underline{\quad} \qquad \text{Subtract} \underline{\quad} \text{from both sides.}$$

$$4x = 8 \qquad \text{Find the differences.}$$

$$\frac{4x}{\underline{\quad}} = \frac{8}{\underline{\quad}} \qquad \text{Divide both sides by} \underline{\quad}.$$

$$x = \underline{\quad} \qquad \text{Find the quotients.}$$

Now we check our solution in the original equation.

$$\text{CHECK:} \quad 11 + 4x = 19$$

$$11 + 4(\underline{\quad\quad}) \stackrel{?}{=} 19$$

$$11 + \underline{\quad\quad} \stackrel{?}{=} 19$$

$$\underline{\quad\quad} = 19$$

Check your work on page 680. ▶

Example 2. Solve $8.3 = -1.2x - 3.7$.

Solution

$$8.3 = -1.2x - 3.7$$

$$8.3 + 3.7 = -1.2x - 3.7 + 3.7 \qquad \text{Add } 3.7 \text{ to both sides.}$$

$$12 = -1.2x \qquad \text{Find the sums.}$$

$$\frac{12}{-1.2} = \frac{-1.2x}{-1.2} \qquad \text{Divide both sides by } -1.2.$$

$$-10 = x \qquad \text{Find the quotients.}$$

Example 3. Solve $\dfrac{x}{7} + 9 = 9$.

Solution

$$\frac{x}{7} + 9 = 9$$

$$\frac{x}{7} + 9 - 9 = 9 - 9 \qquad \text{Subtract } 9 \text{ from both sides.}$$

$$\frac{x}{7} = 0 \qquad \text{Find the differences.}$$

$$7 \cdot \frac{x}{7} = 7 \cdot 0 \qquad \text{Multiply both sides by } 7.$$

$$\frac{\overset{1}{7}}{1} \cdot \frac{x}{\underset{1}{7}} = 0 \qquad \text{Find the products.}$$

$$x = 0$$

⟫ Trial Run

Solve each equation and check.

———— **1.** $2x - 3 = 7$ ———— **2.** $4a + 5 = 17$

———— **3.** $\dfrac{n}{3} + 4 = 7$ ———— **4.** $12 - 5x = 22$

———— **5.** $9.4 = 3.6x - 8.6$ ———— **6.** $\dfrac{a}{5} - 3 = 4$

Answers are on page 680.

Simplifying an Equation

Sometimes an equation will contain **like terms** that should be combined or parentheses that should be dealt with before we can begin to find the solution. When this happens, we must *simplify* the equation before we can solve it. For instance,

$$3x + x = 16$$

$$3x + 1x = 16 \qquad \text{Rewrite } x \text{ as } 1x.$$

$$4x = 16 \qquad \text{Combine like terms.}$$

$$\frac{4x}{4} = \frac{16}{4} \qquad \text{Divide both sides by } 4.$$

$$x = 4 \qquad \text{Find the quotients.}$$

This solution should be checked in the *original* equation.

$$3x + x = 16 \qquad \text{Write the original equation.}$$

$$3 \cdot 4 + 4 \stackrel{?}{=} 16 \qquad \text{Substitute } 4 \text{ for each } x.$$

$$12 + 4 \stackrel{?}{=} 16 \qquad \text{Find the product.}$$

$$16 = 16 \qquad \text{Find the sum.}$$

Example 4. Solve $2x + 3 + 9x - 1 = 13$.

Solution

$$2x + 3 + 9x - 1 = 13$$

$$\underbrace{2x + 9x} + \underbrace{3 - 1} = 13 \qquad \text{Rearrange terms.}$$

$$11x \quad + \quad 2 \quad = 13 \qquad \text{Combine like terms.}$$

$$11x + 2 - 2 = 13 - 2 \qquad \text{Subtract } 2 \text{ from both sides.}$$

$$11x = 11 \qquad \text{Find the differences.}$$

$$\frac{11x}{11} = \frac{11}{11} \qquad \text{Divide both sides by } 11.$$

$$x = 1 \qquad \text{Find the quotients.}$$

You complete Example 5.

Example 5. Solve $8x + 13 - 3x - 4 = 49$ and check.

Solution

$$8x + 13 - 3x - 4 = 49$$
$$8x - 3x + 13 - 4 = 49$$
$$5x + \underline{\hphantom{xx}} = 49$$
$$5x + 9 - \underline{\hphantom{xx}} = 49 - \underline{\hphantom{xx}}$$
$$5x = \underline{\hphantom{xx}}$$
$$x = \underline{\hphantom{xx}}$$

CHECK: $8x + 13 - 3x - 4 = 49$

$$8(\underline{\hphantom{xx}}) + 13 - 3(\underline{\hphantom{xx}}) - 4 \stackrel{?}{=} 49$$
$$\underline{\hphantom{xx}} + 13 - \underline{\hphantom{xx}} - 4 \stackrel{?}{=} 49$$
$$\underline{\hphantom{xx}} - \underline{\hphantom{xx}} \stackrel{?}{=} 49$$
$$\underline{\hphantom{xx}} = 49$$

Check your work on page 680. ▶

If an equation contains **parentheses,** they should be dealt with before the equation is solved.

$$2(3x + 5) = 28$$
$$2 \cdot 3x + 2 \cdot 5 = 28 \qquad \text{Use distributive property to remove parentheses.}$$
$$6x + 10 = 28 \qquad \text{Find the products.}$$
$$6x + 10 - 10 = 28 - 10 \qquad \text{Subtract 10 from both sides.}$$
$$6x = 18 \qquad \text{Find the differences.}$$
$$\frac{6x}{6} = \frac{18}{6} \qquad \text{Divide both sides by 6.}$$
$$x = 3 \qquad \text{Find the quotients.}$$

Let's be sure this solution checks in the original equation.

$$2(3x + 5) = 28 \qquad \text{Write the original equation.}$$
$$2(3 \cdot 3 + 5) \stackrel{?}{=} 28 \qquad \text{Substitute 3 for } x.$$
$$2(9 + 5) \stackrel{?}{=} 28 \qquad \text{Find the product.}$$
$$2(14) \stackrel{?}{=} 28 \qquad \text{Find the sum within parentheses.}$$
$$28 = 28 \qquad \text{Find the product.}$$

You complete Example 6.

Example 6. Solve $3(x + 1) - 1 = 3$.

Solution

$$3(x + 1) - 1 = 3$$
$$3 \cdot x + 3 \cdot 1 - 1 = 3 \qquad \text{Use distributive property to remove parentheses.}$$
$$3x + 3 - 1 = 3 \qquad \text{Find the products.}$$
$$3x + \underline{\hphantom{xx}} = 3 \qquad \text{Combine like terms.}$$
$$3x + 2 - \underline{\hphantom{xx}} = 3 - \underline{\hphantom{xx}} \qquad \text{Subtract } \underline{\hphantom{xx}} \text{ from both sides.}$$
$$3x = \underline{\hphantom{xx}} \qquad \text{Find the differences.}$$
$$\frac{3x}{3} = \frac{1}{3} \qquad \text{Divide both sides by } \underline{\hphantom{xx}}.$$
$$x = \underline{\hphantom{xx}} \qquad \text{Find the quotients.}$$

Check your work on page 680. ▶

Example 7. Solve $-5(4x + 1) + 7x = 34$.

Solution

$$-5(4x + 1) + 7x = 34$$

$$-5 \cdot 4x - 5 \cdot 1 + 7x = 34 \qquad \text{Use distributive property to remove parentheses.}$$

$$-20x - 5 + 7x = 34 \qquad \text{Find the products.}$$

$$-20x + 7x - 5 = 34 \qquad \text{Rearrange terms.}$$

$$-13x - 5 = 34 \qquad \text{Combine like terms.}$$

$$-13x - 5 + 5 = 34 + 5 \qquad \text{Add 5 to both sides.}$$

$$-13x = 39 \qquad \text{Find the sums.}$$

$$\frac{-13x}{-13} = \frac{39}{-13} \qquad \text{Divide both sides by } -13.$$

$$x = -3 \qquad \text{Find the quotients.}$$

⫸ Trial Run

Solve and check.

———— 1. $4x + 5 + 3x - 2 = 17$ ———— 2. $9x + 15 - 3x - 3 = -18$

———— 3. $5(a + 3) - 6 = 34$ ———— 4. $4(2x - 3) - 9x = 1$

———— 5. $-4(x + 5) + 3(7 - x) = 15$ ———— 6. $0.5(2x - 6.4) = 3.2$

Answers are on page 680.

In solving equations we have now learned to use the following steps to get the variable by itself.

Steps for Solving Equations

1. Use the distributive property to deal with any parentheses.
2. Combine like terms if possible.
3. Do any necessary additions or subtractions to get the variable term alone on one side of the equation.
4. Do any necessary multiplications or divisions to solve for the variable.
5. Check your solution in the original equation.

▶ Examples You Completed

Example 1. Solve $11 + 4x = 19$ and check.

Solution

$$11 + 4x = 19$$

$$11 - 11 + 4x = 19 - 11 \qquad \text{Subtract 11 from both sides.}$$

$$4x = 8 \qquad \text{Find the differences.}$$

$$\frac{4x}{4} = \frac{8}{4} \qquad \text{Divide both sides by 4.}$$

$$x = 2 \qquad \text{Find the quotients.}$$

CHECK: $11 + 4x = 19$

$$11 + 4(2) \overset{?}{=} 19$$

$$11 + 8 \overset{?}{=} 19$$

$$19 = 19$$

Example 5. Solve $8x + 13 - 3x - 4 = 49$ and check.

Solution

$$8x + 13 - 3x - 4 = 49$$

$$8x - 3x + 13 - 4 = 49$$

$$5x + 9 = 49$$

$$5x + 9 - 9 = 49 - 9$$

$$5x = 40$$

$$\frac{5x}{5} = \frac{40}{5}$$

$$x = 8$$

CHECK:

$$8x + 13 - 3x - 4 = 49$$

$$8(8) + 13 - 3(8) - 4 \overset{?}{=} 49$$

$$64 + 13 - 24 - 4 \overset{?}{=} 49$$

$$77 - 28 \overset{?}{=} 49$$

$$49 = 49$$

Example 6. Solve $3(x + 1) - 1 = 3$.

Solution

$$3(x + 1) - 1 = 3$$

$$3 \cdot x + 3 \cdot 1 - 1 = 3 \qquad \text{Use distributive property to remove parentheses.}$$

$$3x + 3 - 1 = 3 \qquad \text{Find the products.}$$

$$3x + 2 = 3 \qquad \text{Combine like terms.}$$

$$3x + 2 - 2 = 3 - 2 \qquad \text{Subtract 2 from both sides.}$$

$$3x = 1 \qquad \text{Find the differences.}$$

$$\frac{3x}{3} = \frac{1}{3} \qquad \text{Divide both sides by 3.}$$

$$x = \frac{1}{3} \qquad \text{Find the quotient.}$$

Answers to Trial Runs

page 677   **1.** $x = 5$ **2.** $a = 3$ **3.** $n = 9$ **4.** $x = -2$ **5.** $x = 5$ **6.** $a = 35$

page 679   **1.** $x = 2$ **2.** $x = -5$ **3.** $a = 5$ **4.** $x = -13$ **5.** $x = -2$ **6.** $x = 6.4$

EXERCISE SET 15.3

Solve each equation.

_____ 1. $3x + 5 = 26$ _____ 2. $4x + 6 = 26$

_____ 3. $2x - 9 = 11$ _____ 4. $7x - 9 = 5$

_____ 5. $5y + 3 = -22$ _____ 6. $2y + 8 = -12$

_____ 7. $0.8x - 1.5 = -0.7$ _____ 8. $0.6x + 2.3 = 1.1$

_____ 9. $-4a + 9 = 45$ _____ 10. $-5a + 10 = 55$

_____ 11. $13 + 2m = 0$ _____ 12. $20 + 3m = 0$

_____ 13. $-x - 7 = 6$ _____ 14. $-x - 8 = 9$

_____ 15. $7\frac{1}{2} - 2x = 13\frac{1}{2}$ _____ 16. $6\frac{3}{4} - 5x = 41\frac{3}{4}$

_____ 17. $24 = 3x - 15$ _____ 18. $17 = 6x - 7$

_____ 19. $4y + \frac{2}{5} = \frac{2}{5}$ _____ 20. $7y + \frac{1}{9} = \frac{1}{9}$

_____ 21. $-23 = 9 - 4a$ _____ 22. $-21 = 15 - 9a$

_____ 23. $-1.7x + 3.4 = 0$ _____ 24. $-0.4x + 3.2 = 0$

_____ 25. $\frac{x}{6} + 3 = 5$ _____ 26. $\frac{x}{7} + 4 = 9$

_____ 27. $\frac{x}{5} - 9 = -9$ _____ 28. $\frac{x}{8} - 7 = -7$

_____ 29. $13 = \frac{y}{4} + 9$ _____ 30. $17 = \frac{y}{5} + 8$

_____ 31. $6 - \frac{x}{4} = 7$ _____ 32. $9 - \frac{x}{2} = 12$

_____ 33. $0.7 + \frac{a}{5} = -0.1$ _____ 34. $0.6 + \frac{a}{3} = -0.2$

_____ 35. $4x + 3x = 56$ _____ 36. $7x + 2x = 72$

_____ 37. $x + 3 + 4x = -2$ _____ 38. $x + 5 + 7x = -3$

_____ 39. $3x + 4 - x + 2 = 15$ _____ 40. $4x + 7 - x + 5 = 21$

_____ 41. $(x + 3) + (x - 5) = 8$ _____ 42. $(x + 5) + (x - 6) = 9$

_____ 43. $2(3x - 4) = 16$

_____ 44. $3(4x - 1) = 18$

_____ 45. $12 = 4\left(\dfrac{1}{2}x - 7\right) + 8$

_____ 46. $12 = 6\left(\dfrac{1}{2}x - 6\right) + 9$

_____ 47. $2(x - 8) - (x + 3) = -22$

_____ 48. $3(x - 4) - (2x + 7) = -26$

_____ 49. $3 = 4(x - 5) - 2(x + 6)$

_____ 50. $2 = 5(x - 5) - 3(x - 2)$

☆ Stretching the Topics ————————————

Solve each equation.

_____ 1. $2.3 = -0.9 - (9.4x + 1.6) - 0.2x$

_____ 2. $2(-x - 6) + 3x - (-x - 4) - 3(x - 2) = 0$

_____ 3. $4[x - 3(x - 2)] + [5 - (x - 1)] = 9$

Check your answers in the back of your book.

If you can solve the equations in **Checkup 15.3,** you are ready to go on to Section 15.4.

✔ CHECKUP 15.3

Solve each equation.

_____ 1. $3x - 7 = 8$

_____ 2. $0.7 = 0.8x + 1.5$

_____ 3. $-2y - 5 = 7$

_____ 4. $\dfrac{x}{2} - 1 = 5$

_____ 5. $8 + \dfrac{x}{3} = 9$

_____ 6. $6x - 3x - 12 = 6$

_____ 7. $2x - 6 - 4x - 2 = 2$

_____ 8. $5(x - 3.2) = -16$

_____ 9. $5y - 3(3y - 1) = 31$

_____ 10. $4(x - 3) + 2(4x - 3) = 6$

Check your answers in the back of your book.

If You Missed Problems:	You Should Review Examples:
1–3	1, 2
4, 5	3
6, 7	4, 5
8–10	6, 7

15.4 Switching from Words to Equations

We have practiced switching from words to numbers throughout the chapters of this book. Now that we have learned some basic ideas of algebra, we will find that we can use variables to switch more complicated word statements into equations that we can solve.

Switching from Words to Algebraic Expressions

If 1 credit hour at a university costs a student $27, what would be the cost for

2 hours?	$27 \cdot 2 = $54
9 hours?	$27 \cdot 9 = $243
h hours?	$27h

Notice that we used an algebraic expression ($27h$) to stand for the cost of *any* unknown number of hours.

Suppose the enrollment at a college is 9240 students. What would be the enrollment if it increased by

200 students?	$9240 + 200 = 9440$
83 students?	$9240 + 83 = 9323$
s students?	$9240 + s$

Once again, the algebraic expression $9240 + s$ allows us to describe the final enrollment after *any* increase of s students.

Suppose an ice cream distributor must equally divide a week's supply of ice cream among 7 different stores. How many gallons will each store receive if the total is

63 gallons?	$\dfrac{63}{7} = 9$
98 gallons?	$\dfrac{98}{7} = 14$
G gallons?	$\dfrac{G}{7}$

Using the algebraic expression $\dfrac{G}{7}$ allows us to describe the amount received by each store for *any* total amount G.

When we use a variable to stand for an unknown number, it is important that we write down what we are letting the variable represent. That will help you and anyone looking at your work to follow what has been done.

Example 1. Sam has a gross monthly income of $2030. Write an expression for his take-home pay after total deductions, *d*, have been made.

Solution. Let d = amount of total deductions. Then Sam's take-home pay is $2030 - d$ dollars.

Example 2. Tickets for a rock concert are selling for $12 each. Write an expression for the money received from the sale of *x* tickets.

Solution. Let x = number of tickets sold. Then the money received is $12x$ dollars.

You complete Example 3.

Example 3. Regina weighed 172 pounds before starting her diet. Write an expression for her weight after she has lost p pounds.

Solution. Let p = number of pounds lost. Then Regina's weight is _____ pounds.

Check your work on page 689. ▶

Example 4. Suppose that Alex has $1460 in his savings account. If he deposits $30 each week into this account, write an expression for the amount he will have after w weeks.

Solution. Let w = number of weeks. Alex will save an additional $30w$ dollars after w weeks. His *total* will be $1460 + 30w$ dollars.

Example 5. In a certain triangle, the second side is twice as long as the first side. The third side is 5 times as long as the first side. If the first side is s feet long, write an expression for the perimeter of this triangle.

Solution

$$\text{Let } s = \text{length of first side}$$

$$\text{then } 2s = \text{length of second side}$$

$$\text{and } 5s = \text{length of third side}$$

We know the perimeter of a triangle is found by *adding* the lengths of its 3 sides, so the perimeter of this triangle is

$$s + 2s + 5s = 8s \text{ feet.}$$

Changing from Words to Algebraic Expressions

1. Write down what the variable represents.
2. Change word expressions to algebraic expressions containing the variable.

⫸ Trial Run

_____ **1.** If Tyrone drives the speed limit, 55 miles per hour, write an expression for the distance he could travel in h hours.

_____ **2.** There are 1252 students enrolled in Central High School. Write an expression for the number enrolled after n students drop out.

_____ **3.** The cost of renting a garden tiller is $25 plus $10 for each day it is used. Write an expression for the cost of renting a tiller for x days.

_____ **4.** In a certain rectangle, the length is twice the width. If the width is x feet, write an expression for the perimeter.

Answers are on page 689.

Switching from Words to Equations

Having learned to switch from words to algebraic expressions containing variables, we are now ready to solve word problems using equations.

Example 6. Suppose tickets for a rock concert sell for $12 each. How many tickets must be sold to have a total of $13,800 in sales?

Solution. The quantity we must find is the total number of tickets.

$$\text{Let} \quad t = \text{number of tickets}$$

$$\text{then} \quad 12t = \text{total dollars from tickets}$$

Since we want a total of $13,800 in sales, we can write an equation and solve it by the usual methods.

$$12t = 13{,}800$$

$$\frac{12t}{12} = \frac{13{,}800}{12}$$

$$t = \frac{13{,}800}{12}$$

$$t = 1150$$

$$\begin{array}{r} 1150 \\ 12\overline{)13800} \\ \underline{12} \\ 18 \\ \underline{12} \\ 60 \\ \underline{60} \\ 00 \\ \underline{0} \\ 0 \end{array}$$

To have $13,800 in sales, 1150 tickets must be sold.

Switching problems from words to equations should be done in an orderly way.

Switching from Words to Equations

1. Write down what you want the variable to represent.
2. Write any expressions containing the variable.
3. Write an equation from the information in the word problem.
4. Solve the equation by the usual methods.
5. Write your conclusion in a sentence, using appropriate units.

Example 7. Alex has $1460 in his savings account and he plans to save $30 each week. For how many weeks must he continue this plan if he wishes to have a total of $2000 in his savings account?

Solution. The quantity we must find is the number of weeks.

$$\text{Let} \quad w = \text{number of weeks}$$

$$\text{then} \quad 30w = \text{amount saved in } w \text{ weeks}$$

$$\text{and} \quad 1460 + 30w = \text{total amount in account}$$

Since Alex wants a certain total ($2000), we can write our equation.

$$1460 + 30w = 2000$$

$$1460 - 1460 + 30w = 2000 - 1460$$

$$30w = 540$$

$$\frac{30w}{30} = \frac{540}{30}$$

$$w = 18$$

Alex must continue saving $30 each week for 18 weeks.

Now let's solve the problem stated at the beginning of the chapter.

Example 8. Suppose Bradley has 42 yards of fencing to use around his rectangular garden. If he wants the garden to be twice as long as it is wide, find the length and width of the garden.

Solution. Let's draw the garden and label its dimensions, remembering that it is twice as long as it is wide. In other words, 1 length is equal to *2 widths*.

To find the amount of fencing we must find the *perimeter* of the garden.

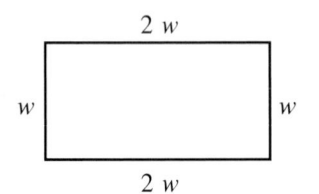

Our equation becomes

$$2w + w + 2w + w = 42$$

$$6w = 42$$

$$\frac{6w}{6} = \frac{42}{6}$$

$$w = 7$$

The width of the garden is 7 yards. The length is *twice* the width, so the length is 14 yards.

As you can see, algebra is useful in solving many problems that would be difficult to solve without using a variable. Although the basic skills of arithmetic are an important foundation for all mathematics, algebra provides the key to understanding higher level mathematics.

⏵ Trial Run

Use an equation to solve each problem.

_____ 1. One credit hour at a university costs $27. For how many hours may Shawna enroll if she has saved $297?

_____ 2. The cost of renting a garden tiller is $25 plus $10 per day. If Jake was charged $185 when he returned his tiller, for how many days did he use it?

_____ 3. Manuel has received grades of 83 and 75 on his first 2 English tests. What grade must he earn on his third test so that his average for the 3 tests will be 80?

Answers are on page 689.

▶ **Example You Completed** ─────────────────────────

Example 3. Regina weighed 172 pounds before starting her diet. Write an expression for her weight after she has lost p pounds.

Solution. Let p = number of pounds lost. Then Regina's weight is $172 - p$ pounds.

Answers to Trial Runs ─────────────────────────

page 686   **1.** $55h$ **2.** $1252 - n$ **3.** $\$25 + \$10x$ **4.** $6x$

page 688   **1.** Shawna may enroll for 11 hours.   **2.** Jake used the tiller for 16 days.
3. Manuel must earn a grade of 82 on his third test.

EXERCISE SET 15.4

Write an algebraic expression for each word expression.

_____ 1. Hazel has a part-time job paying $4.15 per hour. Write an expression for her earnings after she has worked x hours.

_____ 2. The Deltas are selling lottery tickets for a video cassette recorder at $2.50 per ticket. Write an expression for the receipts from the sale of n tickets.

_____ 3. The cost of a chartered bus to the ball game is $850, to be divided equally among the students who ride the bus. Write an expression for the cost to each passenger if x students ride the bus.

_____ 4. Write an expression for the number of liters remaining in a 75-liter fuel tank if y liters have already been used.

_____ 5. Jamal earns $350 per week plus $7.25 per hour of overtime. Write an expression for Jamal's total pay for a week in which he worked x hours of overtime.

_____ 6. Tammy is saving money for a vacation during spring break. If she saves $50 and then saves $15 per week for n weeks, write an expression for the amount she will have at the end of n weeks.

_____ 7. The cost of the annual Christmas party is to be shared equally by 52 factory employees. Write an expression for the amount each employee will pay if the party costs x dollars.

_____ 8. An oil tank that contains 300 barrels of oil begins to leak at a rate of 5 barrels a day. Write an expression for the amount of oil in the tank after n days.

_____ 9. The winning candidate in an election received 75 votes more than twice the number of votes her opponent received. If her opponent received x votes, write an expression for the number of votes received by the winner.

_____ 10. The balance owed on Chico's charge account is $683.74. If he pays x dollars per month for 5 months, write an expression for the balance. (Assume no interest is charged.)

Use an equation to solve each problem.

_____ 11. If Carlotta finds she can save $12.50 a week, how many weeks will it take her to save $300?

_____ 12. If Herbert drives at a rate of 60 miles per hour, how long will it take him to travel 315 miles?

_____ 13. Courtney bought a shirt and a blazer. The cost of the blazer was $15 less than 3 times the cost of the shirt. If the blazer cost $47.97, what was the cost of the shirt?

_____ 14. Daphne and Sarah are selling tickets for a concert to benefit Special Olympics. Sarah has sold 4 times as many tickets as Daphne. Together they have sold 65 tickets. How many tickets has each sold?

_____ 15. Mr. Henshaw has 120 feet of fencing with which to build a loading pen for cattle. If the pen must be 3 times as long as it is wide, find the dimensions of the pen.

_____ 16. André is buying framing for a rectangular picture that is twice as long as it is wide. The sales person tells him that he needs 72 inches of framing. What are the dimensions of the frame?

_____ 17. Bernice is saving money for a vacation. She deposits $50 and then plans to save a certain amount each week for the next 8 weeks. If she estimates she will need $600 for her vacation, how much must she save each week?

_____ 18. The restaurant manager tells the chairperson of the banquet committee that she will charge $200 for the use of the room plus $8.75 per person for the dinner. If the total cost of the banquet is $900, how many persons will be attending?

_____ 19. The width of a room is 5 feet more than half its length. If a paperhanger needs 70 feet of ceiling border, find the dimensions of the room.

_____ 20. Irma has scored 80, 62, 75, and 69 on her first 4 mathematics tests. What must she score on her fifth test in order to have an average of 70?

☆ Stretching the Topics _____

Use an equation to solve each problem.

_____ 1. A board 50 inches in length is to be sawed into 4 pieces so that each piece is 1 inch shorter than the preceding piece. Find the length of each piece.

_____ 2. A manufacturing company has decreased its inventory by 8 percent. If it now has 2070 items in stock, how many items did it have before the decrease?

Check your answers in the back of your book.

If you can complete **Checkup 15.4,** you are ready to do the **Review Exercises** for Chapter 15.

✓ CHECKUP 15.4

Write an algebraic expression for each word expression.

_____ 1. A hiker left Horseshoe Cave, walking along the trail at 3 miles per hour. Write an expression for the distance traveled after h hours.

_____ 2. Aunt Patsy left her entire estate to her 3 nieces. If the estate is valued at x dollars, write an expression for each niece's share.

_____ 3. The cost of renting a cabin at the lake is $200 for the first week and $19.75 for each additional day. Write an expression for the total cost of the cabin after one week and n additional days.

Use an equation to solve each problem.

_____ 4. Martin sells garden tillers. He earns $75 for each tiller he sells. How many tillers must he sell each month so that he will have a monthly income of $1425?

_____ 5. Tickets to Wonder World are $8.25 each. How many members of the Senior Citizens' Club can go if the organization has $198?

_____ 6. Martina buys a lawnmower for $500. She makes a down payment of $150 and pays the balance by making payments of $25 a week. For how many weeks will she be making payments? (Assume no interest is charged.)

Check your answers in the back of your book.

If You Missed Problems:	You Should Review Examples:
1, 2	2
3	4
4, 5	6
6	7

Summary

In this chapter we carried our understanding of arithmetic to the next level of mathematics: the study of basic algebra. We learned to use **variables** to stand for unknown numbers and to write **algebraic expressions** containing constant terms and variable terms. In our work with algebraic expressions we learned to combine like terms and simplify expressions involving parentheses.

If We Wish to	We Must	Examples
Evaluate an expression	Substitute a particular value for the variable.	Evaluate $3 - 2x$ when $x = 4$. $3 - 2x = 3 - 2(4)$ $= 3 - 8$ $= -5$
Combine like terms	Combine numerical coefficients and keep the same variable part.	$5x - 3 + 2x - 8$ $= 5x + 2x - 3 - 8$ $= 7x - 11$
Multiply a constant times a variable term	Multiply the constant times the coefficient and keep the same variable part.	$-3(10x)$ $= -30x$
Divide a variable term by a constant	Divide the coefficient by the constant and keep the same variable part.	$\dfrac{35x}{7} = 5x$ $\dfrac{-18a}{3} = -6a$
Remove parentheses	Use the distributive property.	$3(x - 2)$ $= 3 \cdot x - 3 \cdot 2$ $= 3x - 6$

Throughout the chapter we followed the rules for the order of operations in evaluating and simplifying expressions.

We also learned to solve **equations** by getting the variable by itself on one side of the equation. To accomplish this, we made use of several properties of equations.

Property of Equations	Examples
We may *add* the same number to both sides of an equation.	$x - 10 = 5$ $x - 10 + 10 = 5 + 10$ $x = 15$
We may *subtract* the same number from both sides of an equation.	$7 + x = 1$ $7 - 7 + x = 1 - 7$ $x = -6$
We may *multiply* both sides of an equation by the same nonzero number.	$\dfrac{a}{4} = -1$ $\dfrac{4}{1} \cdot \dfrac{a}{4} = 4(-1)$ $a = -4$
We may *divide* both sides of an equation by the same nonzero number.	$3x = 21$ $\dfrac{3x}{3} = \dfrac{21}{3}$ $x = 7$

Then we developed the method for solving equations containing parentheses and/or more than one variable term. This was accomplished by using the rules for simplifying algebraic expressions and the properties of equations, as the next example illustrates.

$$3 - 2(x + 5) + 7x = -11$$

$$3 - 2x - 10 + 7x = -11 \qquad \text{Use distributive property.}$$

$$5x - 7 = -11 \qquad \text{Combine like terms.}$$

$$5x - 7 + 7 = -11 + 7 \qquad \text{Add 7 to both sides.}$$

$$5x = -4 \qquad \text{Find the sums.}$$

$$\frac{5x}{5} = \frac{-4}{5} \qquad \text{Divide both sides by 5.}$$

$$x = -\frac{4}{5} \qquad \text{Find the quotient.}$$

Finally, we practiced switching word problems to algebraic equations containing a variable. We solved such equations by the methods already described.

❏ Speaking the Language of Mathematics

Complete each sentence with the appropriate word or phrase.

1. In algebra we work with known numbers, called _____ , and unknown numbers, called _____ .

2. Combinations of constants and variables involving the operations of addition, subtraction, multiplication, and division are called _____ _____ .

3. When we substitute a particular value for the variable in an expression, we are _____ the expression.

4. In the expression $-12x$, we call -12 the _____ _____ of x.

5. To find the product $3(x + 2)$ we must use the _____ property.

6. The value of the variable that makes an equation a true statement is called the _____ for the equation.

△ Writing About Mathematics

Write your response to each question in complete sentences.

1. Write an algebraic expression that contains 2 different variable terms and a constant term. Then identify the variables, their numerical coefficients, and the constant term.

2. Show and explain the steps you would use to solve the equation

$$3x + 2(x + 3) = 4$$

3. Make up a word problem that would lead you to write the equation

$$5x + 30 = 120$$

REVIEW EXERCISES for Chapter 15

Write an algebraic expression for each word statement.

_____ 1. Add -10 to x.

_____ 2. Multiply $\frac{3}{4}$ times y.

_____ 3. Subtract 3.5 from a.

_____ 4. Find the sum of x and -2, then divide by 5.

Evaluate each expression.

_____ 5. $9 - y$, when $y = 10$

_____ 6. $2a - 5b$, when $a = -2$ and $b = -1$

_____ 7. $a - 2(3a - 5)$, when $a = 4$

_____ 8. $y^2 - 12$, when $y = 3$

Combine like terms.

_____ 9. $5x - 3x + 2x$

_____ 10. $7a - 3b - 9a + b$

_____ 11. $2x - 3y + 6 - 4x + 3y - 5$

_____ 12. $3x - 2y + z - 3x + y - 2z$

Perform the indicated operation.

_____ 13. $5(-7x)$

_____ 14. $-\dfrac{1}{5}(15x)$

_____ 15. $\dfrac{20a}{-4}$

_____ 16. $\dfrac{-45x}{-18}$

_____ 17. $5(2x - 3)$

_____ 18. $-3(4x - 5y)$

_____ 19. $\dfrac{1}{4}(-8x + 16y)$

_____ 20. $0.2(-4m + 3n)$

Solve each equation.

_____ 21. $x + 7 = 18$

_____ 22. $x - 10 = 21$

_____ 23. $y + 5 = -7$

_____ 24. $3.2 + a = 5.7$

_____ 25. $y - 12 = -4$

_____ 26. $-9 + a = 7$

_____ 27. $-\dfrac{2}{3} = m + \dfrac{5}{3}$

_____ 28. $3x = 27$

_____ 29. $-2a = 54$

_____ 30. $-4.2 = -6x$

_____ 31. $0.3x = 0$

_____ 32. $-x = 7.2$

_____ 33. $\dfrac{x}{3} = -2$

_____ 34. $\dfrac{x}{12} = 0$

_____ 35. $\dfrac{x}{-5} = 2$

_____ 36. $\dfrac{1}{4} = \dfrac{x}{-8}$

_____ 37. $4x - 7 = 31$ _____ 38. $8 - 2x = 6$

_____ 39. $-15 = 12 + 9x$ _____ 40. $\dfrac{x}{4} - 2 = -1$

_____ 41. $7x - 5 - 3x = 12$ _____ 42. $4x - 5 + 2x + 3 = -8$

_____ 43. $3(y - 4) - 7 = -4$ _____ 44. $7(2 - 3x) + x = 6$

_____ 45. $3(x - 7) - (x + 1) = 0$ _____ 46. $2(x + 2) - 5(2 + x) = -6$

Write an algebraic expression for each word expression.

_____ 47. A clothing store makes x dollars profit on each sweater sold. If the store wishes to have $600 profit from the sale of sweaters, write an expression for the number of sweaters the store must sell.

_____ 48. A building contractor must pay $2500 plus an additional $500 a day for each day over the deadline for completion of a building. Write an expression for the amount of penalty after n days.

_____ 49. In a certain rectangle, the width is 2 inches less than half the length. If the length is n inches, write an expression for the width.

Use an equation to solve each problem.

_____ 50. Clara has 5 times as many cassettes as Chris. Together they have 48 cassettes. How many cassettes does Clara have?

_____ 51. The length of a rectangle is 3 times its width. If the perimeter of the rectangle is 32 feet, find its dimensions.

_____ 52. Jesse earns $56 a day plus $8.75 an hour for each hour he works overtime. If Jesse earned $82.25 last Tuesday, how many hours did he work overtime?

Check your answers in the back of your book.

If You Missed Exercises:	You Should Review Examples:	
1–4	Section 15.1	1, 2
5–8		3–6
9–12		8–12
13–16		13–17
17–20		18–21
21–27	Section 15.2	2–5
28–32		6–8
33–36		9, 10
37–40	Section 15.3	1–3
41–46		4–7
47–49	Section 15.4	1–5
50–52		6–8

If you have completed the **Review Exercises** and corrected your errors, you are ready to take the **Practice Test** for Chapter 15.

PRACTICE TEST for Chapter 15

		SECTION	EXAMPLES

_____ 1. Write an algebraic expression for the sum of x and -5, divided by 3.   15.1   2

_____ 2. Evaluate $x - 3(2x - 3)$ when $x = -1$.   15.1   5

Combine like terms.

_____ 3. $5n - 6n - 3n$ 15.1   8

_____ 4. $-5x - 2y + 3 + 5x - 2y - 1$ 15.1   12

Find the product or quotient.

_____ 5. $-5(4n)$ 15.1   13

_____ 6. $\dfrac{-9x}{3}$ 15.1   16

_____ 7. $\dfrac{2}{3}(9x - 3)$ 15.1   20

Solve each equation.

_____ 8. $x + 12 = 5$ 15.2   2

_____ 9. $15 + n = 21$ 15.2   3

_____ 10. $a - 3.2 = -7.5$ 15.2   5

_____ 11. $8x = -96$ 15.2   6

_____ 12. $\dfrac{x}{7} = -9$ 15.2   9

_____ 13. $\dfrac{x}{-1.3} = -5$ 15.2   10

_____ 14. $4x - 9 = 3$ 15.3   1

_____ 15. $\dfrac{x}{5} - 9 = -14$ 15.3   3

_____ 16. $4x - 5 - 7x + 3 = 46$ 15.3   5

_____ 17. $4(x - 3) - (6x + 1) = 1$ 15.3   6, 7

_____ 18. A cable TV subscription salesperson is paid $250 a week plus $4.35 for every new subscriber. Write an expression for the weekly pay of a salesperson enrolling n new subscribers.   15.4   4

	SECTION	EXAMPLES

_____ 19. A limousine rental agency charges customers $55 per day plus $2.15 per mile driven. If Armand was charged $297.95 for 1 day's rental, how far did he drive?   15.4   7

_____ 20. The length of a rectangular painting is twice its width. If the perimeter of the painting is 36 inches, find the dimensions of the painting.   15.4   8

Check your answers in the back of your book.

SHARPENING YOUR SKILLS after Chapters 1–15

SECTION

_____ 1. Change $\frac{89}{12}$ to a mixed number. 3.1

_____ 2. Change $\frac{7}{8}$ to an equivalent fraction with a denominator of 72. 5.2

_____ 3. Use $<$, $>$, or $=$ to compare $\frac{19}{38}$ and $\frac{7}{12}$. 5.4

_____ 4. Round 78.549 to the tenths place. 6.1

_____ 5. Change $\frac{421}{38}$ to a decimal number and round to the hundredths place. 7.3

_____ 6. Change $12\frac{1}{4}$ percent to a fraction. 9.1

_____ 7. Change 0.036 to a percent. 9.1

_____ 8. Change 0.35 liter to milliliters. 12.3

_____ 9. Change $2\frac{1}{2}$ gallons to cups. 12.3

_____ 10. Compare -3.054 and -3.54 using $<$ or $>$. 14.1

Perform the indicated operations.

_____ 11. $509 \times 17 \times 200$ 2.1

_____ 12. $7\frac{2}{7} \times 3\frac{1}{9}$ 4.1

_____ 13. $\dfrac{3\frac{5}{9}}{\frac{3}{8}}$ 4.2

_____ 14. $12\frac{1}{2}\%$ of \$3980 9.2

_____ 15. $[(3.6)(-2.5)](-1.2)$ 14.3

_____ 16. $(7-9)^3$ 14.4

_____ 17. Perry's bank balance is \$2378. He writes checks for \$419 and \$87. He also makes deposits of \$325 and \$78. What is his new balance? 1.3

SECTION

———— 18. On a scale drawing of a house plan, 1 inch represents 4 feet. How many inches on the   2.4
scale drawing would be used to represent the width of the front of the house if the
actual width is 83 feet?

———— 19. Steve's diet allows him to have 1400 calories per day. If he uses 425 calories for   3.1
breakfast, what fractional part of his day's allowance is left for the rest of the day?

———— 20. If $\frac{3}{7}$ of 245 students surveyed had jobs to help pay college expenses, how many   4.3
students had jobs?

———— 21. Lenora uses $\frac{1}{4}$ of her salary for rent, $\frac{1}{6}$ for food, and $\frac{1}{9}$ for utilities. What fractional part   5.5
of her salary is left for other things?

———— 22. In 3 weeks of driving Harry used 15.4, 18.3, and 22.5 gallons of gasoline. If his   6.3
budget allows him to average 20 gallons per week for a four-week month, how many
gallons can he use during the fourth week?

———— 23. Each month, Dewayne's spending and saving are in the ratio of 9 to 2. If he saves $76   8.3
this month, how much will he spend?

———— 24. A survey showed that 3100 of 5000 eligible voters voted in the last presidential   9.2
election. What percent of those surveyed voted?

———— 25. A VCR that regularly sells for $389 is on sale for $330.65. Find the amount of mark-   10.1
down. Then find the percent of markdown.

———— 26. Shandra will receive a bonus from her employer if the mean of her sales for 6 months   11.1
is at least $1200. What must be her sales for the sixth month if the total of her sales
for the first 5 months is $5950?

———— 27. A certain solution is purchased in a bottle which contains 1 liter 600 milliliters. If   12.3
the solution is divided equally among 8 beakers, how many milliliters will be in
each beaker?

———— 28. Find the volume of a cylindrical container that has a diameter of 14 inches and a   13.3
height of 18 inches. Use $\pi \doteq \frac{22}{7}$.

———— 29. Write an algebraic expression for the sum of x and -9, multiplied by 7.   15.1

———— 30. Evaluate $x - 4(3x - 5)$ when $x = -2$.   15.1

Check your answers in the back of your book.

CHAPTER

16

Working
with a Calculator

> On a map, City A is 42 miles directly north of City C. City B is 79
> miles due east of City C. Find the shortest distance from City A to
> City B.

In recent years, the handheld calculator has become a popular tool for performing calculations in arithmetic and other branches of mathematics and science. Even people who enjoy doing computations in arithmetic prefer to turn to a calculator when doing lengthy calculations involving large numbers, fractional numbers, and decimal numbers.

In this chapter we learn to

1. Use a calculator to add, subtract, multiply, and divide.
2. Use a calculator to work with fractions.
3. Use a calculator to solve percent problems.
4. Use a calculator to work with measurement and geometry.

A calculator can be used to work even the simplest arithmetic problems, but it should *not* be used as a crutch. The basic number facts reviewed in Chapters 1 and 2 *must* be learned, even if you expect to use a calculator frequently. Without an understanding of the basic facts, you will become overly dependent upon the accuracy of your calculator and will not even recognize when its answer is not reasonable. Moreover, a calculator can only do what you tell it to do. You must have an understanding of the basic operations of arithmetic and the rules for the order of those operations if you hope to arrive at the correct answer for a problem involving more than one operation.

16.1 Using a Calculator to Add, Subtract, Multiply, and Divide

There are many different calculators on the market today. The calculator illustrated here is a typical example.

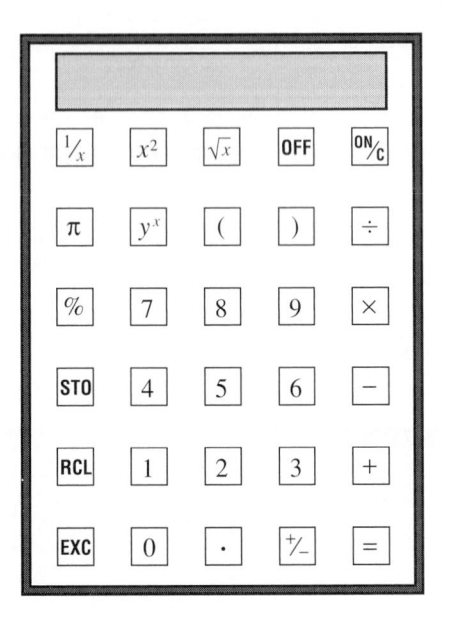

As you can see, this calculator has 30 **keys**, some of which perform calculations that we are not concerned with here. The instructions for using this calculator will be useful when working with any calculator that has an **algebraic operating system**. A calculator that uses such a system performs operations according to the rules for the order of operations that we learned in Chapter 14.

(1) First, multiplications and divisions are done in order from left to right.

(2) Then additions and subtractions are done in order from left to right.

In other words, a calculator using an algebraic operating system performs its calculations in the *same* order that you would perform those calculations by hand. If you own a calculator that does *not* use this system, you must carefully read the instructions for operating your calculator.

To begin working with your calculator, you must press the $\boxed{ON/C}$ key. A zero (**0.**) will appear in the **display**. When you finish a calculation, you should press the $\boxed{OFF}$ key to avoid unnecessary drain on the battery. If you plan to do another calculation immediately, you may press the $\boxed{ON/C}$ key twice to **clear** the display, instead of turning the calculator $\boxed{OFF}$.

Doing One Operation

To **add** two numbers using a calculator, we must enter the first number, press the $\boxed{+}$ key, enter the next number, and press the $\boxed{=}$ key. The sum appears in the display. For example, we find 37 + 42 as follows:

$$37 \boxed{+} 42 \boxed{=}$$

The answer, 79, will appear in the display.

If you wish to find the sum of several numbers, just repeat the $\boxed{+}$ step as many times as necessary. Let's find the sum 83 + 27 + 92 + 68.

Press: (ON/C) 83 (+) 27 (+) 92 (+) 68 (=)

Display: 270

Press: (OFF)

By the way, you may look at your subtotal along the way by pressing (=). Then you may finish adding the rest of the numbers as usual. Watch.

Press: (ON/C) 83 (+) 27 (=)

Display: 110

Press: (+) 92 (=)

Display: 202

Press: (+) 68 (=)

Display: 270

Subtotals can be useful when you are trying to stay "under" a certain sum while totaling dollars spent, or calories consumed, and so on. Speaking of dollars, notice that the calculator contains a decimal point key (·) which should be pressed when you enter a decimal number.

| EXC | 0 | · | +/− | = |

To find the sum $83.92 + $29.89, we proceed with the following steps.

Press: (ON/C) 83 (·) 92 (+) 29 (·) 89 (=)

Display: 113.81

Press: (OFF)

Our answer is $113.81. Notice that the calculator has no $ sign. You must attach it to your answer.

To **subtract** two numbers, we enter the first number, press the (−) key, and then press the (=) key. The difference will appear in the display. Let's find the difference 6391 − 2499.

Press: (ON/C) 6391 (−) 2499 (=)

Display: 3892

Press: (OFF)

To add and subtract within the same problem, we enter the numbers and press the appropriate (+) and (−) keys. Finally, we press the (=) key.

Example 1. Find $83.20 − $27.06 − $11.30 + $159.

Solution. Press: (ON/C) 83 (·) 20 (−) 27 (·) 06 (−) 11 (·) 30 (+) 159 (=)

Display: 203.84

Press: (OFF)

The answer is $203.84.

You complete Example 2.

Example 2. Find $0.0387 - 0.001 + 0.6298$.

Solution. Be careful with decimal points here.

Press: (ON/C) ⬭ ⎯⎯ ⬭⬭ ⎯⎯ ⬭⬭ ⎯⎯ (=)

Display: ⎯⎯

The answer is ⎯⎯ .

Check your work on page 712. ▶

As you surely have guessed, we **multiply** and **divide** on the calculator by using the (×) and (÷) keys.

Example 3. Find the product $16 \times 20 \times 9$. Find the quotient $8436 \div 12$.

Solution To find the product	To find the quotient
Press: (ON/C) 16 (×) 20 (×) 9 (=)	Press: (ON/C) 8436 (÷) 12 (=)
Display: 2880	Display: 703
The product is 2880.	The quotient is 703.

Example 4. Find $28 \div 3$.

Solution. Without a calculator, we would find

$$28 \div 3 = \frac{28}{3} = 9\frac{1}{3}$$

Let's see what the calculator comes up with.

Press: (ON/C) 28 (÷) 3 (=)

Display: 9.3333333

Our answer is the repeating decimal number $9.\overline{3}$.

From this example we can see that a calculator does *not* display fractional numbers. Every fractional number will appear as a decimal number in the display.

Example 5. Find $62.317 \div 0.84$.

Solution. Press: (ON/C) 62 (·) 317 (÷) (·) 84 (=)

Display: 74.186905

Our calculator will only display 8 places, so we cannot assume that this answer is exact. Let's round the quotient to the thousandths place.

$$62.317 \div 0.84 \doteq 74.187$$

To perform multiplications and divisions several times within the same problem, we must remember that our calculator will do those operations *in order from left to right*. If you do not wish them done in that order, you must use **parentheses** to tell the calculator how the operations are to be done. We discuss the use of parentheses in the next section.

Example 6. Find $0.83 \times 1.19 \div 3$ and round to the hundredths place.

Solution. Press: (ON/C) (·) 83 (×) 1 (·) 19 (÷) 3 (=)

Display: 0.32923333

Therefore, $0.83 \times 1.19 \div 3 \doteq 0.33$.

⟩ Trial Run

Perform the operations.

_____ 1. $74.38 − $29.13 − $6 + $117.25

_____ 2. 0.0429 − 0.003 + 0.7284

_____ 3. 51 ÷ 9

_____ 4. 0.723 × 5.32

_____ 5. 48.723 ÷ 0.39 (Round to the thousandths place.)

_____ 6. 0.293 × 1.76 ÷ 6 (Round to the hundredths place.)

Answers are on page 712.

Using Parentheses to Do Several Operations

If you wish to calculate

$$13 + 7 - 8 \div 2 + 3 \times 4$$

you may enter the numbers and press the operation keys just as they appear. Your calculator will obey the usual rules for the order of operations.

Press: (ON/C) 13 (+) 7 (−) 8 (÷) 2 (+) 3 (×) 4 (=)

Display: 28

Perhaps you would like to check this answer by hand. Remember to use the rules for the order of operations.

$$
\begin{aligned}
&13 + 7 - \underbrace{8 \div 2} + 3 \times 4 \\
=\; &13 + 7 - \quad 4 \quad + \underbrace{3 \times 4} \\
=\; &\underbrace{13 + 7} - \quad 4 \quad + \quad 12 \\
=\; &\quad \underbrace{20 \quad - \quad 4} \quad + \quad 12 \\
=\; &\qquad\qquad 16 \qquad + \quad 12 \\
=\; &28
\end{aligned}
$$

Suppose that you wish to find the **mean** of the two numbers 37 and 42. We know that the mean is found by adding the numbers and then dividing that sum by 2. We *cannot* enter this problem as 37 + 42 ÷ 2 or the calculator will divide 42 by 2 and then add that quotient (21) to 37. Instead, we must use **parentheses** to show that we are to *add first and then divide*. Parentheses always say ''do this first''. We write the problem as

$$(37 + 42) \div 2$$

Find the parentheses keys on your calculator.

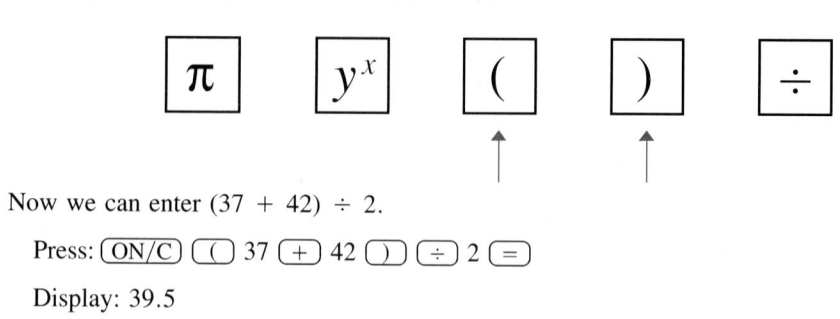

Now we can enter $(37 + 42) \div 2$.

Press: ON/C (37 + 42) ÷ 2 =

Display: 39.5

The mean of 37 and 42 is 39.5.

If you were watching the display as you closed your parentheses, you saw the sum 79 appear. The calculator was following your instructions. You should always *write down* a problem in its correct form *before* you enter it into the calculator. A calculator cannot be accurate if it does not receive information in the proper form.

Example 7. Find the mean of the following weekly grocery bills: $75.63, $82.27, $91.13, and $69.75. (Round your answer to the nearest cent.)

Solution. First we write the problem in its correct form.

$$\frac{75.63 + 82.27 + 91.13 + 69.75}{4}$$

$$= (75.63 + 82.27 + 91.13 + 69.75) \div 4$$

Press: ON/C (75.63 + 82.27 + 91.13 + 69.75) ÷ 4 =

Display: 79.695

Rounding to the nearest cent, our mean grocery bill is $79.70. By the way, since the calculator did not display all 8 places, we know this decimal number *terminated* at 79.695.

Example 8. Mario weighed 187 pounds before he started his diet. Now he weighs 149 pounds. His sister Helen lost *twice* as much weight as Mario lost. Find Helen's weight loss.

Solution. First we must write down the problem. To find Helen's loss, we first find Mario's loss by subtracting. Then we multiply Mario's loss by 2.

$$(187 - 149) \times 2$$

Press: ON/C (187 − 149) × 2 =

Display: 76

Helen lost 76 pounds.

Remember that your calculator will not keep track of *units* for you. You must decide from the problem what kind of units to give to your answer.

Example 9. Find $(3 + 14) \times (10 - 4)$.

Solution. Press: ON/C (3 + 14) × (10 − 4) =

Display: 102

$$(3 + 14) \times (10 - 4) = 102$$

In problems of this kind, parentheses are very important. Leaving them out will give an entirely different (and incorrect) answer.

⫸ Trial Run

Perform the operations.

_____1. (29.72 + 38.65 + 78.32 + 53.31) ÷ 4

_____2. (293 − 126) × 3

_____3. (0.372 + 0.124) × 3.2

_____4. (13 + 42) × (29 − 15)

_____5. (8.3 + 7.4) × 5 + (3.6 + 9.4) ÷ 2

_____6. (93 + 84 + 75 + 36) ÷ (6.2 − 4.6)

Answers are on page 712.

Finding Powers on the Calculator

At several times throughout this course, we have looked at the **squares** of numbers. Recall that we said that

$$3^2 = 3 \times 3 = 9$$
$$5^2 = 5 \times 5 = 25$$

Because the need for squaring a number comes up often in mathematics, there is a key on your calculator that does the squaring in one step. Find the $\boxed{x^2}$ key.

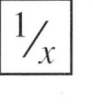 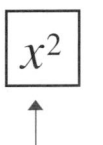    

This is called the **squaring key**. The letter x stands for any number that you might like to square. To square a number using your calculator, enter the number and press the $\boxed{x^2}$ key. The square will appear immediately in the display. Let's use the calculator to find 5^2.

Press: $\boxed{ON/C}$ 5 $\boxed{x^2}$

Display: 25

As we already knew, $5^2 = 25$.

Example 10. Find $(1.238)^2$.

Solution. Parentheses need not be entered in the calculator. They are used here to make the problem clearer.

Press: $\boxed{ON/C}$ 1.238 $\boxed{x^2}$

Display: 1.532644

$(1.238)^2 = 1.532644$

By the way, unless parentheses say otherwise, your calculator will do squarings *first*, then multiplications and divisions, and then additions and subtractions.

Example 11. Find 4×9^2.

Solution. Press: (ON/C) 4 (×) 9 (x^2) (=)

Display: 324

$4 \times 9^2 = 324$

Example 12. Find $7^2 + 12^2$.

Solution. Press: (ON/C) 7 (x^2) (+) 12 (x^2) (=)

Display: 193

$7^2 + 12^2 = 193$

If parentheses appear around a *quantity* to be squared, those parentheses must be entered in the calculator. Otherwise the calculator will only square the number entered just before you press the (x^2) key.

Example 13. Find $(3 + 5)^2$ and check by hand.

Solution. On the calculator   CHECK:

Press: (ON/C) (( ) 3 (+) 5 ( )) (x^2) $(3 + 5)^2 = 8^2$

Display: 64 $= 64$

$(3 + 5)^2 = 64$

Example 14. Find $15^2 - 4 \times 11 \times 2$.

Solution. Press: (ON/C) 15 (x^2) (−) 4 (×) 11 (×) 2 (=)

Display: 137

$15^2 - 4 \times 11 \times 2 = 137$

⟫ Trial Run

Perform the operations.

———— 1. $(2.376)^2$ ———— 2. $5 \times (0.063)^2$

———— 3. $15^2 + 9^2$ ———— 4. $(8.7 + 3.2)^2$

———— 5. $(6.67 - 3.25)^2$ ———— 6. $43^2 - 9 \times 7 \times 15$

Answers are on page 712.

When we first learned about finding a square such as 5^2, we labeled each part of the expression

$$\quad \overset{\quad\quad}{\underset{\text{base}}{\longrightarrow}} 5^2 \leftarrow \text{exponent}$$

and we said that the exponent (2) told us to use the base (5) as a factor in a product 2 times. We read 5^2 as "5 squared," but we may also read 5^2 as "the second power of 5".

In a similar way, we can find other powers of the base 5. Using an exponent to tell us how many times to use the base as a factor in a product, we may find the

third power of 5: $5^3 = 5 \times 5 \times 5$

fourth power of 5: $5^4 = 5 \times 5 \times 5 \times 5$

seventh power of 5: $5^7 = 5 \times 5 \times 5 \times 5 \times 5 \times 5 \times 5$

Each of these powers can be found on a calculator (or by hand) by multiplying the factors.

Are you ready for an easier method for finding powers with a calculator? On your calculator, notice the $\boxed{y^x}$ key.

$$\boxed{\pi} \quad \boxed{y^x} \quad \boxed{(} \quad \boxed{)} \quad \boxed{\div}$$

This is the **power key**. To find a power, we enter the base, press the $\boxed{y^x}$ key, enter the exponent, and press $\boxed{=}$. The display will flicker a bit as the calculator cranks out the answer, but it will appear in a few seconds. Let's try using the power key to find 5^7.

Press: $\boxed{ON/C}$ 5 $\boxed{y^x}$ 7 $\boxed{=}$

Display: 78125

We find that $5^7 = 78,125$.

You try Example 15.

Example 15. Find 2^9.

Solution. Press: $\boxed{ON/C}$ 2 $\boxed{\;}$ 9 $\boxed{=}$

Display: _____

Therefore $2^9 = $ _____ .

Check your work on page 712. ▶

Example 16. Find $(0.123)^5$.

Solution. Imagine finding this power by hand!

Press: $\boxed{ON/C}$ $\boxed{\cdot}$ 123 $\boxed{y^x}$ 5 $\boxed{=}$

Display: .00002815

Therefore $(0.123)^5 \doteq 0.00002815$.

As we said earlier, your calculator will do powers, then multiplications and divisions, then additions and subtractions, unless parentheses tell it to do otherwise.

Example 17. Find $7^4 + 829$.

Solution. Press: $\boxed{ON/C}$ 7 $\boxed{y^x}$ 4 $\boxed{+}$ 829 $\boxed{=}$

Display: 3230

$7^4 + 829 = 3230$

Example 18. Find $(3.27 + 8.09)^4 + 1.52^3$. Round your answer to the hundredths place.

Solution. We must use parentheses around the first sum.

Press: $\boxed{ON/C}$ $\boxed{(}$ 3.27 $\boxed{+}$ 8.09 $\boxed{)}$ $\boxed{y^x}$ 4 $\boxed{+}$ 1.52 $\boxed{y^x}$ 3 $\boxed{=}$

Display: 16657.311

Therefore $(3.27 + 8.09)^4 + 1.52^3 \doteq 16657.31$.

Be sure to give your calculator time to find the powers along the way before you enter the next number.

⫸ Trial Run

Perform the operations.

_____ **1.** 8^3

_____ **2.** $(0.235)^4$ (Round to the hundred thousandths place.)

_____ **3.** $4^5 + 938$

_____ **4.** $(4.25 + 7.83)^4 + 2.35^5$ (Round to the hundredths place.)

_____ **5.** $1.5^5 \times 2.1^3 - 8.3^2$ (Round to the tenths place.)

_____ **6.** $2.9^4 \div (3.6 + 2.4)^3$ (Round to the ten thousandths place.)

Answers are given below.

▶ Examples You Completed

Example 2. Find $0.0387 - 0.001 + 0.6298$.

   Solution. Be careful with decimal points here.

   Press: $\boxed{\text{ON/C}}$ $\boxed{\cdot}$ 0387 $\boxed{-}$ $\boxed{\cdot}$ 001 $\boxed{+}$ $\boxed{\cdot}$ 6298 $\boxed{=}$
   Display: 0.6675

The answer is 0.6675.

Example 15. Find 2^9.

   Solution. Press: $\boxed{\text{ON/C}}$ 2 $\boxed{y^x}$ 9 $\boxed{=}$
   Display: 512

Therefore $2^9 = 512$.

Answers to Trial Runs

page 707 **1.** $156.50   **2.** 0.7683   **3.** $5.\overline{6}$ **4.** 3.84636   **5.** 124.931   **6.** 0.09

page 709 **1.** 50   **2.** 501   **3.** 1.5872   **4.** 770   **5.** 85   **6.** 180

page 710 **1.** 5.645376   **2.** 0.019845   **3.** 306   **4.** 141.61   **5.** 11.6964   **6.** 904

page 712 **1.** 512   **2.** 0.00305   **3.** 1962   **4.** 21366.18   **5.** 1.4   **6.** 0.3274

EXERCISE SET 16.1

Perform the operations.

_____ 1. $94.20 − $36.07 − $12.35 + $88

_____ 2. $76.25 − $29.08 − $16.29 + $67

_____ 3. 0.386 − 0.002 + 0.3975

_____ 4. 0.485 − 0.005 + 0.3865

_____ 5. 1.37 − 0.035 − 0.9

_____ 6. 1.76 − 0.058 − 0.7

_____ 7. 26 × 75 × 83

_____ 8. 39 × 52 × 37

_____ 9. 3.72 × 82.53

_____ 10. 5.76 × 28.37

_____ 11. 36 ÷ 8

_____ 12. 78 ÷ 8

_____ 13. 56 ÷ 9

_____ 14. 44 ÷ 6

_____ 15. 0.382 × 6.71

_____ 16. 0.543 × 8.23

_____ 17. 36.072 ÷ 0.29
(Round answer to thousandths place.)

_____ 18. 48.039 ÷ 0.34
(Round answer to thousandths place.)

_____ 19. 0.79 × 1.32 ÷ 15
(Round answer to hundredths place.)

_____ 20. 0.89 × 2.34 ÷ 19
(Round answer to hundredths place.)

_____ 21. 3.724 ÷ 0.4 × 12
(Round answer to tenths place.)

_____ 22. 4.631 ÷ 0.8 × 17
(Round answer to tenths place.)

_____ 23. (16.32 + 29.76 + 35.61 + 32.31) ÷ 6

_____ 24. (19.32 + 52.31 + 18.75 + 17.62) ÷ 9

_____ 25. (1037 − 879) × 12

_____ 26. (1237 − 978) × 15

_____ 27. 8.735 − (0.25 × 9)

_____ 28. 9.386 − (0.39 × 6)

_____ 29. (0.546 + 0.135) × 2.4

_____ 30. (0.782 + 0.183) × 3.6

_____ 31. (17 + 59) × (37 − 19)

_____ 32. (92 + 13) × (45 − 37)

_____ 33. (9.4 + 7.3) × 6 + (8.1 + 9.3) ÷ 3

_____ 34. (8.2 + 7.6) × 7 + (8.3 + 7.2) ÷ 5

_____ 35. (87 + 65 + 39 + 57) ÷ (9.8 − 3.6)

_____ 36. (98 + 53 + 28 + 47) ÷ (6.2 − 3.7)

_____ 37. 572^2

_____ 38. 398^2

_____ 39. $(3.784)^2$

_____ 40. $(1.583)^2$

_____ 41. $9 × (7.32)^2$

_____ 42. $7 × (8.35)^2$

_____ 43. $13^2 + 7^2$ _____ 44. $19^2 + 15^2$

_____ 45. $(3.6 + 7.9)^2$ _____ 46. $(8.4 + 9.7)^2$

_____ 47. $(13.65 - 7.28)^2$ _____ 48. $(15.76 - 9.37)^2$

_____ 49. $(4.3 \times 7.9)^2$ _____ 50. $(5.7 \times 8.4)^2$

_____ 51. $24^2 - 6 \times 8 \times 12$ _____ 52. $37^2 - 9 \times 9 \times 13$

_____ 53. $(48^2 \div 64) \times (8.34 - 2.75)$ _____ 54. $(35^2 \div 49) \times (7.36 - 3.48)$

_____ 55. $(7.3 - 5.8)^2 + (2.9)^2$ _____ 56. $(8.4 - 3.7)^2 + (3.7)^2$

_____ 57. 9^3 _____ 58. 6^4

_____ 59. $6^5 + 3428$ _____ 60. $7^4 + 8379$

_____ 61. $(0.372)^4$ _____ 62. $(0.583)^3$
          (Round answer to hundred   (Round answer to hundred
          thousandths place.)   thousandths place.)

_____ 63. $(3.72 + 4.59)^3 + 8.54^4$ _____ 64. $(1.72 + 3.13)^4 + 2.74^5$
          (Round answer to hundredths place.)   (Round answer to hundredths place.)

_____ 65. $2.3^4 \times 3.4^3 - 9.14^2$ _____ 66. $12.3^2 \times 1.5^5 - 2.72^4$
          (Round answer to tenths place.)   (Round answer to tenths place.)

_____ 67. $(4.6 - 2.3)^4 - (9.3 - 8.5)^5$ _____ 68. $(8.9 - 7.4)^5 - (3.6 - 2.7)^4$
          (Round answer to thousandths place.)   (Round answer to thousandths place.)

_____ 69. $(4.7)^3 \div (1.6 + 3.9)^2$ _____ 70. $(5.2)^4 \div (2.3 + 1.5)^3$
          (Round answer to ten   (Round answer to ten
          thousandths place.)   thousandths place.)

☆ Stretching the Topics _____

_____ 1. $5.3 \times (1.7 + 3.2)^2 \div (1.2 - 0.3)^3$ (Round answer to thousandths place.)

_____ 2. $(0.003 \times 0.007^2) \div 0.05^3$ (Round answer to ten thousandths place.)

_____ 3. $(3.75^2 + 0.32^2 - 1.75^2)^3 \div (3.75^3 + 0.32^3 - 1.75^3)^2$ (Round answer to
          hundredths place.)

Check your answers in the back of your book.

If you can do the problems in **Checkup 16.1**, you are ready to go on to Section 16.2.

✔ # CHECKUP 16.1

Perform the operations.

_____ 1. $93.78 − $42.36 − $24.75 + $115.50

_____ 2. 57 ÷ 9

_____ 3. 0.79 × 1.37 ÷ 5.3
(Round answer to hundredths place.)

_____ 4. ($26.79 + $83.72 + $15.28) ÷ 3

_____ 5. (24 + 32) × (15 − 9)

_____ 6. $(2.374)^2$

_____ 7. $(12.7 − 8.3)^2$

_____ 8. $26^2 − 71 × 12 × 7$

_____ 9. $(0.379)^4$
(Round answer to thousandths place.)

_____ 10. $(2.37 + 9.08)^4 + 2.51^3$
(Round answer to hundredths place.)

Check your answers in the back of your book.

If You Missed Problems:	You Should Review Examples:
1	1
2, 3	4–6
4, 5	7–9
6–8	10–14
9, 10	16–18

16.2 Using a Calculator to Work with Fractions

We mentioned earlier that a calculator deals with decimal numbers rather than fractional numbers. When we work fraction problems with a calculator, we cannot expect answers in fractional form. The answers will appear in decimal number form.

Changing Fractional Numbers to Decimal Numbers

The calculator is very handy for changing fractional numbers to decimal numbers. Remember that the **fraction bar** is a symbol for the operation of **division**.

$$\frac{2}{3} \text{ means } 2 \div 3$$

$$\frac{1}{9} \text{ means } 1 \div 9$$

Therefore, to change a fraction to a decimal number, we enter the fraction as a division problem. For example, suppose we wish to change $\frac{2}{3}$ to a decimal number.

$$\frac{2}{3} \text{ means } 2 \div 3$$

Press: (ON/C) 2 (÷) 3 (=)

Display: 0.6666667

$$\frac{2}{3} = 0.\overline{6}$$

Example 1. Change $\frac{163}{107}$ to a decimal number. (Round to the thousandths place.)

Solution. $\frac{163}{107}$ means $163 \div 107$.

Press: (ON/C) 163 (÷) 107 (=)

Display: 1.5233645

Therefore $\frac{163}{107} \doteq 1.523$.

You complete Example 2.

Example 2. Change $\frac{1}{9}$ to a decimal number.

Solution. $\frac{1}{9}$ means _____ .

Press: (ON/C) 1 (　) 9 (=)

Display: _____

Therefore $\frac{1}{9}$ = _____ .

Check your work on page 724. ▶

There is another way to use the calculator to find the decimal value of a fraction with a numerator of 1 (such as $\frac{1}{9}$). Find the (1/x) key on your calculator.

We can use this key to find the decimal value of a fraction with a numerator of 1 and *any* denominator. The letter x stands for any denominator you wish to enter. To use this key we enter the denominator and press the $\boxed{1/x}$ key. The answer appears in the display. Let's try using this key to change $\frac{1}{9}$ to a decimal number.

Press: $\boxed{\text{ON/C}}$ 9 $\boxed{1/x}$

Display: 0.1111111

$$\frac{1}{9} = 0.\overline{1}$$

Notice that we did not need to press $\boxed{=}$ because the calculator displays $1/x$ immediately.

Example 3. Find $\dfrac{1}{6395}$. (Round to the hundred thousandths place.)

Solution. Press: $\boxed{\text{ON/C}}$ 6395 $\boxed{1/x}$

Display: 0.00015637

$$\frac{1}{6395} \doteq 0.00016$$

Suppose we wish to write the fraction $\dfrac{2 + 3}{8 + 1}$ as a decimal number. To change this fraction to a decimal number, we must use **parentheses** around the numerator and denominator.

> When a numerator or denominator of a fraction contains a sum or difference, we must use parentheses to enter the fraction in a calculator.

As before, we write our problem before we try to enter it in the calculator to be certain that we have used parentheses where needed.

To change $\dfrac{2 + 3}{8 + 1}$ to a decimal number, we write

$$\frac{2 + 3}{8 + 1} = (2 + 3) \div (8 + 1)$$

Press: $\boxed{\text{ON/C}}$ $\boxed{(}$ 2 $\boxed{+}$ 3 $\boxed{)}$ $\boxed{\div}$ $\boxed{(}$ 8 $\boxed{+}$ 1 $\boxed{)}$ $\boxed{=}$

Display: 0.55555556

If you do *not* use parentheses, your calculator will do the operations according to the rules for order of operations. Your answer will be incorrect. (Try it.) Then complete Example 4.

Example 4. Find $\dfrac{7 - 4}{9 - 2}$. (Round to the thousandths place.)

Solution. First we write the problem with necessary parentheses.

$$\frac{7 - 4}{9 - 2} = (7 - 4) \div (9 - 2)$$

Press: $\boxed{\text{ON/C}}$ $\boxed{\ }$ 7 $\boxed{\ }$ 4 $\boxed{\ }$ $\boxed{\ }$ $\boxed{\ }$ 9 $\boxed{\ }$ 2 $\boxed{\ }$ $\boxed{=}$

Display: _____

Therefore $\dfrac{7 - 4}{9 - 2} \doteq$ _____ .

Check your work on page 725. ▶

Example 5. Find $\dfrac{1}{7 \times 6 - 2}$.

Solution. This looks like a problem in which we can use the $\boxed{1/x}$ key. But we must use parentheses around the denominator as we enter it.

$$\frac{1}{7 \times 6 - 2} = \frac{1}{(7 \times 6 - 2)}$$

Press: $\boxed{\text{ON/C}}$ $\boxed{(}$ 7 $\boxed{\times}$ 6 $\boxed{-}$ 2 $\boxed{)}$ $\boxed{1/x}$

Display: 0.025

Therefore $\dfrac{1}{7 \times 6 - 2} = 0.025$. Notice that this answer is exact.

⇒ Trial Run

_____ 1. Change $\dfrac{235}{178}$ to a decimal number. (Round to the hundredths place.)

_____ 2. Change $\dfrac{13}{99}$ to a decimal number.

_____ 3. Change $\dfrac{1}{375}$ to a decimal number. (Round to the ten thousandths place.)

_____ 4. Change $\dfrac{7}{8}$ to a decimal number.

_____ 5. Find $\dfrac{9 - 4}{15 - 9}$.

_____ 6. Find $\dfrac{5^2}{3 \times 7 - 13}$. (Round to the hundredths place.)

Answers are on page 725.

Multiplying and Dividing Fractions on a Calculator

You may use your calculator to perform the operations of addition, subtraction, multiplication, and division with fractions as long as you understand that the answers will appear as **decimal numbers**.

For instance, to **multiply fractions** we enter each fraction as a division problem, with multiplication between. Let's try an example that can be checked easily by hand.

Multiply $\dfrac{2}{3} \times \dfrac{9}{2}$.

Press: $\boxed{\text{ON/C}}$ 2 $\boxed{\div}$ 3 $\boxed{\times}$ 9 $\boxed{\div}$ 2 $\boxed{=}$

Display: 3

CHECK: $\dfrac{2}{3} \times \dfrac{9}{2}$

$= \dfrac{\overset{1}{\cancel{2}}}{\underset{1}{\cancel{3}}} \times \dfrac{\overset{3}{\cancel{9}}}{\underset{1}{\cancel{2}}}$

$= 3$

Actually, you would probably have done this little problem by hand instead of using your calculator. We chose to do it both ways to convince ourselves that the calculator steps we used were correct.

You complete Example 6.

Example 6. Find $\dfrac{13}{83} \times \dfrac{19}{17}$. (Round to the thousandths place.)

Solution. Press: $\boxed{\text{ON/C}}$ 13 $\boxed{}$ 83 $\boxed{}$ 19 $\boxed{}$ 17 $\boxed{=}$

Display: _____

$\dfrac{13}{83} \times \dfrac{19}{17} \doteq$ _____

Check your work on page 725. ▶

If you prefer a fractional answer to Example 6, you may multiply the numerators on the calculator and multiply the denominators on the calculator.

$$\frac{13}{83} \times \frac{19}{17} = \frac{247}{1411}$$

Of course we have not tried to reduce this fraction, but you could have checked for common factors *before* multiplying. There are none. As you expect, if you now change $\frac{247}{1411}$ to a decimal number on the calculator, the display reads 0.17505315.

Example 7. Find $\dfrac{15}{16} \times \dfrac{3}{11} \times \dfrac{1}{8}$. Write your answer as a decimal number. (Round to the thousandths place.)

Solution. Press: $\boxed{\text{ON/C}}$ 15 $\boxed{\div}$ 16 $\boxed{\times}$ 3 $\boxed{\div}$ 11 $\boxed{\times}$ 8 $\boxed{1/x}$ $\boxed{=}$

Display: 0.031960203

$$\frac{15}{16} \times \frac{3}{11} \times \frac{1}{8} \doteq 0.032$$

Notice that we used the $\boxed{1/x}$ key when multiplying by $\frac{1}{8}$. Your calculator knows to use $\boxed{1/x}$ only on the number you enter just before pressing the $\boxed{1/x}$ key. To complete this calculation with several operations, we pressed the $\boxed{=}$ key at the end.

Example 8. Find $\dfrac{15}{16} \times \dfrac{3}{11} \times \dfrac{1}{8}$. Write your answer as a *fraction*.

Solution. We see no common factors, so we use the calculator to multiply numerators and multiply denominators.

Numerator	*Denominator*
Press: $\boxed{\text{ON/C}}$ 15 $\boxed{\times}$ 3 $\boxed{\times}$ 1 $\boxed{=}$	Press: $\boxed{\text{ON/C}}$ 16 $\boxed{\times}$ 11 $\boxed{\times}$ 8 $\boxed{=}$
Display: 45	Display: 1408

Therefore $\dfrac{15}{16} \times \dfrac{3}{11} \times \dfrac{1}{8} = \dfrac{45}{1408}$.

To **divide fractions** we must be more careful as we enter the problem in the calculator.

There are several approaches that we may use, but the easiest approach requires that we *use parentheses around the divisor*. Recall the parts of a division problem.

$$\frac{2}{3} \div \frac{5}{7} \qquad = \; ?$$

dividend⎤   divisor   ⎣quotient

We can certainly rewrite this problem using parentheses around the divisor without changing the meaning of the problem.

$$\frac{2}{3} \div \frac{5}{7} = \frac{2}{3} \div \left(\frac{5}{7}\right)$$

Here's how we enter this problem in the calculator.

Press: ON/C 2 ÷ 3 ÷ (5 ÷ 7) =

Display: 0.93333333

Therefore $\frac{2}{3} \div \frac{5}{7} = 0.9\overline{3}$.

As long as we remember to put parentheses around a fraction that is acting as a divisor we will be able to divide fractions correctly on the calculator.

Example 9. Find $\frac{17}{18} \div \frac{12}{7} \times \frac{1}{5}$. (Round to the ten thousandths place.)

Solution. Remember to put parentheses around the divisor fraction.

$$\frac{17}{18} \div \left(\frac{12}{7}\right) \times \frac{1}{5}$$

Press: ON/C 17 ÷ 18 ÷ (12 ÷ 7) × 5 1/x =

Display: 0.11018519

Therefore $\frac{17}{18} \div \frac{12}{7} \times \frac{1}{5} \doteq 0.1102$.

You complete Example 10.

Example 10. Find $\frac{5}{16} \div \frac{2}{3} \div \frac{10}{11}$.

Solution. We must use parentheses around each of the fractions that are acting as divisors.

$$\frac{5}{16} \div \left(\frac{2}{3}\right) \div \left(\frac{10}{11}\right)$$

Press: ON/C 5 ÷ 16 ÷ (2 ÷ 3) ÷ (10 ÷ 11) =

Display: _____

Therefore $\frac{5}{16} \div \frac{2}{3} \div \frac{10}{11} =$ _____ .

Check your work on page 725. ▶

Example 11. Find $\dfrac{5}{16} \div \left(\dfrac{2}{3} \div \dfrac{10}{11}\right)$. (Round to the thousandths place.)

Solution. We see parentheses around the big divisor, but inside those parentheses there is another little divisor. We must use parentheses *twice* here and write

$$\frac{5}{16} \div \left(\frac{2}{3} \div \frac{10}{11}\right) = \frac{5}{16} \div \left[\frac{2}{3} \div \left(\frac{10}{11}\right)\right]$$

Brackets help us organize the problem, but we can only use *parentheses* on the calculator. We will enter 2 pairs of parentheses.

Press: $\boxed{\text{ON/C}}$ 5 $\boxed{\div}$ 16 $\boxed{\div}$ $\boxed{(}$ $\boxed{(}$ 2 $\boxed{\div}$ 3 $\boxed{\div}$ $\boxed{(}$ $\boxed{(}$ 10 $\boxed{\div}$ 11 $\boxed{)}$ $\boxed{)}$ $\boxed{=}$

Display: 0.42613636

$$\frac{5}{16} \div \left(\frac{2}{3} \div \frac{10}{11}\right) \doteq 0.426$$

Let's work Example 11 by hand to convince ourselves that the double parentheses do what they are supposed to do.

$$\frac{5}{16} \div \left(\frac{2}{3} \div \frac{10}{11}\right)$$

$$= \frac{5}{16} \div \left(\frac{2}{3} \times \frac{11}{10}\right) \qquad \text{Inside parentheses, invert the divisor and multiply.}$$

$$= \frac{5}{16} \div \left(\frac{11}{15}\right) \qquad \text{Find the product in parentheses.}$$

$$= \frac{5}{16} \times \frac{15}{11} \qquad \text{Invert the divisor and multiply.}$$

$$= \frac{75}{176} \qquad \text{Find the product.}$$

$$= 75 \div 176 \qquad \text{Change the fraction to a division problem.}$$

$$= 0.42613636 \qquad \text{Find the quotient on the calculator.}$$

$$\doteq 0.426 \qquad \text{Round to the thousandths place.}$$

Indeed both methods gave the same answer. It is easy to get confused when working with more than one pair of parentheses. That is why it is important to write down the problem using the necessary parentheses and/or brackets *before* you enter it in the calculator.

⫸ Trial Run

Perform the operations.

_____ 1. $\dfrac{4}{5} \times \dfrac{15}{6}$ _____ 2. $\dfrac{2}{3} \times \dfrac{4}{5}$ _____ 3. $\dfrac{9}{20} \times \dfrac{3}{5} \times \dfrac{1}{4}$

_____ 4. $\dfrac{7}{8} \div \dfrac{2}{3}$ _____ 5. $\dfrac{11}{15} \div \left(\dfrac{3}{4} \times \dfrac{2}{7}\right)$ _____ 6. $\left(\dfrac{5}{3} \div \dfrac{4}{5}\right) \div \dfrac{5}{16}$

Answers are on page 725.

Adding and Subtracting Fractions on a Calculator

Because our calculator performs its operations according to the usual rules for the order of operations, adding and subtracting fractions using the calculator will not be difficult. Let's practice with a sum that we can check easily by hand. Find the sum $\frac{2}{3} + \frac{7}{3}$.

We enter the numbers as they appear.

Press: $\boxed{\text{ON/C}}$ 2 $\boxed{\div}$ 3 $\boxed{+}$ 7 $\boxed{\div}$ 3 $\boxed{=}$

Display: 3

$$\frac{2}{3} + \frac{7}{3} = 3$$

CHECK:
$$\frac{2}{3} + \frac{7}{3} = \frac{2 + 7}{3}$$
$$= \frac{9}{3}$$
$$= 3$$

The *advantage* of adding (or subtracting) fractions on a calculator is that we do not need to worry about finding a lowest common denominator (LCD). The *disadvantage* of using a calculator is that our answers will be decimal numbers instead of fractional numbers.

Example 12. Find $\frac{3}{17} + \frac{9}{11} + \frac{82}{45}$. (Round to the hundredths place.)

Solution. Press: $\boxed{\text{ON/C}}$ 3 $\boxed{\div}$ 17 $\boxed{+}$ 9 $\boxed{\div}$ 11 $\boxed{+}$ 82 $\boxed{\div}$ 45 $\boxed{=}$

Display: 2.8168746

$$\frac{3}{17} + \frac{9}{11} + \frac{82}{45} \doteq 2.82$$

You complete Example 13.

Example 13 Find $\frac{43}{51} + \frac{3}{10} - \frac{18}{19}$. (Round to the thousandths place.)

Solution. Press: $\boxed{\text{ON/C}}$ 43 $\boxed{\div}$ 51 $\boxed{+}$ 3 $\boxed{\div}$ 10 $\boxed{-}$ 18 $\boxed{\div}$ 19 $\boxed{=}$

Display: _____

$$\frac{43}{51} + \frac{3}{10} - \frac{18}{19} \doteq \text{_____}$$

Check your work on page 725 ▶.

If *parentheses* appear in an addition or subtraction problem, they must be entered in the calculator where they appear.

Example 14. Find $\frac{73}{82} - \left(\frac{19}{39} - \frac{2}{43} \right)$. (Round to the ten thousandths place.)

Solution. Press: $\boxed{\text{ON/C}}$ 73 $\boxed{\div}$ 82 $\boxed{-}$ $\boxed{(}$ 19 $\boxed{\div}$ 39 $\boxed{-}$ 2 $\boxed{\div}$ 43 $\boxed{)}$ $\boxed{=}$

Display: 0.44957604

$$\frac{73}{82} - \left(\frac{19}{39} - \frac{2}{43} \right) \doteq 0.4496$$

Suppose we try a fraction problem that asks us to do all 4 operations. We must pay close attention to parentheses.

Example 15. Find $\left(\dfrac{3}{8} + \dfrac{5}{2} \times \dfrac{1}{3}\right) \div \left(\dfrac{11}{15} - \dfrac{2}{9}\right)$. (Round to the hundredths place.)

Solution.   Press: $\boxed{\text{ON/C}}$ $\boxed{(\,}$ 3 $\boxed{\div}$ 8 $\boxed{+}$ 5 $\boxed{\div}$ 2 $\boxed{\times}$ 1 $\boxed{\div}$ 3 $\boxed{\,)}$
$\boxed{\div}$ $\boxed{(\,}$ 11 $\boxed{\div}$ 15 $\boxed{-}$ 2 $\boxed{\div}$ 9 $\boxed{\,)}$ $\boxed{=}$

Display: 2.3641304

The rounded answer is 2.36.

⫸ Trial Run

Perform the operations.

_____ 1. $\dfrac{2}{3} + \dfrac{3}{8}$ (Round to the thousandths place.)

_____ 2. $\dfrac{7}{8} + \dfrac{3}{5} + \dfrac{3}{4}$

_____ 3. $\dfrac{9}{10} + \dfrac{13}{16} - \dfrac{5}{6}$ (Round to the hundredths place.)

_____ 4. $\dfrac{17}{20} - \left(\dfrac{15}{32} - \dfrac{4}{9}\right)$ (Round to the thousandths place.)

_____ 5. $\dfrac{3}{4} \times \left(\dfrac{9}{25} + \dfrac{7}{16}\right)$

_____ 6. $\left(\dfrac{3}{7} + \dfrac{4}{5}\right) \div \left(\dfrac{7}{8} - \dfrac{2}{3}\right)$ (Round to the hundredths place.)

Answers are on page 725.

▶ Examples You Completed

Example 2. Change $\dfrac{1}{9}$ to a decimal number.

Solution. $\dfrac{1}{9}$ means $1 \div 9$.

Press: $\boxed{\text{ON/C}}$ 1 $\boxed{\div}$ 9 $\boxed{=}$

Display: 0.1111111

Therefore $\dfrac{1}{9} = 0.\overline{1}$.

Example 4. Find $\dfrac{7-4}{9-2}$. (Round to the thousandths place.)

Solution. First we write the problem with necessary parentheses.

$$\frac{7-4}{9-2} = (7-4) \div (9-2)$$

Press: (ON/C) (( ) 7 (−) 4 ( )) (÷) (( ) 9 (−) 2 ( )) (=)

Display: 0.4285714

Therefore $\dfrac{7-4}{9-2} \doteq 0.429$.

Example 6. Find $\dfrac{13}{83} \times \dfrac{19}{17}$. (Round to the thousandths place.)

Solution. Press: (ON/C) 13 (÷) 83 (×) 19 (÷) 17 (=)

Display: 0.17505315

$$\frac{13}{83} \times \frac{19}{17} \doteq 0.175$$

Example 10. Find $\dfrac{5}{16} \div \dfrac{2}{3} \div \dfrac{10}{11}$.

Solution. We must use parentheses around each of the fractions that are acting as divisors.

$$\frac{5}{16} \div \left(\frac{2}{3}\right) \div \left(\frac{10}{11}\right)$$

Press: (ON/C) 5 (÷) 16 (÷) (( ) 2 (÷) 3 ( )) (÷) (( ) 10 (÷) 11 ( )) (=)

Display: 0.515625

Therefore $\dfrac{5}{16} \div \dfrac{2}{3} \div \dfrac{10}{11} = 0.515625$.

Example 13. Find $\dfrac{43}{51} + \dfrac{3}{10} - \dfrac{18}{19}$. (Round to the thousandths place.)

Solution. Press: (ON/C) 43 (÷) 51 (+) 3 (÷)10 (−) 18 (÷) 19 (=)

Display: 0.19576883

$$\frac{43}{51} + \frac{3}{10} - \frac{18}{19} \doteq 0.196$$

Answers to Trial Runs

page 719   1. 1.32   2. $0.\overline{13}$ 3. 0.0027   4. 0.875   5. $0.8\overline{3}$ 6. 3.125

page 722   1. 2   2. $0.5\overline{3}$ 3. 0.0675   4. 1.3125   5. $3.4\overline{2}$ 6. $6.\overline{6}$

page 724   1. 1.041   2. 2.225   3. 0.88   4. 0.826   5. 0.598125   6. 5.90

EXERCISE SET 16.2

_____ 1. Change $\dfrac{13}{25}$ to a decimal number.

_____ 2. Change $\dfrac{15}{16}$ to a decimal number.

_____ 3. Change $\dfrac{8}{9}$ to a decimal number.

_____ 4. Change $\dfrac{7}{11}$ to a decimal number.

_____ 5. Change $\dfrac{9}{23}$ to a decimal number.
(Round to hundredths place.)

_____ 6. Change $\dfrac{16}{29}$ to a decimal number.
(Round to hundredths place.)

_____ 7. Change $\dfrac{476}{245}$ to a decimal number.
(Round to thousandths place.)

_____ 8. Change $\dfrac{576}{341}$ to a decimal number.
(Round to thousandths place.)

_____ 9. Find $\dfrac{8 - 3}{31 - 25}$.

_____ 10. Find $\dfrac{12 - 7}{21 - 12}$.

_____ 11. Find $\dfrac{7^2}{5 \times 9 - 23}$.
(Round to hundredths place.)

_____ 12. Find $\dfrac{13^2}{12 \times 7 - 43}$.
(Round to hundredths place.)

Perform the operations. Round answers to the hundredths place.

_____ 13. $\dfrac{3}{5} \times \dfrac{7}{8}$

_____ 14. $\dfrac{5}{7} \times \dfrac{8}{13}$

_____ 15. $\dfrac{3}{4} \times \dfrac{8}{5}$

_____ 16. $\dfrac{2}{3} \times \dfrac{9}{4}$

_____ 17. $\dfrac{3}{5} \times \dfrac{7}{8} \times \dfrac{1}{4}$

_____ 18. $\dfrac{5}{7} \times \dfrac{5}{6} \times \dfrac{1}{3}$

_____ 19. $\dfrac{5}{8} \times \dfrac{7}{9}$

_____ 20. $\dfrac{3}{7} \times \dfrac{4}{5}$

_____ 21. $\left(\dfrac{4}{5} \times \dfrac{9}{8}\right) \div \dfrac{2}{3}$

_____ 22. $\left(\dfrac{3}{4} \times \dfrac{7}{6}\right) \div \dfrac{4}{5}$

_____ 23. $\dfrac{13}{8} \div \left(\dfrac{1}{2} \times \dfrac{3}{4}\right)$

_____ 24. $\dfrac{19}{15} \div \left(\dfrac{2}{3} \times \dfrac{4}{5}\right)$

_____ 25. $\left(\dfrac{5}{2} \div \dfrac{3}{4}\right) \div \dfrac{16}{21}$

_____ 26. $\left(\dfrac{7}{3} \div \dfrac{2}{3}\right) \div \dfrac{18}{25}$

_____ 27. $\dfrac{4}{5} + \dfrac{7}{8}$

_____ 28. $\dfrac{2}{3} + \dfrac{3}{4}$

_____ 29. $\dfrac{2}{7} + \dfrac{3}{5} + \dfrac{2}{3}$

_____ 30. $\dfrac{3}{4} + \dfrac{2}{5} + \dfrac{5}{6}$

_____ 31. $\dfrac{8}{9} + \dfrac{5}{12} - \dfrac{3}{8}$

_____ 32. $\dfrac{7}{8} + \dfrac{9}{16} - \dfrac{4}{9}$

_____ 33. $\dfrac{9}{10} - \left(\dfrac{3}{5} - \dfrac{4}{7}\right)$

_____ 34. $\dfrac{7}{8} - \left(\dfrac{3}{4} - \dfrac{1}{3}\right)$

_____ 35. $\dfrac{3}{5} \times \left(\dfrac{7}{8} + \dfrac{2}{3}\right)$

_____ 36. $\dfrac{5}{9} \times \left(\dfrac{3}{4} + \dfrac{1}{5}\right)$

_____ 37. $\left(\dfrac{1}{2} + \dfrac{7}{8}\right) \times \left(\dfrac{4}{5} \times \dfrac{7}{3}\right)$

_____ 38. $\left(\dfrac{1}{3} + \dfrac{5}{6}\right) \times \left(\dfrac{7}{2} \times \dfrac{3}{5}\right)$

_____ 39. $\left(\dfrac{5}{8} + \dfrac{3}{4}\right) \div \left(\dfrac{9}{2} + \dfrac{3}{7}\right)$

_____ 40. $\left(\dfrac{3}{4} + \dfrac{7}{10}\right) \div \left(\dfrac{3}{5} + \dfrac{3}{2}\right)$

☆ Stretching the Topics

Perform the operations. Round answers to the hundredths place.

_____ 1. $\dfrac{4}{5} \times \left(\dfrac{3}{13} + \dfrac{5}{16}\right)^2 \div \left(\dfrac{7}{8} \times \dfrac{9}{11}\right)^3$

_____ 2. $\dfrac{7}{8} \times \left(\dfrac{2}{3}\right)^2 - \left(\dfrac{4}{7}\right)^3 \div \dfrac{2}{9}$

_____ 3. $\dfrac{3}{5} \times \left(\dfrac{3}{4} - \dfrac{3}{8}\right)^3 + \dfrac{5}{6} \div \left(\dfrac{1}{2} + \dfrac{3}{5}\right)^3$

Check your answers in the back of your book.

If you can complete **Checkup 16.2**, you are ready to go on to Section 16.3.

 **CHECKUP 16.2**

_____ 1. Change $\dfrac{125}{84}$ to a decimal number, rounded to the tenths place.

_____ 2. Change $\dfrac{17}{36}$ to a decimal number.

_____ 3. Find $\dfrac{3^2 + 5}{7 \times 4 - 12}$.

Perform the operations. Round answers to the hundredths place.

_____ 4. $\dfrac{14}{15} \times \dfrac{3}{4} \times \dfrac{1}{2}$

_____ 5. $\dfrac{15}{16} \div \dfrac{5}{7} \times \dfrac{8}{9}$

_____ 6. $\dfrac{10}{11} \div \dfrac{2}{3} \div \dfrac{5}{16}$

_____ 7. $\dfrac{15}{16} \div \left(\dfrac{3}{8} \times \dfrac{5}{4} \right)$

_____ 8. $\dfrac{3}{4} + \dfrac{7}{9} - \dfrac{5}{18}$

_____ 9. $\dfrac{9}{8} - \left(\dfrac{15}{16} - \dfrac{7}{8} \right)$

_____ 10. $\left(\dfrac{4}{5} + \dfrac{2}{3} \right) \div \left(\dfrac{5}{12} - \dfrac{1}{3} \right)$

Check your answers in the back of your book.

If You Missed Problems:	You Should Review Examples:
1, 2	1–3
3	4, 5
4	7, 8
5, 6	9, 10
7	11
8	12, 13
9, 10	14, 15

16.3 Using a Calculator to Solve Percent Problems

In Chapter 9 we learned to work several kinds of percent problems. We found that we could use the **percent proportion** to solve any kind of percent problem.

$$\frac{\text{numerator}}{\text{denominator}} = \frac{p}{100}$$

Remember that the denominator is always the number that follows the word "of" in the statement of our problem. The letter p always stands for the percent.

Now let's see how the calculator can simplify the arithmetic we must do to solve percent problems.

Finding Some Percent of a Number (Alternate Method)

In Section 9.2 we discussed an alternate way to solve problems such as

Find 18% of 922.

What is 94% of $1825?

Find the $5\frac{1}{2}$% sales tax on a $21 shirt.

In each of these problems we are asked to find some percent *of* a number. In Chapter 9 we decided that we could work such problems by remembering that

18% of 922 means 18% × 922

94% of $1825 means 94% × $1825

$5\frac{1}{2}$% of $21 means $5\frac{1}{2}$% × $21

Most calculators have a ⟨%⟩ key which allows us to work these kinds of problems just as we have always worked multiplication problems on the calculator.

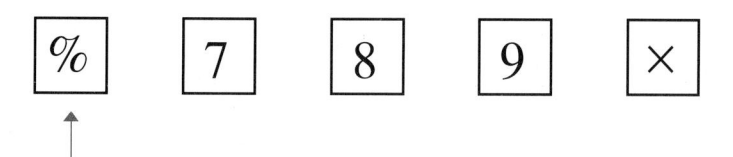

To find 18% × 922 we enter the number 18, press the ⟨%⟩ key, press the ⟨×⟩ key, enter 922, and press the ⟨=⟩ key.

Press: ⟨ON/C⟩ 18 ⟨%⟩ ⟨×⟩ 922 ⟨=⟩

Display: 165.96

Therefore, 18% of 922 = 165.96.

Did you see what happened when you entered 18 and then pressed the ⟨%⟩ key?

Press: ⟨ON/C⟩ 18 ⟨%⟩

Display: 0.18

When you press the ⟨%⟩ key, the calculator changes the percent to a decimal number. If you had done this problem by hand, *you* would have changed the percent to a decimal number before multiplying. The calculator now makes that change for you.

Example 1. Change 127% to a decimal number.

Solution. This is a quickie.

Press: (ON/C) 127 (%)

Display: 1.27

127% = 1.27

You complete Example 2.

Example 2. What is 94% of $1825?

Solution. 94% of $1825 = 94% × $1825

Press: (ON/C) 94 (%) (×) 1825 (=)

Display: _____

94% of $1825 = $_____

Check your work on page 733. ▶

Example 3. Find the $5\frac{1}{2}$ percent sales tax on a $21 shirt.

Solution. The percent here is a mixed number that must be expressed as a decimal number. Recalling that $5\frac{1}{2}$ means $5 + \frac{1}{2}$, we can have the calculator make the necessary change. However, parentheses will be needed. Let's write the problem before we enter it.

$$5\frac{1}{2}\% \text{ of } 21 = \left(5 + \frac{1}{2}\right) \% \times 21$$

Press: (ON/C) (( ) 5 (+) 1 (÷) 2 ( )) (%) (×) 21 (=)

Display: 1.155

The $5\frac{1}{2}$ % tax on a $21 shirt is $1.16.

Example 4. Find $16\frac{2}{3}\%$ of 950.

Solution. $16\frac{2}{3}\% \text{ of } 950 = \left(16 + \frac{2}{3}\right) \% \times 950$

Press: (ON/C) (( ) 16 (+) 2 (÷) 3 ( )) (%) (×) 950 (=)

Display: 158.3333

$16\frac{2}{3}\% \text{ of } 950 = 158.\overline{3}$

Example 5. In Kentucky, the automobile sales tax is 5 percent of 90 percent of the car's "sticker price." Find the sales tax for an automobile whose "sticker price" is $9495.

Solution. 5% of 90% of $9495 = 5% × 90% × $9495

Press: (ON/C) 5 (%) (×) 90 (%) (×) 9495 (=)

Display: 427.275

The tax will be $427.28.

‖➡ **Trial Run**

_____ 1. Change 87% to a decimal number.

_____ 2. Change 175% to a decimal number.

_____ 3. What is 38% of 455?

_____ 4. Find $4\frac{1}{2}$ % of $18.98.

_____ 5. Find $33\frac{1}{3}$ % of $270.

_____ 6. Find 3% of 85% of $7265.

Answers are on page 733.

Solving Percent Proportions Using the Calculator

Recall how we used the percent proportion to solve a problem such as this. If 17 of the 29 teenagers in Amy's class listen to radio station WBGN, what percent of the class listens to that station? (Round the percent to the tenths place.)

$$\frac{17}{29} = \frac{p}{100} \qquad \text{Write the percent proportion.}$$

$$17 \times 100 = 29 \times p \qquad \text{Write the cross products.}$$

$$\frac{17 \times 100}{29} = \frac{29 \times p}{29} \qquad \text{Divide both sides by 29.}$$

$$\frac{17 \times 100}{29} = p$$

Now we can use our calculator to do the arithmetic on the left-hand side. The steps are the usual ones for multiplication and division.

Press: (ON/C) 17 (×) 100 (÷) 29 (=)

Display: 58.62069

Rounding to the tenths place, we conclude that 58.6 percent of the students in Amy's class listen to WBGN.

To use a calculator to solve percent problems, *you* must set up the percent proportion and work it out to the point at which the unknown quantity is by itself on one side of the equation. Remember that your calculator can only find an answer if you are accurate when entering the information.

Example 6. Of the 650 seats in Municipal Auditorium, 420 were filled on Saturday night. What percent of the seats were filled? (Round to the tenths place.)

Solution. Our percent proportion becomes

$$\frac{420}{650} = \frac{p}{100}$$

$$420 \times 100 = 650 \times p$$

$$\frac{420 \times 100}{650} = \frac{650 \times p}{650}$$

$$\frac{420 \times 100}{650} = p$$

Press: (ON/C) 420 (×) 100 (÷) 650 (=)

Display: 64.615385

On Saturday night about 64.6 percent of the seats were filled.

Example 7. After lunch, Mary discovered she had already consumed 80 percent of her total calories for the day. If she had consumed 900 calories, what is her total calorie allowance for 1 day?

Solution. Here we know that 900 is 80 percent *of* some unknown number. We must find the denominator in the percent proportion.

$$\frac{900}{d} = \frac{80}{100}$$

$$900 \times 100 = d \times 80$$

$$\frac{900 \times 100}{80} = \frac{d \times 80}{80}$$

$$\frac{900 \times 100}{80} = d$$

Press: $\boxed{\text{ON/C}}$ 900 $\boxed{\times}$ 100 $\boxed{\div}$ 80 $\boxed{=}$

Display: 1125

Mary's daily allowance is 1125 calories.

Example 8. Martin was charged $2.94 in sales tax when he had his car tuned. If the sales tax rate is 6 percent, what was he charged for the tune-up before the tax was added? What was Martin's total bill?

Solution. We know that 6 percent *of* some number is $2.94. We must find the denominator in the percent proportion.

$$\frac{2.94}{d} = \frac{6}{100}$$

$$2.94 \times 100 = d \times 6$$

$$\frac{2.94 \times 100}{6} = \frac{d \times 6}{6}$$

$$\frac{2.94 \times 100}{6} = d$$

Press: $\boxed{\text{ON/C}}$ 2.94 $\boxed{\times}$ 100 $\boxed{\div}$ 6 $\boxed{=}$

Display: 49

Before tax, Martin was charged $49. To find his total bill we must add the tax.

Press: $\boxed{\text{ON/C}}$ 49 $\boxed{+}$ 2.94 $\boxed{=}$

Display: 51.94

Martin's total bill was $51.94.

As you can see, a calculator can certainly help with the arithmetic involved in percent problems. However, a calculator cannot tell you *how* to solve a problem. *You* must set up the problem and then use the calculator to do the arithmetic.

➠ Trial Run

_____ 1. If 5 of every 8 persons are overweight, what percent of the population is over-weight?

_____ 2. Of the 720 students living in Bates Hall, 576 are freshmen. What percent of the residents are freshmen?

_____ 3. At the end of 3 weeks, Tyrone discovered he had already spent 85 percent of the money he had budgeted for food that month. If Tyrone had spent $168.30, how much had he budgeted for food?

_____ 4. Yolanda was charged $8.46 sales tax on her new winter coat. If the sales tax rate is $4\frac{1}{2}$ percent, what was the price of Yolanda's coat before the tax was added?

Answers are given below.

▶ **Example You Completed** ─────────────────────────

Example 2. What is 94% of $1825?

Solution. 94% of $1825 = 94% × $1825

Press: (ON/C) 94 (%) (×) 1825 (=)

Display: 1715.5

94% of $1825 = $1715.50

Answers to Trial Runs ─────────────────────────

page 730   **1.** 0.87   **2.** 1.75   **3.** 172.9   **4.** $0.85   **5.** $90   **6.** $185.26

page 732   **1.** 62.5%   **2.** 80%   **3.** $198   **4.** $188

EXERCISE SET 16.3

_____ 1. Change 83 percent to a decimal number.

_____ 2. Change 58 percent to a decimal number.

_____ 3. Change 125 percent to a decimal number.

_____ 4. Change 250 percent to a decimal number.

_____ 5. Change 24.5 percent to a decimal number.

_____ 6. Change 74.2 percent to a decimal number.

_____ 7. What is 55 percent of 783?

_____ 8. What is 72 percent of 387?

_____ 9. Find $6\frac{1}{4}$ percent of 350.

_____ 10. Find $8\frac{3}{4}$ percent of 585.

_____ 11. Find $15\frac{2}{3}$ percent of 600. (Round answer to hundredths place.)

_____ 12. Find $12\frac{1}{3}$ percent of 800. (Round answer to hundredths place.)

_____ 13. Find $21\frac{3}{5}$ percent of 775.

_____ 14. Find $22\frac{3}{4}$ percent of 380.

_____ 15. Find 6 percent of 80 percent of $8725.

_____ 16. Find 8 percent of 60 percent of 9500.

_____ 17. Oliver's salary last year was $28,800. His salary for this year has been raised to $29,880. Find the percent increase in Oliver's salary. (Round answer to the tenths place.)

_____ 18. Last year a lawn mower sold for $189.98. This year the same mower sells for $198.99. Find the percent change in the price. (Round answer to the tenths place.)

_____ 19. The enrollment in the adult literacy program increased by $33\frac{1}{3}$ percent this year. If the enrollment last year was 72 persons, how many were enrolled this year?

_____ 20. If prices at the sporting goods store have kept up with the $5\frac{3}{4}$ percent rate of inflation, what will be the new price of a $39.95 pair of tennis shoes?

_____ 21. Tuition per semester at Central University last year was $5685. If there will be a 6.25 percent increase in tuition for next semester, how much should a student expect to pay?

_____ 22. The budget of the Health Department will be cut by 15.75 percent next year. If the budget this year is $225,784, how much money should the department plan to spend next year?

_____ 23. If Elsie paid $25.50 for a souvenir that was priced at $24, find the sales tax rate in the state where she was taking her vacation.

_____ 24. When Edith ordered a meal priced on the menu at $14, she paid $1.05 sales tax. Find the sales tax rate.

_____ 25. Mr. Hoheimer has 15,685 bushels of corn in storage. This is 85 percent of the capacity of his storage bins. Find how many bushels of corn he could store. (Round to the nearest bushel.)

_____ 26. Victoria's salary this year is $23,256. If she received a 5.3 percent raise for this year, what was her salary (to the nearest dollar) last year? (Hint: Victoria's current salary is 105.3 percent of her salary last year.)

_____ 27. Miranda was charged $80.13 in sales tax when she purchased a freezer. If the sales tax rate is 5.5 percent, what was the price of the freezer?

_____ 28. Stokeley paid $3.38 in tax on a concert ticket. If the entertainment tax rate is 13.5 percent, what was the price (to the nearest dollar) of the concert ticket (before tax)?

_____ 29. If a coat regularly selling for $95.98 is marked down 25 percent, what will be the sale price of the coat?

_____ 30. A VCR that had been selling for $329.95 has been marked down $33\frac{1}{3}$ percent. Find the new selling price of the VCR.

☆ Stretching the Topics _____

_____ 1. A clothing store advertises a sale of $33\frac{1}{3}$ percent off on all merchandise. Tammie purchased a sweater that regularly sells for $59.95. She also received a student discount of $15\frac{1}{2}$ percent off the sale price. Then she paid a 6.25 percent sales tax. How much change did Tammie receive if she gave the clerk a fifty-dollar bill?

_____ 2. If Roland deposits $12,500 in his credit union account, paying 7.25 percent interest compounded semiannually, how much will he have on deposit at the end of 2 years?

Check your answers in the back of your book.

If you can do the problems in **Checkup 16.3,** you are ready to go on to Section 16.4.

✓ CHECKUP 16.3

_____ 1. Change $12\frac{1}{2}$ percent to a decimal number.

_____ 2. Change 225 percent to a decimal number.

_____ 3. What is 75 percent of $3568?

_____ 4. Find $9\frac{3}{4}$ percent of $1375.

_____ 5. Find $6\frac{1}{2}$ percent of 85 percent of $15,345. (Round to the nearest cent.)

_____ 6. Of the 853 employees at a factory, 325 were laid off for 3 months. What percent of the work force was laid off? (Round to the tenths place.)

_____ 7. After 3 days of her 10-day vacation Teresa discovered she had spent 65 percent of her total vacation savings. If she had spent $487.50, how much had she saved for her vacation?

_____ 8. Ray was charged $5.78 sales tax on a sweeper he purchased at Save Mart. If the sales tax in his state is $6\frac{1}{2}$ percent, what was the price of the sweeper (before tax)?

Check your answers in the back of your book.

If You Missed Problems:	You Should Review Examples:
1, 2	1
3	2
4, 5	3, 4
6	6
7	7
8	8

16.4 Using a Calculator for Measurement and Geometry

We did a great deal of arithmetic when we worked with measurement and geometry in Chapters 12 and 13. Wouldn't it be nice to let the calculator do some of that arithmetic for us?

Converting Units of Measure on the Calculator

In Chapter 12, we learned to use **conversion factors** to change one American unit to another or to change one metric unit to another. Let's summarize the most important units again in a table that we can use for reference.

American Units			
Length	Area	Weight	Capacity
1 ft = 12 in. 1 yd = 3 ft 1 mi = 5280 ft	1 sq ft = 144 sq in. 1 sq yd = 9 sq ft	1 lb = 16 oz 1 ton = 2000 lb	1 T = 3 t 1 pt = 2 c 1 qt = 2 pt 1 gal = 4 qt
Metric Units			
Length	Area	Weight	Capacity
1 cm = 10 mm 1 m = 100 cm 1 m = 1000 mm 1 km = 1000 m	1 sq m = 10,000 sq cm 1 sq km = 1,000,000 sq m	1 g = 1000 mg 1 g = 100 cg 1 kg = 1000 g	1 ℓ = 1000 ml 1 kl = 1000 ℓ 1 ml = 1 cu cm (cc) 1 ℓ = 1000 cu cm (cc)

A calculator can help us make conversions, but we must still set up the conversion product just as though we were doing the problem by hand. For instance, to change 150 inches to feet, we write

$$150 \text{ in.} = 150 \times \overset{1}{\cancel{1 \text{ in.}}} \times \frac{1 \text{ ft}}{\underset{12}{\cancel{12 \text{ in.}}}}$$

$$= \frac{150}{12} \times 1 \text{ ft}$$

Then we can use the calculator to do the actual division.

Press: (ON/C) 150 (÷) 12 (=)

Display: 12.5

So we know that 150 in. = 12.5 ft.

Example 1. Change 9 yards to inches.

Solution. $9 \text{ yd} = 9 \times \overset{1}{\cancel{1 \text{ yd}}} \times \dfrac{\overset{3}{\cancel{3 \text{ ft}}}}{\underset{1}{\cancel{1 \text{ yd}}}} \times \dfrac{12 \text{ in.}}{\underset{1}{\cancel{1 \text{ ft}}}}$

$= 9 \times 3 \times 12 \text{ in.}$

Press: $\boxed{\text{ON/C}}$ 9 $\boxed{\times}$ 3 $\boxed{\times}$ 12 $\boxed{=}$

Display: 324

Therefore 9 yd = 324 in.

Example 2. How many ounces are in $\frac{1}{9}$ ton? (Round to whole ounces.)

Solution. We must change $\frac{1}{9}$ ton to ounces.

$$\frac{1}{9} \text{ ton} = \frac{1}{9} \times \overset{1}{\cancel{1 \text{ ton}}} \times \dfrac{\overset{2000}{\cancel{2000 \text{ lb}}}}{\underset{1}{\cancel{1 \text{ ton}}}} \times \dfrac{16 \text{ oz}}{\underset{1}{\cancel{1 \text{ lb}}}}$$

$$= \frac{1 \times 2000 \times 16 \text{ oz}}{9}$$

Press: $\boxed{\text{ON/C}}$ 2000 $\boxed{\times}$ 16 $\boxed{\div}$ 9 $\boxed{=}$

Display: 3555.5556

Therefore $\dfrac{1}{9}$ ton $\doteq$ 3556 oz.

Example 3. Change 0.3 millimeters to meters.

Solution. $0.3 \text{ mm} = 0.3 \times \overset{1}{\cancel{1 \text{ mm}}} \times \dfrac{1 \text{ m}}{\underset{1000}{\cancel{1000 \text{ mm}}}}$

$= \dfrac{0.3}{1000} \times 1 \text{ m}$

Press: $\boxed{\text{ON/C}}$ .3 $\boxed{\div}$ 1000 $\boxed{=}$

Display: 0.0003

Therefore 0.3 mm = 0.0003 m.

⚞ Trial Run

_____ 1. Change 15 feet to inches.

_____ 2. Change 8 yards to inches.

_____ 3. How many ounces are in $\frac{1}{5}$ of a ton?

_____ 4. Change 235 millimeters to meters.

_____ 5. Change 780 centimeters to meters.

_____ 6. Change 0.7 meter to millimeters.

Answers are on page 753.

Converting back and forth between American units and metric units can also be done using conversion factors. We learned a few such conversions in Chapter 12, but the arithmetic involved is so messy in some cases that we have waited until we could use the calculator to try them. Here is a table of some approximate equivalent measurements in American and metric units.

	Metric Units		American Units
Length	1 m	$\doteq$	3.28 ft
	1 m	$\doteq$	1.09 yd
	1 km	$\doteq$	0.62 mi
	1 cm	$\doteq$	0.39 in.
Weight	1 g	$\doteq$	0.035 oz
	1 kg	$\doteq$	2.2 lb
Capacity	1 ℓ	$\doteq$	1.06 qt
	1 kl	$\doteq$	264.18 gal

Each of these equivalences gives us a conversion factor that we can use to change measurements from either system to the other system.

Suppose we wish to change 17 meters to feet. As usual we set up our conversion product, choosing the conversion factor that will let us eliminate the unwanted unit (m) and keep the unit (ft) that we want.

$$17 \text{ m} \doteq 17 \times \overset{1}{\cancel{1 \text{ m}}} \times \frac{3.28 \text{ ft}}{\underset{1}{\cancel{1 \text{ m}}}}$$

$$\doteq 17 \times 3.28 \text{ ft}$$

Using our calculator to multiply, we find that

$$17 \text{ m} \doteq 55.76 \text{ ft}$$

Example 4. Mary Jane ran in a 10-kilometer race last weekend. How many miles did she run?

Solution. Find the relationship between kilometers and miles in the table.

$$1 \text{ km} \doteq 0.62 \text{ mi}$$

Then set up the conversion.

$$10 \text{ km} \doteq 10 \times \overset{1}{\cancel{1 \text{ km}}} \times \frac{0.62 \text{ mi}}{\underset{1}{\cancel{1 \text{ km}}}}$$

$$\doteq 10 \times 0.62 \text{ mi}$$

$$\doteq 6.2 \text{ mi}$$

Mary Jane ran about 6.2 miles.

You complete Example 5.

Example 5. Find the weight in kilograms of a football player who weighs 215 pounds. (Round to the tenths place.)

Solution. The relationship between kilograms and pounds is

$$1 \text{ kg} \doteq \underline{\hspace{1cm}} \text{ lb}$$

Now set up the conversion.

$$215 \text{ lb} \doteq 215 \times 1 \text{ lb} \times \frac{\text{kg}}{\text{lb}}$$

$$\doteq \frac{215}{2.2} \text{ kg}$$

$$\doteq \underline{\hspace{1cm}} \text{ kg}$$

The football player weighs about _____ kilograms.

Check your work on page 752. ▶

Example 6. If Max needs 18 gallons of gas to fill his tank, how many liters of gas must he buy?

Solution. We know

$$1 \text{ gal} \doteq 4 \text{ qt}$$

$$1 \ \ell \doteq 1.09 \text{ qt}$$

so we must use two conversion factors to make our change.

$$18 \text{ gal} \doteq 18 \times 1 \text{ gal} \times \frac{4 \text{ qt}}{1 \text{ gal}} \times \frac{1 \ \ell}{1.09 \text{ qt}}$$

$$\doteq \frac{18 \times 4}{1.09} \ \ell$$

$$\doteq 66.055046 \ \ell$$

Max must buy about 66 liters of gas.

⫸ Trial Run

_____ 1. If Marquita jogs 5 kilometers each day, how many miles does she jog?

_____ 2. Find the weight in kilograms of an 850-pound beef calf. (Round to tenths place.)

_____ 3. Sheila needs 12 gallons of punch for her party. How many liters does she need? (Round to the hundredths place.)

_____ 4. If Fran needs 9 yards of border for a bedspread, how many meters should she buy? (Round to the hundredths place.)

_____ 5. Jake's recipe from a French cookbook calls for 1.5 kilograms of ground beef. How many pounds of ground beef should he buy?

_____ **6.** If Rick's living room is 7.2 meters long, find its length to the nearest foot.

Answers are on page 753.

Finding Perimeters, Areas, and Volumes

In Chapter 13 we learned the formulas for the perimeter, circumference, area, and volume of some basic geometric figures. Once again, calculators can help us do the arithmetic *after* we have decided what numbers to use in the correct formula.

Example 7. Use the calculator to find the feet of baseboard needed for a room whose floor plan is illustrated here.

Solution. We need the *perimeter* of this room, so we *add* the measurements.

$$P = 1 + 2\frac{7}{8} + 6 + 2\frac{7}{8} + 8\frac{1}{4} + 10 + 7\frac{1}{4}$$

$$= 1 + 2 + \frac{7}{8} + 6 + 2 + \frac{7}{8} + 8 + \frac{1}{4} + 10 + 7 + \frac{1}{4}$$

Remember that we do not need to find a common denominator to add fractional numbers on a calculator. We simply enter each fraction as division.

Press: $\boxed{\text{ON/C}}$ 1 $\boxed{+}$ 2 $\boxed{+}$ 7 $\boxed{\div}$ 8 $\boxed{+}$ 6 $\boxed{+}$ 2 $\boxed{+}$ 7
$\boxed{\div}$ 8 $\boxed{+}$ 8 $\boxed{+}$ 1 $\boxed{\div}$ 4 $\boxed{+}$ 10 $\boxed{+}$ 7 $\boxed{+}$ 1 $\boxed{\div}$ 4 $\boxed{=}$

Display: 38.25

We need 38.25 feet of baseboard.

To find the **circumference** or the **area** of a circle, recall that our calculations required the use of the constant π. On your calculator, locate the $\boxed{\pi}$ key.

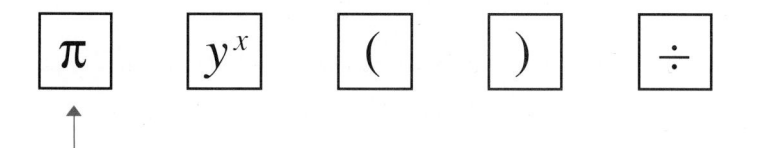

Turn on your calculator and press the $\boxed{\pi}$ key.

$$\pi \doteq 3.1415927$$

Recall that we used 3.14 or $\frac{22}{7}$ as approximations for π in our earlier calculations. On the calculator, find the value of $\frac{22}{7}$.

Press: $\boxed{\text{ON/C}}$ 22 $\boxed{\div}$ 7 $\boxed{=}$

Display: 3.1428571

Notice that, rounded to the hundredths place, $\dfrac{22}{7} \doteq 3.14$.

Remember that both 3.14 and $\frac{22}{7}$ are just *approximations* for π. In using our calculator, the $\boxed{\pi}$ key will give us an even better approximation for π to use in finding circumferences and areas of circles. However, it is still not *exact*, because π is not a terminating or repeating decimal number.

To find the circumference (C) of a circle with radius (r) and diameter (d), we learned to use either of these formulas.

Circumference of a Circle
$$C = \pi \cdot d$$
$$C = 2 \cdot \pi \cdot r$$

Example 8. Find the circumference of a circular table whose diameter is 60 inches. (Round to the tenths place.)

Solution. $C = \pi \cdot d$

$\qquad = \pi \cdot 60$ in.

Press: $\boxed{\text{ON/C}}$ $\boxed{\pi}$ $\boxed{\times}$ 60 $\boxed{=}$

Display: 188.49556

The circumference is about 188.5 inches.

⫸ Trial Run

_____ 1. Find the circumference of a circle that has a radius of 3.8 meters. (Round answer to the tenths place.)

_____ 2. Find the perimeter of a rectangle with a length of $6\frac{1}{2}$ feet and a width of $9\frac{3}{4}$ feet.

_____ 3. Find the perimeter of a triangle with sides of 3.29, 4.61, and 5.92 meters.

_____ 4. Find the perimeter of the figure illustrated.

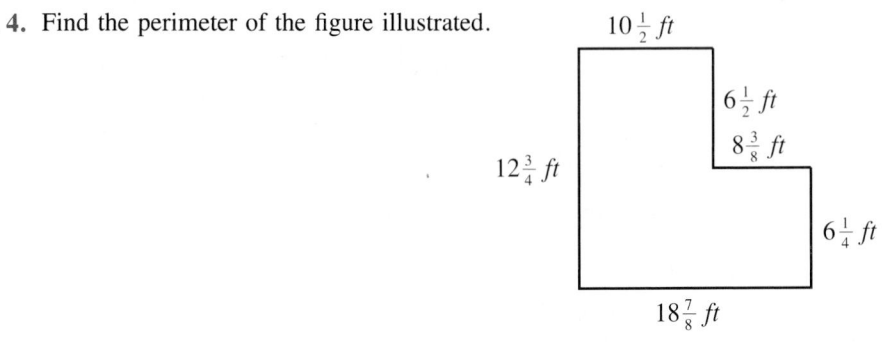

Answers are on page 753.

We learned several formulas for **areas** of geometric figures in Chapter 13.

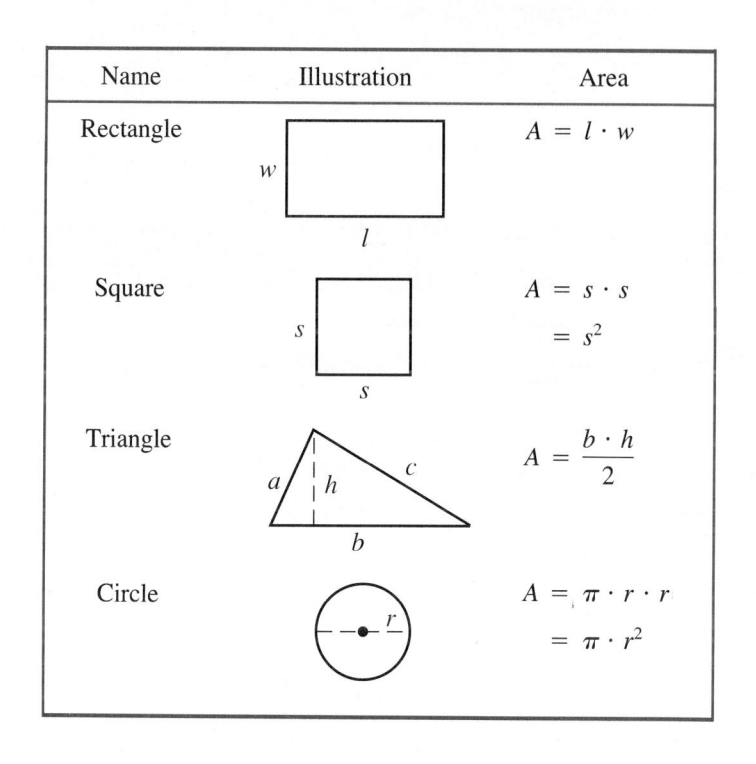

Name	Illustration	Area
Rectangle		$A = l \cdot w$
Square		$A = s \cdot s$ $= s^2$
Triangle		$A = \dfrac{b \cdot h}{2}$
Circle		$A = \pi \cdot r \cdot r$ $= \pi \cdot r^2$

Now we will use the calculator to perform the necessary arithmetic in finding areas. Remember that area is always expressed in **square units**.

Example 9. Find the area of a triangular patio with a 3.7-meter base and a 4.1-meter height.

Solution. Our figure is a triangle in which $b = 3.7$ meters and $h = 4.1$ meters.

$$A = \frac{b \cdot h}{2}$$

$$A = \frac{3.7 \times 4.1}{2} \text{ sq m}$$

Press: $\boxed{\text{ON/C}}$ 3.7 $\boxed{\times}$ 4.1 $\boxed{\div}$ 2 $\boxed{=}$

Display: 7.585

The area is 7.585 square meters.

Example 10. How many square yards of carpeting are needed to carpet a rectangular room that is 17 feet long and $14\frac{1}{2}$ feet wide?

Solution. Our figure is a rectangle in which $l = 17$ feet and $w = 14\frac{1}{2}$ feet.

$$A = l \cdot w$$

$$A = 17 \cdot 14\frac{1}{2} \text{ sq ft}$$

$$= 17 \cdot \frac{29}{2} \text{ sq ft}$$

But we want our area in **square** yards. Let's include the necessary conversion factor in our calculation.

$$A = \frac{17 \cdot 29}{2} \text{ sq ft} \times \frac{1 \text{ sq yd}}{9 \text{ sq ft}}$$

$$= \frac{17 \cdot 29}{2 \cdot 9} \text{ sq yd}$$

We will use parentheses here around the numerator and denominator.

Press: (ON/C) (( ) 17 (×) 29 ( )) (÷) (( ) 2 (×) 9 ( )) (=)

Display: 27.388889

We'll need about 27.4 square yards of carpeting.

Example 11. A rectangle is 2.3 meters long and 1.6 meters wide. If 1.76 meters are added to the length and to the width, find the area of the enlarged rectangle.

Solution. Let's use a drawing to help us here.

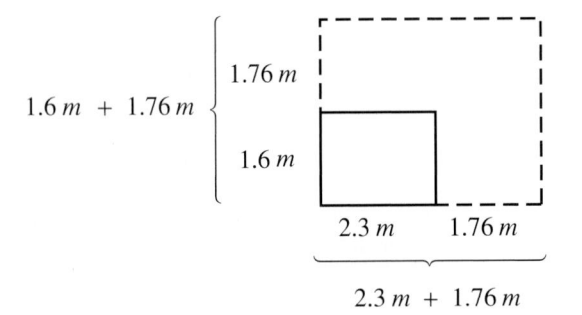

In finding the area of the enlarged rectangle, we must use the *new* length and the *new* width.

$$\text{new length} = 2.3 + 1.76$$

$$\text{new width} = 1.6 + 1.76$$

$$A = l \cdot w$$

$$A = (2.3 + 1.76)(1.6 + 1.76)$$

Press: (ON/C) (( ) 2.3 (+) 1.76 ( )) (×) (( ) 1.6 (+) 1.76 ( )) (=)

Display: 13.6416

The area of the enlarged rectangle is 13.6416 square meters.

Remember that your calculator will not keep track of units for you. *You* must decide what kind of units to give to your answer.

Example 12. If the radar equipment at a weather station surveys the conditions within a 15-kilometer radius of the station, what is the area of the region surveyed? (Round to the hundredths place.)

Solution. The region surveyed is a circle with a radius of 15 kilometers.

$$A = \pi \cdot r^2$$

$$A = \pi \cdot 15^2 \text{ sq km}$$

Here we'll have a chance to use the $\boxed{\pi}$ key and the $\boxed{x^2}$ key.

Press: $\boxed{\text{ON/C}}$ $\boxed{\pi}$ $\boxed{\times}$ 15 $\boxed{x^2}$ $\boxed{=}$

Display: 706.85835

The radar surveys an area of about 706.86 square kilometers.

Example 13. The bottom part of a church window is a square measuring 4 feet on each side. The top part of the window is half a circle. Find the area of the window. (Round to the hundredths place.)

Solution. Let's draw a picture of the window.

We must find the area of the square and *add* it to the area of the half-circle.

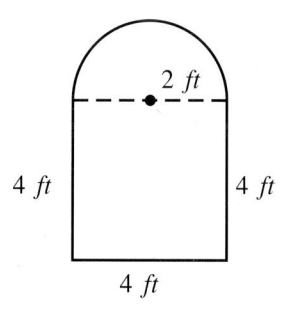

For the square with side $s = 4$ ft

$$A = s \cdot s$$
$$= s^2$$
$$A = 4^2 \text{ sq ft}$$

For the *half*-circle with radius $r = 2$ ft

$$A = \frac{1}{2} \cdot \pi \cdot r^2$$

$$A = \frac{1}{2} \times \pi \times 2^2 \text{ sq ft}$$

The total area of the window is the *sum* of these areas. Let's use the calculator.

Press: $\boxed{\text{ON/C}}$ 4 $\boxed{x^2}$ $\boxed{+}$ 1 $\boxed{\div}$ 2 $\boxed{\times}$ $\boxed{\pi}$ $\boxed{\times}$ 2 $\boxed{x^2}$ $\boxed{=}$

Display: 22.283185

The area of the window is about 22.28 square feet.

⫸ Trial Run

_____ 1. Find the area of a triangular lot with a 152.5-foot base and a 238.2-foot height.

_____ 2. Find the area of a rectangle that is 7.8 meters long and 15.3 meters wide.

_____ 3. Find the area covered by a circular oriental rug that has a radius of 4.8 meters. (Round to hundredths place.)

_____ 4. Find the area of a flower bed that has the shape and dimensions of the figure illustrated. (Round to tenths place.)

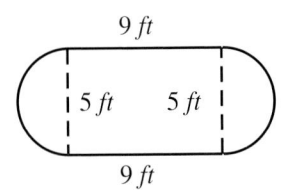

Answers are on page 743.

To find **volumes** (*V*) of geometric solids, we used the formulas summarized below.

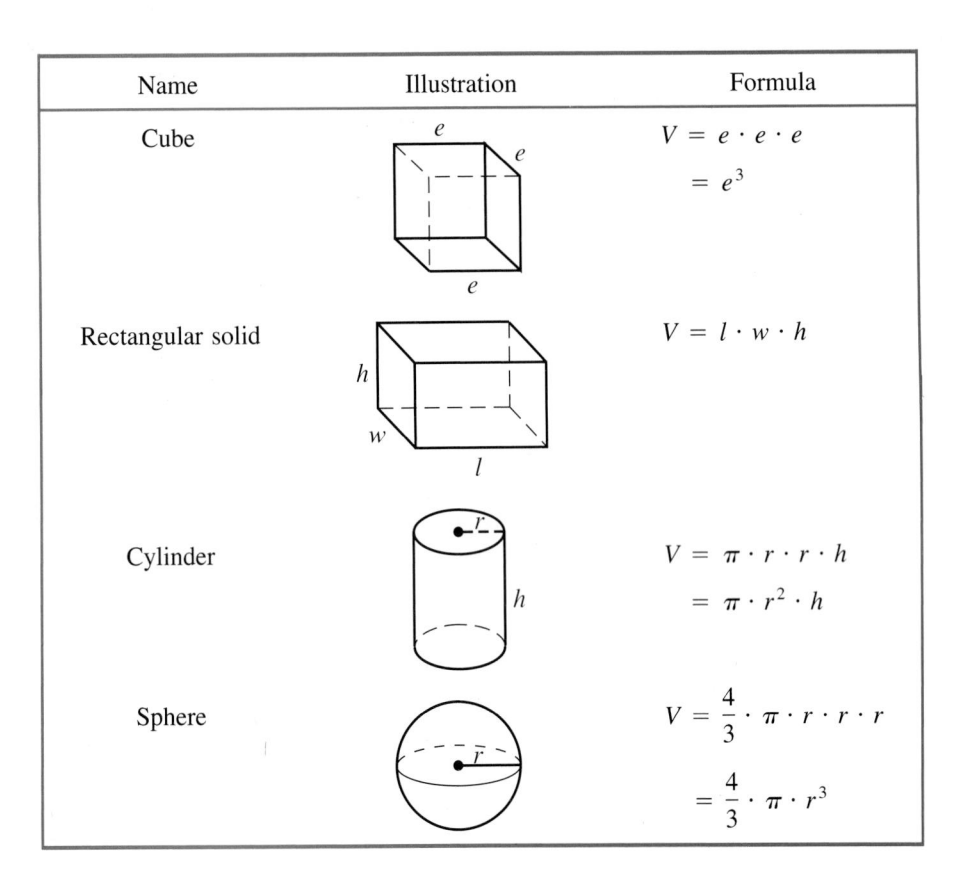

Name	Illustration	Formula
Cube		$V = e \cdot e \cdot e$ $= e^3$
Rectangular solid		$V = l \cdot w \cdot h$
Cylinder		$V = \pi \cdot r \cdot r \cdot h$ $= \pi \cdot r^2 \cdot h$
Sphere		$V = \dfrac{4}{3} \cdot \pi \cdot r \cdot r \cdot r$ $= \dfrac{4}{3} \cdot \pi \cdot r^3$

Let's use the calculator to help us compute volume. Remember that volume is always measured in **cubic units**.

Example 14. A storage shed is 12.5 feet long, 11.3 feet wide, and 8.2 feet high. Find the volume of this shed.

Solution. The shed is a rectangular solid in which $l = 12.5$ ft, $w = 11.3$ ft, and $h = 8.2$ ft.

$$V = l \cdot w \cdot h$$

$$V = 12.5(11.3)(8.2) \text{ cu ft}$$

Multiplying on the calculator, we find

$$V = 1158.25 \text{ cu ft}$$

Example 15. Find the amount of plastic needed to make a pair of dice, if the edge of each die measures 1.35 centimeters. (Round answer to the ones place.)

Solution. Each die is a *cube* for which we need the *volume*. We know that $e = 1.35$ cm.

$$\text{For 1 die: } V = (1.35)^3 \text{ cu cm}$$

$$\text{For 2 dice: } V = 2 \cdot (1.35)^3 \text{ cu cm}$$

Let's enter this product in our calculator, remembering to use the power key.

Press: (ON/C) 2 (×) 1.35 (yˣ) 3 (=)

Display: 4.92075

The amount of plastic needed is about 5 cubic centimeters.

You complete Example 16.

Example 16. Find the volume of air that will fill a spherical balloon with a radius of 10.5 meters. (Round to the ones place.)

Solution. The balloon is a sphere with $r = $ _____ m.

$$V = \frac{4}{3} \cdot \pi \cdot r^3$$

$$V = \frac{4}{3} \cdot \pi \cdot (\underline{\hspace{1cm}})^3 \text{ cu m}$$

Press: (ON/C) 4 (÷) 3 (×) (π) (×) 10.5 (yˣ) 3 (=)

Display: _____

The volume of air that will fill the balloon is about _____ cubic meters.

Check your work on page 752. ▶

⟫ Trial Run

_____ 1. A box is 1.2 meters long, 0.5 meter wide, and 0.4 meter high. Find the volume of the box.

_____ 2. Find the volume of a cereal box that is 24 centimeters long, 6 centimeters wide, and 30 centimeters tall.

_____ 3. Find the volume of a toy box that is cube whose edge is 1.5 feet.

_____ 4. Find the volume of a sphere with a radius of 7.5 meters. (Round to the nearest cubic meter.)

Answers are on page 753.

Finding Parts of Right Triangles

In *any* triangle, we know that the sum of the measures of all 3 angles must equal 180 degrees. Using the symbols discussed in Chapter 13, we write

$$\angle A + \angle B + \angle C = 180°$$

If we know the measures of 2 of the angles, we can find the measure of the third angle. For instance, if we know

$$\angle A = 37° \quad \text{and} \quad \angle B = 69°$$

we can find $\angle C$ in 2 steps. First we *add* the measures of the two known angles, and then we *subtract* this sum from the total of 180 degrees.

$$\angle C = 180° - (37° + 69°)$$

We can do this problem in 1 step with the calculator if we use *parentheses* around our sum.

Press: (ON/C) 180 (−) (() 37 (+) 69 ()) (=)

Display: 74

The measure of the third angle is 74°. Notice that we had to remember to attach the symbol for degrees (°) to our answer because the calculator will not carry units for us.

Example 17. In triangle *ABC*, suppose that $\angle A = 10.7°$ and $\angle C = 116.5°$. Find $\angle B$.

Solution. We must add the measures of the known angles and subtract that sum from 180 degrees.

$$\angle B = 180° - (10.7° + 116.5°)$$

Press: (ON/C) 180 (−) (() 10.7 (+) 116.5 ()) (=)

Display: 52.8

Therefore the measure of $\angle B$ is 52.8°.

The **right triangle** is a special triangle with a 90-degree angle at one vertex. We have agreed that in right triangle ABC, we will always label the right angle as $\angle C$.

Example 18. In right triangle *ABC*, suppose that $\angle B = 63.6°$. Find $\angle A$.

Solution. Since this is a *right* triangle, we know that $\angle C = 90°$. Therefore

$$\angle A = 180° - (63.6° + 90°)$$

Press: (ON/C) 180 (−) (() 63.6 (+) 90 ()) (=)

Display: 26.4

Therefore the measure of $\angle A$ is 26.4°.

⫸ Trial Run

———— **1.** In triangle *ABC*, suppose ∢*A* = 15.3° and ∢*C* = 113.8°. Find ∢*B*.

———— **2.** In triangle *ABC*, suppose ∢*A* = 37.5° and ∢*B* = 83.5°. Find ∢*C*.

———— **3.** In right triangle *ABC*, ∢*A* = 68.3°. Find ∢*B*.

———— **4.** In right triangle *ABC*, ∢*B* = 78.8°. Find ∢*A*.

Answers are on page 753.

In Chapter 13, we discussed the **Pythagorean Formula** for right triangles. At that time we discovered that in right triangle *ABC*, with legs *a* and *b*, and hypotenuse *c*,

$$a^2 + b^2 = c^2$$

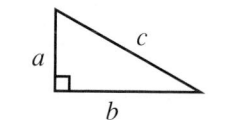

We used a table of squares to find the squares of legs *a* and *b*. Then we used the table again to find *c* when we knew its square. The calculator allows us to work with the Pythagorean Formula without using a table of squares.

Suppose we know the length of the two legs of a right triangle and wish to find the length of the hypotenuse. For instance, suppose *a* = 3, *b* = 4, and we wish to find *c*. We know that

$$a^2 + b^2 = c^2$$
$$3^2 + 4^2 = c^2$$
$$9 + 16 = c^2$$
$$25 = c^2$$

We are looking for a number, *c*, whose *square* is 25. If *c* is a number whose square is 25, we say that *c* is the **square root** of 25. You may have already concluded that 5 is the square root of 25 because you know that $5^2 = 5 \times 5 = 25$. Therefore,

$$c = 5$$

The mathematical symbol that is used to stand for the square root of a number is $\sqrt{}$.

> We say that a positive number, *c*, is the principal square root of a positive number *n* if $c^2 = n$.
>
> $$c = \sqrt{n} \quad \text{if} \quad c^2 = n$$

Thus we may write

$$\sqrt{25} = 5 \quad \text{because} \quad 5^2 = 25$$
$$\sqrt{100} = 10 \quad \text{because} \quad 10^2 = 100$$
$$\sqrt{9} = 3 \quad \text{because} \quad 3^2 = 9$$

If you cannot figure out the square root of a number in your head, you may again turn to your calculator for help. Locate the $\sqrt{x}$ key.

$$\boxed{1/x} \qquad \boxed{x^2} \qquad \boxed{\sqrt{x}} \qquad \boxed{\textbf{OFF}} \qquad \boxed{\textbf{ON}/\textbf{C}}$$

$\uparrow$

This is called the **square root key** and can be used to find the square root of any number. To find the square root of a number, we enter the number, press the $\sqrt{x}$ key, and wait a second for the answer to appear in the display. Let's practice using the $\sqrt{x}$ key and find $\sqrt{25}$.

Press: $\boxed{\text{ON/C}}$ 25 $\boxed{\sqrt{x}}$

Display: 5

You complete Example 19.

Example 19. Find $\sqrt{841}$.

Solution. Press: $\boxed{\text{ON/C}}$ 841 $\boxed{\sqrt{x}}$

Display: _____

$\sqrt{841}$ = _____

Check your work on page 752. ▶

Example 20. In right triangle *ABC*, $a = 15$ feet and $b = 36$ feet. Find the hypotenuse.

Solution. Here we must practice setting up the problem so that we can transfer it correctly to our calculator.

$$a^2 + b^2 = c^2 \qquad \text{Write the Pythagorean Formula.}$$
$$15^2 + 36^2 = c^2 \qquad \text{Substitute 15 for } a \text{ and 36 for } b.$$
$$\sqrt{15^2 + 36^2} = c \qquad \text{Find the square root of the sum.}$$

The calculator will only find the square root of the last number entered, so we must use *parentheses* around our sum.

Press: $\boxed{\text{ON/C}}$ $\boxed{(}$ 15 $\boxed{x^2}$ $\boxed{+}$ 36 $\boxed{x^2}$ $\boxed{)}$ $\boxed{\sqrt{x}}$

Display: 39

The hypotenuse is 39 feet.

A whole number whose square root is a whole number is called a **perfect square**.

841 is a perfect square because $\sqrt{841} = 29$

25 is a perfect square because $\sqrt{25} = 5$

100 is a perfect square because $\sqrt{100} = 10$

There are many numbers that are *not* perfect squares. For instance

2 is not a perfect square

3 is not a perfect square

20 is not a perfect square

Square roots of numbers that are not perfect squares can be estimated by hand, using a complicated process. We will not learn that process here. Instead we will estimate such square roots using the calculator.

Example 21. Use the calculator to find: $\sqrt{2}$, $\sqrt{3}$, and $\sqrt{20}$. (Round to the thousandths place.)

Solution

To find $\sqrt{2}$,	To find $\sqrt{3}$,	To find $\sqrt{20}$,
Press: $\boxed{\text{ON/C}}$ 2 $\boxed{\sqrt{x}}$	Press: $\boxed{\text{ON/C}}$ 3 $\boxed{\sqrt{x}}$	Press: $\boxed{\text{ON/C}}$ 20 $\boxed{\sqrt{x}}$
Display: 1.4142136	Display: 1.7320508	Display: 4.472136
$\sqrt{2} \doteq 1.414$	$\sqrt{3} \doteq 1.732$	$\sqrt{20} \doteq 4.472$

You complete Example 22.

Example 22. In right triangle *ABC*, $a = 0.23$ meter and $b = 1.06$ meters. Find the hypotenuse. (Round to the hundredths place.)

Solution. We must use the Pythagorean Formula.

$$a^2 + b^2 = c^2$$
$$0.23^2 + 1.06^2 = c^2$$
$$\sqrt{0.23^2 + 1.06^2} = c$$

Press: $\boxed{\text{ON/C}}$ $\boxed{(}$ .23 $\boxed{x^2}$ $\boxed{+}$ 1.06 $\boxed{x^2}$ $\boxed{)}$ $\boxed{\sqrt{x}}$

Display: _____

The hypotenuse, $c \doteq$ _____ meters.

Check your work on page 753. ▶

Let's work the problem stated at the beginning of this chapter.

Example 23. On a map, City A is 42 miles directly north of City C. City B is 79 miles due east of City C. Find the shortest distance from City A to City B. (Round to the tenths place.)

Solution. A picture will help with this problem.

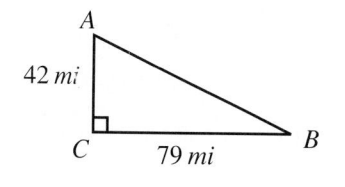

We have a right triangle in which

$$a = 79$$
$$b = 42$$

and the hypotenuse, *c*, is unknown.

Using the Pythagorean Formula,

$$a^2 + b^2 = c^2$$
$$79^2 + 42^2 = c^2$$
$$\sqrt{79^2 + 42^2} = c$$

Press: $\boxed{\text{ON/C}}$ $\boxed{(}$ 79 $\boxed{x^2}$ $\boxed{+}$ 42 $\boxed{x^2}$ $\boxed{)}$ $\boxed{\sqrt{x}}$

Display: 89.470666

The shortest distance from City A to City B is about 89.5 miles.

⮕ Trial Run

———— 1. Find $\sqrt{3364}$.

———— 2. Find $\sqrt{5}$. (Round to thousandths place.)

———— 3. In right triangle ABC, $a = 3.42$ meters and $b = 5.26$ meters. Find the hypotenuse. (Round to hundredths place.)

———— 4. In right triangle ABC, $a = 300$ feet and $b = 720$ feet. Find the hypotenuse.

Answers are on page 753.

▶ Examples You Completed

Example 5. Find the weight in kilograms of a football player who weighs 215 pounds. (Round to the tenths place.)

Solution. The relationship between kilograms and pounds is

$$1 \text{ kg} \doteq 2.2 \text{ lb}$$

Now set up the conversion.

$$215 \text{ lb} \doteq 215 \times \overset{1}{\cancel{\text{lb}}} \times \frac{1 \text{ kg}}{\underset{2.2}{\cancel{2.2 \text{ lb}}}}$$

$$\doteq \frac{215}{2.2} \text{ kg}$$

$$\doteq 97.727273 \text{ kg}$$

The football player weighs about 97.7 kilograms.

Example 16. Find the volume of air that will fill a spherical balloon with a radius of 10.5 meters. (Round to the ones place.)

Solution. The balloon is a sphere with $r = 10.5$ meters.

$$V = \frac{4}{3} \cdot \pi \cdot r^3$$

$$V = \frac{4}{3} \cdot \pi \cdot (10.5)^3 \text{ cu m}$$

Press: $\boxed{\text{ON/C}}$ 4 $\boxed{\div}$ 3 $\boxed{\times}$ $\boxed{\pi}$ $\boxed{\times}$ 10.5 $\boxed{y^x}$ 3 $\boxed{=}$

Display: 4849.0483

The volume of air that will fill the balloon is about 4849 cubic meters.

Example 19. Find $\sqrt{841}$.

Solution. Press: $\boxed{\text{ON/C}}$ 841 $\boxed{\sqrt{x}}$

Display: 29

$$\sqrt{841} = 29$$

Example 22. In right triangle ABC, $a = 0.23$ meters and $b = 1.06$ meters. Find the hypotenuse. (Round to the hundredths place.)

Solution. We must use the Pythagorean Formula

$$a^2 + b^2 = c^2$$

$$0.23^2 + 1.06^2 = c^2$$

$$\sqrt{0.23^2 + 1.06^2} = c$$

Press: ⏺ON/C ⏺(.23 ⏺x^2 ⏺+ 1.06 ⏺x^2 ⏺) ⏺$\sqrt{x}$

Display: 1.0846658

The hypotenuse, $c \doteq 1.08$ meters.

Answers to Trial Runs

page 738   **1.** 180 in.   **2.** 288 in.   **3.** 6400   **4.** 0.235 m   **5.** 7.8 m   **6.** 700 mm

page 740   **1.** 3.1   **2.** 386.4   **3.** 45.28   **4.** 8.26   **5.** 3.3   **6.** 24 ft

page 742   **1.** 23.9 m   **2.** 32.5 ft   **3.** 13.82 m   **4.** 63.25 ft

page 746   **1.** 18,162.75 sq ft   **2.** 119.34 sq m   **3.** 72.38 sq m   **4.** 64.6 sq ft

page 747   **1.** 0.24 cu m   **2.** 4320 cu cm   **3.** 3.375 cu ft   **4.** 1767 cu m

page 749   **1.** 50.9°   **2.** 59°   **3.** 21.7°   **4.** 11.2°

page 752   **1.** 58   **2.** 2.236   **3.** 6.27 m   **4.** 780 ft

EXERCISE SET 16.4

_____ 1. Change 27 feet to inches.

_____ 2. Change 18 feet to inches.

_____ 3. Change 12 yards to inches.

_____ 4. Change 15 yards to inches.

_____ 5. Change 545 millimeters to meters.

_____ 6. Change 680 millimeters to meters.

_____ 7. Change 1285 centimeters to meters.

_____ 8. Change 890 centimeters to meters.

_____ 9. Change $\frac{1}{4}$ of a ton to ounces.

_____ 10. Change $\frac{2}{5}$ of a ton to ounces.

_____ 11. Change 0.92 meter to millimeters.

_____ 12. Change 0.87 meter to millimeters.

_____ 13. Change 8.4 kilometers to meters.

_____ 14. Change 12.5 kilometers to meters.

_____ 15. Change 0.28 kilometer to centimeters.

_____ 16. Change 0.47 kilometer to centimeters.

_____ 17. Find the length in meters of a piece of lumber that is 24 feet long. (Round to the hundredths place.)

_____ 18. If Irma jogged 5 kilometers, how many miles did she jog?

_____ 19. In feet, how high is a stack of 17 books if each book is 4 centimeters thick? (Round to the hundredths place.)

_____ 20. If a box of detergent weighs 45 ounces, what will be the weight, to the nearest gram, of a case of 20 boxes?

_____ 21. One bar of soap weighs 140 grams. What is the weight of a box of 40 bars? (Round to the tenths place.)

_____ 22. Tinker needs $\frac{1}{2}$ pint of wood stain for each chair that he refinishes. If he has 24 chairs to refinish, how many whole liters of wood stain should he buy?

_____ 23. If Hilda needs 75 liters of gasoline to fill her tank, how many gallons of gasoline must she buy? (Round to the tenths place.)

_____ 24. Find the circumference of a circle that has a radius of 5.27 meters. (Round to the hundredths place.)

_____ 25. Find the perimeter of a rectangle with a length of 8.75 feet and a width of 9.87 feet.

_____ 26. Find the perimeter of a square with each side 8.32 meters.

_____ 27. Find the circumference of a circle with a diameter of $8\frac{3}{4}$ feet. (Round to the hundredths place.)

_____ 28. Find the perimeter of a triangle with sides 0.342, 0.756, and 1.225 kilometers.

754

_____ **29.** Find the perimeter of the figure illustrated. (Round answer to the hundredths place.)

$6\frac{1}{2}$ ft $8\frac{1}{4}$ ft

$7\frac{2}{3}$ ft $7\frac{2}{3}$ ft

$11\frac{3}{4}$ ft

_____ **30.** Find the area of a triangle with base 15.37 meters and height 8.72 meters. (Round answer to the hundredths place.)

_____ **31.** Find the area of a triangle with base $17\frac{1}{2}$ feet and height $8\frac{3}{4}$ feet.

_____ **32.** Find the area of a circle with a diameter of 14.35 centimeters. (Round to the hundredths place.)

_____ **33.** Find the area of a circle with a radius of $5\frac{1}{2}$ feet. (Round to the hundredths place.)

_____ **34.** Find the area of a rectangular farm that is 1.43 kilometers by 3.75 kilometers.

_____ **35.** Find the amount of carpeting needed to cover the floor of a rectangular room that is $11\frac{1}{2}$ feet wide by $13\frac{2}{3}$ feet long. (Round answer to the hundredths place.)

_____ **36.** Find the area of the bottom of the illustrated swimming pool. (Round answer to the tenths place.)

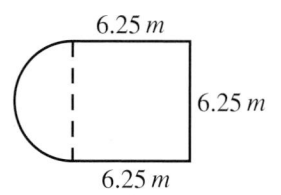

6.25 m

6.25 m

6.25 m

_____ **37.** Find the area of the figure illustrated. (Round answer to the hundredths place.)

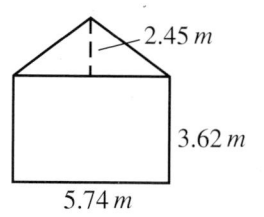

2.45 m

3.62 m

5.74 m

_____ **38.** Find the volume of a box that is a cube with edge 15.25 centimeters. (Round answer to the hundredths place.)

_____ **39.** Find the volume of a carton that is a cube with edge $2\frac{1}{4}$ feet. (Round answer to the hundredths place.)

_____ 40. Find the volume of a sphere with radius 15.25 centimeters. (Round answer to the hundredths place.)

_____ 41. Find the volume of a sphere with radius $2\frac{1}{2}$ inches. (Round answer to the hundredths place.)

_____ 42. An aluminum can in the shape of a cylinder has diameter 4.5 centimeters and height 12.6 centimeters. Find its volume. (Round answer to the tenths place.)

_____ 43. Ms. Briscoe has a grain bin in the shape of a cylinder. The diameter is 25 feet and the height is $42\frac{1}{2}$ feet. Find the volume of the grain bin. (Round answer to the tenths place.)

_____ 44. Find the volume of a rectangular solid with length 8.32 meters, width 3.58 meters, and height 7.96 meters. (Round answer to the hundredths place.)

_____ 45. Find the volume of a storage box with length $5\frac{2}{3}$ feet, width $1\frac{1}{2}$ feet, and height $3\frac{3}{4}$ feet. (Round answer to the hundredths place.)

_____ 46. In triangle ABC, $\angle A = 27.9°$ and $\angle B = 55.7°$. Find $\angle C$.

_____ 47. In triangle ABC, $\angle A = 36.7°$ and $\angle C = 87.9°$. Find $\angle B$.

_____ 48. In right triangle ABC, $\angle A = 78.4°$. Find $\angle B$.

_____ 49. In right triangle ABC, $a = 35$ feet and $b = 84$ feet. Find the hypotenuse.

_____ 50. In right triangle ABC, $a = 45$ feet and $b = 24$ feet. Find the hypotenuse.

_____ 51. In right triangle ABC, $a = 8.47$ meters and $b = 12.65$ meters. Find the hypotenuse. (Round to the hundredths place.)

_____ 52. In right triangle ABC, $a = 23.4$ centimeters and $b = 15.2$ centimeters. Find the hypotenuse. (Round to the tenths place.)

☆ Stretching the Topics _____

_____ 1. A patio has the shape and dimensions illustrated. Find the cost of covering the patio with indoor-outdoor carpeting if the carpet costs $3.95 per square yard.

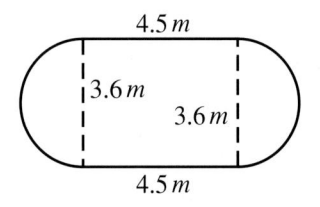

4.5 m

3.6 m

3.6 m

4.5 m

_____ 2. A bottle contains $2\frac{1}{2}$ liters of solution. If the solution is divided equally among 4 beakers, how many ounces of the solution will each beaker contain?

_____ **3.** Find the perimeter and area of triangle *ABC* shown in the illustration. (Round the missing lengths, the perimeter, and the area to the tenths place.)

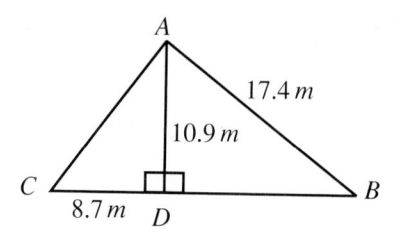

Check your answers in the back of your book.

If you can complete **Checkup 16.4**, you are ready to do the **Review Exercises** for Chapter 16.

✓ **CHECKUP 16.4**

_____ 1. Change 0.74 millimeter to meters.

_____ 2. Change 558 inches to yards.

_____ 3. Paul lives 35 kilometers from the school where he teaches. How many miles is his home from the school?

_____ 4. If Helen's family drinks $2\frac{3}{10}$ gallons of milk in 1 week, how many liters should she buy for the week? (Round answer to the ones place.)

_____ 5. Find the circumference of a circle whose diameter is 48 inches. (Round answer to the hundredths place.)

_____ 6. Find the perimeter of a rectangle with length $7\frac{1}{4}$ feet and width $3\frac{7}{8}$ feet.

_____ 7. Find the area of a triangle with base 8.35 meters and height 4.17 meters. (Round answer to the hundredths place.)

_____ 8. Find the area of a circular rug that has a radius of 4.7 meters. (Round to the hundredths place.)

_____ 9. Find the volume of a storage carton that is 3.5 feet long, 4.7 feet wide, and 2.3 feet high. (Round to the tenths place.)

_____ 10. In right triangle ABC, $a = 15.2$ inches and $b = 24.4$ inches. Find the hypotenuse. (Round to the tenths place.)

Check your answers in the back of your book.

If You Missed Problems:	You Should Review Examples:
1, 2	1–3
3, 4	4–6
5, 6	7, 8
7, 8	9–12
9	15
10	22

Summary

In this chapter we learned to use the calculator to help us do the arithmetic in many kinds of problems. We performed the operations of addition, subtraction, multiplication, and division with whole numbers, fractional numbers, and decimal numbers. Since a calculator using an algebraic operating system obeys the usual rules for order of operations, we were careful to include parentheses wherever necessary.

We learned to change fractional numbers to decimal numbers by treating fractions as quotients. Then we practiced the four operations of arithmetic with fractional numbers, noting that our answers appeared as decimal numbers instead of fractions.

In the section on percents we discovered that the ⟨%⟩ key was very helpful for finding some percent of a number. To solve other kinds of percent problems, however, we set up percent proportions and solved them using the calculator.

We found that the calculator was very useful in converting measurements and in finding perimeters, circumferences, areas, and volumes of geometric figures. Finally, we were able to use the calculator to find missing parts of a right triangle. In applying the Pythagorean Formula, we learned to find the square root of any number with the calculator.

Throughout this chapter it was pointed out that a calculator can do only what the user tells it to do. When working with your calculator you must be careful to enter numbers correctly, press operation keys in the proper order, and use parentheses when necessary. If you write down a problem before entering it in the calculator, you will be less likely to make errors. Finally, in writing the answer, *you* must attach the necessary units.

❑ Speaking the Language of Mathematics

Complete each sentence with the appropriate word or phrase.

1. A calculator that performs multiplications and divisions from left to right and then performs additions and subtractions from left to right is using an _____ _____ system.

2. The screen where the numbers appear on a calculator is called the _____ .

3. To find a power of a number (such as 3^5) on the calculator, we may use the _____ key.

4. When the numerator or denominator of a fraction contains a sum or difference, we must use _____ to enter the fraction in the calculator.

5. To find the square root of a number on the calculator, we should use the _____ key.

△ Writing About Mathematics

Write your response to each question in complete sentences.

1. Show the sequence of calculator keys that you would press to find the following.

$$\frac{2}{3} + \sqrt{7} - (3 + 2)^3 + 4(1.23)$$

2. Describe the method you would use to find the product $\frac{12}{13} \cdot \frac{45}{17}$ with your calculator if you want your answer
   (a) in decimal number form.
   (b) in fractional number form.

3. Express your feelings about using the calculator as a tool in mathematics. At what grade level should the use of the calculator be permitted?

REVIEW EXERCISES for Chapter 16

Perform the indicated operations.

_____ 1. $103.85 - $38.72 + $78.92 - $98.36   _____ 2. 96 ÷ 5

_____ 3. ($78.25 + $95 + $23.79) ÷ 3   _____ 4. 22.365 ÷ 21.3

_____ 5. 1.395 × 0.712 ÷ 8.7 (Round answer to the hundredths place.)

_____ 6. (78 + 39) ÷ (11 × 2)   _____ 7. $(8.72)^2$

_____ 8. $(13.5 - 7.4)^2$ _____ 9. $17^2 - 3^3 \times 7 \times 5$

_____ 10. $(0.475)^4$ _____ 11. $(3.72 - 0.51)^4 + 3.74^3$
(Round to the thousandths place.)   (Round answer to hundredths place.)

_____ 12. Change $\dfrac{384}{179}$ to a decimal number.   _____ 13. Change $\dfrac{19}{36}$ to a decimal number.
(Round to the tenths place.)

_____ 14. Find $\dfrac{8^2 - 50}{9 \times 6 - 38}$.   _____ 15. Find $\dfrac{38 - 6^2}{15 \div 3 + 4}$.

Perform the indicated operations. (In Exercises 16–28, round answers to the hundredths place.)

_____ 16. $\dfrac{9}{10} \times \dfrac{7}{12} \times \dfrac{2}{3}$ _____ 17. $\dfrac{12}{25} \div \dfrac{4}{9} \times \dfrac{5}{8}$

_____ 18. $\dfrac{13}{21} \div \dfrac{4}{7} \div \dfrac{20}{39}$ _____ 19. $\dfrac{8}{15} \div \left(\dfrac{5}{9} \times \dfrac{3}{2}\right)$

_____ 20. $\dfrac{7}{8} + \dfrac{5}{9} - \dfrac{7}{12}$ _____ 21. $\dfrac{15}{7} - \left(\dfrac{8}{9} - \dfrac{3}{4}\right)$

_____ 22. $\left(\dfrac{2}{3} + \dfrac{7}{8}\right) \div \left(\dfrac{1}{2} - \dfrac{2}{7}\right)$ _____ 23. $\left(\dfrac{15}{11} - \dfrac{3}{8}\right) \times \left(\dfrac{5}{3} - \dfrac{5}{7}\right)$

_____ 24. Change $10\frac{3}{4}$ percent to a decimal number.

_____ 25. Change 325 percent to a decimal number.

_____ 26. Find $7\frac{1}{4}$ percent of $3728.

_____ 27. Find 115 percent of 783.

_____ 28. Of the 60 members of Central State University's football team, 28 were from outside the state. What percent of the team's members were from outside the state?

_____ 29. Emile pays $435 per month for rent. If this amount represents 23 percent of his salary, how much does Emile earn per month? (Round to the nearest cent.)

_____ 30. Find the total amount paid for a refrigerator if the selling price is $498.98 and the state sales tax rate is 5.5 percent.

_____ 31. Change 0.39 millimeter to meters.   _____ 32. Change 15 yards to inches.

_____ 33. Change 85 miles to kilometers.   _____ 34. Change 32 quarts to liters.

_____ 35. In triangle ABC, $\angle A = 89.9°$ and $\angle C = 36.7°$. Find $\angle B$.

_____ 36. Find the circumference of a circle that has a diameter of $12\frac{1}{2}$ feet. (Round answer to the hundredths place.)

_____ 37. Find the perimeter of the figure illustrated. (Round answer to the tenths place.)

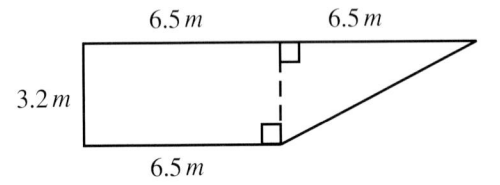

_____ 38. Find the area of a circular swimming pool that has a diameter of $9\frac{1}{2}$ feet. (Round answer to the hundredths place.)

_____ 39. Find the volume of a sphere that has a radius of 3.5 meters. (Round answer to the tenths place.)

_____ 40. Find the volume of a cedar chest that is 4 feet long, 1.75 feet wide, and 2.25 feet deep. (Round volume to the nearest cubic foot.)

Check your answers in the back of your book.

If You Missed Exercises:	You Should Review Examples:	
1–3	Section 16.1	1–6
4, 5		7–9
6–8		10–14
9, 10		16–18
11, 12	Section 16.2	1–3
13, 14		4, 5
16–19		7–11
20–23		12–15
24, 25	Section 16.3	1
26, 27		2–4
28		6
29		7
30		8
31–34	Section 16.4	1–6
35		17
36		7
37		8, 22
38		12
39, 40		14–16

If you have completed the **Review Exercises** and corrected your errors, you are ready to take the **Practice Test** for Chapter 16.

PRACTICE TEST for Chapter 16

Perform the indicated operations.

		SECTION	EXAMPLE
_____	1. $34.83 \times 2.46 \div 5.25$ (Round to the hundredths place.)	16.1	6
_____	2. $42^2 - 2^3 \times 9 \times 15$	16.1	14
_____	3. $(0.187)^5$	16.1	16
_____	4. $(9.35 \times 5.07)^3 - 3.62^4$	16.1	18
_____	5. Change $\dfrac{25}{36}$ to a decimal number.	16.2	2
_____	6. $7 \div 215$ (Round to the thousandths place.)	16.2	3
_____	7. $\dfrac{12}{13} \times \dfrac{3}{7} \times \dfrac{5}{8}$ (Round to the thousandths place.)	16.2	7
_____	8. $\dfrac{7}{8} \div \dfrac{2}{3} \div \dfrac{5}{16}$ (Round to the hundredths place.)	16.2	10
_____	9. $\dfrac{5}{9} + \dfrac{11}{17} + \dfrac{25}{12}$ (Round to the hundredths place.)	16.2	12
_____	10. $\dfrac{85}{93} - \left(\dfrac{2}{3} - \dfrac{3}{5}\right)$ (Round to the ten thousandths place.)	16.2	14
_____	11. What is 75 percent of $2589?	16.3	2
_____	12. Find $17\frac{2}{3}$ percent of 1050. (Round to the hundredths place.)	16.3	4
_____	13. Of the 32 basketball games played by the Blazers, 21 were victories. What percent of their games did the Blazers win?	16.3	6
_____	14. By the end of April, Mr. MacDonald had planted 80 percent of his crop acreage. If he had planted 875 acres, how many acres of crops did he plan to plant?	16.3	7
_____	15. Ms. Zagiba was charged $3.18 in sales tax for parts when she had some electrical repairs done. If the sales tax rate is 6.5 percent, what was the charge for the parts?	16.3	8
_____	16. Change 0.85 millimeter to meters.	16.4	3
_____	17. Find the weight in kilograms of 5 pounds of hamburger.	16.4	5
_____	18. Find the circumference of a circular tablecloth whose diameter is 48 inches. (Round answer to the tenths place.)	16.4	8

		SECTION	EXAMPLE
_____	19. Find the area of a rectangle that is 8.25 meters long and 12.35 meters wide. (Round answer to the hundredths place.)	16.4	11
_____	20. In right triangle *ABC*, $a = 19.2$ m and $b = 25.3$ m. Find the hypotenuse. (Round answer to the tenths place.)	16.4	20

Check your answers in the back of your book.

SHARPENING YOUR SKILLS after Chapters 1–16

SECTION

_____ 1. Write the word form for 1,302,019.

1.1

_____ 2. Change $\frac{5}{9}$ to a fraction with a denominator of 108.

5.2

_____ 3. Write 0.0485 as a fraction.

6.1

_____ 4. Change $\frac{96}{15}$ to a decimal number.

7.3

_____ 5. Use cross products to decide whether the ratios $\frac{4.5}{0.3}$ and $\frac{31.5}{2.1}$ are equal.

8.2

_____ 6. Change 225 percent to a decimal number.

9.1

_____ 7. Change $\frac{118}{354}$ to a percent. (Round answer to the tenths place.)

9.1

_____ 8. Change 15,540 grams to kilograms.

12.2

_____ 9. Write an algebraic expression for 5 more than twice x.

15.1

_____ 10. Combine the like terms: $5x - 7 - 2x + 3y + 2$.

15.1

Perform the indicated operations.

_____ 11. $468,780 \div 156$

2.2

_____ 12. $12\frac{2}{3} \times 5 \times \frac{15}{19}$

4.1

_____ 13. $2\frac{7}{8} - \frac{7}{12} - 1\frac{3}{4}$

5.3

_____ 14. $8 + 2.035 + 0.02$

6.2

_____ 15. $\$50 - \$29.98$

6.2

_____ 16. 31.005×0.083

7.1

_____ 17. $-5.2(8.7 - 10.03)$

14.3

_____ 18. $\dfrac{3(-5) - 2(-7)}{(6)(-4) + (12)(2)}$

14.4

_____ 19. Solve $5x - 8 = 12$

15.3

_____ 20. What is $12\frac{1}{2}$ percent of 3895? (Round to the nearest cent.)

16.3

_____ 21. Find the missing part in the proportion $\dfrac{5}{8} = \dfrac{n}{72}$.   8.2

_____ 22. If a pair of tennis shoes that regularly sells for $39.95 has been marked down 20   10.1
percent, find the sale price of the shoes.

_____ 23. For the first basketball game of the season the arena which seats 27,000 was sold out.   10.2
After the team lost the first game, the attendance decreased by 3132. Find the percent
of decrease in attendance.

_____ 24. Midterm examination scores for 10 students in an algebra class are given as follows.   11.1
Find the mean score to the nearest whole number.
  55, 63, 78, 48, 65, 84, 93, 88, 78, 85

_____ 25. If 148 of the 200 students enrolled in an English class were ranked below him, what   11.3
was Ted's percentile rank in the class?

_____ 26. Joy cut a piece of rope 8.4 meters long into 6 equal pieces. What is the length of each   12.1
_____ piece in meters? in centimeters?

_____ 27. A nurse plans to administer flu shots to 1600 patients. If each shot he administers is   12.3
0.75 milliliter, how many liters of medication does he need?

_____ 28. Find the area (in square feet) of a rectangle that is 6 yards long and $12\frac{2}{3}$ feet wide.   13.2

_____ 29. If a car travels 24 miles due north and then 7 miles due west, how far is the car from   13.4
its starting point?

_____ 30. Georgia has $1590 in her savings account and she plans to save $30 each week. For   15.4
how many weeks must she continue this plan if she wishes to have a total of $3000 in
her savings account?

_____ 31. The length of a rectangle is 3 less than twice the width. If the perimeter of the   15.4
rectangle is 30 inches, find its dimensions.

_____ 32. Find the volume of a rectangular solid with length 8.42 meters, width 5.28 meters, and   16.4
height 9.85 meters. (Round answer to the hundredths place.)

Check your answers in the back of your book.

Answers

This section includes the answers to Odd-Numbered Exercises, Stretching the Topics, Checkups, Speaking the Language of Algebra, Review Exercises, Practice Tests, and Sharpening Your Skills.

CHAPTER 1

Exercise Set 1.1 (page 10)
1. 12 **3.** 9 **5.–8.**

```
+---+------------+------------+---+-->
0   1            5            12
```

9. 2 **11.** 9 **13.** <
15. forty-seven thousand, five hundred ninety-two
17. 56,704
19. 9 ten thousands and 3 thousands and 5 ones
21. 1 million and 3 thousands and 5 hundreds and 7 tens and 6 ones **23.** 3 **25.** 2
27. five thousand, two hundred thirty-six
29. seven thousand, one hundred seventy-five **31.** 40
33. 570 **35.** 1500 **37.** 3000 **39.** 30,000
41. $600,000

Stretching the Topics (page 12)
1. eleven million, one hundred one thousand, one
2. 2 hundred millions and 5 millions and 3 hundred thousands and 6 ten thousands and 7 ones **3.** 1246

Checkup 1.1 (page 13)
1. 12 **2.** 65 **3.** thirty thousand, one hundred seventy
4. 26,000,002
5. 7 ten thousands and 3 thousands and 4 hundreds and 5 tens and 2 ones
6. 1 hundred thousand 5 ten thousands **7.** 7 **8.** 0
9. forty-two million, four hundred seven thousand, nine hundred forty-three **10.** 5400

Exercise Set 1.2 (page 20)
1. 8 **3.** 15 **5.** 11 **7.** 13 **9.** 19 **11.** 14
13. 18 **15.** 17 **17.** 19 **19.** 38 **21.** 119
23. 1098 **25.** 1099 **27.** 298 **29.** 787
31. 679 **33.** 126 **35.** 289 **37.** 7768
39. 9859 **41.** 23 **43.** 61 **45.** 166 **47.** 898
49. 238 **51.** 11,987 **53.** 11,000 **55.** 9025

57. 43,605 **59.** 129,677 **61.** 2591 **63.** 21,659

Stretching the Topics (page 21)
1. 764,911 **2.** 631 **3.** 111,231,105

Checkup 1.2 (page 22)
1. 17 **2.** 18 **3.** 69 **4.** 1290 **5.** 1565
6. 14,500 **7.** 13,090 **8.** 91,368 **9.** 5961
10. 2040

Exercise Set 1.3 (page 29)
1. 8 **3.** 6 **5.** 5 **7.** 5 **9.** 9 **11.** 4
13. 8 **15.** 19 **17.** 41 **19.** 6 **21.** 197
23. 118 **25.** 542 **27.** 329 **29.** 15,778
31. 478 **33.** 23,399 **35.** 142,999 **37.** 4695
39. 260,709

Stretching the Topics (page 30)
1. 13,925,204 **2.** 120,624 **3.** 5,431,674

Checkup 1.3 (page 31)
1. 8 **2.** 5 **3.** 55 **4.** 16 **5.** 38 **6.** 1124
7. 17,848 **8.** 564,778

Exercise Set 1.4 (page 38)
1. 6605 **3.** 262,870 **5.** $9285 **7.** 4955 ft
9. $514 **11.** Plaza II; 6 **13.** 107,933 **15.** 47,150
17. 2252 **19.** 75 units

Stretching the Topics (page 39)
1. 70 **2.** $350; $135; $215

Checkup 1.4 (page 40)
1. 200 **2.** 3989 **3.** 1187 lb **4.** 533 **5.** 120

Speaking the Language of Mathematics (page 42)
1. whole numbers **2.** right **3.** digits

4. hundreds; tens; ones   **5.** sum; difference
6. $17 = 8 + 9$ **7.** $10 + 11$ **8.** perimeter

Review Exercises for Chapter 1 (page 43)
1. >   **2.** <
3. 3 thousands and 7 hundreds and 8 tens and 4 ones
4. 10,061   **5.** 2; 750,200   **6.** 8; 6,590,000
7. two hundred eight thousand, four hundred seven
8. 22   **9.** 77   **10.** 799   **11.** 491   **12.** 11,936
13. 26,450   **14.** 126,854   **15.** 3192   **16.** 18
17. 12   **18.** 8   **19.** 12   **20.** 38   **21.** 23
22. 526   **23.** 664   **24.** 6402   **25.** 2947
26. 28,108   **27.** 95,324   **28.** 194,607
29. $232,653   **30.** $1726

Practice Test for Chapter 1 (page 45)
1. >
2. three hundred six thousand, seven hundred nineteen
3. 3; 78,400   **4.** 595   **5.** 2110   **6.** 11,053
7. 21,757   **8.** 196,308   **9.** 15   **10.** 412
11. 267   **12.** 1389   **13.** 6985   **14.** 1392
15. 1157   **16.** 86   **17.** 526   **18.** 2-yard gain

CHAPTER 2

Exercise Set 2.1 (page 55)
1. 56   **3.** 42   **5.** 8   **7.** 49   **9.** 0   **11.** 68
13. 150   **15.** 266   **17.** 1932   **19.** 40,500
21. 3397   **23.** 24,168   **25.** 97,265   **27.** 1,374,540
29. 540   **31.** 560   **33.** 5040   **35.** 4280
37. 4228   **39.** 325,600   **41.** 279,744
43. 2,177,420   **45.** 336,000   **47.** 7,770,000

Stretching the Topics (page 58)
1. 29,754,000   **2.** 147,836,224   **3.** 439,296,000

Checkup 2.1 (page 59)
1. 63   **2.** 408   **3.** 4200   **4.** 320   **5.** 2376
6. 945   **7.** 2314   **8.** 136,422   **9.** 41,122
10. 204,800

Exercise Set 2.2 (page 67)
1. 9   **3.** 5   **5.** 0   **7.** 8   **9.** 8   **11.** 7 R5
13. 6 R6   **15.** 7 R1   **17.** 6 R1   **19.** 14
21. 12 R5   **23.** 54   **25.** 79 R4   **27.** 458
29. 856 R3   **31.** 7
33. It is not possible to divide by 0.   **35.** 37   **37.** 405
39. 90 R2   **41.** 204 R5   **43.** 346 R4   **45.** 6 R29
47. 12   **49.** 23 R5   **51.** 29

Stretching the Topics (page 67)
1. 27,950   **2.** 3,019 R832   **3.** 784 R100

Checkup 2.2 (page 68)
1. 9   **2.** 16 R2   **3.** 72 R3   **4.** 429 R1   **5.** 25
6. 0   **7.** 403   **8.** 60 R15   **9.** 656   **10.** 62

Exercise Set 2.3 (page 73)
1. composite; 3×5 **3.** composite; 3^3 **5.** prime

7. composite; $2^2 \times 3$ **9.** composite; $2^3 \times 5^3$
11. composite; 3×61 **13.** composite; 7×13
15. prime   **17.** composite; $3 \times 5 \times 7$
19. composite; $2^3 \times 3^3 \times 7$

Stretching the Topics (page 73)
1. 2   **2.** $2^5 \times 3^3 \times 7$ **3.** 13

Checkup 2.3 (page 74)
1. composite; 2×5^2 **2.** composite; $3^2 \times 5$
3. composite; $2^2 \times 17$ **4.** composite; $3^2 \times 19$
5. prime   **6.** composite; $2^3 \times 13$
7. composite; $2^3 \times 3^3$ **8.** composite; 7×19
9. composite; $2^2 \times 3 \times 59$
10. composite; $2^3 \times 3 \times 5 \times 7$

Exercise Set 2.4 (page 85)
1. 42   **3.** 18   **5.** 612 mi   **7.** 368 mi   **9.** 61
11. 22 sec   **13.** 8   **15.** 2704 sq ft   **17.** 672 sq ft
19. 230

Stretching the Topics (page 86)
1. $29,000   **2.** 210 sq in.

Checkup 2.4 (page 87)
1. $1656   **2.** 120   **3.** 16,500 sq ft   **4.** 8483
5. 18

Speaking the Language of Mathematics (page 90)
1. factors; product   **2.** 0
3. dividend; divisor; quotient   **4.** $4 \times 6 = 24$
5. remainder   **6.** $(6 \times 3) + 2 = 20$
7. prime; prime or composite   **8.** area   **9.** add; divide

Review Exercises for Chapter 2 (page 93)
1. 635   **2.** 4812   **3.** 6258   **4.** 11,088
5. 237,654   **6.** 1512   **7.** 417,200   **8.** 56,400
9. 173,262   **10.** 112,179,000   **11.** 8
12. It is not possible to divide by 0.   **13.** 13 R3
14. 823 R1   **15.** 372   **16.** 480   **17.** 3007
18. 1300   **19.** 1349 R4   **20.** 356 R300
21. composite; $2 \times 7 \times 13$ **22.** prime
23. composite; $2^3 \times 3^2$ **24.** composite; $2^3 \times 3 \times 5^2$
25. $84   **26.** 117 sq in.   **27.** $26   **28.** 265 mi
29. $4   **30.** $150

Practice Test for Chapter 2 (page 95)
1. 30,456   **2.** 19,342   **3.** 316,506   **4.** 270
5. 982,800   **6.** 2,460,691,620   **7.** 4 R5   **8.** 803
9. 4720 R3   **10.** 23,277 R5   **11.** 1234
12. 500 R36   **13.** composite; $2^2 \times 3 \times 7$ **14.** prime
15. composite; $3^2 \times 7^2$ **16.** $48   **17.** 270 sq ft
18. 336 mi   **19.** 6560 sq ft   **20.** 4

CHAPTER 3

Exercise Set 3.1 (page 109)
1. $\frac{5}{9}$ **3.** $\frac{2}{5}$ **5.** $\frac{486}{625}$ **7.** $\frac{385}{1283}$ **9.** $\frac{13}{28}$ **11.** $\frac{123}{250}$

13. 7   **15.** 56   **17.** 9   **19.** 6   **21.** 12
23. $\frac{95}{12}$ or $7\frac{11}{12}$ **25.** $\frac{138}{25}$ or $5\frac{13}{25}$ **27.** $\frac{101}{12}$ or $8\frac{5}{12}$
29. $8\frac{7}{10}$ **31.** $19\frac{3}{4}$ **33.** $2\frac{4}{5}$ **35.** $31\frac{2}{3}$ **37.** $\frac{43}{5}$
39. $\frac{5}{3}$ **41.** $\frac{197}{10}$ **43.** $\frac{183}{4}$

Stretching the Topics (page 111)
1. $\frac{13}{150}$ **2.** (a) $\frac{27,349}{100}$ (b) $329\frac{21}{25}$ **3.** $\frac{161}{30}$ or $5\frac{11}{30}$

Checkup 3.1 (page 112)
1. $\frac{5}{6}$ **2.** $\frac{23}{38}$ **3.** $\frac{3}{5}$ **4.** 5   **5.** 7   **6.** $\frac{257}{36}$ or $7\frac{5}{36}$
7. $7\frac{1}{7}$ **8.** $3\frac{3}{4}$ **9.** $\frac{38}{5}$ **10.** $\frac{76}{9}$

Exercise Set 3.2 (page 119)
1. $\frac{2}{3}$ **3.** $\frac{5}{32}$ **5.** $\frac{1}{3}$ **7.** $\frac{1}{8}$ **9.** $\frac{2}{3}$ **11.** $\frac{3}{4}$
13. $\frac{1}{4}$ **15.** $\frac{1}{3}$ **17.** $\frac{5}{9}$ **19.** $\frac{28}{45}$ **21.** $\frac{5}{21}$ **23.** $\frac{4}{11}$
25. $\frac{7}{9}$ **27.** $\frac{49}{100}$ **29.** $\frac{3}{11}$ **31.** $\frac{2}{3}$ **33.** $\frac{4}{9}$
35. $\frac{165}{364}$ **37.** $\frac{35}{33}$ **39.** $\frac{9}{11}$

Stretching the Topics (page 120)
1. $\frac{10}{33}$ **2.** $\frac{5}{8}$ **3.** $\frac{2}{3}$

Checkup 3.2 (page 121)
1. $\frac{1}{6}$ **2.** $\frac{4}{7}$ **3.** $\frac{1}{5}$ **4.** $\frac{3}{11}$ **5.** $\frac{2}{3}$ **6.** $\frac{10}{21}$ **7.** $\frac{4}{9}$
8. $\frac{34}{39}$ **9.** $\frac{75}{98}$ **10.** $\frac{2}{3}$

Exercise Set 3.3 (page 126)
1. $\frac{1}{3}$ **3.** $\frac{5}{9}$ **5.** $\frac{1}{3}$ **7.** $\frac{2}{5}$ **9.** $\frac{13}{30}$ **11.** $\frac{2}{3}$
13. $\frac{5}{2}$ or $2\frac{1}{2}$ **15.** $\frac{22}{15}$ or $1\frac{7}{15}$ **17.** $\frac{3}{4}$ **19.** $\frac{3}{5}$

Stretching the Topics (page 127)
1. $\frac{62}{101}$ **2.** $\frac{15}{29}$ **3.** $\frac{5}{2}$ or $2\frac{1}{2}$

Checkup 3.3 (page 129)
1. $\frac{2}{5}$ **2.** $\frac{12}{37}$ **3.** $\frac{2}{7}$ **4.** $\frac{5}{8}$ **5.** $\frac{5}{2}$ or $2\frac{1}{2}$

Speaking the Language of Mathematics (page 131)
1. numerator; denominator   **2.** proper   **3.** improper
4. mixed numbers   **5.** common factors   **6.** $\frac{5}{13}$; $\frac{8}{13}$

Review Exercises for Chapter 3 (page 133)
1. $\frac{5}{6}$ **2.** $\frac{139}{378}$ **3.** $\frac{59}{112}$ **4.** $\frac{2}{3}$ **5.** 54   **6.** 4
7. 6   **8.** 405   **9.** $\frac{97}{4}$ or $24\frac{1}{4}$ **10.** $\frac{223}{20}$ or $11\frac{3}{20}$
11. $\frac{136}{25}$ or $5\frac{11}{25}$ **12.** $6\frac{1}{4}$ **13.** $9\frac{7}{8}$ **14.** $\frac{38}{5}$
15. $\frac{17}{3}$ **16.** $\frac{7}{19}$ **17.** $\frac{3}{8}$ **18.** $\frac{4}{63}$ **19.** $\frac{3}{10}$ **20.** $\frac{3}{4}$
21. $\frac{4}{15}$ **22.** $\frac{7}{10}$ **23.** $\frac{12}{35}$ **24.** $\frac{49}{81}$ **25.** $\frac{17}{61}$
26. $\frac{8}{15}$ **27.** $\frac{4}{7}$ **28.** $\frac{7}{4}$ or $1\frac{3}{4}$

Practice Test for Chapter 3 (page 135)
1. $\frac{286}{335}$ **2.** $\frac{336}{515}$ **3.** 135   **4.** 24   **5.** $\frac{88}{15}$ or $5\frac{13}{15}$
6. $6\frac{3}{8}$ **7.** $3\frac{4}{5}$ **8.** $\frac{79}{8}$ **9.** $\frac{71}{3}$ **10.** $\frac{5}{6}$ **11.** $\frac{10}{13}$
12. $\frac{4}{5}$ **13.** $\frac{38}{51}$ **14.** $\frac{5}{9}$ **15.** $\frac{4}{9}$ **16.** $\frac{18}{55}$ **17.** $\frac{3}{20}$
18. $\frac{5}{9}$ **19.** $\frac{1}{3}$ **20.** $\frac{9}{10}$

Sharpening Your Skills after Chapters 1–3 (page 137)
1. seven hundred three thousand, four hundred fifty
2. composite; $2^2 \times 3^3 \times 7$ **3.** $8\frac{2}{7}$ **4.** $\frac{117}{8}$

5. 12,797   **6.** 153,407   **7.** 15   **8.** 17,778
9. 6451   **10.** 315,146   **11.** 2268   **12.** 873
13. 1005   **14.** 85 R9   **15.** $\frac{3}{5}$ **16.** $\frac{17}{19}$ **17.** 46
18. $\frac{46}{77}$ **19.** 150,000 sq ft   **20.** 143

Chapter 4

Exercise Set 4.1 (page 150)
1. 9   **3.** $\frac{15}{2}$ or $7\frac{1}{2}$ **5.** $\frac{8}{5}$ or $1\frac{3}{5}$ **7.** 7
9. $\frac{125}{8}$ or $15\frac{5}{8}$ **11.** 48   **13.** $\frac{1}{64}$ **15.** $\frac{8}{13}$ **17.** $\frac{1}{5}$
19. $\frac{16}{27}$ **21.** $\frac{1}{3}$ **23.** $\frac{1}{55}$ **25.** $\frac{3}{8}$ **27.** $\frac{3}{10}$ **29.** $\frac{3}{28}$
31. $\frac{1}{5}$ **33.** $\frac{2}{5}$ **35.** $\frac{2}{3}$ **37.** $\frac{5}{12}$ **39.** $\frac{3}{8}$
41. $\frac{8}{3}$ or $2\frac{2}{3}$ **43.** $\frac{15}{4}$ or $3\frac{3}{4}$ **45.** 10   **47.** $\frac{37}{3}$ or $12\frac{1}{3}$
49. $\frac{7}{4}$ or $1\frac{3}{4}$ **51.** $\frac{2}{3}$ **53.** $\frac{16}{5}$ or $3\frac{1}{5}$ **55.** $\frac{22}{3}$ or $7\frac{1}{3}$
57. 8   **59.** 1   **61.** $\frac{1}{4}$ **63.** $\frac{1}{27}$ **65.** 3

Stretching the Topics (page 151)
1. 4   **2.** 1   **3.** $\frac{2}{5}$

Checkup 4.1 (page 152)
1. 49   **2.** 9   **3.** $\frac{7}{10}$ **4.** $\frac{8}{35}$ **5.** $\frac{3}{5}$ **6.** $\frac{5}{6}$
7. $\frac{61}{2}$ or $30\frac{1}{2}$ **8.** $\frac{5}{2}$ or $2\frac{1}{2}$ **9.** 1   **10.** 11

Exercise Set 4.2 (page 160)
1. 41   **3.** $\frac{3}{5}$ **5.** $\frac{7}{9}$ **7.** $\frac{1}{39}$ **9.** $\frac{3}{5}$ **11.** $\frac{1}{5}$
13. $\frac{9}{25}$ **15.** $\frac{5}{7}$ **17.** $\frac{40}{9}$ or $4\frac{4}{9}$ **19.** $\frac{5}{2}$ or $2\frac{1}{2}$
21. 189   **23.** $\frac{1}{3}$ **25.** $\frac{23}{2}$ or $11\frac{1}{2}$ **27.** $\frac{33}{20}$ or $1\frac{13}{20}$
29. $\frac{16}{15}$ or $1\frac{1}{15}$ **31.** $\frac{3}{14}$ **33.** $\frac{21}{8}$ or $2\frac{5}{8}$ **35.** $\frac{5}{7}$
37. $\frac{7}{6}$ or $1\frac{1}{6}$ **39.** 10   **41.** $\frac{1}{36}$ **43.** $\frac{1}{3}$ **45.** $\frac{5}{4}$ or $1\frac{1}{4}$
47. $\frac{4}{5}$ **49.** 4

Stretching the Topics (page 161)
1. $\frac{35}{4}$ or $8\frac{3}{4}$ **2.** $\frac{2}{9}$ **3.** $\frac{225}{128}$ or $1\frac{97}{128}$

Checkup 4.2 (page 162)
1. $\frac{3}{4}$ **2.** $\frac{33}{35}$ **3.** $\frac{1}{4}$ **4.** $\frac{1}{30}$ **5.** $\frac{4}{7}$ **6.** $\frac{45}{2}$ or $22\frac{1}{2}$
7. $\frac{1}{6}$ **8.** $\frac{48}{49}$ **9.** $\frac{14}{5}$ or $2\frac{4}{5}$ **10.** $\frac{9}{16}$

Exercise Set 4.3 (page 166)
1. 135   **3.** 60 min   **5.** 60   **7.** 16   **9.** $450
11. 32   **13.** $8\frac{3}{4}$ **15.** $3\frac{1}{3}$¢   **17.** $8
19. $\frac{1}{8}$ c flour; $\frac{3}{4}$ c milk; $\frac{1}{3}$ c cooking oil   **21.** $215\frac{1}{9}$ sq ft
23. $1\frac{1}{3}$ mi   **25.** 32

Stretching the Topics (page 167)
1. $\frac{3}{4}$ lb   **2.** 12

Checkup 4.3 (page 168)
1. $13\frac{1}{3}$ **2.** 36   **3.** $12\frac{3}{8}$ in.   **4.** 6   **5.** $503\frac{1}{4}$ sq ft

Speaking the Language of Mathematics (page 170)
1. numerators; denominators   **2.** reciprocal
3. $\frac{11}{17}$; $\frac{17}{11} \times \frac{11}{17} = 1$ **4.** invert; multiply
5. complex fraction   **6.** division

Review Exercises for Chapter 4 (page 171)
1. 14   **2.** $\frac{21}{2}$ or $10\frac{1}{2}$ **3.** $\frac{7}{20}$ **4.** $\frac{2}{21}$ **5.** $\frac{4}{15}$
6. $\frac{1}{10}$ **7.** $\frac{3}{22}$ **8.** $\frac{29}{2}$ or $14\frac{1}{2}$ **9.** 2   **10.** $16\frac{1}{4}$

11. $8\frac{3}{4}$ **12.** 1   **13.** $\frac{50}{27}$ or $1\frac{23}{27}$ **14.** $\frac{55}{6}$ or $9\frac{1}{6}$
15. $\frac{3}{26}$ **16.** $\frac{5}{8}$ **17.** 10   **18.** $\frac{5}{8}$ **19.** $3\frac{3}{8}$
20. $7\frac{41}{252}$ **21.** $\frac{7}{4}$ or $1\frac{3}{4}$ **22.** 45   **23.** $1\frac{5}{12}$
24. $3\frac{1}{36}$ **25.** 21   **26.** \$36   **27.** $\frac{1}{2}$ acre   **28.** 16
29. 6   **30.** 343 sq ft

Practice Test for Chapter 4 (page 173)
1. $\frac{15}{2}$ or $7\frac{1}{2}$ **2.** $\frac{5}{9}$ **3.** $\frac{5}{7}$ **4.** $\frac{22}{45}$ **5.** $\frac{38}{3}$ or $12\frac{2}{3}$
6. $\frac{9}{2}$ or $4\frac{1}{2}$ **7.** $146\frac{2}{3}$ **8.** $\frac{5}{2}$ or $2\frac{1}{2}$ **9.** $\frac{6}{5}$ or $1\frac{1}{5}$
10. $\frac{2}{5}$ **11.** $5\frac{1}{3}$ **12.** $1\frac{1}{2}$ **13.** $6\frac{2}{3}$ **14.** $1\frac{1}{5}$
15. $\frac{2}{5}$ **16.** $7\frac{1}{2}$ **17.** \$208   **18.** $73\frac{1}{8}$ sq in.
19. 64   **20.** $1\frac{5}{8}$ mi

Sharpening Your Skills after Chapters 1–4 (page 175)
1. 7   **2.** prime   **3.** $16\frac{2}{5}$ **4.** 10,851   **5.** 23
6. 421   **7.** $\frac{13}{21}$ **8.** 362,946   **9.** 555,000
10. 3645   **11.** 40   **12.** $\frac{2}{7}$ **13.** 28   **14.** 329
15. 380 R102   **16.** $2\frac{1}{2}$ **17.** $6\frac{1}{2}$ **18.** $\frac{12}{25}$
19. 5100 sq ft   **20.** 27

CHAPTER 5
Exercise Set 5.1 (page 190)
1. $\frac{1}{2}$ **3.** $\frac{3}{2}$ or $1\frac{1}{2}$ **5.** $\frac{13}{15}$ **7.** 3   **9.** $\frac{8}{3}$ or $2\frac{2}{3}$
11. $\frac{11}{7}$ or $1\frac{4}{7}$ **13.** $\frac{1}{3}$ **15.** $\frac{6}{13}$ **17.** 2   **19.** $\frac{9}{5}$ or $1\frac{4}{5}$
21. 2   **23.** $\frac{16}{9}$ or $1\frac{7}{9}$ **25.** $\frac{3}{2}$ or $1\frac{1}{2}$ **27.** $\frac{17}{16}$ or $1\frac{1}{16}$
29. 1   **31.** 0   **33.** $\frac{1}{2}$ **35.** 2   **37.** $5\frac{6}{7}$ **39.** $9\frac{3}{8}$
41. $6\frac{4}{7}$ **43.** $5\frac{1}{4}$ **45.** $\frac{1}{2}$ **47.** $6\frac{3}{5}$ **49.** $4\frac{5}{8}$
51. 5   **53.** 0   **55.** $2\frac{7}{8}$ **57.** $2\frac{1}{13}$ **59.** $\frac{3}{4}$

Stretching the Topics (page 191)
1. 36   **2.** $8\frac{7}{9}$ **3.** $952\frac{49}{50}$

Checkup 5.1 (page 192)
1. $\frac{1}{2}$ **2.** $\frac{17}{8}$ or $2\frac{1}{8}$ **3.** 2   **4.** $\frac{13}{3}$ or $4\frac{1}{3}$
5. $\frac{7}{6}$ or $1\frac{1}{6}$ **6.** 0   **7.** 3   **8.** $17\frac{1}{2}$ **9.** $2\frac{2}{9}$
10. $3\frac{1}{3}$

Exercise Set 5.2 (page 198)
1. $\frac{8}{12}$ **3.** $\frac{15}{36}$ **5.** $\frac{21}{35}$ **7.** $\frac{28}{49}$ **9.** $\frac{48}{75}$ **11.** $\frac{81}{36}$
13. $\frac{90}{100}$ **15.** $\frac{1300}{1000}$ **17.** $\frac{115}{100}$ **19.** $\frac{24}{4}$ **21.** $\frac{16}{16}$
23. $\frac{65}{5}$ **25.** $\frac{36}{8}$ **27.** $\frac{39}{36}$ **29.** $\frac{250}{30}$

Stretching the Topics (page 198)
1. $\frac{42}{108}$ **2.** $\frac{570}{675}$ **3.** $\frac{2022}{396}$

Checkup 5.2 (page 199)
1. $\frac{9}{12}$ **2.** $\frac{36}{18}$ **3.** $\frac{70}{18}$ **4.** $\frac{28}{35}$ **5.** $\frac{135}{100}$ **6.** $\frac{57}{6}$
7. $\frac{32}{16}$ **8.** $\frac{30}{2}$

Exercise Set 5.3 (page 209)
1. $\frac{3}{2}$ or $1\frac{1}{2}$ **3.** $\frac{3}{2}$ or $1\frac{1}{2}$ **5.** $\frac{3}{16}$ **7.** $\frac{3}{2}$ or $1\frac{1}{2}$ **9.** $\frac{2}{9}$
11. $10\frac{7}{8}$ **13.** $5\frac{11}{12}$ **15.** $\frac{53}{30}$ or $1\frac{23}{30}$ **17.** $\frac{89}{100}$
19. $\frac{61}{105}$ **21.** $\frac{83}{28}$ or $2\frac{27}{28}$ **23.** $\frac{79}{90}$ **25.** $\frac{19}{12}$ or $1\frac{7}{12}$
27. $\frac{413}{312}$ or $1\frac{101}{312}$ **29.** $\frac{359}{340}$ or $1\frac{19}{340}$ **31.** $14\frac{1}{6}$ **33.** $3\frac{3}{10}$
35. $2\frac{15}{28}$ **37.** $4\frac{1}{30}$ **39.** $2\frac{5}{6}$

Stretching the Topics (page 210)
1. $\frac{73}{63}$ or $1\frac{10}{63}$ **2.** $\frac{3001}{2310}$ or $1\frac{691}{2310}$ **3.** $\frac{399}{100}$ or $3\frac{99}{100}$

Checkup 5.3 (page 211)
1. 2   **2.** $\frac{5}{8}$ **3.** $1\frac{2}{3}$ **4.** $\frac{11}{6}$ or $1\frac{5}{6}$ **5.** $\frac{61}{40}$ or $1\frac{21}{40}$
6. $10\frac{19}{36}$ **7.** $4\frac{9}{40}$ **8.** $4\frac{5}{24}$

Exercise Set 5.4 (page 216)
1. >   **3.** <   **5.** <   **7.** >   **9.** >   **11.** >
13. =   **15.** <   **17.** <   **19.** >   **21.** Sarah
23. Dantley

Stretching the Topics (page 216)
1. $\frac{3}{8}, \frac{5}{9}, \frac{4}{7}$ **2.** =   **3.** $3\frac{1}{8}, 3\frac{3}{16}, \frac{7}{2}, \frac{15}{4}$

Checkup 5.4 (page 217)
1. >   **2.** <   **3.** <   **4.** >   **5.** =   **6.** >

Exercise Set 5.5 (page 222)
1. $10\frac{1}{2}$ lb   **3.** $\frac{3}{10}$ **5.** $3\frac{5}{12}$ c   **7.** $5\frac{23}{24}$ ft   **9.** 793
11. $25\frac{5}{6}$ **13.** $\frac{3}{8}$ dollar decrease   **15.** $\frac{7}{8}$ in.   **17.** $8\frac{7}{12}$
19. $3\frac{23}{24}$

Stretching the Topics (page 223)
1. \$875   **2.** $2\frac{11}{20}$ mi; $1\frac{4}{5}$ mi   **3.** $1\frac{3}{16}$ in.; $2\frac{1}{2}$ in.

Checkup 5.5 (page 224)
1. $8\frac{11}{12}$ **2.** $2\frac{3}{8}$ ft   **3.** Bunzer Road   **4.** $2\frac{1}{12}$ c
5. $\frac{7}{20}$

Speaking the Language of Mathematics (page 226)
1. numerators; denominator   **2.** numerators
3. lowest common denominator   **4.** building
5. whole number; fractional

Review Exercises for Chapter 5 (page 227)
1. $\frac{1}{2}$ **2.** $\frac{17}{13}$ or $1\frac{4}{13}$ **3.** $\frac{4}{9}$ **4.** $\frac{29}{25}$ or $1\frac{4}{25}$
5. $\frac{31}{16}$ or $1\frac{15}{16}$ **6.** $5\frac{3}{8}$ **7.** $\frac{20}{25}$ **8.** $\frac{72}{42}$ **9.** $\frac{108}{9}$
10. $\frac{48}{48}$ **11.** $\frac{29}{14}$ or $1\frac{15}{14}$ **12.** $4\frac{1}{4}$ **13.** $\frac{1}{35}$
14. $\frac{5}{3}$ or $1\frac{2}{3}$ **15.** $18\frac{2}{21}$ **16.** $11\frac{11}{36}$ **17.** $5\frac{1}{16}$
18. $5\frac{1}{12}$ **19.** $8\frac{17}{24}$ **20.** $3\frac{17}{24}$ **21.** >   **22.** <
23. >   **24.** <   **25.** $6\frac{1}{4}$ **26.** $7\frac{13}{30}$ ft
27. $6\frac{7}{8}$ points   **28.** $1\frac{1}{2}$ ton   **29.** $63\frac{3}{40}$ **30.** $180\frac{1}{8}$ lb

Practice Test for Chapter 5 (page 229)
1. $\frac{37}{16}$ or $2\frac{5}{16}$ **2.** $\frac{3}{5}$ **3.** $\frac{19}{4}$ **4.** $5\frac{1}{3}$ **5.** $\frac{35}{63}$
6. $\frac{96}{24}$ **7.** 1   **8.** $4\frac{1}{3}$ **9.** $14\frac{1}{2}$ **10.** $\frac{113}{60}$ or $1\frac{53}{60}$
11. $\frac{61}{72}$ **12.** $\frac{991}{420}$ or $2\frac{151}{420}$ **13.** $2\frac{2}{5}$ **14.** $12\frac{5}{4}$
15. $2\frac{11}{20}$ **16.** >   **17.** >   **18.** $8\frac{1}{15}$ **19.** $11\frac{9}{20}$
20. $\frac{1}{15}$

Sharpening Your Skills after Chapters 1–5 (page 231)
1. twenty thousand, three hundred nine   **2.** prime
3. $\frac{253}{8}$ **4.** $\frac{2}{3}$ **5.** $\frac{18}{48}$ **6.** >   **7.** 5925
8. 107,615   **9.** $\frac{38}{15}$ or $2\frac{8}{15}$ **10.** $\frac{19}{10}$ or $1\frac{9}{10}$ **11.** $16\frac{13}{24}$

12. 16,768   **13.** 371   **14.** $\frac{3}{5}$ **15.** $3\frac{7}{8}$ **16.** $21\frac{2}{5}$
17. 1995   **18.** 1,843,535   **19.** $\frac{48}{5}$ or $9\frac{3}{5}$
20. $\frac{93}{4}$ or $23\frac{1}{4}$ **21.** 32   **22.** 10   **23.** 6
24. $5\frac{1}{2}$ mi   **25.** $\frac{2}{5}$

CHAPTER 6

Exercise Set 6.1 (page 240)

1. $\frac{3}{100}$ **3.** $\frac{73}{10,000}$ **5.** 0.345   **7.** 0.00056
9. five and thirty-eight thousandths
11. seventy-five ten thousandths   **13.** 6.09   **15.** 8.700
17. 0.3000   **19.** >   **21.** <   **23.** <   **25.** >
27. <   **29.** 0.0367, 0.052, 0.351
31. 2.072, 2.3, 2.36   **33.** 8.4   **35.** 15.0   **37.** 0.9
39. 3.0   **41.** 52.753   **43.** 6.040   **45.** 0.043
47. 200.244   **49.** 7.03   **51.** 25.33   **53.** 0.87
55. 154.00   **57.** 0.7385   **59.** 12.0439   **61.** 0.0037
63. 1356.7295

Stretching the Topics (page 242)

1. 0.04783 > 0.04779
2. 1.0173, 1.0598, 1.1001, 1.101, 1.103, 1.1102
3. 3.86060

Checkup 6.1 (page 243)

1. $\frac{39}{1000}$ **2.** 0.0008
3. eight hundred sixty-five thousandths   **4.** 0.380
5. 0.73 > 0.0839   **6.** 3.0934, 3.672, 3.76   **7.** 9.3
8. 0.050   **9.** 2.37   **10.** 0.0004

Exercise Set 6.2 (page 251)

1. 0.15   **3.** 1.2   **5.** 24.2   **7.** 4.34   **9.** 7.24
11. 18.91   **13.** 12.626   **15.** 3.118   **17.** 22.018
19. 18.642   **21.** 17.9765   **23.** $274.73   **25.** 0.42
27. 3.421   **29.** 0.1244   **31.** 3.237   **33.** 0.6388
35. $286.45   **37.** $401.89   **39.** 4.86   **41.** 5.47
43. 63.6176   **45.** 8.356   **47.** $65.21   **49.** 8.238
51. 5.797   **53.** $9.39   **55.** 4.47   **57.** $58.57
59. 2.675

Stretching the Topics (page 252)

1. 5.3045   **2.** 125.3616   **3.** 165.04   **4.** 5271.28

Checkup 6.2 (page 253)

1. 1.73   **2.** 52.602   **3.** 23.164   **4.** $25.80
5. 2.69   **6.** 5.285   **7.** 8.861   **8.** $4.22
9. 1.8535   **10.** $22.69

Exercise Set 6.3 (page 262)

1. Four hundred twenty-three and $\frac{78}{100}$
3. Two thousand thirty-six and $\frac{no}{100}$ **5.** Fifteen and $\frac{50}{100}$
7. $59.75   **9.** $3009.00   **11.** $320.07
13. $284.47   **15.** $331.13   **17.** 18.3   **19.** 3.4
21. 16.6   **23.** 1.56 m   **25.** 1.08

Stretching the Topics (page 264)

1. $4.63; $5.20; Carlos; $0.57; $0.17   **2.** 9.4
3. Tuesday; Thursday; Wednesday and Thursday; 1.1; 1.8

Checkup 6.3 (page 265)

1. Seventy-three and $\frac{85}{100}$ **2.** $506.03   **3.** $200.55
4. 2.342 in.   **5.** 37.931 mph

Speaking the Language of Mathematics (page 268)

1. fractional   **2.** tenths; hundredths   **3.** decimal points
4. thousandths   **5.** 100; one

Review Exercises for Chapter 6 (page 269)

1. $\frac{43}{1000}$ **2.** $\frac{31}{100}$ **3.** 0.028   **4.** 0.0009
5. seven hundred eighty-three ten thousandths
6. seven and eight hundred two thousandths   **7.** 9.300
8. 1.20   **9.** >   **10.** <   **11.** <   **12.** <
13. 3.029, 3.09, 3.26   **14.** 16.1   **15.** 0.654
16. 189.68   **17.** 0.0467   **18.** 27.3   **19.** 14.239
20. 15.82   **21.** 35.144   **22.** 19.962   **23.** 40.689
24. $783.64   **25.** $3235.72   **26.** 0.34   **27.** 2.427
28. 1.7617   **29.** $103.34   **30.** 1.42   **31.** 69.2518
32. 8.4764   **33.** $50.84   **34.** 4.81   **35.** $41.80
36. 7.58   **37.** $338.40
38. Three hundred twenty-nine and $\frac{7}{100}$
39. Seven thousand six and $\frac{25}{100}$ **40.** $408.06
41. $7368.25   **42.** $985.83   **43.** 316.8 m
44. 2113.3 mi   **45.** $23.79

Practice Test for Chapter 6 (page 271)

1. $\frac{397}{10,000}$ **2.** 0.08
3. five hundred two and six hundredths   **4.** 8.2300
5. <   **6.** 8.3   **7.** 0.570   **8.** 27.8   **9.** 6.78
10. 12.27   **11.** 45.923   **12.** $143.34   **13.** 2.251
14. $89.86   **15.** 5.446   **16.** 10.366   **17.** $36.08
18. 23.874   **19.** 14.355   **20.** $39.65
21. Ninety-eight and $\frac{89}{100}$ **22.** 3.74 m   **23.** 14.2
24. 6 m   **25.** Grove City; 1.9 in.

Sharpening Your Skills after Chapters 1–6 (page 273)

1. five hundred seven thousand, two hundred nine
2. composite; $2^2 \times 3^2 \times 7$ **3.** $\frac{74}{3}$ **4.** $\frac{38}{51}$ **5.** <
6. $\frac{423}{10,000}$ **7.** 7.10   **8.** 69,107   **9.** 1214   **10.** $3\frac{1}{3}$
11. $\frac{77}{120}$ **12.** $10\frac{5}{12}$ **13.** 13.12   **14.** $36.02
15. 4.68   **16.** 75,000   **17.** 171
18. 1021 R214 or $1021\frac{214}{407}$ **19.** $\frac{45}{2}$ or $22\frac{1}{2}$ **20.** $6\frac{1}{8}$
21. $630   **22.** 295 mi   **23.** $\frac{5}{6}$ **24.** $245\frac{4}{9}$ sq ft
25. $\frac{23}{60}$

CHAPTER 7

Exercise Set 7.1 (page 282)

1. 0.02448   **3.** 0.58968   **5.** 290.088   **7.** 0.80127

9. 0.0031004   **11.** 13.704   **13.** 0.59   **15.** 440
17. 20.208   **19.** 0.00388   **21.** 36.95   **23.** 8625
25. 230   **27.** 0.487   **29.** 38,720   **31.** 0.72
33. 8,853,600   **35.** 7   **37.** 3745   **39.** 8.8786
41. 0.384128   **43.** 23.10924   **45.** 40,708.2
47. 12,816   **49.** 14.8393

Stretching the Topics (page 282)
1. 13,517.7984   **2.** 15,159.42   **3.** 641.03276

Checkup 7.1 (page 283)
1. 0.02106   **2.** 184.356   **3.** 0.02616   **4.** 97.4
5. 1520   **6.** 2.38   **7.** 0.8909   **8.** 16.171
9. 17.91495   **10.** 19,493

Exercise Set 7.2 (page 292)
1. 1.3   **3.** 120   **5.** 0.142   **7.** 300   **9.** 0.25
11. 3.6   **13.** 2200   **15.** 1.02   **17.** 12
19. 0.035   **21.** 286.$\overline{6}$ **23.** 0.1$\overline{90}$ **25.** 80.84$\overline{7}$
27. 13.4$\overline{8}$ **29.** 0.00$\overline{6}$ **31.** 11,111.$\overline{1}$ **33.** 23.9$\overline{18}$
35. 55.0   **37.** 0.6   **39.** 63.689   **41.** 23.388
43. 6   **45.** 649   **47.** 109.14   **49.** 56.67

Stretching the Topics (page 293)
1. 10,035   **2.** 4.9   **3.** 30,100

Checkup 7.2 (page 294)
1. 3.9   **2.** 0.203   **3.** 600   **4.** 35   **5.** 12.9$\overline{62}$
6. 2736.$\overline{6}$ **7.** 116.$\overline{8}$ **8.** 1.26   **9.** 7.0   **10.** 8

Exercise Set 7.3 (page 302)
1. 0.879   **3.** 0.0072   **5.** 0.09   **7.** 3.78
9. 23.00074   **11.** 0.076   **13.** 3.97   **15.** 0.00035
17. 0.004296   **19.** 0.000039   **21.** 74.0   **23.** 32.7
25. 24.667   **27.** 0.001   **29.** 9.38   **31.** 120.29
33. 0.0004   **35.** 0.1667   **37.** 90   **39.** 616

Stretching the Topics (page 303)
1. 35.57   **2.** 12.462

Checkup 7.3 (page 304)
1. 0.00048   **2.** 3.0023   **3.** 87.65   **4.** 0.129
5. 0.008   **6.** 36.5   **7.** 0.000456   **8.** 0.03563
9. 0.004   **10.** 56.1

Exercise Set 7.4 (page 308)
1. $21.25   **3.** $118.63   **5.** $16.68   **7.** $802.75
9. $70.02   **11.** $2.95   **13.** $0.20   **15.** 2.82 hr
17. 11.5   **19.** 166

Stretching the Topics (page 309)
1. 0.85 mi   **2.** $10.65   **3.** $107.59

Checkup 7.4 (page 310)
1. $4.50   **2.** 80   **3.** $0.51   **4.** 376   **5.** 500

Speaking the Language of Mathematics (page 312)
1. three   **2.** whole number   **3.** 3; right   **4.** 2; left

5. repeating   **6.** hundredths
7. denominator; numerator   **8.** 60; 60

Review Exercises for Chapter 7 (page 313)
1. 0.02072   **2.** 246.806   **3.** 5.796   **4.** 0.00258
5. 9342   **6.** 0.83   **7.** 15.0144   **8.** 0.187085
9. 98,688.6   **10.** 2303.22   **11.** 216   **12.** 0.021
13. 1100   **14.** 6180   **15.** 858   **16.** 7250
17. 12.0   **18.** 79.78   **19.** 196.667   **20.** 2
21. 0.00089   **22.** 5.007   **23.** 50.3   **24.** 0.001
25. 5.43   **26.** 0.4444   **27.** 20   **28.** $1093.76
29. $14.30   **30.** 172.66 sq cm   **31.** 200   **32.** 839
33. $4.49

Practice Test for Chapter 7 (page 315)
1. 0.02604   **2.** 480.624   **3.** 207.12   **4.** 1.932276
5. 4938.468   **6.** 0.39$\overline{594}$ **7.** 2300   **8.** 12.$\overline{48}$
9. 19.03   **10.** 9.2   **11.** 23.71   **12.** 0.00075
13. 7.4   **14.** 6.82   **15.** 0.714   **16.** 27.625
17. $320.47   **18.** $0.07   **19.** $5.85
20. 1.87 min; Mark, by 0.10 min

Sharpening Your Skills after Chapters 1–7 (page 317)
1. 0   **2.** 13$\frac{2}{3}$ **3.** $\frac{49}{56}$
4. eighty-five and three hundred sixty-five thousandths
5. <   **6.** 0.0429   **7.** 24,528   **8.** 93   **9.** $\frac{113}{120}$
10. 20$\frac{19}{24}$ **11.** $168.67   **12.** 44.275   **13.** 1234
14. 173,880   **15.** $\frac{25}{21}$ or 1$\frac{4}{21}$ **16.** $\frac{98}{9}$ or 10$\frac{8}{9}$
17. 3$\frac{1}{3}$ **18.** 0.01387   **19.** 3107.07   **20.** 2.2
21. 7000   **22.** 336 mi   **23.** $8   **24.** 50$\frac{11}{24}$
25. $618.92

CHAPTER 8

Exercise Set 8.1 (page 327)
1. $\frac{2}{3}$ **3.** $\frac{5}{36}$ **5.** $\frac{46}{81}$ **7.** $\frac{1}{3}$ **9.** $\frac{3}{2}$ **11.** $\frac{5000}{583}$
13. $\frac{1}{8}$ **15.** $\frac{22}{5}$ **17.** $\frac{4}{3}$ **19.** $\frac{42}{55}$ **21.** $\frac{8}{5}$ **23.** $\frac{5}{2}$

Stretching the Topics (page 328)
1. $\frac{30}{43}$ **2.** Dominic $\frac{21}{1}$; Tory $\frac{18}{1}$; Dominic's car   **3.** $\frac{7}{11}$

Checkup 8.1 (page 329)
1. $\frac{1}{5}$ **2.** $\frac{3}{7}$ **3.** $\frac{2}{3}$ **4.** $\frac{4}{3}$ **5.** $\frac{3}{40}$

Exercise Set 8.2 (page 338)
1. no   **3.** yes   **5.** no   **7.** no   **9.** yes   **11.** no
13. yes   **15.** yes   **17.** yes   **19.** no   **21.** 12
23. 12   **25.** 9   **27.** 19$\frac{1}{2}$ **29.** 6$\frac{2}{7}$ **31.** 1$\frac{5}{9}$
33. 12   **35.** $1.68   **37.** 18   **39.** 2.88

Stretching the Topics (page 339)
1. 1$\frac{2}{13}$ **2.** 3.75   **3.** $0.80

Checkup 8.2 (page 340)
1. yes   **2.** no   **3.** yes   **4.** yes   **5.** 21   **6.** 12
7. 39   **8.** 1   **9.** $13.05   **10.** 2.4

Exercise Set 8.3 (page 344)
1. 19   **3.** 291   **5.** 57   **7.** 100   **9.** 23

11. $640   **13.** $4\frac{3}{4}$ **15.** 315   **17.** $13\frac{1}{2}$ in.
19. 24,580   **21.** 400   **23.** 30

Stretching the Topics (page 345)
1. 40,000   **2.** 2412   **3.** 19

Checkup 8.3 (page 346)
1. 38   **2.** 120   **3.** 35   **4.** 2.8   **5.** $154\frac{2}{3}$

Speaking the Language of Mathematics (page 348)
1. ratio   **2.** numerator; denominator; 10
3. cross products   **4.** proportion   **5.** right side

Review Exercises for Chapter 8 (page 349)
1. $\frac{3}{8}$ **2.** $\frac{1}{18}$ **3.** $\frac{2267}{400}$ **4.** $\frac{207}{5}$ **5.** $\frac{5}{11}$ **6.** yes
7. no   **8.** yes   **9.** no   **10.** yes   **11.** no
12. $2\frac{2}{3}$ **13.** 10   **14.** $8\frac{1}{2}$ **15.** $2\frac{2}{9}$ **16.** $16\frac{7}{8}$
17. $11\frac{1}{3}$ **18.** 1.25   **19.** 21.5   **20.** 23
21. $15,000   **22.** 6500; 18,200   **23.** 675   **24.** 15
25. 20 in.

Practice Test for Chapter 8 (page 351)
1. $\frac{2}{17}$ **2.** $\frac{23}{52}$ **3.** $\frac{15}{1}$ **4.** $\frac{1}{5}$ **5.** $\frac{23}{8}$ **6.** yes
7. no   **8.** yes   **9.** yes   **10.** 27   **11.** 75
12. 9   **13.** 75   **14.** 6.5   **15.** 186.92   **16.** 2 c
17. $400   **18.** $2\frac{3}{4}$ **19.** $6.25

Sharpening Your Skills after Chapters 1–8 (page 353)
1. composite; $2^2 \times 3^2 \times 5 \times 13$ **2.** $\frac{4}{15}$ **3.** =
4. 0.750   **5.** Fifty-seven and $\frac{89}{100}$ **6.** 6.56   **7.** 7.9
8. yes   **9.** 6124   **10.** 258,923   **11.** 323   **12.** $\frac{7}{9}$
13. 75   **14.** $42\frac{3}{4}$ **15.** $\frac{1}{5}$ **16.** $\frac{27}{64}$ **17.** $\frac{47}{72}$
18. $7\frac{29}{40}$ **19.** 30.174   **20.** 16.32   **21.** 203.44
22. 87   **23.** 27   **24.** $\frac{1}{3}$ **25.** 10.6   **26.** $8.10

CHAPTER 9

Exercise Set 9.1 (page 368)
1. $\frac{79}{100}$ **3.** $\frac{2}{25}$ **5.** $\frac{2}{5}$ **7.** $\frac{5}{2}$ **9.** $\frac{5}{1}$ **11.** $\frac{1}{200}$
13. $\frac{3}{1000}$ **15.** $\frac{13}{150}$ **17.** $\frac{231}{200}$ **19.** $\frac{1}{3}$ **21.** $\frac{17}{400}$
23. $\frac{439}{1000}$ **25.** $\frac{83}{400}$ **27.** $\frac{1}{200}$ **29.** $\frac{3401}{2000}$ **31.** 0.23
33. 0.623   **35.** 0.8   **37.** 0.06   **39.** 0.0003
41. 3.75   **43.** 0.1023   **45.** 0.0025   **47.** 0.003
49. 0.055   **51.** 0.1013   **53.** 9%   **55.** 3.6%
57. 27.5%   **59.** 0.8%   **61.** 380%   **63.** 132%
65. 253.5%   **67.** 35%   **69.** 8%   **71.** 145%
73. 40%   **75.** 0.9%   **77.** 37.5%   **79.** 31.3%
81. 55.6%   **83.** 48%   **85.** 187.5%   **87.** 28.8%

Stretching the Topics (page 369)
1. $\frac{1}{20,000}$ **2.** 344.23%   **3.** $\frac{3}{4}$

Checkup 9.1 (page 370)
1. $\frac{17}{20}$ **2.** $\frac{3}{400}$ **3.** $\frac{13}{200}$ **4.** 0.225   **5.** 0.035
6. 450%   **7.** 62.3%   **8.** 83%   **9.** 80%
10. 66.7%

Exercise Set 9.2 (page 384)
1. 30.5%   **3.** 25%   **5.** 17.2%   **7.** 620
9. $22,590   **11.** $540.93; $10,375.93   **13.** 44.5
15. 160 lb; 24   **17.** 93%   **19.** 15,000   **21.** 49%
23. 274,176; 2,558,976   **25.** $5.33; $30.17

Stretching the Topics (page 385)
1. 451,500   **2.** 32 per min   **3.** 2228

Checkup 9.2 (page 387)
1. 7%   **2.** 74%   **3.** $48,500   **4.** $3.97; $30.42
5. 6510

Speaking the Language of Mathematics (page 388)
1. per hundred   **2.** $\frac{p}{100}$ **3.** 2; left; drop
4. 2; right; attach   **5.** percent proportion

Review Exercises for Chapter 9 (page 391)
1. $\frac{19}{50}$ **2.** $\frac{5}{4}$ **3.** $\frac{91}{200}$ **4.** $\frac{7}{30}$ **5.** $\frac{287}{1000}$ **6.** $\frac{11}{2000}$
7. $\frac{61}{400}$ **8.** $\frac{1}{800}$ **9.** 0.82   **10.** 0.357   **11.** 0.17
12. 0.0005   **13.** 2.53   **14.** 0.4575   **15.** 8%
16. 3.9%   **17.** 37.5%   **18.** 350%   **19.** 168%
20. 432.5%   **21.** 84%   **22.** 5%   **23.** 235%
24. 62.5%   **25.** 52%   **26.** 400%   **27.** 72%
28. 115%   **29.** $52,000; $55,120
30. 36,300; 41,745   **31.** 222   **32.** $238.25

Practice Test for Chapter 9 (page 393)
1. $\frac{9}{100}$ **2.** $\frac{7}{4}$ **3.** $\frac{11}{80}$ **4.** $\frac{9}{400}$ **5.** 0.23
6. 0.752   **7.** 0.0005   **8.** 3.45   **9.** 50%
10. 4.8%   **11.** 39%   **12.** 125%   **13.** 83%
14. 60%   **15.** 53.3%   **16.** 33.3%   **17.** 76%
18. 9065; 1813   **19.** 2,975,037   **20.** 12.54 in.

Sharpening Your Skills after Chapters 1–9 (page 395)
1. twenty-eight thousand, three hundred seventy   **2.** $\frac{127}{8}$
3. $\frac{10}{13}$ **4.** $\frac{105}{135}$ **5.** <   **6.** $\frac{379}{1000}$ **7.** 8.7   **8.** no
9. $\frac{49}{80}$ **10.** 120%   **11.** $\frac{98}{3}$ or $32\frac{2}{3}$ **12.** $17\frac{1}{2}$
13. $1\frac{1}{5}$ **14.** $\frac{254}{105}$ or $2\frac{44}{105}$ **15.** $4\frac{3}{4}$ **16.** 246.84
17. 4.894   **18.** 168.26   **19.** 45   **20.** $91
21. 676   **22.** $\frac{59}{80}$ **23.** $6\frac{1}{2}$ ft   **24.** $\frac{1}{40}$
25. $1390.80   **26.** $408.33

CHAPTER 10

Exercise Set 10.1 (page 406)
1. $4; 50%   **3.** $130; $455   **5.** 65¢   **7.** 60%
9. $736.30   **11.** $4.29; 10%   **13.** $11.79; $47.16
15. $16.73   **17.** $669.80   **19.** 200%

Stretching the Topics (page 407)
1. $780; $702   **2.** $48.96; $39.17; $30.78; 44%

Checkup 10.1 (page 408)
1. $67.50; 75%   **2.** $243.75; $438.75
3. $88.11; 33%   **4.** $25.60; $102.40   **5.** $37.70

Exercise Set 10.2 (page 417)
1. 40% increase   **3.** 8% increase   **5.** 150% increase
7. 25% increase   **9.** 10% decrease   **11.** 1139
13. $2.28   **15.** $1156.05   **17.** 827,410
19. $3460.40

Stretching the Topics (page 418)
1. $1111   **2.** $14,500

Checkup 10.2 (page 419)
1. 33% increase   **2.** $7.13   **3.** 51.6 persons
4. $1183   **5.** yes

Exercise Set 10.3 (page 427)
1. $65.15   **3.** 5%   **5.** $141.53   **7.** $79.50
9. $2187.50   **11.** $2917.50   **13.** line 4: $333.00
15. $11,644.50   **17.** $8032.50

Stretching the Topics (page 428)
1. hometown   **2.** yes; no; yes

Checkup 10.3 (page 429)
1. $837.90   **2.** 6%   **3.** $117.05   **4.** $703.50
5. $6352.50

Exercise Set 10.4 (page 436)
1. $360   **3.** $125   **5.** $337.50; $3337.50   **7.** $410
9. $11,925.18

Stretching the Topics (page 436)
1. $1350; $1407.99; Mercantile

Checkup 10.4 (page 437)
1. $1500   **2.** $260   **3.** $6998.40; $998.40

Speaking the Language of Mathematics (page 439)
1. markup   **2.** markdown   **3.** percent proportion
4. change; old   **5.** years   **6.** compound

Review Exercises for Chapter 10 (page 441)
1. $36; 40%   **2.** 30%   **3.** $74; $259   **4.** $1.49
5. $18; 40%   **6.** 33.$\bar{3}$%   **7.** $210.45
8. 12% decrease   **9.** 4.3% increase
10. 33.9% decrease   **11.** $7.67   **12.** 22,230
13. $1.95   **14.** yes   **15.** $1041.22   **16.** $104.45
17. $113.60   **18.** 1.3%   **19.** line 4: $1075.25
20. $19,518.10   **21.** $2457.75
22. $937.50; $2437.50   **23.** $183.75; $5433.75
24. $7581.60; $1081.60   **25.** $2894.06; $394.06

Practice Test for Chapter 10 (page 445)
1. $64.68; 33%   **2.** $150.40; 40%   **3.** $3.89
4. 40% increase   **5.** 8% increase   **6.** $287.50
7. 550   **8.** $10.44   **9.** no   **10.** $75.08
11. $75.58   **12.** line 4: $1489.00   **13.** $3487.50
14. $25,616.50   **15.** $306; $1156
16. $2970.25; $470.25

Sharpening Your Skills after Chapters 1–10 (page 447)
1. 7   **2.** $\frac{109}{126}$ **3.** <   **4.** Thirty-seven and $\frac{85}{100}$
5. >   **6.** 0.570   **7.** 5.8   **8.** 0.00087   **9.** 75
10. $\frac{11}{4}$ **11.** 0.0004   **12.** 5.2%   **13.** 80%
14. 11,024   **15.** 6,897,960   **16.** 170 R460
17. $42\frac{3}{4}$ **18.** $\frac{3}{4}$ **19.** $\frac{991}{420}$ or $2\frac{151}{420}$ **20.** 23.116
21. 133.614   **22.** 40.59   **23.** 432   **24.** 111
25. $12\frac{13}{24}$ **26.** $1043.70   **27.** 3.76¢

CHAPTER 11

Exercise Set 11.1 (page 464)
1. $27,570   **3.** $7\frac{1}{4}$ lb   **5.** 71   **7.** 3   **9.** 3.11
11. 53   **13.** 133,500   **15.** $540,000   **17.** 35
19. $1.85   **21.** 26   **23.** two-bedroom

Stretching the Topics (page 466)
1. 71; no   **2.** 59.3

Checkup 11.1 (page 467)
1. $172   **2.** 13.5   **3.** 384   **4.** $878   **5.** 16
6. 3

Exercise Set 11.2 (page 475)
1. $27,000; $26,500   **3.** 42.212 mph; 140.92 mph
5. 29°; 23.5°   **7.** $225.50; $235; $235 and $195
9. 82; 80; 80; 35; 77.5

Stretching the Topics (page 476)
1. 156.752; no mode; 153.991; 36.708; 159.248

Checkup 11.2 (page 477)
1. 113.75   **2.** $22.97
3. 15.7; 16; 17 and 15; 15.5; 15

Exercise Set 11.3 (page 485)
1. 525   **3.** 64th   **5.** 900   **7.** 50th   **9.** 75th

Stretching the Topics (page 486)
1. Roberto   **2.** 99.3   **3.** Tony

Checkup 11.3 (page 487)
1. 169   **2.** 80th   **3.** 525   **4.** 50th   **5.** June

Speaking the Language of Mathematics (page 489)
1. central tendency   **2.** sum; 15   **3.** order
4. 73%; below

Review Exercises for Chapter 11 (page 491)
1. 105°F   **2.** $374.19   **3.** 76.2   **4.** 28   **5.** 26.3
6. 1925   **7.** $446\frac{2}{5}$ oz   **8.** $452   **9.** 6; 7
10. $60,000   **11.** 77   **12.** 22   **13.** 340 yd; 320 yd
14. 5; 4.5   **15.** $40.62; $40.39; none; $25; $41.22
16. 170   **17.** 60th   **18.** 200   **19.** 50th
20. Roger

Practice Test for Chapter 11 (page 493)
1. 163.9 lb   **2.** 26   **3.** 48   **4.** $2250
5. 4825.5   **6.** 26.5; 25   **7.** 138.5 cm   **8.** $21,627
9. 13,900; 22,478.$\overline{3}$; 22,050; none; 23,200   **10.** 33
11. 40th   **12.** 1100   **13.** 65th   **14.** 10th
15. 120   **16.** 84

Sharpening Your Skills after Chapters 1–11 (page 495)
1. prime   **2.** $\frac{119}{8}$ **3.** $\frac{15}{103}$
4. thirty-four and seventy-two thousandths   **5.** 0.714
6. 48   **7.** 62.5%   **8.** $\frac{63}{400}$ **9.** 481.25
10. 96.43°   **11.** 11   **12.** $4\frac{6}{7}$ **13.** $\overline{2}$ **14.** $\frac{1}{8}$
15. 0.0674   **16.** 3435.49   **17.** 13.4$\overline{8}$
18. $46\frac{1}{5}$ sq cm   **19.** $2\frac{11}{12}$ **20.** $6.14   **21.** $943.75
22. 2044   **23.** 60%   **24.** $216; 30%
25. 16% decrease   **26.** $393.75; $1643.75   **27.** 45
28. 50th

CHAPTER 12

Exercise Set 12.1 (page 509)
1. 84 in.   **3.** $3\frac{1}{3}$ yd   **5.** 13 ft   **7.** 2 yd
9. 288 in.   **11.** 26,400 ft   **13.** $\frac{1}{4}$ mi   **15.** 36 ft
17. 1700 cm   **19.** 3.85 km   **21.** 5.35 m
23. 0.815 m   **25.** 9300 m   **27.** 5000 cm
29. 3400 mm   **31.** $6\frac{1}{2}$ ft; 78 in.   **33.** 9.1 mi
35. 24   **37.** 4500 m; 4.5 km   **39.** $4.53

Stretching the Topics (page 510)
1. 108 m; 0.108 km; 0.0675 mi   **2.** 337 cm   **3.** $\frac{3}{5}$

Checkup 12.1 (page 511)
1. 84 in.   **2.** $\frac{1}{4}$ yd   **3.** $\frac{1}{5}$ mi   **4.** 234 in.
5. 0.07 cm   **6.** 900,000 cm   **7.** 0.000018 km
8. 3600 mm   **9.** 4800 m   **10.** Gary

Exercise Set 12.2 (page 519)
1. $6\frac{1}{2}$ lb   **3.** $4\frac{1}{2}$ ton   **5.** 84 oz   **7.** 4400 lb
9. 1.5 kg   **11.** 1250 cg   **13.** 5 mg   **15.** 5250 g
17. 1.5 g   **19.** 0.0185 kg   **21.** 4 lb 9 oz
23. $33\frac{3}{4}$ lb   **25.** 250 mg   **27.** 4.5 kg   **29.** 2.6 g

Stretching the Topics (page 520)
1. 3.6 g   **2.** 150,000   **3.** 7.26 kg

Checkup 12.2 (page 521)
1. 50,250,000 lb   **2.** 132 oz   **3.** $2\frac{13}{16}$ lb
4. 0.00085 g   **5.** 8500 g   **6.** 0.025 kg   **7.** $9\frac{1}{3}$ oz
8. 337.5 g

Exercise Set 12.3 (page 528)
1. 14 qt   **3.** 11 pt   **5.** $5\frac{1}{2}$ pt   **7.** 60 pt
9. 750 mℓ **11.** 8500 ℓ **13.** 25 ℓ **15.** 5000 mℓ
17. 4 c   **19.** $2\frac{1}{2}$ gal   **21.** 625 mℓ **23.** 6000
25. 250 mℓ

Stretching the Topics (page 529)
1. $1.08   **2.** $16.08

Checkup 12.3 (page 530)
1. 30 qt   **2.** 10 qt   **3.** 48 c   **4.** 430 mℓ
5. 8 kℓ **6.** 8.5 ℓ **7.** $\frac{3}{4}$ gal   **8.** 36 ℓ

Speaking the Language of Mathematics (page 532)
1. American   **2.** metric   **3.** conversion factor
4. capacity   **5.** length

Review Exercises for Chapter 12 (page 533)
1. 216 in.   **2.** 3 mi   **3.** $9\frac{1}{2}$ ft   **4.** 10 yd
5. 7 mi   **6.** $\frac{1}{4}$ mi driveway   **7.** 0.423 m
8. 0.008 km   **9.** 165 mm   **10.** 900,000 cm
11. 864 cm; 8.64 m   **12.** 7.5 m   **13.** 12 lb
14. 6500 lb   **15.** 248 oz   **16.** 36 oz   **17.** 13 oz
18. $7\frac{1}{2}$ ton   **19.** 7500 mg   **20.** 1.2 kg   **21.** 3.5 g
22. 0.025 kg   **23.** 6.475 g   **24.** 16   **25.** 22 qt
26. 8 pt   **27.** 30 pt   **28.** $3\frac{5}{8}$ gal   **29.** 32 qt
30. 4 qt   **31.** 375 mℓ **32.** 4730 ℓ **33.** 32,000 mℓ
34. 85 ℓ **35.** 12,500   **36.** 1.5 kℓ

Practice Test for Chapter 12 (page 535)
1. 29,040 ft   **2.** 33 in.   **3.** $10\frac{1}{2}$ ft   **4.** 0.825 m
5. 120 cm   **6.** Springfield   **7.** 1,450,000 lb
8. 244 oz   **9.** 11 lb   **10.** 0.153 g   **11.** 12.45 kg
12. 3.875 kg   **13.** 24 c   **14.** $7\frac{1}{2}$ qt   **15.** $1\frac{1}{2}$ qt
16. 450 mℓ **17.** 8.4 ℓ **18.** 310 mℓ

Sharpening Your Skills after Chapters 1–12 (page 537)
1. two hundred ninety thousand, seventy-eight
2. composite; $2^3 \times 3^2 \times 13$ **3.** $17\frac{4}{5}$ **4.** $\frac{5}{17}$ **5.** <
6. 8.3   **7.** 0.75   **8.** yes   **9.** $\frac{67}{400}$ **10.** 36%
11. 44,000 ft   **12.** 0.156 g   **13.** 2-yd loss
14. 180 sq ft   **15.** $\frac{9}{20}$ gal   **16.** $\frac{147}{16}$ or $9\frac{3}{16}$ **17.** $\frac{17}{25}$
18. $1\frac{7}{8}$ mi   **19.** 5.8 m   **20.** $380.91   **21.** 5 in.
22. 5.38 in.   **23.** $221.98   **24.** $3.81
25. $210.87   **26.** 9   **27.** 20 and 25
28. 14 lb 7 oz

CHAPTER 13

Exercise Set 13.1 (page 549)
1. parallelogram; 28 ft   **3.** square; 32 in.
5. rectangle; 31 m   **7.** trapezoid; 52 cm
9. triangle; 12 yd   **11.** 110 in.   **13.** 53.38 m
15. 29 m   **17.** 6.9 cm   **19.** 72 in.; 2 yd   **21.** 39 m
23. 4.8 km   **25.** 142   **27.** 1120   **29.** 36.56 m

Stretching the Topics (page 551)
1. $66.55   **2.** L-shaped patio; 2.32 ft

Checkup 13.1 (page 552)
1. parallelogram; 108 in.   **2.** trapezoid; 39 m
3. 49.2 cm   **4.** $20\frac{1}{2}$ ft   **5.** 13.2 cm   **6.** circle

Exercise Set 13.2 (page 561)
1. 88.36 sq m   **3.** $14\frac{7}{8}$ sq ft   **5.** 60 sq m
7. 154 sq cm   **9.** 324   **11.** 6500 sq yd
13. 6600 sq ft   **15.** 379.94 sq cm   **17.** 210 sq in.
19. 27,000   **21.** 45   **23.** 0.3852   **25.** 4,900,000
27. 0.1386

Stretching the Topics (page 562)
1. 2.09 sq mi   **2.** 49,200 sq ft   **3.** 2444

Checkup 13.2 (page 564)
1. 15 sq in.   **2.** $2\frac{1}{4}$ sq ft   **3.** 706.5 sq cm
4. 3600 sq ft   **5.** 360

Exercise Set 13.3 (page 573)
1. $22\frac{1}{2}$ cu ft   **3.** 3581.577 cu cm   **5.** 1125 cu ft
7. 5832 cu in.   **9.** 15 cu ft   **11.** 1695.6 cu cm
13. 137,188.6933 cu ft   **15.** 188.4 cu cm
17. 278.12079 cu m   **19.** 403.8825 cu cm
21. 4.2 mℓ **23.** 300 cc   **25.** 5 ℓ **27.** 42
29. 2

Stretching the Topics (page 574)
1. 823.68 cu in.   **2.** 8624 cu yd   **3.** 1125

Checkup 13.3 (page 575)
1. 512 cu ft   **2.** 1950 cu cm   **3.** 1953 cu cm
4. 141.3 cu in.   **5.** 33,493.$\overline{3}$ cu in.   **6.** 4.5 mℓ
7. 2300 cc   **8.** 2.35

Exercise Set 13.4 (page 583)
1. 79°   **3.** 52°   **5.** 90°; yes   **7.** no   **9.** 17 yd
11. 25 ft   **13.** 17 ft   **15.** 15 ft

Stretching the Topics (page 584)
1. 16 ft   **2.** 80 ft   **3.** 48 in.; 84 sq in.

Checkup 13.4 (page 585)
1. 50°   **2.** 41°   **3.** 90°; yes   **4.** yes   **5.** 13 ft
6. 25 mi

Speaking the Language of Mathematics (page 589)
1. polygon   **2.** perimeter; area   **3.** square; cubic
4. right triangle; hypotenuse   **5.** Pythagorean

Review Exercises for Chapter 13 (page 591)
1. trapezoid; 44 cm   **2.** triangle; 37 in.
3. parallelogram; 26 ft   **4.** square; 12 m   **5.** 40.82 ft
6. 264 cm   **7.** 122 ft   **8.** 48 in.   **9.** 35.42 yd
10. 256 sq in.   **11.** 21 sq ft   **12.** 204 sq m
13. 640 sq in.   **14.** 1017.36 sq ft   **15.** 72 sq ft
16. 0.5285 sq m   **17.** 113.04   **18.** 0.36
19. 1728 cu cm   **20.** 45 cu ft   **21.** 391.6 cu cm
22. 75.36 cu in.   **23.** $11,498\frac{2}{3}$ cu cm   **24.** 3.8 mℓ
25. 800 cc   **26.** 1.5 ℓ **27.** 0.7   **28.** 29°
29. 27°   **30.** 10 in.

Practice Test for Chapter 13 (page 593)
1. 9.2 mm   **2.** 17 ft   **3.** 50.24 in.   **4.** 18.28 m
5. 30.25 sq m   **6.** 216 sq ft   **7.** 72 sq in.
8. 28.26 sq ft   **9.** 5184 sq in.   **10.** 11,304 sq m
11. 5832 cu in.   **12.** 2   **13.** 3696 cu in.
14. 3052.08 cu cm   **15.** 1.8 ℓ **16.** 3.2   **17.** 30°
18. 31°   **19.** 17   **20.** 25 ft

Sharpening Your Skills after Chapters 1–13 (page 595)
1. one hundred thousand, nine hundred seven
2. composite; $3^2 \times 7 \times 11$ **3.** $\frac{78}{5}$ **4.** $\frac{3}{20}$ **5.** <
6. 0.0156   **7.** 0.84   **8.** One hundred three and $\frac{8}{100}$
9. 0.556   **10.** yes   **11.** 35   **12.** $\frac{3}{80}$ **13.** 53.3%
14. 92 in.   **15.** 0.175 m   **16.** 0.254 g   **17.** $8\frac{1}{2}$ qt
18. $\frac{14}{3}$ or $4\frac{2}{3}$ **19.** $4\frac{6}{7}$ **20.** $\frac{111}{40}$ or $2\frac{31}{40}$ **21.** 17.158
22. 27.20   **23.** 4   **24.** $920.17   **25.** $1126.25
26. 320   **27.** 1625; 4875   **28.** $600; $5600
29. 80th   **30.** 48   **31.** 9938.375 cu cm   **32.** 26 cm

CHAPTER 14

Exercise Set 14.1 (page 607)
1.

3. A: -5; B: -2; C: 0; D: 1.5; E: 3
5. (a) < (b) > (c) > (d) <   **7.** 5   **9.** -8 **11.** 3
13. 6   **15.** 1.1   **17.** 22   **19.** -12 **21.** 12
23. -6 **25.** -7.1 **27.** 1   **29.** 20   **31.** 8
33. -17 **35.** -3.1 **37.** 18   **39.** $-\frac{21}{2}$ **41.** 0
43. -11 **45.** -12 **47.** 0   **49.** -2.5

Stretching the Topics (page 608)
1. 15   **2.** 12   **3.** $\frac{17}{20}$

Checkup 14.1 (page 609)
1. >   **2.** -1.2 **3.** -5 **4.** 8   **5.** -12.2
6. -1 **7.** -1 **8.** 0   **9.** 1   **10.** 9.1

Exercise Set 14.2 (page 614)
1. 3   **3.** -16 **5.** -22 **7.** 15   **9.** -2
11. $-\frac{1}{6}$ **13.** -12.3 **15.** 10   **17.** 25   **19.** $-\frac{5}{9}$
21. -9 **23.** 4   **25.** -12 **27.** -10 **29.** -6
31. -7.4 **33.** 1   **35.** 8   **37.** -26 **39.** 0
41. 0   **43.** 3   **45.** $-\frac{22}{5}$ **47.** -7 **49.** 2

Stretching the Topics (page 615)
1. -22 **2.** 28   **3.** $\frac{43}{2}$

Checkup 14.2 (page 616)
1. 7   **2.** 21   **3.** -22 **4.** -10 **5.** 1
6. $-\frac{8}{5}$ **7.** $-\frac{25}{12}$ **8.** -15.2

Exercise Set 14.3 (page 626)
1. -30 **3.** 42   **5.** -24 **7.** 1.35   **9.** $-\frac{5}{8}$
11. 150   **13.** 0   **15.** 45   **17.** -80 **19.** 4.54
21. -162 **23.** 4   **25.** 64   **27.** -27 **29.** $\frac{4}{9}$

31. -1.728 **33.** -288 **35.** -4 **37.** -2
39. 5   **41.** -2 **43.** $-\frac{11}{6}$ **45.** -7.2
47. 1.47   **49.** $\frac{2}{5}$

Stretching the Topics (page 627)
1. 63   **2.** -244 **3.** -34

Checkup 14.3 (page 628)
1. -30 **2.** $\frac{5}{8}$ **3.** -24 **4.** 23.5   **5.** 10.24
6. -2 **7.** -5 **8.** 7   **9.** -2.8 **10.** $\frac{2}{9}$

Exercise Set 14.4 (page 636)
1. -2 **3.** 9   **5.** 2   **7.** 9   **9.** 45   **11.** 11.75
13. -3 **15.** -17.8 **17.** 60   **19.** -2
21. -6 **23.** -3 **25.** 24   **27.** $\frac{1}{3}$ **29.** 1.47
31. -1 **33.** undefined   **35.** 1   **37.** -7
39. -3 **41.** 9   **43.** -27 **45.** $\frac{1}{32}$ **47.** 29
49. -9

Stretching the Topics (page 637)
1. $\frac{16}{25}$ **2.** -0.0000154 **3.** $-\frac{7}{48}$

Checkup 14.4 (page 638)
1. 2   **2.** -25 **3.** -5 **4.** $-\frac{1}{2}$ **5.** 36
6. -3 **7.** 3   **8.** $-\frac{8}{27}$ **9.** -75 **10.** 0

Exercise Set 14.5 (page 641)
1. 78   **3.** 1718   **5.** \$14.87   **7.** 346.5 mi   **9.** $9\frac{1}{3}$
11. won 75¢   **13.** \$15.70   **15.** owes her \$14
17. 780   **19.** \$32.25   **21.** \$0.17; \$3.90; \$6.10
23. 13

Stretching the Topics (page 642)
1. loss of \$61.59 each   **2.** \$12,078.17

Checkup 14.5 (page 643)
1. \$720.24   **2.** 48 ft   **3.** \$203,603   **4.** $47\frac{79}{180}$ ft
5. loss of \$1.22

Speaking the Language of Mathematics (page 646)
1. left   **2.** absolute value   **3.** opposites
4. negative; positive   **5.** base; exponent
6. distributive

Review Exercises for Chapter 14 (page 647)
1. (a) >   (b) <   (c) >   (d) >
2. (a) 3 (b) 2 (c) -3.9 (d) -9 **3.** -14 **4.** -3
5. -4.5 **6.** -10 **7.** $-\frac{1}{8}$ **8.** 4   **9.** 24
10. 6   **11.** -9 **12.** 2.5   **13.** $-\frac{1}{2}$ **14.** -22
15. 24   **16.** -15 **17.** -13 **18.** 45
19. -6.5 **20.** $-\frac{3}{2}$ **21.** -27 **22.** 10.35
23. -11 **24.** $\frac{1}{3}$ **25.** 16.81   **26.** 0.2   **27.** -5
28. 6   **29.** -3.2 **30.** 6   **31.** -6 **32.** -23
33. -3 **34.** $\frac{16}{9}$ **35.** -42 **36.** -7 **37.** 4
38. $-\frac{27}{8}$ **39.** 40   **40.** 0   **41.** $-5°$
42. \$35.93   **43.** 43 ft   **44.** \$99.65   **45.** \$49

Practice Test for Chapter 14 (page 649)
1. (a) <   (b) >   **2.** -8 **3.** -20 **4.** $-\frac{8}{5}$
5. 4   **6.** -18 **7.** 4.4   **8.** 84   **9.** 42
10. 9.5   **11.** 9   **12.** -140 **13.** $\frac{5}{8}$ **14.** 9.5
15. $-\frac{8}{9}$ **16.** -3.2 **17.** 0   **18.** -125
19. lost $1\frac{3}{4}$ lb   **20.** won \$9.20

Sharpening Your Skills after Chapters 1–14 (page 651)
1. 6   **2.** $\frac{6}{7}$ **3.** >   **4.** $\frac{17}{1}$ **5.** 175%
6. 11,880 ft   **7.** 150 cm   **8.** 136 oz   **9.** 2.5 ℓ
10. <   **11.** 376   **12.** 9   **13.** $2\frac{15}{28}$ **14.** \$333.13
15. 348.56   **16.** -26 **17.** 0   **18.** 480 sq ft
19. $\frac{13}{16}$ **20.** $7\frac{1}{2}$ **21.** $746\frac{2}{3}$ **22.** \$0.07
23. 15 in.   **24.** \$29,000; \$30,305   **25.** \$2128.13
26. 23.61 in.   **27.** \$4400   **28.** 113.04 ft   **29.** 30
30. \$19,719

CHAPTER 15

Exercise Set 15.1 (page 662)
1. $x + 12$ **3.** $\frac{1}{3}n$ **5.** $x - 6.5$
7. $\dfrac{x + (-3)}{7}$ or $\dfrac{x - 3}{7}$ **9.** $\frac{1}{6}n + 9$ **11.** 7
13. -5 **15.** -1 **17.** -12 **19.** -6
21. $-2a$ **23.** $3x$ **25.** $-8a + 10b + 7$
27. $11x - 2y + 7$ **29.** $a - 11b + 4c$ **31.** $-21x$
33. $5a$ **35.** $-8x$ **37.** $3a$ **39.** $\dfrac{3y}{2}$ **41.** $4x - 8$
43. $-2x + 4y$ **45.** $3x + y$ **47.** $9x - 3y$
49. $-18m - 27n$

Stretching the Topics (page 663)
1. $-\frac{77}{9}$ **2.** $-\frac{7}{10}x - \frac{17}{3}y + \frac{7}{4}z$
3. $7.2x + 7.7y + 0.3$

Checkup 15.1 (page 664)
1. $-12 + y$ **2.** $\dfrac{x + 9}{5}$ **3.** -23 **4.** 2
5. $-a$ **6.** $3a - 10b + 7$ **7.** $-4x$ **8.** $-\dfrac{3x}{2}$
9. $-3x + 21$ **10.** $2x - y$

Exercise Set 15.2 (page 672)
1. $x + 5 = -7$ **3.** $-4x = 24$ **5.** $3(x - 2) = 12$
7. $x = 7$ **9.** $x = 12$ **11.** $y = -8$ **13.** $y = 2.6$
15. $m = 5$ **17.** $a = 10$ **19.** $k = 6$ **21.** $x = 0$
23. $m = -\frac{5}{2}$ **25.** $y = -8$ **27.** $x = 7$
29. $x = -\frac{35}{2}$ **31.** $x = 0.6$ **33.** $x = 0$
35. $x = -6.5$ **37.** $x = 12$ **39.** $x = -12$
41. $x = -56$ **43.** $x = 0$ **45.** $x = -40$
47. $x = 63$ **49.** $x = 6$

Stretching the Topics (page 673)
1. $x = \frac{7}{12}$ **2.** $x = -1.5$ **3.** $y = d - c$ **4.** $x = \dfrac{m}{a}$

Checkup 15.2 (page 674)

1. $5[x + (-3)] = 11$ or $5(x - 3) = 11$
2. $x = -1$ **3.** $x = 7$ **4.** $x = 4.8$ **5.** $x = 0$
6. $x = 9$ **7.** $x = 0$ **8.** $x = 24$ **9.** $x = 9.6$
10. $x = -60$

Exercise Set 15.3 (page 681)

1. $x = 7$ **3.** $x = 10$ **5.** $y = -5$ **7.** $x = 1$
9. $a = -9$ **11.** $m = -\frac{13}{2}$ **13.** $x = -13$
15. $x = -3$ **17.** $x = 13$ **19.** $y = 0$ **21.** $a = 8$
23. $x = 2$ **25.** $x = 12$ **27.** $x = 0$ **29.** $y = 16$
31. $x = -4$ **33.** $a = -4$ **35.** $x = 8$
37. $x = -1$ **39.** $x = \frac{9}{2}$ **41.** $x = 5$ **43.** $x = 4$
45. $x = 16$ **47.** $x = -3$ **49.** $x = \frac{35}{2}$

Stretching the Topics (page 682)

1. $x = -0.5$ **2.** $x = -2$ **3.** $x = \frac{7}{3}$

Checkup 15.3 (page 683)

1. $x = 5$ **2.** $x = -1$ **3.** $y = -6$ **4.** $x = 12$
5. $x = 3$ **6.** $x = 6$ **7.** $x = -5$ **8.** $x = 0$
9. $y = -7$ **10.** $x = 2$

Exercise Set 15.4 (page 690)

1. $\$4.15x$ **3.** $\dfrac{\$850}{x}$ **5.** $\$350 + \$7.25x$

7. $\dfrac{x}{52}$ dollars   **9.** $2x + 75$ **11.** 24   **13.** $\$20.99$

15. 15 ft by 45 ft   **17.** $\$68.75$ **19.** 15 ft by 20 ft

Stretching the Topics (page 691)

1. 11 in.; 12 in.; 13 in.; 14 in.   **2.** 2250

Checkup 15.4 (page 692)

1. $3h$ **2.** $\dfrac{x}{3}$ dollars   **3.** $\$200 + \$19.75n$ **4.** 19

5. 24   **6.** 14

Speaking the Language of Mathematics (page 695)

1. constants; variables   **2.** algebraic expressions
3. evaluating   **4.** numerical coefficient   **5.** distributive
6. solution

Review Exercises for Chapter 15 (page 697)

1. $x + (-10)$ or $x - 10$ **2.** $\frac{3}{4}y$ **3.** $a - 3.5$
4. $\dfrac{x + (-2)}{5}$ or $\dfrac{x - 2}{5}$ **5.** -1 **6.** 1   **7.** -10
8. -3 **9.** $4x$ **10.** $-2a - 2b$ **11.** $-2x + 1$
12. $-y - z$ **13.** $-35x$ **14.** $-3x$ **15.** $-5a$
16. $\dfrac{5x}{2}$ **17.** $10x - 15$ **18.** $-12x + 15y$
19. $-2x + 4y$ **20.** $-0.8m + 0.6n$ **21.** $x = 11$
22. $x = 31$ **23.** $y = -12$ **24.** $a = 2.5$
25. $y = 8$ **26.** $a = 16$ **27.** $m = -\frac{7}{3}$
28. $x = 9$ **29.** $a = -27$ **30.** $x = 0.7$
31. $x = 0$ **32.** $x = -7.2$ **33.** $x = -6$

34. $x = 0$ **35.** $x = -10$ **36.** $x = -2$
37. $x = \frac{19}{2}$ **38.** $x = 1$ **39.** $x = -3$
40. $x = 4$ **41.** $x = \frac{17}{4}$ **42.** $x = -1$
43. $y = 5$ **44.** $x = \frac{2}{5}$ **45.** $x = 11$ **46.** $x = 0$
47. $\dfrac{600}{x}$ **48.** $\$2500 + \$500n$ **49.** $(\frac{1}{2}n - 2)$ in.
50. 40   **51.** 4 ft by 12 ft   **52.** 3

Practice Test for Chapter 15 (page 699)

1. $\dfrac{x + (-5)}{3}$ or $\dfrac{x - 5}{3}$ **2.** 14   **3.** $-4n$
4. $-4y + 2$ **5.** $-20n$ **6.** $-3x$ **7.** $6x - 2$
8. $x = -7$ **9.** $n = 6$ **10.** $a = -4.3$
11. $x = -12$ **12.** $x = -63$ **13.** $x = 6.5$
14. $x = 3$ **15.** $x = -25$ **16.** $x = -16$
17. $x = -7$ **18.** $\$250 + \$4.35n$ **19.** 113 mi
20. 6 in. by 12 in.

Sharpening Your Skills after Chapters 1–15 (page 701)

1. $7\frac{5}{12}$ **2.** $\frac{63}{72}$ **3.** $<$ **4.** 78.5   **5.** 11.08
6. $\frac{49}{400}$ **7.** 3.6%   **8.** 350 mℓ **9.** 40 c   **10.** $>$
11. 1,730,600   **12.** $22\frac{2}{3}$ **13.** $\frac{256}{27}$ or $9\frac{13}{27}$
14. $\$497.50$ **15.** 10.8   **16.** -8 **17.** $\$2275$
18. $20\frac{3}{4}$ **19.** $\frac{39}{56}$ **20.** 105   **21.** $\frac{17}{36}$ **22.** 23.8
23. $\$342$ **24.** 62%   **25.** $\$58.35$; 15%   **26.** $\$1250$
27. 200   **28.** 2772 cu in.
29. $7[x + (-9)]$ or $7(x - 9)$ **30.** 42

Chapter 16

Exercise Set 16.1 (page 713)

1. $\$133.78$ **3.** 0.7815   **5.** 0.435   **7.** 161,850
9. 307.0116   **11.** 4.5   **13.** $6.\overline{2}$ **15.** 2.56322
17. 124.386   **19.** 0.07   **21.** 111.7   **23.** 19
25. 1896   **27.** 6.485   **29.** 1.6344   **31.** 1368
33. 106   **35.** 40   **37.** 327,184   **39.** 14.318656
41. 482.2416   **43.** 218   **45.** 132.25   **47.** 40.5769
49. 1153.9609   **51.** 0   **53.** 201.24   **55.** 10.66
57. 729   **59.** 11,204   **61.** 0.01915   **63.** 5892.87
65. 1016.3   **67.** 27.656   **69.** 3.4322

Stretching the Topics (page 714)

1. 174.558   **2.** 0.0012   **3.** 0.61

Checkup 16.1 (page 715)

1. $\$142.17$ **2.** 6.3   **3.** 0.20   **4.** $\$41.93$
5. 336   **6.** 5.635876   **7.** 19.36   **8.** -5288
9. 0.021   **10.** 17,203.68

Exercise Set 16.2 (page 726)

1. 0.52   **3.** $0.\overline{8}$ **5.** 0.39   **7.** 1.943   **9.** $0.8\overline{3}$
11. 2.23   **13.** 0.53   **15.** 1.20   **17.** 0.13
19. 0.49   **21.** 1.35   **23.** 4.33   **25.** 4.38
27. 1.68   **29.** 1.55   **31.** 0.93   **33.** 0.87
35. 0.93   **37.** 2.57   **39.** 0.28

Stretching the Topics (page 727)
1. 0.64   **2.** −0.45   **3.** 0.66

Checkup 16.2 (page 728)
1. 1.5   **2.** 0.47$\overline{2}$ **3.** 0.875   **4.** 0.35   **5.** 1.17
6. 4.36   **7.** 2   **8.** 1.25   **9.** 1.06   **10.** 17.60

Exercise Set 16.3 (page 734)
1. 0.83   **3.** 1.25   **5.** 0.245   **7.** 430.65
9. 21.875   **11.** 94   **13.** 167.4   **15.** $418.80
17. 3.8%   **19.** 96   **21.** $6040.31   **23.** 6.25%
25. 18,453   **27.** $1456.91   **29.** $71.98

Stretching the Topics (page 735)
1. $14.12   **2.** $14,413.46

Checkup 16.3 (page 736)
1. 0.125   **2.** 2.25   **3.** $2676   **4.** $134.06
5. $847.81   **6.** 38.1%   **7.** $750   **8.** $88.92

Exercise Set 16.4 (page 754)
1. 324 in.   **3.** 432 in.   **5.** 0.545 m   **7.** 12.85 m
9. 8000 oz   **11.** 920 mm   **13.** 8400 m
15. 28,000 cm   **17.** 7.32 m   **19.** 2.23 ft
21. 12 lb   **23.** 19.9   **25.** 37.24 ft   **27.** 27.49 ft
29. 41.83 ft   **31.** 76.5625 sq ft   **33.** 95.03 sq ft
35. 157.17 sq ft   **37.** 27.81 sq m   **39.** 11.39 cu ft
41. 65.45 cu in.   **43.** 20,862.1 cu ft   **45.** 31.88 cu ft
47. 55.4°   **49.** 91 ft   **51.** 15.22 m

Stretching the Topics (page 756)
1. $123.77   **2.** 21.2   **3.** 53.6 m; 121.5 sq m

Checkup 16.4 (page 758)
1. 0.00074 m   **2.** 15.5 yd   **3.** 21.7   **4.** 9
5. 150.80 in.   **6.** 22.25 ft   **7.** 17.41 sq m
8. 69.40 sq m   **9.** 37.8 cu ft   **10.** 28.7 in.

Speaking the Language of Mathematics (page 759)
1. algebraic operating   **2.** display   **3.** y^x
4. parentheses   **5.** $\sqrt{x}$

Review Exercises for Chapter 16 (page 761)
1. $45.69   **2.** 19.2   **3.** $65.68   **4.** 1.05
5. 0.11   **6.** 5.3$\overline{18}$ **7.** 76.0384   **8.** 37.21
9. −656   **10.** 0.051   **11.** 158.49   **12.** 2.1
13. 0.52$\overline{7}$ **14.** 0.875   **15.** 0.$\overline{2}$ **16.** 0.35
17. 0.68   **18.** 2.11   **19.** 0.64   **20.** 0.85
21. 2.00   **22.** 7.19   **23.** 0.94   **24.** 0.1075
25. 3.25   **26.** $270.28   **27.** 900.45   **28.** 46.7%
29. $1891.30   **30.** $526.42   **31.** 0.00039 m
32. 540 in.   **33.** 137.1 km   **34.** 30.2 l **35.** 53.4°
36. 39.27 ft   **37.** 29.9 m   **38.** 70.88 cu ft
39. 179.6 cu m   **40.** 16 cu ft

Practice Test for Chapter 16 (page 763)
1. 16.3$\underline{2}$ **2.** 684   **3.** 0.000229   **4.** 2826.717589
5. 0.69$\overline{4}$ **6.** 0.033   **7.** 0.247   **8.** 4.2   **9.** 3.29
10. 0.8473   **11.** $1941.75   **12.** 185.5   **13.** 65.6%
14. 1093.75   **15.** $48.92   **16.** 0.00085 m
17. 2.$\overline{27}$ kg   **18.** 150.8 in.   **19.** 101.8875 sq m
20. 31.8 m

Sharpening Your Skills after Chapters 1–16 (page 765)
1. one million, three hundred two thousand, nineteen
2. $\frac{60}{108}$ **3.** $\frac{97}{2000}$ **4.** 6.4   **5.** yes   **6.** 2.25
7. 33.3%   **8.** 15.54 kg   **9.** $2x + 5$
10. $3x + 3y - 5$ **11.** 3005   **12.** 50   **13.** $\frac{13}{24}$
14. 10.055   **15.** $20.02   **16.** 2.573415   **17.** 6.916
18. undefined   **19.** $x = 4$ **20.** $486.88   **21.** 45
22. $31.96   **23.** 11.6%   **24.** 74   **25.** 74th
26. 1.4 m; 140 cm   **27.** 1.2   **28.** 228 sq ft
29. 25 mi   **30.** 47   **31.** 6 in. by 9 in.
32. 437.91 cu m

Index